High-brightness Metal Vapour Lasers

Vol. 1 Physical Fundamentals and Mathematical Models

High-brightness Metal Vapour Lasers

Vol. 1 Physical Fundamentals and Mathematical Models

V. M. Batenin, V. V. Buchanov, A. M. Boichenko
M. A. Kazaryan, I. I. Klimovskii, E. I. Molodykh

CISP

CRC Press is an imprint of the
Taylor & Francis Group, an **informa** business

CRC Press
Taylor & Francis Group
6000 Broken Sound Parkway NW, Suite 300
Boca Raton, FL 33487-2742

First issued in paperback 2020

© 2017 by CISP
CRC Press is an imprint of Taylor & Francis Group, an Informa business

No claim to original U.S. Government works

ISBN 13: 978-0-367-57404-8 (pbk)
ISBN 13: 978-1-4822-5004-6 (hbk)

Visit the Taylor & Francis Web site at
http://www.taylorandfrancis.com

and the CRC Press Web site at
http://www.crcpress.com

Contents

Foreword

The lasing at self-contained transitions of metal atoms was achieved for the first time in lead vapours [1] in 1965, at the transition between the resonant level $6p7s\ ^3P_1^0$ and one of the levels $6p^2\ ^1D_2$ of the basic configuration, at a wavelength of 722.9 nm. Since then, many studies have been published that contain the results of research and development of lasers on self-contained transitions of metal atoms. Moreover, in the meantime numerous attempts have been made (see [2–7]) in some way to summarize the results of studies carried out by different groups, primarily in the USA and the former USSR.

12 years has passed since the publication of the monograph [6], which made at that time a significant contribution to the development of research and development of the self-contained metal vapour lasers. During this time the investigations were almost completed of the physical processes taking place in the active medium of self-contained lasers and determining their lasing parameters; mathematical models of these lasers were developed with different degrees of accuracy so that it was possible to transfer from physical investigations to numerical experiments in order to determine the optimum operating conditions as regards the efficiency or the mean lasing power of the self-contained lasers; new methods of development of these lasers have been formulated. Constant interest in the lasers at self-contained transitions of metal vapours has been maintained by the extensive possibilities of using them in solving various applied problems.

The last and relatively successful attempt for the collective generalization of the results of investigations of self-contained lasers is a collection of the Proceedings of the International Scientific School: Pulsed metal vapour lasers, organised in 1995 at the St. Andrews University (Scotland) under the supervision of NATO. The proceedings, published for this scientific school [8], provides quite complete information on the state and priority directions of current investigations and applications of self-contained lasers. However, the absence in [8] of a united approach typical of monographs which would make it possible to determine the level and mutual correspondence of the results of different investigations again indicates that it is necessary to write and publish a monograph generalizing from a single viewpoint the results of long-term investigations of self-contained metal vapour lasers.

In Russia there are several teams capable of generalizing the results of more than 30 years of investigations of self-contained metal vapour lasers. However, it may be concluded that the team of the authors of this monograph was quite lucky. They obtained funds from the Russian Fund of Fundamental Investigations and this support played a significant role in the publication of this monograph.

The monograph concerned with the self-contained lasers should be started with the retrospective analysis of the history of development of these lasers. This view indicates that in the initial stage of this history the strongest effect was exerted, in addition to [1], also by the studies [9–12]. In [9] the authors formulated the condition of producing effective lasing at self-contained transitions which determined for a long period of time the strategy of finding working metals and investigations of lasers based on them.

In [10] experiments were carried out with the self-heating (as a result of the energy generated in the discharge) operating mode of the copper vapour lasers previously used in ionic lasers [13]. The application of this operating mode has demonstrated the practical possibilities of creating highly efficient repetitively pulsed (pulsed periodic) lasers on metal vapours referred to subsequently as self-heating lasers.

In [11] the double excitation pulses were used for the first time to produce lasing in the vapours of a chemical compound of copper – copper chloride. In fact, the study [11] is the initial study of lasers referred to at present as metal halide vapour lasers.

In the theoretical study [12] concerned with the calculation of the copper vapour laser the authors made the principal step in understanding the lasing kinetics of these lasers and showed that the single-pulse copper vapour laser (the best of the metal vapour lasers) cannot compete in the specific lasing energy with the single pulse molecular high-pressure lasers.

After publishing the studies in [10, 12] it became clear that the most promising direction of the development of the self-contained metal vapour lasers is the construction of effective repetitively pulsed lasers. In particular, the construction of this type of laser (including self-heating lasers) was the subject for a long period of time of theoretical and experimental studies with the results published in the main text of this monograph. It should be noted that the successes achieved in these investigations have been supported to a large extent by the large amount of experience obtained in previous studies of the gas discharge physics, the physics and chemistry of low-temperature plasma, atomic physics, collected by Soviet and foreign scientists and generalized in a number of monographs.

In studies of self-contained lasers the lasing was produced both at neutral atoms and ions in the lasing wavelength range from 312.2 nm (gold vapours) to 645.16 nm (strontium vapours). The most powerful and effective are the copper vapour lasers ($\lambda_1 = 510.5; 570.82$ nm). For this reason, these lasers are used most widely in the industry. In the copper vapour lasers with the average lasing power from several watts to several tens of watts there are sealed-off emitters with the guaranteed operating life of ~1000 hours [14]. Considerable successes have also been achieved with copper bromide vapour lasers. The operating life of the sealed-off emitters of these lasers is at present 500 hours [15] but with their low price the copper bromide lasers are capable of competition in a number of applications with the copper vapour lasers.

The maximum average lasing powers are obtained at present using laser systems consisting of a master oscillator and several amplifiers [16, 17]. The study [16] describes a laser system based on an oscillator and three amplifiers. The power of each amplifier is 250 W at the general output power of the laser system higher than 750 W. The laser system, described in [17], consists of an oscillator and four amplifiers. The maximum average lasing power, taken from a single amplifier, is 560 W with the total power of the laser system of 1902 W. [1]

A significant special feature of the development of self-contained lasers is that regardless of more than 30 years history these development has not as yet been completed and has both short-term and long-term prospects. The short-term prospects include the construction, firstly, of the effective repetitively pulsed copper halide vapour lasers working in the oscillator mode, with the mean lasing power of several hundred of watts and with the practical efficiency of several percent. Secondly, it is the development of highly efficient copper and copper halide vapour lasers with the practical efficiency of 5–8% and the average lasing power in the range from several watts to several tens of watts. This is confirmed by the results of [18, 19]. The work [18] reports on the construction of a laboratory prototype of the CuBr laser formed

---

[1]We intentionally avoid discussing the problems associated with different applications of self-contained layers. This discussion would require writing a separate monograph. The authors.

as a result of the interaction of solid copper with gaseous HBr in the discharge (Cu/HBr laser) with the average lasing power of 200 W and the efficiency of 1.9%, and the study [19] is concerned with the construction of also a prototype of a copper vapour laser with the physical efficiency of 9%. It is possible that in near future the operating mode of the lasers on other metal vapours will also be optimized, not only on copper vapours, with the industrial production of different modifications.

The long-term prospects include evidently the construction of powerful and effective metal vapour lasers and, in particular, copper vapour lasers with a large beam aperture excited by electron beams with the energy of 1–10 keV [20, 21]. At present, considerable successes have been achieved using the so-called open electrical discharge [20] between a coaxial cylindrical cathode and a mesh anode situated in the region of the cathodic voltage drop. A similar excitation method was used for the first time in [22] in the experiments with a continuous helium–neon laser. However, a special feature of the open discharge, used in [20], is that it was possible to use it in pulses with the duration of approximately 1 ns to obtain extremely high densities of the of the electron beam current ($\sim$50 A/cm^2) with the energy on the level of several keV which also determines its potential as an excitation source of self-contained metal vapour lasers.

Finally, it is hoped to use high-voltage repetitively pulsed and open discharges employed at present for the excitation of self-contained lasers in plasma chemistry and in particular in solving problems such as the cleaning of fumes to remove oxides of nitrogen and sulphur and also for the elimination of toxic and odoriferous substances. The experience with the lasers on chemical compound vapours of metals shows that as regards their capacity to dissociate different chemical compounds, the high-voltage repetitively pulsed discharges evidently are not inferior to the periodic corona discharge [23] which is at present one of the most promising methods of cleaning fumes gases to remove oxides of nitrogen and sulphur.

The further prospects for the development and application of both self-contained lasers and the discharges used for the excitation of active media of such lasers are also confirmed by the fact that this monograph is directed not only to the generalization of the results of experimental and theoretical investigations of the energy characteristics of the self-contained lasers but also to the generalization of the results of investigations of high-voltage repetitively pulsed periodic discharges used for the excitation of these lasers. It is hoped that the monograph will be of interest to both experts the area of lasers and quantum electronics and those working in the area of plasma chemistry, plasma physics and gas discharge.

V.M. Batenin
V.V. Buchanov
M.A. Kazaryan
I.I. Klimovskii

References

1. Fowles G.R., Silfvast W.T., Appl. Phys. Lett. 1965. Vol. 6. No. 12. P. 236–237.
2. Vernyi E.A., Self-terminating lasers. In: L. Allen, D. Jones, Fundamentals of physics of gas lasers, translated from English. Moscow, Nauka. 1970. P. 78–81.
3. Petras G.G., Usp. Fiz. Nauk, 1971. Vol .105. Issue 4. P. 645–676.
4. Isaev A.A., Petrash G.G., Research of pulsed gas lasers at atomic transitions. Pulsed gas discharge lasers on transitions of atoms and molecules (Proceedings of the Lebedev Physical Institute. Vol. 81). - Moscow, Nauka. 1975. P. 3–87.
5. Buzhinskiy O.I., Evolution of research of Cu-laser and the possibility of practical application (overview). Moscow, Kurchatov Institute of Atomic Energy. 1983.
6. Soldatov A.N., Solomonov V.I., Gas discharge lasers at self-contained transitions in metal vapor. Novosibirsk, Nauka. 1985.
7. The metal vapor lasers and their halides (Proceedings of the Lebedev Physical Institute. Vol .181). Moscow, Nauka. 1987.
8. Pulsed Metal Vapour Lasers, Proceedings of the NATO Research Advanced Workshop on Pulsed Metal Vapor Lasers, Physic and Emerging Applications in Industry, Medicine and Science, St. Andrews, U.K., Aug. 6-10, 1995: C.E. Little and N.V. Sabotinov, Eds. Dordrecht: NATO ASI Series, Kluwer Academic Publishers. 1996.
9. Walter W.T., Solimene N., Piltch M., Gould G., IEEE J. Quantum Electronics. 1966. Vol. QE-2.No. 9. P. 474–479.
10. Isaev A.A., Kazaryan M.A., Petras G.G., Pis'ma ZhETF. 1972. Vol. 16. Issue 1. P. 40–42.
11. Chen C.J., Nerheim N.M., Russell G.R., Appl. Physics Letters. 1973. Vol. 23, No. 9. P. 514–515.
12. Eletskii A.V., et al., Dokl. AN SSSR. 1975. Vol. 220. No. 2. P. 318–321.
13. Silfvast V.T., Metal-Vapor Lasers. Sci. Amer. 1973. Vol. 228. No. 2. P. 89–97.
14. Zubov V.V., et al., Copper-vapor lasers with sealed-off active elements. Metal Vapor Lasers and Their Applications. CIS Celected Papers, G.G. Petrash, editor. Proc. SPIE 2110. 1993. P .78–89.
15. Marazov O.R., Manev L.G., Optics Communications. 1990. Vol. 78. No. 1. P. 63-66 (Bulgarian Patent No. 9751 / 27.06.1995, Spectronica Ltd.- Bulgaria).
16. Hackel R.P., Warner B.E. The copper-pumped dye laser system at LawrenceLivermore National Laboratory. Laser Isotope Separation: CIS Selected Papers, J.A. Paisner, Editor. Proc. SPIE. 1993. Vol. 859. P. 2–13.
17. Konagai C., Aoki N., Ohtani R., Kobayashi N., Kimura H. Copper Vapor Laser System Development. Proc. 6th International Symposium on Advanced Nuclear Energy Research. P. 637-642.
18. Jones D.R., Maitland A. Little C.E., IEEE J. Quantum Electronics. 1994. Vol. 30. No. 10. P. 2385–2390.
19. Soldatov A.N., et al., Kvant. Elektronika. 1994. Vol. 21. No. 8. P. 733–734.

20. Bohan P.A., Sorokin A.R., Zh. Tekh. Fiz. 1985. Vol. 55. No. 1. P. 88–95.
21. Kolbychev G.V., Optika atmosfery i okeana, 1993. Vol. 6. No. 6. P. 635–649.
22. Klimkin V.I., et al., Helium-neon laser at a wavelength of 1.15 microns with a discharge in the hollow cathode. Conference Proceedings in Electronic Engineering. Gas Optical Quantum Generators. 1970. No. 2 (18). P. 29.
23. Amirov R.Kh., et al., Modelling of processes of purification of the flue gases, initiated by periodic corona discharge. Preprint IVTAN No. 1-403. 1997.

Notations, indexes and abbreviations

A, B – Einstein coefficients for spontaneous and stimulated transitions;

C_S – electric capacity of the storage capacitor;

C_p – electric capacity of the peaking capacitor;

D_a – ambipolar diffusion coefficient;

D_{Cu} – diffusion coefficient of copper atoms;

D_{Ne} – diffusion coefficient of neon atoms;

d_b – the inner diameter of the return current conductor, located coaxially with the discharge;

d_d – the diameter of the discharge;

d_p – inner diameter of the discharge tube (GDT);

E – electric field in the discharge;

E_i – the energy of the i-th level of the atom;

e – the electron charge;

f_{ij} – oscillator strength for absorption in the transition from the i-th (bottom) to the j-th (upper) level;

f – lasing (excitation) pulsed repetition frequency

G – the specific energy in the discharge of the excitation pulse;

G_d – full energy deposited in the discharge of the excitation pulse;

G_D – specific energy input during the dissociation of the pulse, discharge current calculated or the momentum and tension, or energy stored in the storage capacitor;

g_i – statistical weight of i-th level (i = 1, 2, n, a, g, m, r...);

I – intensity of radiation;

I_d – instant discharge current;

I_i the ionization energy of the atom;

I_l – the intensity of the laser radiation;

I_r – current in the rectifier;

j – the instantaneous current density;

k – Boltzmann constant;

L_C – inductance of the discharge circuit;

L_d – discharge inductance;

L_{ch} – charging inductance;

L_S – inductance, shunting the discharge;

l – length;

l_a – the length of the laser active medium;

l_d – the distance between the electrodes (length of discharge);

l_p — the length of the discharge tube;

l_r — the distance between the mirrors of the resonator (cavity length);

M_j — the mass of the heavy particles ($j = a, M, i$, Ne, He, etc., where a – atom, M – molecule, i – ion, Ne – neon atom, He – helium atom);

m — mass of the electron;

n_e — the concentration of electrons;

n_g — concentration of metal atoms in the ground state;

n_i — concentration of ions;

n_j ($j = 0, 1, 2, n......$) – atom concentration on j-th level;

n_M — the concentration of the working of metal atoms, the concentration of the molecules;

n_m — the concentration of atoms in the metastable level (metastable atoms);

np — photon density;

n_r — the concentration of atoms in the resonance level (resonance atoms);

P — instantaneous specific lasing power, the rate of formation of the induced photons;

$\bar{P}$ — time-averaged power density of the laser;

P_l — the instantaneous power of the laser;

$\bar{P}_l$ — time-averaged power of the laser;

$\bar{P}_{ll}$ — time-average laser power per unit length;

P_{lpic} — the peak power of the laser;

P_{pic} — peak power density of the laser;

p — pressure;

p_a — buffer gas pressure ($a = a$, Ne, He, etc.);

Q — spectral density of radiation, specific electrical power supplied to the discharge

Q_d — the instantaneous value of the electrical power input to the discharge during the excitation pulse;

$\bar{Q}_d$ — time-averaged power supplied to the discharge;

$\bar{Q}_{dl}$ — time-averaged power per unit length introduced into the discharge;

Q_r — power consumed from the rectifier;

R — characteristic impedance of the cable line of the excitation unit,

R_{ch} — charge resistance;

R_d — discharge resistance;

R_s — the shunt resistance;

r — radius;

r_d — discharge radius;

r_p — inner radius of the discharge tube;

S_d — cross-sectional area of the discharge;

$T = 1/f$ – lasing pulse repetition frequency (excitation);

T_a — temperature of the atoms;

T_e — the electron temperature;

T_g — gas temperature;

T_i — ion temperature;

T_w — wall temperature of the discharge tube (cell);

t — current time;

U_c — instantaneous voltage on the storage capacitor, the instantaneous voltage at the output of the cable line, the cable of the transformer;

U_d — the instantaneous voltage on the electrodes of the discharge;

U_l — the instantaneous voltage across the inductor;

U_R — the instantaneous voltage on the active resistance of the discharge;

U_r — the output voltage of the rectifier;

U_0 — prepulse voltage on the storage capacitor, amplitude of the voltage pulse on the cable line input (cable inlet transformer);

V_p — GDT volume;

v — the thermal velocity of the electrons;

v_d — electron drift velocity;

W — specific energy (specific laser energy);

W_l — the total energy of the laser.

α_i — ionization rate constant;

α_{ij}, α_{ji} — the rate constants of excitation and de-excitation of the atom (ions, molecules), respectively, by electron impact from level i to level j and the level j to level i;

α_j, ($j = r$, 1, 2, $n...$) – total (complete) constant of the rate of destruction of the level j by electron impact;

β — recombination coefficient;

β_e — the three-body recombination coefficient of electrons;

β_{Cu} — recombination coefficient of copper atoms;

$\Delta v, \Delta \lambda$ — width at half maximum of the spectral lines for frequency and wavelength;

ε — the energy of the electron;

$\overline{\varepsilon}$ — the average energy of electrons;

η_d — efficiency of excitation unit, determining the efficiency of transfer of energy, stored in a storage device, to the discharge;

η_f — total coefficient of efficiency of the laser with all the energy on the creation of metal vapors and their excitement taken into account;

η_p — the physical factor of laser efficiency, defining the energy conversion efficiency, delivered to the discharge during the excitation pulse to generate energy (in some cases the physical efficiency is called the efficiency of the discharge);

$\eta_p(t)$ — the physical coefficient of laser efficiency, defining the energy conversion efficiency, delivered to the discharge in given time (current time, the end time of the lasing pulse, etc.) to lasing energy;

η_r — practical laser efficiency ratio, defining the efficiency of energy conversion, stored in the storage device at the start of discharge to lasing energy;

η_t — dynamic (instant) efficiency of the laser determining the conversion efficiency of the instantaneous electrical power input to the discharge in the laser power;

κ, κ_0 – the amplification factor and the amplification factor at the centre of the line;

λ – radiation wavelength;

λ_g – thermal conductivity of the medium;

λ_l – wavelength of laser radiation;

μ – reduced mass of the colliding particles;

μ_0 – magnetic constant;

μ_i – ion mobility;

ν – frequency of radiation;

ν_{ej} ($j = a, i, m, e$) – the frequency of elastic collisions of electrons with atoms, ions, molecules, electrons;

ν_l – the frequency of laser radiation;

ρ – reflection coefficient;

σ – cross-section of the elementary process, plasma conductivity;

σ_{ij}, σ_{ji} – excitation and de-excitation sections of the atoms (ions, molecules), respectively, by electron impact from level i to level j and level j to level i;

τ_D – characteristic diffusion time;

τ_d – the time delay between the pulses of the excitation and dissociation in lasers excited by two pulses;

τ_{ex} – the duration of the excitation pulse of the laser active medium;

τ_g – lasing pulse duration of the laser;

τ_R – typical response time of a switch in the circuit with direct discharge of the storage capacitor;

τ_r – the characteristic relaxation time of a parameter;

τ_{rel} – the duration of the relaxation of the active medium;

τ_U – voltage pulse duration at the output of a voltage pulse generator (VPG) to matched load.

Indexes

a, e, i, g, m, r – refer to an atom, electron, ion, ground, metastable and resonant levels of the atom;

st, fin – refer to the beginning and the end of the laser pulse (excitation)

opt – refers to the parameters of the laser, optimum for a given characteristic of laser radiation.

Abbreviations and symbols in the text and figures

AZ   – active laser zone;

v|pg   – voltage pulse generator;

GDC   – gas discharge chamber;

GDT   – gas discharge tube;

D   – diode;

PCT   – pulse cablr transformer;

MVPL   – pulsed metal vapor lasers;

PT   – pulse transformer;

S	– switch;
CVL	– copper vapor laser;
RSD	– reversed switched dynistor;
DG	– discharge gap;
TGI	– thyratron;
TC	– tacitron;
EB	– electron beam.
D_{ch}	– charging diode;
Tr	– transformer;
C_S	– storage capacitor;
C_c	– correction capacitor;
L_c	– correction inductance;
L_{conn}	– connection inductance;
L_{ch}	– charging choke, charging inductance;
L_S	– inductance, shunting the discharge;
R_{ch}	– charge resistance;
R_S	– shunting resistance

Introduction

I.1. What is the 'self-contained' transition?

The group of self-contained lasers includes a large number of various gas lasers the common hallmark feature of which is the self-contained lasing (induced transitions) over time, due to the properties of the laser (working) levels, with the schematic representation shown in Fig. I.1. The upper (2) level is efficiently populated as a result of some process (for example, in collisions with electrons or excited molecules) and the lower level (1) is metastable and populated only by the spontaneous decay of the upper level by the laser (working) transition with A_{21} probability.

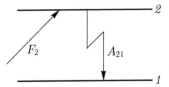

Figure I.1. Ideal scheme of levels with 'self-contained transition" between them.

Assuming that at the initial time ($t = 0$) the concentration of the atoms in the upper n_2 and lower levels n_1 is zero and that the speed of excitation of the upper level F (t) [cm^{-3} s^{-1}] = const, by solving the unsteady balance equations for the concentrations of atoms n_1 and n_2 we easily obtain relations that determine the dependence of these concentrations on time

$$n_2 = \frac{F_2\left(e^{A_{21}t} - 1\right)}{A_{21}e^{A_{21}t}}, \quad n_1 = F_2 t - \frac{F_2\left(e^{A_{21}t} - 1\right)}{A_{21}e^{A_{21}t}}. \tag{I.1}$$

These relationships show that taking into account the statistical weights g_1 and g_2 of the laser levels the inversion of their populations will exist only in the range of a limited time period τ_{inv}, determined by the equation

$$\tau_{inv}A_{21} = \left(1 + g_1/g_2\right)\cdot(1 - 1/e^{A_{21}t_{inv}}). \tag{I.2}$$

At $g_1 = g_2$

$$\tau_{inv} \approx 1.6 A_{21}^{-1}. \qquad (I.3)$$

One of the most important characteristics of the self-contained transition is the limiting energy conversion coefficient (limiting efficiency of the laser transition) η_{lim}, used for the excitation of the upper laser level, to the induced radiation energy

$$\eta_{lim} = \frac{g_1}{g_1 + g_2} \cdot \frac{h\nu_l}{E_2}, \qquad (I.4)$$

where h is the Planck constant; ν_l is the laser radiation frequency; E_2 is the excitation energy of the upper laser level counted from the ground state of the atom, the ion or the molecule.

Coefficient η_{lim} was used for the first time in [1] by analogy with the quantum efficiency of laser transitions taking into account the fact that in the self-contained lasers at the moment of equalisation of the populations of the working levels and interruption of lasing only the fraction equal to $(n_{2\Sigma})g_1/(g_1+g_2)$ takes part in lasing of the total number of atoms excited to the upper level. It should be noted that the relationships (I.1)–(I.4) were derived for the ideal self-contained transition. In practice there is a large number of reasons leading to both a decrease of the duration of existence of the inversion of the populations of the laser levels and a decrease of the efficiency of the lasers in comparison with η_{lim}.

I.2. The conditions of producing effective lasing in self-contained lasers

In 1966, Walter et al [1], analysing the conditions of producing effective lasing at self-contained transitions, formulated five criteria for laser levels, with the fulfilment of these criteria resulting in high efficiency of operations of self-contained lasers.

1. The upper laser level should be resonant, strongly bonded with the ground level by the radiation transition. Since the excitation cross section in the Brownian approximation is proportional to the matrix element of the electrical dipole transition, the fulfillment of this condition ensures a high rate of excitation of the upper level in the presence of a sufficiently high concentration of the electrons in the discharge.

2. The lower laser level should be metastable, not connected with the ground level by the electriuc dipole transition. In most cases this

conditions is associated with the requirement that the ground and lower laser levels have the same parity, and the upper laser level the parity opposite to them. In this case, the excitation cross section of the lower laser level is small and is only slightly populated in collisions of the electrons with the atoms in the ground state.

3. It is desirable for the upper laser level to be optically connected only with the ground and lower laser levels. The electric dipole matrix elements for any other transitions from the upper laser level should be small in comparison with these two levels in order to ensure that there are no shunting transitions of the atoms from the upper laser level. To decrease the probability of the transition from the upper laser level to the ground level, the concentration of the atoms should be sufficiently high (of the order of 10^{13} cm^{-3} in a tube with a diameter of 1 cm) and the effective capture of the resonant radiation should take place. The ground state of the atom should be unique. In the presence of several levels, they should be situated quite closely and have the population, ensuring the capture of the resonant radiation.

The probability of the laser transition should be lower than the probability of the resonant transition which is usually equal to 10^8 s^{-1} but greater than the probability of the relaxation transition, reaching in some cases the value of 1 s^{-1}. The central region of the values of the population A of the laser transition should be in the range $10^7 > A > 10^4$ s^{-1}. If the probability of transition is higher than the rate of excitation of the upper laser level, the spontaneous radiation destructs the upper level prior to reaching the sufficiently high value of the inversion of the population. On the other hand, if the probability of the laser transition is very small, to obtain the adequate amplification it is necessary to ensure the unobtainable density of the population inversion.

To ensure the high efficiency of the laser rates it is also necessary to make sure that the efficiency coefficient (quantum efficiency) of the laser transition, i.e., the ratio of the quantum energy to the difference of the energies of the ground and upper laser levels, is sufficiently high. This requirement restricts the upper position of the lower laser level. On the other hand, the lower laser level should not be populated because of the Boltzmann distribution at the working temperature. It may be assumed that the thermal population of the lower laser level should not exceed 0.1% of the total number of the atoms. This means that the lower laser level should not be higher than the ground level by 6000–18000 cm^{-1} (0.74–2.23 eV).

If all the criteria formulated in [1] are fulfilled, we can expect the lasers with the efficiency on the level of several tens of percent. However, regardless of the fact that the structure of the levels with the self-contained transitions is characteristic of a large number of atoms and ions of metals, with lasing in vapours of these metals, the atoms and ions whose laser levels would fully satisfy the above five conditions have not been found. These criteria are satisfied most efficiently by the laser levels of the copper atom (see Figure I. 2 and Table I. 1). It should be noted that the lasing at the radiation lines of the copper atoms at 510.5 and 578.2 nm was reported for the first time in [4].

Table I.1. Characteristics of laser transitions of the copper atom

Transition	Oscillator strength	Transition probability, s^{-1}
$4p\ ^2P^0_{3/2} \rightarrow 4s\ ^2S_{1/2}$	0.43 [3]	$1.37 \cdot 10^8$
$4p\ ^2P^0_{3/2} \rightarrow 4s\ ^2D_{5/2}$	0.0051 [3]	$1.97 \cdot 10^6$
$4p\ ^2P^0_{1/2} \rightarrow 4s\ ^2D_{3/2}$	0.0011 [3]	$2.3 \cdot 10^5$
$4p\ ^2P^0_{1/2} \rightarrow 4s\ ^2S_{1/2}$	0.22 [3]	$1.38 \cdot 10^8$
$4p\ ^2P^0_{3/2} \rightarrow 4s\ ^2D_{3/2}$	0.042 [3]	$1.68 \cdot 10^6$

Analysis of the reasons ensuring to various degrees the correspondence of the characteristics of the laser levels of the copper atoms to the criteria formulated in [1] was carried out in [2]. The most significant results of this analysis may be described as follows. The basic configuration of the copper atom $3d^{10}4s$ corresponds to the single level $^2S_{1/2}$, strongly bonded with the nearest excited state of the opposite parity $3d^{10}4p$ by the electrical dipole transition. Since the $3d$-electrons in the copper atoms are not bonded strongly, the states, corresponding to the excitation of one of these electrodes are situated at a considerable depth. The deepest from them $3d^94s^2$ has two metastable levels $^2D_{3/2}$ and $^2D_{5/2}$ with the same parity as the ground state (Fig. I.2). The transitions from the $3d^{10}4p$ configuration levels to the $3d^94s^2$ configuration levels are no forbidden, but they are associated with the simultaneous change of the orbital quantum number of two electrons. Therefore, these transitions are less likely to take place than the single-electron transitions. The quantum efficiency of the transition $^2P^0_{3/2} \rightarrow {}^2D_{5/2}$ (λ_1 = 510.554 nm) is more than 60% because the lower level is quite deep. At the same time, the thermal population of the level at the working temperature is not large. A shortcoming of the scheme of the levels of the copper atom is the presence of two laser transitions to the levels $^2D_{5/2}$ and $^2D_{3/2}$ (the latter, $^2P^0_{1/2} \rightarrow {}^2D_{3/2}$, corresponds to the lasing line 578.213 nm), and also

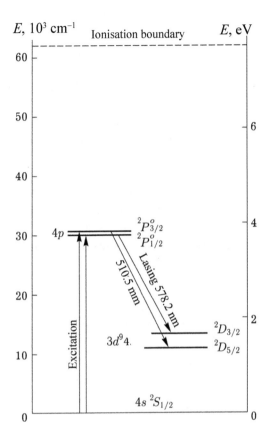

Figure I.2. The diagram of the lower levels of the copper atom.

the relatively high probability of population of other states which do not take part in the lasing.

Regardless of the certain discrepancy in the parameters of the laser transitions of the copper atoms with respect to the parameters of the ideal self-contained transition, the characteristics of the laser levels of the copper atom, as already mentioned, correspond quite satisfactorily in comparison with the characteristics of the laser levels of other atoms to the above-mentioned criteria, and this also explains the fact that in comparison with other lasers the copper vapour self-contained lasers have the highest lasing power and efficiency.

I.3. Lasing characteristics of self-contained lasers

The previously mentioned restricted duration of existence of the innovation of the populations of the working levels in the self-contained lasers predetermines the pulsed methods of excitation (pumping) of these lasers. In order to ensure during the excitation pulse the preferential population of the Apple laser level, the average energy (temperature) of the electrons and the strength of the electric field which determine their values in the discharge should be sufficiently high and, consequently, the voltage applied for a short period of time to the electrodes of the GDT should also be high. In the majority of cases, the excitation pulse (Figure I.3) of the active medium of the self-contained lasers is produced by the discharge of a storage device (for example a capacitor) using some switch (for example, thyratron) through the active medium of the laser having the form of a mixture of the vapours of the working metal and the inert gas and filling either the gas discharge tube (GDT) at a longitudinal electrical discharge, or a gas discharge chamber (GDC) at a transverse electrical discharge. The duration of the excitation pulses τ_{ex} is selected from the condition

$$\tau_{ex} < A_{21}^{-1} \tag{I. 5}$$

The lasers which use the single excitation pulses are referred to as the single-pulse lasers, and the lasers in which the excitation pulses follow with a regular frequency f are referred to as repetitively pulsed. In these

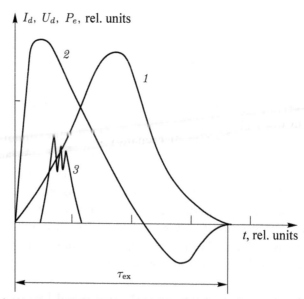

Figure I.3. Typical pulses of discharge current (1), voltage at the electrodes of the GDT (2) and the induced radiation (lasing) (3) during the excitation pulse of the active medium of the self-contained lasers.

lasers, the excitation pulses of the active medium of the laser alternate with the periods of relaxation of the medium. In the stationary operating mode of the repetitively pulse lasers the period of the excitation pulses is characterised by an increase of the electron concentration n_e and the working atoms in the metastable n_1 and the resonant n_2 states and by a decrease in the concentration n_g of the working atoms in the ground state. In the relaxation period n_e, n_1, n_2 decrease and n_g increases to the pre-pulse values.

The energy characteristics of the lasing pulses of the self-contained lasers include a large number of parameters, especially: instantaneous intensity I_l, power P_l, and specific lasing power D of the laser, referred to later in the majority cases simply as intensity, power and specific power. In the case of a longitudinal discharge with a cylindrical GDT I_l, P_l and P are linked together by the following relationships:

$$P_l = 2\pi \int_0^{r_p} r I_l(r)\,dr,$$

$$(\text{I. }6)$$

$$P_l = 2\pi \cdot l_p \int_0^{r_p} r P(r)\,dr,$$

where r, r_p, l are respectively the actual radius, the radius and the length of the GDT. Here and later it is assumed that the active medium fills at the entire volume of the GDT.

The relationship of the specific lasing power P and the lasing power P_l of the laser with the specific lasing energy W and the lasing energy W_l, recorded in a single pulse, is determined by the equation:

$$W = \int_0^{\tau_g} P\,dt, \quad W_l = \int_0^{\tau_g} P_l\,dt. \qquad (\text{I.7})$$

The peak (maximum power in the lasing pulse) power P_{lpic} and specific peak lasing power P_{pic} are linked with W_1 and W by the relationships:

$$W_l \approx P_{lpic} \cdot \tau_{g/2}, \quad W \approx P_{pic} \cdot \tau_{g/2}, \qquad (\text{I.8})$$

where $\tau_{g/2}$ is the duration of the lasing pulse at half height.

The mean (with respect to time) lasing power E_l and also specific P and the unit $\bar{P}_{ll}$ lasing powers of the repetitively pulsed self-contained lasers are:

$$\bar{P}_l = W_l \cdot f, \quad \bar{P} = W \cdot f, \quad \bar{P}_{ll} = \bar{P}_l / l_p. \qquad (\text{I.9})$$

The efficiency of operation of the self-contained lasers is characterised by means of several efficiency coefficients.

The physical efficiency coefficients $\eta_p(t)$ and η_p determine the efficiency of conversion of the electrical energy supplied to the discharge during the excitation pulse, to the lasing energy at the given moment of time t, counted from the start of the excitation pulse (actual time, the time of completion of the lasing pulse) and at the end of the excitation pulse:

$$\eta_p(t) = \frac{W_l(t)}{G_d(t)},$$

$$\eta_p = \frac{W_l}{G_d} = \frac{\bar{P}_l}{\bar{Q}_d} = \frac{\bar{P}_{ll}}{\bar{Q}_{dl}}, \qquad (I.10)$$

$$\bar{P}_{ll} = \bar{P}_l / L_d, \bar{Q}_d = G_d \cdot f, \bar{Q}_{dl} = \bar{Q}_d / l_d;$$

Here, in addition to the well-known notations, $W_l(t)$ and $G_d(t)$ are respectively the lasing energy and the energy input to the discharge during time t; G_d is the energy supplied to the discharge during the excitation pulse; $\bar{Q}_d$ and $\bar{Q}_{dl}$ are the time-averaged electrical power and the unit electrical power, added to the discharge from the excitation unit.

The dynamic efficiency of the laser characterises the efficiency of conversion of electrical power, added to the discharge, to the laser radiation power

$$\eta_t = \frac{P_l}{Q_d}, \qquad (I.11)$$

where Q_d is the instantaneous value of the electrical power, supplied to the discharge during the excitation pulse.

The practical efficiency of the laser η_r determines the efficiency of conversion of the energy, stored in the storage device at the start of the discharge, to the lasing energy. For the excitation circuit with the storage capacitor

$$\eta_r = \frac{2W_l}{C_S U_0^2}, \qquad (I.12)$$

where C_S is the electrical capacitance of the storage capacitor, U_0 is the voltage at the capacitor prior to the start of the discharge (pre-pulse voltage).

The practical efficiency for the repetitively pulsed lasers is often determined as

$$\eta_r = \frac{\overline{P}_l}{Q_r}, \qquad (I.13)$$

here Qr is the power required from the rectifier.

Finally, when determining the general efficiency η_f we take into account the energy losses not only for the excitation of the discharge but also for all auxiliary operations: heating of the thyratrons, maintenance of the working temperature of the GDT or GDC, etc. The determination of the general efficiency has any meaning only for the repetitively pulse lasers in the form of the ratio of the mean lasing power to the total power required from the electrical circuit and used not only for the excitation of the laser but also, for example, for heating of the thyratrons and the external heating of the GDT or GDC.

If the self-contained lasers generate radiation at more than one half length, all the above characteristics can be determined for both the single wavelength or for several wavelength. In the latter case, they are referred to as summary and will be denoted by the subscript Σ.

References

1.   Walter W.T., et al., IEEE J. Quantum Electronics. 1966. Vol. QE-2. No. 9. P. 474–479.
2.   Vernyi E.A., Self-contained lasers. In: L. Allen and D. Jones: Fundamentals of the physics of gas lasers. Translated from English, ed. E.A. Vernyi. Moscow, Nauka, 1970. 78–81.
3.   Kasabov G.A., Eliseev V.V., Spectroscopic tables for low-temperature plasma. Moscow, Atomizdat, 1973.
4.   Walter W.T., et al., Bull. Amer. Phys. Society. 1966. Vol. 11. No. 1, 113.

Devices and methods for producing metal vapours

1.1. First designs of metal vapour lasers

The first mention of the early works with metal vapour lasers generated at the atomic transitions from the resonant to metastable levels can be found in [1–3], which relate to 1965 and 1966. At that time there were great difficulties in the design of high-temperature cavities containing vapours of high-temperature metals, particularly copper and gold. As shown in [4], for the efficient lasing at self-contained transitions of metal atoms it is necessary to achieve densities of 10^{14}–10^{15} cm^{-3}, which requires reaching temperatures of 1500–1700°C for gas-discharge tubes.

Although by this time there were high-vacuum furnaces in which higher temperatures could be achieved, the direct use of this technique was very difficult due to the specifics of the construction of special laser gas-discharge tubes.

This was due to the fact that, firstly, for laser tubes it is essential to supply working electrodes into the high-temperature area, and secondly, there is a need for the withdrawal of optical radiation from the high-temperature zone of the discharge tube. The first condition in those days did not have technical solutions, and the second condition was due to the lack of high-temperature optical materials for manufacturing of optical output windows to output the laser radiation to the visible spectrum.

Thus, there was a need for the construction of new laser tubes for use with low-volatility metal atoms. In [3] this problem was solved as follows. The laser tube was divided into two zones, high-temperature and low-temperature. The central zone of the laser tube where the gas discharge took place was placed in a high-temperature furnace with a platinum–rhodium alloy wire, and the electrode zones and output

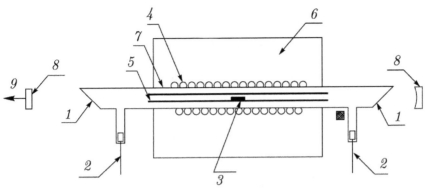

Figure 1.1. Design of the first laser systems for lasing in manganese and copper vapours [5]. 1) the output laser windows; 2) electrodes of the gas-discharge tube; 3) a piece of working substance (copper); 4) 60% Pt–40% Rh (heater); 5) ceramic tube based on aluminium; 6) high-temperature furnace; 7) vacuum tube; 8) the mirror of the resonator; 9) output radiation.

window zones were located outside the high-temperature zone and had low temperature – up to room temperature. The buffer gas – helium – was used to initiate the discharge in the cool areas. The same buffer gas that fills the entire laser tube protected the cold windows from contamination with the working metal as well as some other substances formed during the sputtering of the electrodes when a high-voltage high-power discharge formed. In this design, the working substance was located in the central zone of the third part of the laser tube and in the process continuously diffused to the cold end of the tube. This process, along with the phenomenon of cataphoresis, apparently limited the actual life of such lasers. In these early works the buffer gas pressure was chosen relatively low and, therefore, the diffusion rate was high, which severely limited the service life of lasers and reduced their practical value.

Figure 1.1 [5] shows such a laser tube in which lasing in copper vapours was realized for the first time.

1.2. Self-heated lasers

1.2.1. Thermal regime of discharge laser tubes

To achieve high efficiency in practice it is necessary to reduce the energy consumption for heating the tubes. In this regard, we discuss briefly the thermal regime of the tube when it is heated by the heat generated during a pulsed discharge.

Consider a stationary mode for thermal processes. In this case, the power converted to heat per unit length of the tube is $Q_{dl}(1-\eta)$, where η is the lasing efficiency. In this case, we assume that the tube walls are impervious to radiation from the plasma. In the steady state the heat transfer from the walls of the tube is equal to the heat flux arising in it. Hence

$$2\pi r q(T_w) = Q_{dl}(1-\eta) \qquad (1.1)$$

where $q(T)$ is the heat flux per unit area of the tube at the tube wall temperature T_w. (For simplicity, we neglect the wall thickness of the tube). The value of heat transfer $q(T_w)$ for a given T_w obviously depends on the method of heat removal and usually increases rapidly with increasing temperature T_w. For temperatures $T > 1000°C$, unless special measures are taken, the dominant role is usually played by heat radiation. In this case,

$$q(T_w) = \varepsilon_w \sigma T_w^4, \qquad (1.2)$$

where ε_w is the absorption coefficient of the tube walls (emissivity), T_w is absolute temperature, $\sigma_T = 5.7 \cdot 10^{-12} = W/(cm \cdot K^4)$. At high temperatures for the materials of interest to us $\varepsilon_w \approx 0.5$ and often more.

When working with metal vapours, the discharge tube wall temperature is determined by the need to maintain a specific density of working atoms. To illustrate the examples here and below we will usually use a copper vapour laser. For this type of laser, according to the available data, the operating temperature, as already mentioned, is 1500–1600°C. In this case, the heat transfer by radiation for tubes 1 cm in diameter will be approximately 100 W/cm or 10 kW for a tube 1 m long. With increasing diameter of the tube the heat sink increases in proportion to the diameter. Thus, at these temperatures high powers can be introduced to the tube of a relatively modest size without significant overheating of the walls.

Obviously, when working with lower average powers, if heating of the tube by discharge is considered, it is necessary to insulate the working tubes.

The question arises whether we can introduce the power obtained by calculations into the tubes of a reasonable size. The answer to this question depends, of course, also on the possibility of heat transfer of such powers from the inner wall of the tube to the external one and on further heat removal from the outer surface of the tube. If the heat sink from the outer surface of the wall, depending on its temperature, can generally be produced in different ways (cooling with water, heat removal by radiation, flow of liquid metals), the heat transfer through the walls of the tube, which is impervious to plasma radiation, is given by

$$Q_{dl} = \frac{\pi(T_1 - T_2)}{\frac{1}{2\lambda_{T_1}} \ln \frac{d_2}{d_1}}, \quad (1.3)$$

where T_1 and T_2 are temperatures on the inner and outer surfaces of the tube wall, with $T_1 > T_2$, d_1 and d_2 are respectively the inner and outer diameters of the tube, λ_{T_1} is the thermal conductivity of the tube material at temperature T_1, Q_{dl} is the power transferred by heat transfer. Estimates based on formula (1.3) for one of the materials (BeO) with the highest thermal stability at such high temperatures, which has high thermal conductivity, showed that for the tubes of a reasonable size it is possible to use such high input powers. In some cases, heat removal from the outer wall of the tube can be carried out by radiation, which is very convenient.

As an example, the following section gives the parameters of a BeO tube for a copper vapour laser, which allows the tube to receive without overheating, for example, a power of 850 W/cm and dissipate it by heat radiation: $d_1 = 8.8$ cm, $d_2 = 10$ cm, $T_1 - T_2 = 150$ K at $\lambda_{T_1} = 17$ W/m·K [6]. From the above example it follows that the temperature difference is small and the ceramic material mentioned is quite strong and at this temperature gradient can easily survive the stresses due to the different thermal expansion of the inner and outer walls of the tube.

Thus, the estimates show that a fairly high power can be introduced into the laser tube. When the efficiency of these lasers is the order of several percent, we can obtain high average radiation powers in the visible spectrum.

1.2.2. Design of structures with self-heated laser tubes

As mentioned above, to ensure heating of the working tubes by the heat generated in the discharge, in practical terms it is required to reduce the heat transfer from the tube. A lot of types of self-heated tubes with different sizes, design features and methods of insulation have been studied. In the first versions, insulation with a simple vacuum jacket was used. The design of one of the options for the laser tube is shown in Fig. 1.2. The tubes were made of quartz and the vacuum jacket surrounded

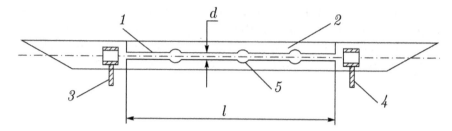

Figure 1.2. Design of the discharge tube with thermal insulation by a vacuum jacket [7]. 1) the working section of the tube; 2) vacuum jacket; 3, 4) electrodes; 5) expanded area for placing the working substance.

the working part of the tube. Cold electrodes were placed in wide end parts. The working medium was placed in the expansion of the active part of the tube. In such tubes it can be fairly easy to raise the temperature of the working section to 800–1000°C and produce, for example, pulsed lasing at the transitions of thallium and lead [7].

However, a further increase of the operating temperature of the quartz tube is not possible. In this regard, to work with manganese, gold and copper vapours it is recommended to use discharge tubes where the working part is made of alundum. At temperatures above 1000°C major heat transfer takes place by radiation and therefore insulation by means of a simple vacuum jacket is not efficient. In principle, heat shields can be used. However, this method of thermal insulation is relatively complex to implement. One can use a simpler method of thermal insulation by means of a powder with a low thermal conductivity and impervious to thermal radiation. The main problem that arises here is that, on the one hand, it is necessary to ensure free thermal expansion of the working tube during its heating to temperatures of 1500°C, but on the other hand – vacuum tightness of the whole structure must be maintained. The problem can be solved quite easily if vacuum-tight connections of the working tube with the heat-insulating powder are not disrupted.

In this case, however, it remains unclear how much gas emission from the powder, which will necessarily occur under heating, may affect the lasing conditions. This question was clarified experimentally. After testing several variants [8] a discharge tube of simple design and very simple to make was proposed, which allowed to obtain very good lasing performance.

The design of the discharge tube, with which the author of [8] first obtained lasing in the self-heated mode, at the 510.5 and 578.2 nm lines in a copper atom, is shown in Fig. 1.3. The working part of the discharge tube was made for the first time from alundum tubes with the

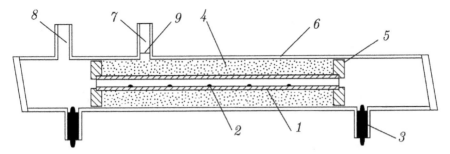

Figure 1.3. Design of the discharge tube with Cu vapours [8]. 1) alundum tube; 2) working substance; 3) electrodes; 4) heat-insulating layer; 5) ring-shaped gaskets; 6) quartz jacket; 7, 8) outputs to the vacuum system; 9) fine metallic mesh.

Al_2O_3 content of not less than 50%. The heat resistance of such tubes is 1700°C, while the normal operating temperature for which they are calculated is about 1350°C which is not enough to work with copper at a temperature of 1500–1600°C. In this regard, after a certain number of hours of operation of the laser the tube were destroyed – especially rapidly if the work was done in the forced mode at which the lasing power was maximum. When working at a power level of approximately half the maximum such tubes could work for a total of 20–30 hours.

Almost all alundum tubes broke down not as a result of the departure of copper from the active zone but as a result of failure due to overheating when trying to get maximum power. The latest versions of the discharge tubes use corundum tubes with the Al_2O_3 content ≥ 98%. This material is good enough to withstand heating to operating temperatures, without showing any signs of damage.

The ceramic discharge tube is mounted on the axis of the quartz tube with inner liners made of lightweight distensilimanite which is quite refractory and easily handled. The tube with some friction could expand during heating by moving into the liner. The thermal insulation of the working tube was provided by zirconia powder, which was chosen because of its high thermal stability and low thermal conductivity. When assembling the discharge tube the annular insert and the alundum tube are first inserted. The zirconia powder is placed in the gap between the alundum tube and the outer quartz tube and thoroughly rammed. After this the second liner is inserted. When evacuating the tube, the cavity filled with zirconium dioxide is initially pumped. To stop the powder from being drawn into the pumping system, the latter is closed with a fine metal mesh through which the evacuation took place. Pumping of the rest of the tube begins only after the forevacuum is reached in the cavity with zirconium dioxide. Otherwise, due to the pressure difference the powder is ejected to the tube window and the area of the electrodes. After preliminary evacuation the tube is treated with a discharge in an inert gas (usually neon) with successive evacuation by a diffusion pump and with a gradual increase in the average power level and the tube temperature. During this treatment zirconium dioxide and other parts of the tube are degassed. It was observed that the degassing process is greatly accelerated if a pre-calcined powder is sued. After treatment, the tube was filled with a buffer gas at a pressure 0.13–2.7 kPa.

If the finished quartz blank is available, the entire process of the preparation of the tube together with treatment takes 1–2 days. In the process, it became clear that, after appropriate treatment, it is possible to ensure relatively stable lasing conditions. This suggests that the generation of a small amount of the gas in the tube and impurities of other gases (e.g. air) has no strong effect on the lasing characteristics. Neon is typically used as the buffer gas.

Lasing was also observed with helium, argon, xenon and air used as a buffer gas. Full degassing of the discharge tube in the laboratory could not be achieved. Heating of the discharge tube above the temperature at which it was degassed led to gas evolution and, consequently, to an increase of the buffer gas pressure.

For the sake of simplicity cold thoriated tungsten electrodes were used in the tubes which were produced from IFP-2000 or other similar quartz lamps. In some cases, when working on significant levels of average power, these electrodes were rapidly heated up which sometimes resulted in cracking and destruction of the electrodes. It was noticed that heating can be reduced if a low-melting metal such as tin is placed in the electrodes. This increased the service life of the electrodes.

Thus, the results were used to create a laser tube, working with self-heating, for the gas lasers on vapours of low-volatility components that require high temperature in the gas-discharge chamber [9,10]. In this laser design, there is no heating furnace, and the discharge tube is heated automatically by the energy released in a pulsed discharge with a high repetition rate [11].

To eliminate possible leakage of impurity gases into the discharge channel the study [12] describes the design of the laser tube in which the heat insulation is provided by heat shields, and the region of the discharge channel can be separated from the rest, in particular, by silicone rubber gaskets and suspended using welded stainless steel bellows. With this method of fixing the tube can freely expand and contract during heating and cooling. The area with the heat shield can be operated in atmospheric conditions, although it is preferred to work in an inert atmosphere or vacuum as in these conditions the possibility of deterioration of the laser due to leakage is minimized.

Fundamentally different structures, so-called 'coaxial tubes', of metal vapour lasers have been considered in [13–16], which led to the specific lasing characteristics comparable to the characteristics of lasers of small transverse dimensions of the gas-discharge tubes. In [13], when working with a copper vapour laser with longitudinal discharge, it was shown that when the ratio of the radii of coaxial tubes, forming the annular working volume, tends to unity the energy input [17, 18] increases tens of times (reaching 50 kW/m) compared with the chamber of the same size, but without the coaxial inserts. In the real conditions, for example at the diameters of the external and internal tubes of, respectively, 20 mm and 3 mm, the calculated limiting energy input was approximately twice as high; in another case, respectively, at 45 mm and 30 mm, the experiments showed a 75% increase in power density [14].

The design of the laser with two cylindrical electrodes placed concentrically in a coaxial line, where the outer electrode is connected to the metal case of the laser, is described in [16]. Lasing on transitions of the copper atom in a pulsed radial–transverse discharge is reported in this study. This design is very compact and can be used to work with other metals.

1.3. The explosive method for producing metal vapours

Explosion of a wire made from the working medium was used for the creation of the active medium of lasers at self-contained transitions of metal atoms for the first time in [19]. This method is proposed, first, to form a sufficient density of copper vapours in transient conditions, and secondly, it was hoped to lengthen the lasing pulse due to the removal of metastable atoms from the active medium because of the relatively high speed of movement of copper atoms. A vacuum chamber was constructed for this purpose in which there were two devices for exploding copper or gold wires, made in the form of an atomic gun. Some time after the explosion when the copper atoms reached the axis of the gas-discharge tube a high voltage of 20 kV was applied to the electrodes. Lasing was observed at the 5105 Å copper line and the 6278 Å gold line.

The lasing power depended on the time delay between the pulse to explode the wire and the pumping pulse. Another feature was that it was noticed that the lasing pulse duration significantly lengthened

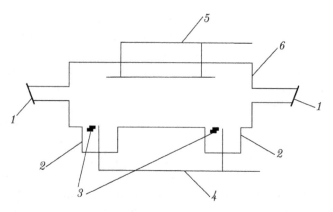

Figure 1.4. Design of a tube for explosive production of the active medium [19]. 1) output windows of the laser tube; 2) the section of the atomic gun; 3) wires of working substance; 4) the cathode; 5) the anode; 6) the vacuum chamber.

compared with the pulse duration in the ordinary copper vapour laser with the stationary thermal method of producing vapours. This was followed by the construction of other similar devices for pulsed vapour generation [20, 21]. However, in all these experiments the lasing pulse time does not exceed the radiative decay time of the upper laser level, so this method can hardly be regarded as a fundamental solution of the problem concerning the control of the lasing time of the copper vapour laser. In all likelihood, the change in the lasing pulse time must be considered as a change in the kinetics of excitation and relaxation due to using the combined method of creating an active environment.

Figure 1.4 shows the design of a vacuum explosive and gas-discharge chamber for creating an active environment.

A simpler device is considered in [22], where copper vapours were produced by electrical explosion of copper wires directly in the discharge chamber. Excitation was conducted at a fixed delay relative to the explosion of the wire (300–700 μs), within which the density of copper vapours and the laser pulse energy were practically unchanged. The density can vary from $4 \cdot 10^{17}$ to $8 \cdot 10^{17}$ cm, depending on the weight of the exploded wire.

The methods proposed in [23, 24] are similar in the physical nature to the method of obtaining the vapours of the working metal by the electric explosion of conductors. It should be noted that the work towards the creation of different methods for obtaining the vapour of the working metal contributed significantly to improving the understanding of the physical processes determining the limiting parameters of metal vapour lasers.

1.4. Pumping laser systems

Due to the fact that when the average pumping power is above some critical value the active medium is overheated [17] and the lasing parameters begin to deteriorate rapidly, it is necessary to produce a sufficiently rapid flow of the working mixture in a closed working volume of the laser. In [25] the authors proposed a basic design of a pumping laser, which consists of two coaxial cylinders–electrodes between which an electrical discharge takes place. The length of the active zone is usually an order of magnitude larger than the size of the cylindrical tubes. The required vapour density is achieved by heating the outer cylindrical tube with a low-voltage source. An external longitudinal magnetic field is applied to the discharge in a direction perpendicular to the vector of the strength of the electric field. In this case the active medium is rotated inside the discharge chamber under the influence of the ponderomotive force, which in principle leads to an increase in the heat transfer rate of gas and in the annihilation rate of the electrons. In certain experimental conditions, the application of the magnetic field led to a marked increase in the output lasing power. In some cases, the rotation of the irregularities at several revolutions per second was observed visually.

In another study [26], attention was given to a physical model of self-pumping of the working medium through the gas discharge gap, based on the results of experimental work [25]. The case where after the acceleration in the discharge chamber the working medium penetrates into a diffuser, a refrigerator and an accelerating nozzle, moving in a closed circuit, was investigated. The working mixture consists of neon or helium with copper vapours, at concentrations $n_{Cu} = 10^{15}$–10^{16} cm^{-3}. The concentration of the buffer gas is typically two orders of magnitude higher. The corresponding closed loop is shown schematically in Fig. 1.5, which is conventionally developed along the x axis so that the section 4 coincides with section 4'. Laser radiation propagates along the z axis.

The results of numerical experiments show that at the most common parameters of the pumping pulses of the copper vapour laser – current density ~100 A/cm^2 – the light He–Cu working mixture can be accelerated in the closed loop to speeds of ~300 m/s in strong magnetic fields of ~5 T. At the same time, the energy consumption for pumping the working environment in these cases is small and constitutes 1–10% of the fraction of the energy used to excite the working medium of the laser. The estimates made in the study show that the pulse repetition rate can be increased up to ~50 kHz, while providing a complete change

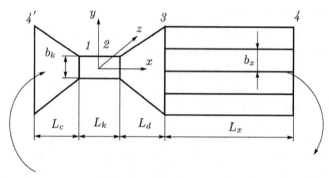

Figure 1.5. Detailed scheme of the circuit [26]: section 1–2 – channel, 2–3 – diffusor, 3 –4 – refrigerator, 4' – 1 – nozzle.

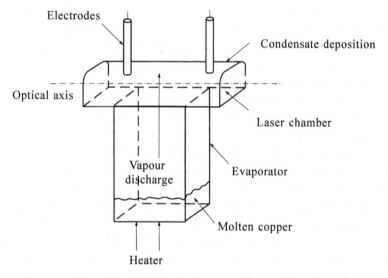

Figure 1.6. Schematic of the system [27].

of the working mixture in the chamber, which can lead to a significant increase in the average lasing power.

Among the many method of pumping a gas flow through the active zone, the simplest method is probably the one based on the use of gravitational forces to create a closed loop [27]. The design of such a tube is shown in Fig. 1.6. Working metal vapours travel into the active zone directly from the evaporator made of boron nitrite. Deposited on the relatively cooler parts of the device, the working substance in the liquid phase is returned to the area of the evaporator. These devices allow to realize a high repetition frequency in copper and lead vapour lasers, ≥ 50 kHz. Moreover, in the case of the lead vapour laser it is not essential to use a buffer gas. This method is very useful for constructing a sealed-off laser, since there is virtually no loss of the working substance in the operation of the laser.

Pulsed lasing in a modified version in the flow of copper vapours, produced in a plasma accelerator, was reported in [28]. Copper vapours formed by erosion of the electrodes of a pulsed plasma accelerator. The distance from the outlet of the accelerator to the lasing axis was 6 cm. The length of the excitation zone was chosen to be 10 cm, and the distance between the electrodes in the transverse dimension was 1 cm.

The method of supersonic discharge of metal vapours to create an active medium, proposed in [29], was developed to improve the performance of copper vapour lasers. It was assumed that the lasing pulse time can be greatly increased, although the system can in principle also increase the pulse repetition frequency.

Figure 1.7 shows the simplified scheme of a laser cell with the supersonic outflow of gas.

A mixture of argon and helium, flowing into the cell 2, is heated to high temperatures and, entering the mixing chamber, entrains the finely divided copper powder. Then, after evaporation of copper the supersonic outflow of the mixture of the buffer gas and the copper atoms through a nozzle into vacuum forms. The electric discharge, commonly used for excitation of copper vapour lasers, results in the generation of laser radiation on green and yellow lines of the copper atom. At the same time, this method is characterized by a relatively long lasing time, ~185 ns. However, the specific characteristics, such as peak power and pulse energy, are inferior to the characteristics of conventional gas-discharge lasers. The study concludes that limiting of the duration of the lasing pulse time is determined by a sharp increase in the rate of arrival of the populations of atomic levels to the equilibrium population,

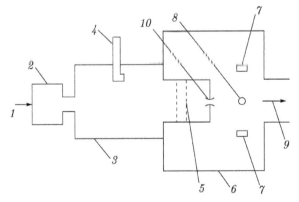

Figure 1.7. The laser cell [29]. 1) gas inlet; 2) arc heating chamber; 3) mixing chamber; 4) system for blowing in copper powder; 5) evaporator; 6) vacuum chamber; 7) gas discharge electrodes; 8) resonator mirror; 9) gas pumping; 10) supersonic nozzle.

as well as by a decrease in the relative pumping rate of laser levels from the ground state due to the fall of the electronic temperature.

1.5. The flow of fine particles as the active medium of metal vapour lasers

The methods of introduction of vapours into the gas discharge tube in metal vapour lasers, discussed in the previous sections, have a common drawback which does not allow high densities of working atoms to be reached. For example, obtaining high-density vapours by thermal heating [5] has, firstly, a limitation associated with considerable difficulties in establishing reliable laser vacuum cavities for operation at temperatures above 1500°C, which is necessary for work with the vapours of many refractory metals and compounds; secondly, because the effective lasing on atomic transitions in metal vapours takes place at self-contained transitions and the lower level is metastable lying close enough to the ground level, and when temperature becomes too high the inverse population may be affected by the population of the lower level and this may lead to reduced lasing generation or even to its breakdown.

Another method – the explosion of wires [21] – of course produces high concentrations of atoms (e.g. copper, 10^{17}–10^{18} cm^{-3}), however, depending on the power consumed for the explosion of a wire, the lasing can be obtained only with a delay of the excitation pulse and, in addition to this, there may form undesirable density inhomogeneities in the expanding stream of particles that can affect the optical quality of the laser beam. Another important fact is that the repetition frequency of lasing in such systems will be severely limited, due to the rate of mechanical change of the wire after an explosion by a new one.

It should also be noted that the physical processes occurring in the explosion of a wire in vacuum depend to a large extent upon the circumstances of the explosion. All this complicates the kinetics of examination of electronic processes in the plasma and introduces an element of instability in the laser output characteristics both in terms of lasing power and beam divergence.

And, finally, the gas-dynamic method for producing atoms of the working medium [29] also fails to achieve relatively high densities of atoms due to the fact that the rate of diffusion 'departure' of atoms is high, and although the atoms concentration can be increased at high discharge speeds, this is accompanied by a large increase of the consumption of copper.

To some extent, these shortcomings can be overcome if fine particles, composed of the working substance, are supplied to the working cavity, and the particle size must be smaller than the wavelength of laser radiation [30].

The concept of the method described in [30] is that the evaporation in a stream of an inert gas produces quite monodispersed particle with the size of 10^{-5}–10^{-6} cm. The size of the particles depends on many parameters, such as the evaporator temperature, flow rate, density of inert gas, and others.

Getting into the laser cavity, these particles evaporate under the action of a high-frequency pulsed gas discharge. The characteristics of the pulsed discharge are selected in such a way that the discharge effectively excites the atoms of the working substance, such as copper, and leads to the creation of an inverse population. Here, however, we must take into account the fact that the fine particles in transit can coalesce into larger conglomerates [31] and, therefore, it is necessary to design the system in such a way that the fine particles form in the immediate vicinity of the discharge chamber. Otherwise, the situation in the formation of large conglomerates may be close to that which occurred in Ref. 24 which used relatively large particles of copper oxide (CuO and Cu_2O) with dimensions of 0.1 mm, with the latter placed in the form of powder in the discharge tube. However, in this case full or partial vapourization of large particles requires several orders of magnitude more energy than that required to excite the atoms, which leads to a sharp drop in the efficiency of the system. In addition, the low rate of renewal of the medium leads to a low repetition rate, and thus, ultimately, to a low average power.

In contrast to conventional designs of gas-discharge laser tubes, lasers on the flow of fine particles contain an additional site in which the metal is evaporated as atoms, and further mixing with a relatively cold stream of the carrier gas results in the formation of fine particles due to condensation.

Structurally, this unit consists of a thermal evaporator of metal of various designs, with an annular input for the supply of the carrier gas. Figure 1.8 shows several designs of laser tubes.

For simplicity, the housing for the evaporator is made of glass or quartz, but due to partial settling of the metal vapours on the walls of the housing the latter is rapidly heated during work in some cases, resulting in a loss of vacuum insulation (the cover and housing the evaporator are sealed using a flat section and vacuum grease). Therefore, in some designs the evaporators were made of steel and cooled with water.

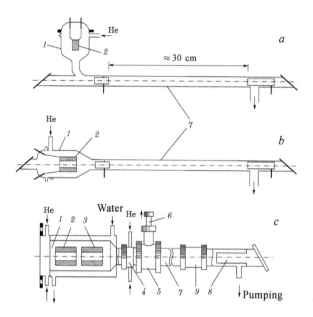

Figure 1.8. Design of laser tubes [30] (*a, b, c*). 1) evaporator housing; 2,3) evaporators; 4) supply of the passing gas flow; 5) device for removing fine-dispersion particles; 6) detachable rod; 7) glass gas discharge tube; 8) electrode section with output window; 9) adapter from the gas discharge tube to the electrode section.

Another problem is to somehow reduce the losses due to atomic diffusion and deposition of particles on the walls of the evaporator. From this point of view the design of evaporators, coaxial with the discharge tube, has been more successful. The amount of metal such as copper, placed in the evaporator, was usually a few grams, and time – about ten seconds.

At a flow rate of ~1 m/s the steady-state concentration of copper in the zone of the pulsed periodic discharge was ~10^{11} cm^{-3}, which is 1–2 orders of magnitude higher than concentrations that occur in the simple thermal creation of vapours. In transverse designs of the gas-discharge tubes with the fine metal feed the pumping rate can be reduced to 1 cm/s, which can lead to an increase in the concentration of copper atoms by more than one order of magnitude.

In these constructions, it is easy to realize the flow of fine particles of, for example, two species, such as copper, manganese or copper and lead. One needs special evaporator designs in which different, e.g. heat-insulating materials, are used to separate the two metals so that two flows of various fine particles can be simultaneously supplied.

1.6. Methods for affecting the output characteristics of lasers

Here we must say that although in many cases lasing occurs at the same transition of an atom of a substance, the initial composition of the substances supplied into the laser cavity (gaseous, solid and liquid) can in principle differ quite significantly. However, during operation of the laser tube the gas-phase, non-stationary plasma-chemical reactions may cause a change in both the physical state and the composition of the substances contained in the laser tube.

When creating laser tubes suitable for use in industrial types of gas discharge lasers there are important issues such as the erosion of electrode materials, the walls of the discharge cavity, the interaction with thermal insulating materials, as well as the accumulation of various substances produced during laser operation. In addition, these effects can be a hindrance even in pumping laser systems.

In some cases, the introduction of additives into the laser cavity is aimed at both improving the output characteristics of lasers and pursuing other goals, such as the transition to the mode of quasi-stationary or stationary lasing.

One of these promising substances is Cs as the cesium atom has the active level very close in energy to the metastable level of the copper atom. Let us now consider the experimental results related to the problem of creating a pulsed Cu–Cs–Ne laser [32,33].

As usual, the gas-discharge tube [32,33] contains a quartz vacuum envelope with natural convection cooling, with welded-in electrodes and a special appendage, located at one end of the discharge tube and used for downloading vials with cesium. The electrodes were made of niobium or tantalum foil connected to the cold electrode sections and placed both inside a discharge ceramic insert and near its ends.

The liner was a BeO ceramic tube 30 cm long with an inner diameter of 10 mm. The insulator, filling the space between the ceramic and the quartz tube of about the same length, was the high-temperature ZrO_2 powder. Copper pieces were located along the entire length inside the ceramic tube.

The vial with metallic cesium was opened with a striker driven by a magnet, and the vial was broken only after degassing of the laser tube and the transfer of the laser tube to the operating mode. Cesium vapours penetrated into the discharge chamber through a process of diffusion and cataphoresis. To prevent the deposition of Cs on the inner surface of the shell, an external furnace was placed near the main ridge around the shell and used to maintain the higher temperature of the walls in this area. The buffer (Ne) gas pressure was typically 2–40 kPa.

Before activating the Cu–Cs laser its discharge tube was usually degassed to a vacuum of 1 Pa and then filled with an inert gas neon to a pressure of 7 kPa, after which the discharge was ignited.

The temperature of the appendage with Cs ranged from 50 to 300°C. The arrival of Cs in the discharge gap increased the rate of development of the discharge, the maximum current in the pulse at the same time increased as a rule. This was accompanied by a simultaneous decrease of the electric field, especially for large values of the pulse repetition rates.

In Ref. 32 the range of optimal concentrations of cesium was shown to be 10^{13}–$5 \cdot 10^{14}$ cm^{-3}, at which there was a noticeable increase in specific energy and laser efficiency. However, with increasing concentration of cesium $>5 \cdot 10^{14}$ cm^{-3} there was a decrease in efficiency and average power. The maximum positive effect when adding the cesium impurities was observed in the repetition frequency range 1–10 kHz and for gas-discharge tubes of small size ≤ 1 cm also at buffer gas pressures <13 kPa.

If the addition of cesium generally did not justify the hopes of obtaining high power and efficiency, a very different effect resulted from additions of hydrogen (H_2). In [34] it was shown for the first time that the addition of hydrogen to the copper vapour laser under certain experimental conditions leads to a significant increase in the lasing efficiency – up to 3%. Today, lasers, containing, *inter alia*, addition of hydrogen, are the most effective – 1.5–3% at the average power of 100–200 W.

Other studies have shown that small additions of HCl, Cl_2H_2 to the buffer gas (Ne) can lead to an increase in the laser output power. In [35], for example, additions of HCl led to a 15% increase in the power output of a Cu–Cl laser with doubled pulses. It should be noted that the delay between the doubled pulses differed from the optimal delay in the case of pure CuCl vapours.

As shown in [36], the laser output parameters are also affected by the working conditions of the electrode sections. When refractory metals are used as a cathode material the laser efficiency is the highest and the cathode material does not affect the discharge characteristics. Additions to the buffer neon of impurity gases such as He, Ar, Kr, and Xe in the amount of 10–30% lead to a marked reduction in the laser output power.

Using the combined discharge (continuous glow (current 0.5 mA) and pulse discharge), as shown in [37], increases the energy generation of a Cu–Cl laser by 3.5–35% depending on the buffer gas pressure. The authors of [37] suggested that the organization of an additional glow discharge leads to: 1) an increase in electron density in the interval between the pulses of dissociation and excitation, thus increasing the rate of current rise in the excitation pulse; 2) a reduction of the rate of electron–ion recombination and a smaller population of the metastable levels, and 3) the dissociation of copper chloride in the interval between pulses.

As stated in [38], the output characteristics of the Cu–Cl laser, excited by twin pulses, are affected by the weak axial magnetic field. An increase in output is recorded at delays between twin pulses greater than the optimum delay corresponding to the maximum power generation during the second pulse and is equal to 175 μs. At lower delays the output power decreases. In particular, the effect of increasing the output power can be related to the improvement of the conditions of stability of the discharge.

Given the fact that the introduction of additional substances into the discharge tube, as well as the effect of additional external factors lead to multi-parameter changes in the conditions in the gas discharge plasma, we can assume that the influence of the additives is currently under development and, apparently, a lot of work is required to be able to uniquely determine the effect of various factors on the lasing characteristics.

1.7. Laser cells for copper vapour lasers with a transverse discharge

Currently, there are many types of structures of cells for copper vapour lasers excited by a transverse discharge. However, the task of creating such cells was associated with a number of difficulties: the development of high-voltage, high-temperature current conductors, efficient evaporation of the working substance; ensuring the uniformity of the concentration of atoms in the working gap between the electrodes, etc. Nevertheless, the fact that such a mode of operation of metal vapour lasers [39] heralds a new opportunity for powerful lasing ultimately determined the fairly intensive research in the physics and technology of transverse discharge lasers.

Calculations of thermophysical processes that determine the temperature of the laser tubes showed that in certain cases, depending on the design features of the laser cell, the pumping power per unit length can be greater than the limiting specific pumping power in the case of a longitudinal cylindrical discharge.

In [40] the authors proposed a modified design of the laser cell with improved thermal insulation in the form of reflective screens and a fairly uniform temperature field in the gap between the electrodes. In this paper, the thermal calculations were carried out for screen insulation made of molybdenum foil with a rectangular profile with an emissivity of 0.2.

The optimum design according to dimensional considerations and acceptable thermal properties was the system consisting of 5 screens. The design temperature for the final screen of the ohmic heater at 1500°C was 650°C.

The design of the laser tube is shown in Fig. 1.9. The main body 1 of the laser tube was constructed of a metallic material and was equipped with flanges 2 for vacuum-tight connection. Radiation was emitted using quartz flat plates placed at Brewster's angle to the optical axis. The voltage on the section with the electrodes 3 was fed through a high-voltage input 4. The insulator of the high-voltage input was 22KhS ceramics and boron nitride ceramics. Paired shaped electrodes made of an pseudoalloy (porous tungsten impregnated with copper) or sheet molybdenum were used.

As already mentioned, the uniform high-temperature field in the discharge gap was formed by the screen insulation 5, made of molybdenum sheets, with the last row replaced in some experiments by packages of the graphite felt. The concentration of copper atoms in the discharge gap was varied by the temperature of the ohmic heater 6 (pyrographite sheet with a thickness 0.5–0.7 mm or 2 mm graphite sheets), which was also used as a reverse bus system to reduce the inductance of the discharge circuit.

The procedure for degassing of the laser cells is not fundamentally different from the methods that are commonly used in the preparation of the tubes with a longitudinal discharge for operation. However, in order to form a homogeneous discharge over the entire volume of the discharge gap it was necessary to preheat the electrodes.

Temperature was measured and controlled with LSP-72 pyrometers or W–Re thermocouple VR 5/20, measurement error was +25°C.

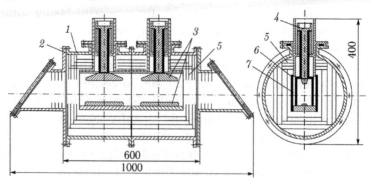

Figure 1.9. Schematic of the laser cell [40].

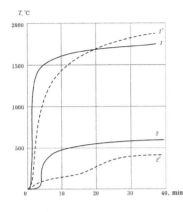

Figure 1.10. Time dependence of the temperature of the heater and the last screen [40].

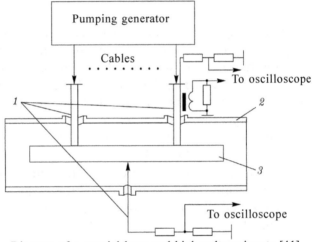

Figure 1.11. Diagram of a coaxial laser and high-voltage inputs [41].

Figure 1.10 shows the results of measurements of the temperature of the heater and the last screen for the five molybdenum screens (curves 1 and 2) and for the combined insulation with packages of graphite fibres (curves 1' and 2'). As can be seen from the figure, the curve goes to the steady state after 15 min. It should be noted that when using the combined insulation for reaching the required operating temperature in the working volume the required heating power was 12 kW instead of 20 kW in the case of the molybdenum screens.

The other – the so-called coaxial – construction of a laser with transverse discharge is shown in Fig. 1.11, where 1 – high-voltage inputs, 2 – anode, 3 – cathode [41]. The discharge zone 40 cm long was formed by the walls of the outer electrode (anode) with a diameter of 6 cm and a 2 cm diameter cathode. The working mixture of the laser consisted of neon and copper vapours.

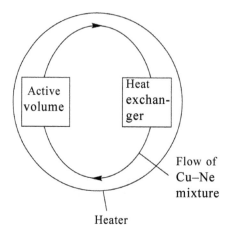

Figure 1.12. Pumping of the Cu–Ne working mixture.

To substantially increase the efficiency of the coaxial laser taking into account the power used to form copper vapour, it is necessary to use evaporators operating without additional heating of the zone. The evaporators were located at the bottom of the anode, the total surface area of molten metal was ~6 cm^2. The temperature of the molten metal in the evaporator could be up to 1600°C, the temperature of the electrodes heated by the heat generated in the evaporators and in the discharge was 1000–1100°C. When increasing the temperature in the evaporators above this temperature range the output power increased, but this dramatically reduced the life of the evaporators, apparently because of the high rate of failure of heating elements.

One of the effective ways of cooling the laser is pumping the mixture in a closed circuit from the active volume to the heat exchanger and back. In [41] experiments were carried out with pumping of the working mixture due to the natural circulation of hot gas in the gravitational field in a direction perpendicular to the optical axis (see Fig. 1.12).

This device operates as follows. The gas heated in the discharge gap travels to the heat exchanger, it is cooled and returns again to the active medium. In order to avoid condensation of metal vapours, the temperature of the electrodes and the heat exchanger is maintained at 1400–1500°C. The gas, heated in the discharge to higher temperatures transferred the heat during motion to the relatively cold heat exchanger and electrodes.

It was shown that application of this method of pumping the mixture in copper vapour lasers and lasers based on other elements is justified by the high-temperature pressure determined by the temperature difference between the gas in the active zone and the heat exchanger, which can be more than 1000 K, although it is difficult to expect to obtain a high-velocity gas flow due to technical difficulties in producing large vertical sizes of the laser and its elements heated to high temperatures. However, in the existing structures, despite low gas flow rates (about 10 cm/s), there was an increase in light output by 1.5 times compared with the laser without pumping with all other things being equal. The maximum pumping power also increased and was about 60 W/cm. It appears that significant progress can be expected with increasing flow rate of gas.

1.8. Laser tubes with a hollow cathode

In contrast to the longitudinal and transverse types of discharge, laser tubes with a hollow cathode have a number of unusual properties which make such structures irreplaceable in many cases. But when it comes to metal vapour lasers operating on self-contained transitions, here we must note that the issue is practically little studied, there are only a few studies [42–44].

Because the operation of the hollow cathode allows for lasing at low temperatures insufficient to achieve due to evaporation the atomic density at which the resonance radiation is completely absorbed at wavelengths characteristic of the size of the active element of the laser, it could be expected to increase the overall efficiency of such lasers in comparison with the self-heated option.

Usually at the same current the discharge in a hollow cathode has a higher concentration of high-energy electrons compared with the discharge of the positive column in a longitudinal discharge. Therefore, the rate of excitation of copper atoms should increase.

Another positive fact is that in the hollow cathode, along with the self-heated mechanism, there also operates another mechanism that leads to the dissipation of the atoms of the working substance, such as copper, from the cathode material determined by ion bombardment of the cathode surface.

In this case, the copper atoms, formed by cathode sputtering, do not suffer large losses because they impact other parts of the cathode surface and re-enter the active medium due to evaporation or sputtering. Moreover, because of the high overpotential the ionization of runaway copper atoms can take place around the cathode gaps, causing them to return to the cathode surface.

Another advantage of the design of the hollow cathode is its special geometry, which allows efficient retention of the heat released during the discharge, due to the fact that the heat radiated by one end of the discharge tube may be reflected or absorbed at the other end of the tube. Here it is also possible to solve the issue of improving the homogeneity of heating the laser tube along its length so that heat pipes can be used, in addition this helps to arrange the return of copper, condensed in the cooler parts of the laser tube.

One of the first tubes with a hollow cathode was a coaxial design similar to those that had previously been used successfully in the He–Cd lasers. The cathode was a molybdenum cylinder with a diameter of about 0.76 cm and length 91 cm, had four 0.25 × 18 cm slits and was uniformly distributed along the entire length. The anode had a diameter of 1.6 cm and a length of 59 cm. Both electrodes had water-cooled shells. The working substance was placed inside the cathode in the form of copper wire.

Lasing was obtained in two different modes of laser tubes.

In the first mode, two types of discharge were used: DC discharge and pulsed-periodic discharge with a frequency of about 8 kHz. The tube was heated to a temperature of 1500°C by the DC discharge, then a continuous discharge was interrupted and a pulsed-periodic one was activated. With the inclusion of such a discharge there appeared initially weak lasing, and within 2 min the power increased to the steady-state level.

Another laser mode was based only on the pulsed-periodic discharge with a high repetition rate of 12 kHz. In this case the average lasing power reached 270 mW (several times higher than that for the first mode), and with further increase in pumping power the average lasing power was reduced. Apparently, to get more power in such a configuration, the principle of self-heating of the discharge is dominant.

A variety of the design of the laser operating at room temperature was proposed in [43] on the basis of the effect of cathode sputtering. The concentration of atoms of the working medium was increased by the gas flow. On this basis, it will be possible to create devices with fast switching, and this technique, it appears, can be practically applied to even the materials with the lowest volatility.

In [43] a laser system was constructed in which the cost of obtaining the atoms of the working medium was minimized due to the fact that a small area (≤10 cm²) of the cathode and a high-speed gas flow were used. Such a system could work with many materials (i.e., to produce multicolour laser radiation) inserted into the working volume together or separately, with the output power at different colours varying with the characteristic time of ~10 ms.

This laser is hardly suitable for generating high-power as there is a need to maintain a low operating temperature, which would be difficult at high pump powers, however, such a system is very attractive

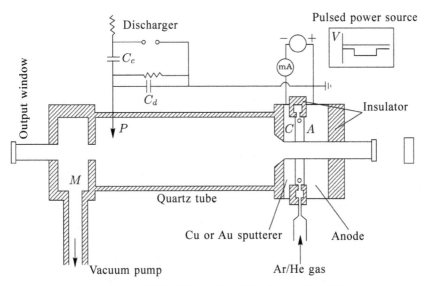

Figure 1.13. Schematic of the laser tube [43].

from the point of view of implementation of efficient lasing at new transitions.

The design of the laser tube is shown in Fig. 1.13. The buffer gas enters the volume through a narrow slit and is initially distributed along the cathode surface. The metal atoms sputtered from the cathode are carried away by the buffer gas, forming a metal vapour jet along the axis of the laser gas-discharge tube. The entrainment process is hampered by the diffusion of metal atoms sputtered back to the cathode or nearby surfaces so that the flow rate of the buffer gas becomes comparable with the average thermal velocity of the metal atoms. This is achieved by selecting a small width of the circular gap (100 mm) and by the buffer gas flow rate of 1 litre/min. By regulating the pumping rate of the gas on the other side of the laser tube the pressure in the tube is maintained from 0.5 to 15 kPa. The main anode is common to both the discharge sputtering of the cathode and to the main pumping discharge. The other electrode for pumping discharge is installed near the end of the stream at the far end of the laser tube. Although the length of the tube is 25 cm and its diameter 44 mm, the optical aperture of the laser tube is limited by the typical size of the sputtered cathode, which was ~8 mm, so the active volume of the laser tube was ~10 cm³.

The modular design of the laser cell allowed the use of different discharge tubes which may contain special branches for spectrochemical studies and measurements of plasma characteristics.

The hydrodynamic stable flow of the mixture of the buffer gas and metal atoms is determined by the length of about 20 cm for one section of the laser tube. With these tubes it was possible to perform lasing in copper and gold vapours, respectively, at a peak power of 6 kW and 1.7 kW. The power ratio of the 510.5 and 578.2 nm green lines of the copper atom was 16:1, and for the 627.8 nm and 312.2 nm gold lines it was 4:1.

In [44] the authors indicate the possibility of a significant increase in the length of the amplifying laser zone, for example, by developing a system with dual cathodes. There are also options of the multicathode tube. Although, as follows from the peculiarities of such laser tubes, one should not expect to obtain high lasing powers, yet they are unique for physics research, including the development of new methods and excitation of metal vapours.

When CuCl was used as the working substance, in [45] lasing was carried out with a power of 400 mW and a useful life of the laser of more than 300 h without recharging the working medium in the laser tube was achieved. This fact may be of interest for the development of the sealed-off lasers.

1.9. Modified cells for gas discharge lasers

As already mentioned, the use of gas discharge in metal vapour lasers has stimulated a substantial upgrading of the conventional gas-discharge technique developed for the room or similar operating temperature. The simplest improvement consisted in the replacement of low-melting electrovacuum glass with quartz or pyroxene operating at up to 800°C and 1100°C, respectively. At present, mainly ceramics and various composite materials are used for this purpose, allowing metal vapours to be introduced at a wall temperature up to 2000°C.

Despite the proliferation of different methods for introducing vapours into the active region, for example, in the form of volatile chemical compounds [46], the explosion of a wire [19] or a thin layer of metal or its compounds deposited on the cell walls [23], cathode sputtering [47], etc., discussed in previous sections, in many cases, particularly during the measurement studies, preference is given to thermal evaporation of the sample material placed in a laser cell. The advantage of this method is, above all, the complete control of the chemical composition of the unexcited mixture. Also important are the purity level of the discharge and the ability to vary the plasma parameters over a wide range.

Generally, the high-temperature gas-discharge device includes a working tube, a heating system (in particular a high-temperature furnace) and a pump generator. In some designs, when the parameters of the mixture and pumping pulse (current) are known in advance, it is advisable to combine the heating of the tube and the generator into a single complex, using the self-heating principle, as proposed in [9–11].

Particularly stringent requirements are imposed on the materials of gas-discharge tubes, directly in contact with metal vapours. The criteria for suitability are: the level of operating temperature, purity of starting material, which guarantees the absence of high-temperature impurities, the given electrical conductivity and dielectric strength, chemical resistance to the vapours of input materials and the environment, adaptability, i.e. the possibility of making long tubes ($l/d \geq 50$, where l is length, d is diameter), preferably vacuum-tight at high temperatures, the ability to produce high-quality joints (junctions) with glass, metals, etc.

There are a large number of compounds with melting points above 2000°C [48, 49], but only few of them are suitable for the manufacture of gas-discharge tubes, which satisfy the above requirements. Even fewer of them have been mastered by the modern industry and are available for wide use. The situation, however, is made significantly easier by the fact that many compounds, especially metal oxides, are produced in the form of pure fine powders, from which it is easy to fabricate ceramics of satisfactory quality in the laboratory.

The list of materials used in studies with the metal vapours is given in Table 1.1. The material used most commonly for research purposes is the beryllium oxide ceramic, satisfying all requirements except chemical stability at high temperatures to the vapours of rare-earth and other elements. In the latter case, tubes of Y_2O_3 or Lu_2O_3 ceramics produced in the laboratory are used.

The production technology [50] was as follows. Cylindrical billets of compacted powders were placed in a flexible mould (PVC, fluoroplastic and other tubes) with a central core of a low-melting material (tin alloys, brass). After hydrostatic pressing at a pressure of $(1–2) \cdot 10^3$ atm and melting of the central core the blank was pre-annealed in vacuum at 1200–1300°C and then machined and sintered at 1600°C. Thus, it was possible to produce products of up to 50 cm long.

The design of the tubes varies considerably depending on the type of heating and cleanliness requirements of the mixture. The best performance is shown by cells with external heating using BeO industrial vacuum-tight ceramics (Fig. 1.14). The basis of the design is the tube (1), coupled with a glass cup (2) using a K-400 adhesive, or through a BeO–Ti–kovar–glass junction, or directly with a specially designed junction of BeO ceramics with glasses of the molybdenum

Table 1.1.

Number	Compound	Melting point, °C	Vapour pressure at 1600°C, Pa		Resistivity at 1600°C, ohm m	Formation energy, kcal/mole
			Compound	Metal		
1	BeO	2500	$1 \cdot 10^{-3}$	$1 \cdot 10^{2}$	$1 \cdot 10^{2}$	−139
2	Al_2O_3	2000	$1 \cdot 10^{-1}$	$1 \cdot 10^{3}$	$1 \cdot 10^{2}$	−126
3	MgO	2800	$1 \cdot 10^{-1}$	$1 \cdot 10^{5}$	10	−136
4	ZrO_2	2700	$1 \cdot 10^{-5}$	$1 \cdot 10^{-5}$	1	−124
5	Y_2O_3	2380		10	$5.4 \cdot 10^{4}$ (1000 K)	− 140
6	Lu_2O_3	2370		$1 \cdot 10^{2}$	10^{6} (1000 K)	−150
7	B	3000	$10^{-5}–10^{-3}$	$1 \cdot 10^{-3}$	$1 \cdot 10^{4}$	−54

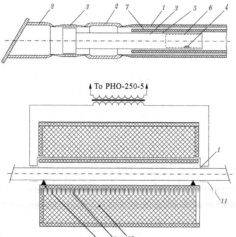

Figure 1.14. Design of the high-temperature gas-discharge tube and the furnace for heating the tube [50]. 1) BeO ceramics tube; 2) glass vessels; 3) electrode; 4) working substance charge; 5,6) bath; 7) insert; 8) muffle; 9) heating spiral; 10) heat insulator; 11) heating spiral.

group S49 or S52. The annealed junction withstands without cracking the operating temperatures up to softening of the glass if, firstly, it contains no less than 1% Na_2O and, secondly, the BeO should contain approximately 0.05–0.15% of sodium oxide. Industrial ceramics with a purity of 99.5% usually does not contain such an amount of Na_2O.

Inside the BeO tube (1) there are metal electrodes (3) and the sample of the working medium (4) placed in cylindrical baths, made, for example, in the form of two layers: a solid foil (5) and a photolithographic grid (6). Due to the capillary effect such a structure ensures the return of the liquid metal, condensed in the cooler parts of the bath, to hotter parts. This technique allows to maintain for a long time a working pressure of the metal vapours up to 1–3 kPa at a relatively small amount of buffer gas (4–10 kPa). In some designs, the electrode and the bath were a single entity. In this case, the grid (6), usually at the hot end of the electrode, was 1–5 cm shorter.

In conducting research with chemically active elements or at high pump power, a cylindrical liner (7) of BeO or chemically more stable Y_2O_3 or Lu_2O_3 ceramics was placed inside the main tube. BeO tubes, coated with a film of a chemically stable oxide (e.g., Y_2O_3, Lu_2O_3, Nd_2O_3 and others) [50] or a refractory metal (Mo, Ta) were also used.

These techniques have been used to introduce into the discharge zone vapours of any metals up to temperatures of 1700°C (beginning of the loss of mechanical strength of the BeO tube, which is expressed in the gradual sagging down to touching the inner surface of the furnace).

The gas discharge cell was heated to the operating temperature in muffle furnaces based of the tube (8) of BeO or Al_2O_3 ceramics. Profiled (for uniform heating along the length) heaters (9) of EI-695 high-temperature alloy (up to 1400°C, without sealing) or molybdenum (up to 1700°C, with argon purging of the furnace) were also used. Insulation in the former case was ensured with foam ceramics (10) based on Al_2O_3 or yttrium oxide powder, which does not interact with the EI-695 alloy. Zirconia powder was used in the second case.

Due to the high thermal conductivity of BeO ceramics this design is characterized by rapid heating of the working ends of the tube, which can lead to depressurization of its connection with the glass. Therefore, the tube is longer than the furnace in which it is heated, and the free end of the ceramic is cooled by the air flow or water. To compensate for the heat losses the end is sometimes heated with a nichrome spiral (11) connected in series with the heater (9) of the muffle furnace.

Testing of the vacuum properties showed that the tubes of the industrial BeO ceramics reached a vacuum of ~10^{-5} Pa at a temperature of 1700°C which enabled the construction of sealed-off copper vapour lasers operating in the self-heated mode [51]. However, this does not mean that the cleanliness of the working mixture is maintained at this level in the working conditions in the cell filled with the buffer gas. Studies have shown that in service the impurities, non-gaseous at room temperature, in particular vapours of Be, BeH, K, Na, etc. [52] accumulate in the tube. Measurements made by the full and linear absorption method [53] showed that in insufficiently trained tubes the

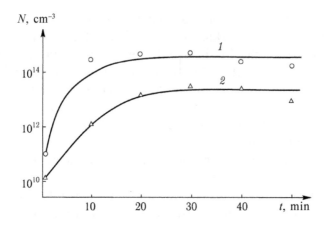

Figure 1.15. Dependence of the concentration of sodium vapours, released from the tube walls, on the duration of preliminary training: for 1 h (1) and 25 h (2) at $T =$ 1400°C [50].

main impurity concentration – sodium vapour – can reach 10^{14} cm^{-3} (Fig. 1.15, curve 1). In cases where especially clean conditions were required, the following techniques were used: long-term training (curve 2, Fig. 1.15), or slow pumping of a gas (usually at the level of 10^{-2} l/ min).

The range of manufactured industrial vacuum-tight tubes made of BeO ceramics is small. Therefore, large discharge tubes use internal thermal insulation. Apparently, this technical innovation has been used widely for spectrochemical and laser research due to the results in [54]. An essential complement to this design is the use of the heating coil of refractory metals (Mo, W–Re alloy, etc.) [50, 55, 56], wound around the outer surface of a high-temperature tube (Fig. 1.16). The pulsed discharge is ignited with a solid spiral, and a DC discharge is produced with a sectioned coil with the length of one section of 5 cm. In the latter case each section is powered by a separate transformer.

The advantage of this design is the ability to produce long tubes of large diameter from short (10–15 cm) segments, the ability to carry out preliminary degassing and the independent regulation of gas-discharge tube wall temperature, etc. The use of BeO (tube), Mo, or W–Re (heater) and ZrO_2 (thermal insulator) as structural materials allows cells to operate at temperatures up to 2000°C. The same approach is possible to create low-inductance high-temperature cells for the excitation of a transverse discharge [57]. It should be noted that abroad similar structures appeared only in 1982 [58].

The characteristic difference and at the same time a shortcoming of the cells with internal thermal insulation is the expansion of the

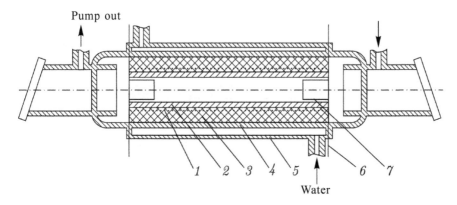

Pump out

Water

Figure 1.16. Design of the high-temperature gas discharge tube with internal insulation:
1) heating coil; 2) ceramic tube; 3) insulator; 4) vacuum shell; 5) water cooling jacket;
6) radial electrical input; 7) electrode [50].

composition of impurities and increase of their concentration, which increases the degassing time from one day up to 1–2 weeks. Therefore, in carrying out the research which required knowledge of the chemical composition of the working mixture in [50] the mixture was constantly renewed. The pumping speed was determined experimentally. The selection criterion was the constancy of the parameters of the active medium, for example, the frequency–energy characteristics of the laser. Compliance with these technological principles has allowed to create sealed-off lasers with internal insulation.

1.10. Laser cells pumped by electron beams

To clarify the possibility of excitation of lasers by electron beams, generated by an open discharge, it is important to establish whether the conditions for the generation of electron beams (EB) in the transition to large cathode surfaces and high pulse repetition frequencies (up to the continuous mode) are preserved.

Figure 1.17 shows a coaxial cell [50] used to investigate the possibility of radial injection of EB into a cylindrical cavity filled with the working mixture. It consists of a dielectric tube 1 with radial inputs 2–4, sealed with the K-400 adhesive, the cathode 5 made of titanium or molybdenum, the anode 6 and the electron collector (EC) of photolithographic Mo-mesh or woven W-cloth, fastened with nickel in an electrolytic bath. The cathode and the anode are separated from each other by a dielectric tube 8. Power is supplied by a thyratron or a valve generator with a peaking tank.

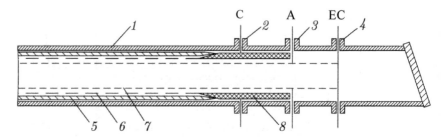

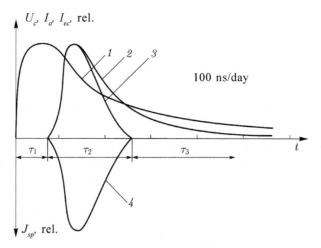

Figure 1.17. Design of a coaxial cell for excitation of gas mixtures with electron beams [50]. 1) dielectric tube; 2–4) radial inputs; 5) cathode; 6) mesh anode; 7) mesh collector (BeO, BN).

Figure 1.18. Oscillograms of voltage on the cathode (1), anode current (2), collector (3) and line luminosity (4) [50].

Study [50] showed that in the above cell one can easily implement the mode of escape of electrons at filling with pure inert gases, mixtures with molecular gases and metal vapours. Figure 1.18 shows typical oscillograms of the voltage at the cathode and the anode and collector currents. It is evident that the stage of the beam corresponds to the interval τ_2 of rapid changes in the anode current and voltage at the cathode. No beam is generated in the final stage τ_3 (apparently, in these conditions the Dreiser criterion). Investigation of the luminosity of spectral lines of different gases excited by direct collisions of electrons with the atoms in the ground state, shows that their emission time is in the interval τ_2, i.e. in the stage of the beam.

As can be seen from the oscillograms, there is quite a long time interval τ_1 during which there is no breakdown of the accelerating gap. The duration of this stage can reach 10^{-5} s. At higher supply voltages close to the limit it is 100–150 ns. This remarkable property of the discharge gap allows for efficient powering of a circuit with the peaking storage, enabling it to be easily charged during the time delay by using a thyratron or an oscillator. The study of this regime has shown that in this case the EBs are more stable and exist in a wider range of conditions. Because the circuit with a peaking storage has very low inductance, the amplitude of the EB current can be three times higher than in the direct discharge of the capacitor through the thyratron and the discharge gap (DG). Accordingly, the pulse time decreases. As a general rule, in [50] the authors used a scheme with a peaking capacity equal to the storage capacity, if the switch was a thyratron. When powered from the oscillator the peaking capacitance value was chosen based on the specific tasks of the experiment.

Using the electron collector for radial injection complicates the installation. Therefore, in the investigation of the active media the collector was removed. The amplitude and shape of the beam current pulse was estimated in [50] based on the following assumptions. In the system with the collector, the electrons, twice passed through the anode grid, enter the retarding field between the cathode and the anode and return back into the cavity of the latter in which they gradually slow down and finally get to the anode. Consequently, from the shunt installed in the anode circuit signals proportional to the sum of the anode current I_a and the current of accelerated electrons are taken. At the same time, the signal, proportional to the current of accelerated electrons I_n is recorded from the shunt of the collector.

The proportion x of the electrons, once intersecting the cavity of the anode and returning to the cavity in reflection from the DG, is

$$x \le \beta_a^3 \beta_{ec}^2$$

where β_a, β_{ec} are respectively the geometric transparency of the anode grid and of the electron collector.

At a typical value of $\beta_a = \beta_{ec} = 0.75$; $x \le 0.24$. Consequently, only 24% of the electrons are theoretically able to return back into the zone of the collector. Thus, the true beam current is in the range

$$\frac{U_{ec}}{(1-\beta_{ec})R_{ec}} > I_n > \frac{U_{ec}}{1.24(1-\beta_{ec})R_{ec}} , \qquad (1.4)$$

where U_{ec} is the signal from the shunt of the collector R_{ec}.

Comparison of the amplitudes of the anode (without the collector) and the collector currents showed that the estimated value of I_n in the optimal conditions is close to

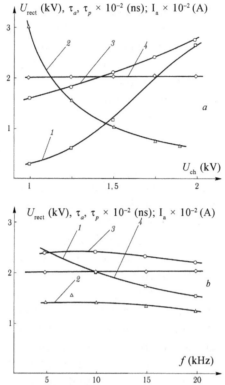

Figure 1.19. Dependence of amplitude I (1) and duration (2) of anode current, voltage U (3) and the delay in development of the discharge (4) on rectifier voltage (a) and repetition frequency of excitation pulses (b) [50].

$$I \cong \beta_a I_a, \qquad (1.5)$$

Consequently, recording the spectral emission lines excited by direct electron impact and measuring the anode current, we can estimate the duration and amplitude of the EB.

Using the developed structure of the cell with radial injection of ED, one can easily initiate lasing in a variety of media [39, 60], in particular, in self-contained lasers (He, Pb); in CW (continuous-wave) lasers with direct excitation of the upper level by electron impact (Xe, $\lambda = 2.02$ μm); in charge-exchange lasers (He + Zn, He + Cd, He + Hg), and others. This confirmed the feasibility of using electron beams, generated by the open discharge, for laser pumping.

The study of the characteristics of an open discharge (OD) in the repetitively pulsed mode was carried with a cell with the length $l = 10$ cm, diameter $d = 2$ cm, the accelerating gap $\delta = 0.5$ mm We used the scheme with the peaking storage and a thyratron as a switch. Figure 1.19 a shows the dependence of the anode current I_a,

EB pulse duration and breakdown voltage on the voltage in the rectifier circuit (resonant charging was carried out using the storage capacitance, a choke coil and a diode). Figure 1.19 b shows the same characteristics, but depending on the pulse repetition frequency f (PRF).

As can be seen from these figures, the acceleration properties of the OD are retained up to PRF of 20 kHz. Reduction of the EB current amplitude with increasing f is entirely due to the voltage drop across the cell caused by the deterioration of the pulse generator. The study of the lasing properties of the He laser ($\lambda_l = 2.02$ μm) also reaffirmed its performance up to $f \approx 10^4$ Hz. Further increase in f is accompanied by a gradual weakening of the energy generation due to the decrease of gas density, i.e., its stopping power, due to heating. In the dual pulse power supply the EB parameters are retained until the delay time between them is 0.5 μs. In the pulse train mode with a duration of 10^{-1} s when using the valve oscillator the EB was generated at up to $f = 300$ kHz, with an average power input capacity of 150 W/cm.

Thus, the periodic-pulse excitation regime does not prevent the generation of EB in the OD. Apparently, the limiting pump power is restricted by overheating of the anode–collector grid. Taking its temperature as equal to 2000°C and the temperature of the helium buffer gas $T_g = 1000°C$, we find that the power dissipated by the grid is

$$P_p = \frac{2\pi\lambda\Delta T}{\ln(d_c / d_{ec})} > 600 \,[\text{W/cm}].$$

where λ is thermal conductivity, d_c is the diameter of the cathode, d_{ec} is the diameter of the electron collector.

For the optimum geometry of the cell, i.e., with full retardation of the beam in the anode–collector cavity, the power fed into it is even higher:

$$P_H - \frac{P_p\beta}{1-\beta}. \tag{1.6}$$

If $\beta = 0.75$, then $P_H > 1.8$ kW/cm, which is probably known to provide the energy needs of any laser media.

Less convincing were the results of the study of the possibility of generating large-amplitude EB current. Thus, under optimal conditions in the above installation the current $I_{a\,max} = 400$ A was obtained, which corresponds to a current density of $J_{a\,max} \approx 6$ A/cm².

To obtain higher current densities, attempts were made to improve the stability of the discharge. In particular, in one experiment an additional discharge was produced in a gap with $\delta = 2$ mm, and the EB of the discharge was injected through the grid in the main acceleration gap with $\delta = 0.5$ mm. This method was used to produce in neon ($p_{Ne} = 1$ kPa) the EB with an amplitude of 300 cm^2 and a duration of 5 ns.

The stability of the generation of EB also improved through the use of polished cathodes. Thus, in a carefully assembled planar electrode system with $s = 1$ cm$^2 = 0.2$ mm, the working pressure of helium was increased to 0.5 atm. At $p_{He} = 30$ kPa and a voltage of 6 kV an EB with an amplitude of 10 A and a duration of 10 µs was produced. The same design was used for continuous generation of the beam in He to $p = 5$ kPa. The maximum power of the EB was limited by the stability of the Mo grid and was 400 W/cm ($p = 2.5$ kPa, $I_n = 0.2$ A, $U_a = 2$ kV).

However, the greatest success in the suppression of instabilities and the stabilization of the generation of the EB was obtained in the selection of semiconducting cathodes with a complex structure. Their use prevents arcing of the discharge due to a sharp decrease in the number of electrons that can go up to the cathode spot in the longitudinal direction. The often used technical approach is to use dielectric materials (e.g. ceramics) with a metallic outer surface and heated to a certain temperature in order to obtain a suitable conductivity. As an example, Fig. 1.20 shows the temperature dependence of the resistivity of the widely used and inexpensive high-alumina ceramics. However, the drawback of dielectric materials is the quantum yield of photoemission. To overcome this effect, the inner surface of the tube is coated (through a mask) by vacuum thermal

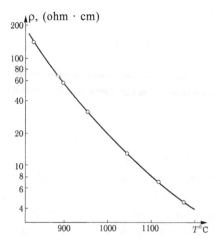

Figure 1.20. Temperature dependence of specific resistivity of high-alumina ceramics [50].

sputtering with a Mo film with a thickness of 1 μm with the size of the elementary area of 3 × 3 mm.

Despite the low resistance at high temperatures, because of the polarization effect, the dielectric materials can not provide high values of $j\tau$. In particular, at a single pulse for high-alumina ceramics at $T = 1300°C$ the value $j\tau < 0.8 \cdot 10^{-6}$ C/cm², where j is current density, τ is pulse duration.

Higher parameters were obtained in [50] using ceramic and glass materials with percolation conductivity, consisting of a mixture of powders of the conductor and the dielectric [61] and having limited conductivity. In the materials used in the study their resistivity was $(10^{-1}-10^3)$ ohm·m. The conductivity of the percolation structures was considered in general in [62]. Up to a certain concentration of the conductor (20–30%), which depends on the method of sample preparation, the material has dielectric conductivity. Above the critical concentration of the conductor, its particles form a network of random connections with each other, and the sample becomes conductive.

The percolation materials have three major advantages:

a) a high content of free metal (40%), which provides sufficient photoemission;

b) the ability to create structures through a combination of a wide range of metals and dielectrics that can satisfy the diverse requirements on the material of the laser cell, in particular, for metal vapour lasers;

c) the weak dependence of the material properties (e.g., conductivity) on the operating temperature.

Application of the percolation materials for manufacturing cathodes for the OD sharply increases the range of operating pressures, voltages and currents of the EB.

Notable successes have also been achieved in the stabilization of the generation of EB by using the discharge through an insulator. The physical basis for stabilization in this case is the same as when using the bulk resistivity. However, given the small magnitude of the dielectric conductivity of the most commonly used ceramics (BeO, Y_2O_3), in these structures it is possible to generate only relatively weak beams. It is also important to cover the inner surface of the tube with a cellular metal film. Such a construction can operate up to the repetition frequency of 3 MHz. The polarization of the dielectric, which limits the maximum repetition rate, is removed in the interval between the pulses due to partial conduction produced by heating the tube. Obviously, this method is especially effective at high temperatures.

1.11. Metal salt vapour lasers

1.11.1. Methods for introducing metal atoms into the working zone

The most common way to introduce the metal atoms into the active zone is, as already noted, the heating of the metal. In this way, it is relatively easy to add to the active volume the atoms of the metal atoms which have significant vapour pressures at temperatures below 1000°C. However, the metals that are attractive from the standpoint of obtaining pulsed lasing in transitions from resonance to metastable levels, have in most cases negligible vapour pressure at this temperature. So to create a pressure of 10 Pa (the saturated vapour pressure of the metal at which lasing is reliably recorded) the temperature of heating most metals should not be less than 700°C. For the group of metals (Cu, Au, Fe, In, Ni, Co, Sn, etc.), the temperature providing this vapour pressure is 1000–2000°C, and for low-volatility metals (La, V, Zr, Mo, Hf , Nb, Ta, W, etc.) it is 2000–3500°C [63].

Therefore, for working with metals it is necessary to construct the working chamber from high-temperature materials. The high-temperature materials based on Al_2O_3 currently used in laser technology can operate at a temperature not higher than 2000°C. For the alkaline earth and rare-earth metals, work in these cells, as already stated, is difficult also at lower temperatures because of the reactivity of the metal. A certain temperature rise was made possible through the use of ceramics based on BeO and the use of refractory metals, for example, in hollow-cathode discharges, but is associated with serious technical difficulties.

Raising the temperature of the metal to achieve a higher density of working atoms creates not only technical difficulties but, as mentioned, a principal difficulty, because at high temperatures metastable lower levels can be significantly populated. This is especially dangerous for lasers on vapours of those metals in which the lower metastable level is not very high above the ground level.

The difficulties associated with the introduction into the active volume of a high-density of working metal atoms by heating, as well as attempts to obtain lasing on atomic transitions of low-volatility metals, necessitate search to find other ways to create a high density of working atoms in the active volume.

Therefore, considerable interest is expressed in the methods based on the creation of a non-equilibrium density of the working metal atoms at a given temperature. They are attractive because the density of the metal atoms, formed in a short time in the working area, has little to do with the temperature of the chamber wall. Due to this, the high density of metal atoms can be obtained at a lower temperature of the working chamber. These methods include the dissociation of chemical compounds of metals in a pulsed discharge, the electric explosion of metal wires, separation of the metal atoms from the wall of the tube and the evaporation of metal particles under the influence of a powerful pulsed discharge (see above).

Of these methods, the currently most developed method is the one based on the dissociation of metal-containing molecules in a pulsed discharge. With its help it is possible not only to implement the lasing on atomic transitions in metals, but, most importantly, produce it in the mode of regular pulses with a sufficiently high performance.

1.11.2. Gas discharge tubes

The plasma of a pulsed discharge in vapours of metal halides is produced using in most cases quartz discharge tubes. For example, in [64] the authors used the simplest design of the gas-discharge tube (Fig. 1.21 a). Typically, the active part of this tube is 40–90 cm long and its internal diameter is 1–3 cm. The longitudinal discharge was excited using two lateral electrodes – electrodes of IFP-2000 lamps.

The copper halide powder is usually distributed in equal portions over the entire length of the working part of the tube. The vapour pressure of the metal halide in the discharge zone depends on the temperature to which the powder is heated, which, in turn, is determined by the temperature of the inner wall of the tube. As the heat removal in

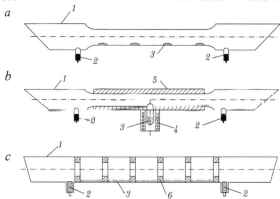

Figure 1.21. Main design features of gas discharge tubes using for lasing in vapours of copper halides [64]: a) with the working substance in the discharge zone; b) with an appendage containing the working substance; c) with restriction of the discharge zone; 1) quartz tube; 2) electrodes; 3) working substance; 4) heating furnace; 5) thermal insulation; 6) restricting diaphragms.

narrow tubes was insufficient to maintain the temperature of the tube necessary for lasing, the lasing in these tubes is usually observed in the superheating mode. The superheating mode is the mode of operation of the laser tube in which the power input to the discharge provides passage through the entire lasing temperature range. In tubes with a large inner diameter (23 cm) the desired temperature is achieved using additional heat insulation. Asbestos cord or asbestos cloth, placed on the outer surface of the tube, is usually used.

The tube of this construction could not meet modern requirements due to the instability and pinching of the discharge. In later studies gas discharge tubes were used in which the working medium was placed in a separate appendage heated by a special oven (Fig. 1.21 b). In this case, the metal halide vapour pressure was determined by the temperature of the appendage, and the temperature of the working part of the discharge tube was always slightly higher.

Despite the fact that the gas discharge tube with the appendage made it possible to obtain a relatively stable lasing for several tens of hours, they did not meet the requirements. Their main drawback was the pinching of the discharge. This effect became stronger with the increase of the diameter of the working part of the tube. Typically, at a diameter greater than 2 cm it was also difficult to obtain a stable discharge cord. So to get the non-constricted and stable plasma in the vapours of metal halides it was necessary to use discharge tubes with limiting apertures (Fig. 1.21 c) [64]. These tubes were made of fused silica. The working part of the tube had a length of 40–60 cm and the inner diameter of 2–6 cm. Inside the working zone of the laser tubes there were inserted the diaphragms limiting the discharge, which were located at equal distances from each other. The diaphragms were made of lightweight fireclay or corundum. The inner diameter of the diaphragms was 1–3 cm. The length of each diaphragm was 1–3 cm, and the total number of diaphragms was usually 5–6. The working medium, as shown in Fig. 1.21 a, was placed in the gap between each pair of the diaphragms. Along with working substance, an amount of copper was placed in the same gap.

When measuring the temperature on the outer surface of the tube [64] it was observed that the temperature along the working part of this design was not the same. This led to local overheating and redistribution of the working substance. External insulation was used to equalize the temperature along the tube. Depending on the material, it was necessary to insulate the tube in the places that housed the diaphragm (in the case of lightweight fireclay) or between the diaphragms (in the case of corundum).

In the tubes of this construction the electrodes were made from copper and cooled with water. All types of discharge tubes had a quartz window at an angle of 5–6° or Brewster angle. Inert gases were used to reduce the diffusion of vapours of metal halides as well as to initiate the discharge.

Before placing the halide powder in the gas discharge tube, the tube was treated with a discharge in neon. Treatment consisted of heating the working zone of the tube to high temperatures (~800°C) and subsequent evacuation. The largest number of heat cycles was applied to tubes with diaphragms.

Spectrally pure inert gases were used in the experiments. The powders of metal halides were usually prepared in the laboratory by conventional methods. In some cases, special purity industrially produced powders were used.

After laying the metal halide powder, the tube was evacuated with a vacuum pump to a vacuum of 1 Pa, and then with a diffusion pump with the use of traps with liquid nitrogen to a pressure of 10^{-2} Pa, after which the tube was filled an inert gas, usually neon, and the discharge was activated. Strong gas evolution occurred before melting of the metal halide powder. This led to a deterioration of the discharge conditions and the discharge spontaneously stopped. The tube was then cooled to a temperature of 100–150°C and evacuated for 30 min. A new portion of the gas was then admitted and all was repeated.

The degree of readiness of the metal halide for lasing was indirectly judged by the colour of the discharge or emission spectrum in the first instants of the discharge: the absence of molecular bands (H_2, O_2) indicated a high purity of the powder. Typically, 6–8 heating cycles of the metal halide to a temperature slightly above the melting point was enough to obtain pure discharge conditions.

It should be pointed out that in cases where the material departure from the working area led to contamination of the windows, special diaphragms, mounted near the windows, were used In most cases, this was enough to make sure the windows were clean when using a laser tube for several hundred hours.

Of all the above structures the best was the one in which the discharge zone was limited and the electrodes were made of copper. The discharge in the tubes of this design was stable which allowed stable lasing with high power and high efficiency.

Great prospects were associated with the creation of lasers on metal halides with a transverse discharge. In this case, large active volumes of laser tubes can be considered. The first experiments [65] demonstrated the ability to scale this type of systems. The active volume of the cell was 700 mm long, 25 mm in height and 5 mm wide consisting of 10 sections with paired electrodes to which voltage was supplied separately. Tubes of a large active volume of 250 cm^3, using CuBr as the working substance, produced high laser energies of 2.5 mJ [66].

1.11.3. Methods for producing vapours for hybrid metal vapour lasers

Due to the fact that the design of laser tubes and the associated methods of introducing the atoms in the active medium can greatly influence the output characteristics of lasers, the question of preparation methods of the environment for the metal vapour lasers becomes very important in some cases .

This situation has already been demonstrated above, when we considered laser media on the vapours of metals and their halides or other metal compounds. In this section we consider the construction of the so-called hybrid metal vapour lasers. Here, the term 'hybrid laser' refers to the laser system that uses a slow flow of the Ne + HBr buffer gas for the generation of the metal bromide in the interaction in the discharge of HBr with the metal placed inside the tube at a temperature of 700°C [67].

As shown by numerous studies of, for example, lasing in the copper atoms, with all other things being equal, the efficiency and output power of the hybrid laser can be usually (depending on various conditions) twice the corresponding values for both the conventional copper vapour lasers and for copper halide vapour lasers. A key fact in this matter is, apparently, the method of producing working atoms in hybrid lasers, which is a hybrid of the technology of conventional metal vapour lasers and copper bromide vapour lasers.

It should be pointed out that HBr in such hybrid systems can be replaced with HBr HCl, Br$_2$Cl$_2$, etc. However, the lasing power increases in the sequence of the employed additional substances: Cl$_2$, HCl, Br$_2$, HBr. Although Br$_2$ provides more power than Cl$_2$, Cl$_2$ is more convenient for study because at room temperature it is a gaseous substance.

However, the most suitable, if not ideal, for efficient lasing is HBr, because besides the fact that it is also a gaseous substance, it delivers bromine and continuously introduces hydrogen into the discharge gap.

This system allows for quick and easy changes of the discharge conditions, which leads to the maximization of output power, in contrast to the usual self-heated of the copper vapour laser, where the density of working atoms is determined by the power generated in the discharge gap.

A hybrid method for introducing the atoms allowed to realize experimentally, except for copper vapour lasers, lasing at 472.2 nm lines for bismuth, iron at 452.9 nm, lead (406.2 nm and 722.9 nm), and manganese (534–554 nm, 1.29–1.40 µm) [68–72].

In these lasers, the metal halides, formed by the reaction of HBr with the metallic working medium in the laser cavity, dissociate in collisions with electrons in a pulsed discharge, following with a high repetition rate, which leads to the formation of free metal atoms. Next, the atoms are excited, transferring to the upper resonant state by electron impacts directly from the ground state of the neutral atom.

Figure 1.22 schematically shows the processes in the hot part of the laser tube, providing the quasi-cyclic mode of producing the working medium. The specified set allows lasing with a wide range of laser tubes with internal diameters ranging from 4.5 to 60 mm and a length of 30 to 200 cm. Because of the relatively low operating temperature and heat capacity the heating time to the steady-state power level is reduced.

Figure 1.23 shows one of the hybrid copper vapour laser structures in which the mean power of 200 W was obtained [72].

One of the important characteristics of the laser tube is the service life. It is known that one of the constraints for metal vapour lasers is the formation in the active zone of the so-called metallic growths

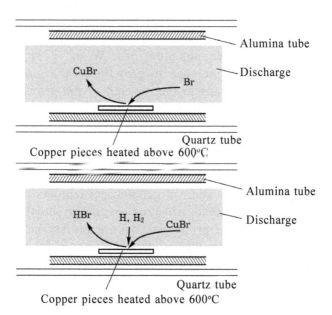

Figure 1.22. Schematic of the departure of copper atoms from the surface of copper pieces and their return [67].

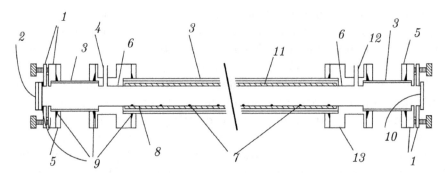

Figure 1.23. 1) nylon; 2) quartz window + reflector; 3) quartz tube; 4) gas pumping; 5) stainless steel; 6) cylindrical copper cathode; 7) copper pieces; 8) cylindrical electrode; 9) vacuum seal; 10) output window; 11) aluminium tube; 12) gas inlet; 13) end stainless steel section with water cooling.

that lead (in operation of the laser) to a gradual narrowing of the gas discharge channel, and, naturally, to a decrease in lasing power.

However, oddly enough, it was observed that the formation of growths does not take place in the cooler parts of the discharge tube, but in the hottest parts.

No growth formation is usually detected in the lead and manganese vapour lasers.

With regard to the laser tubes for generation at the transitions of the copper atom, in order to prevent the formation of growths it is necessary to properly monitor the operating temperature of the laser tube as well as the concentration of hydrogen.

For laser tubes with small diameter, where record lasing characteristics for the copper atom of 2 W/cm^3 [73] were obtained, there is practically no increase in growths as the current density is high enough and this leads to the melting of the growths in the axial part of the tube, thus significantly prolonging the life of these lasers.

1.12. The principles of calculating the thermal regime of emitters

The purpose of this section is the calculation of steady and unsteady two-dimensional (in x, y and r, z geometries) temperature fields in the lasers and structural elements which consist of any number of sites with different thermophysical properties, taking radiant heat transfer into account. Heat transfer conditions must be known for all internal and external boundaries of the region. The temperatures at all points of the region are found by solving the equation [74]:

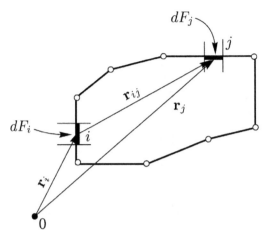

Figure 1.24. Diagram of the surfaces interacting by radiation.

$$C_v \frac{dT}{dt} = \Delta(\lambda \nabla T) + Q(t,\mathbf{r}) - \left\{ \delta_{i,i(\wedge)} \varepsilon_i \sum_j \frac{\delta_{i,j(\wedge)}}{\mathbf{r}_{i,j}} \times \right.$$

$$\left. \times \left[\sigma T_i^4(\mathbf{r}_i) - \int_{F_j} \sigma T_j^4(\mathbf{r}_j) K(\mathbf{r}_i, \mathbf{r}_j) dF_j \right] \right\},$$

where C_v – the bulk heat capacity; λ – the thermal conductivity coefficient; dF_i, dF_j – elements of radiating interacting surfaces that are within the field of research and have calculated temperatures T_i and T_j; $\mathbf{r}_i$, $\mathbf{r}_j$ – radius-vectors determining the locations of dF_i and dF_j, (Fig. 1.24); Q – specific power of heat generation ($Q > 0$) or sink ($Q < 0$) from the internal source; ε – the emissivity of the surface; d – Kronecker symbols, indexes $i(\wedge)$, $j(\wedge)$ correspond to the elements of the surface interacting by radiation; $K(\mathbf{r}_i, \mathbf{r}_j)$ – the function which depends on dF and the position and orientation of the elements dF_i ($\wedge$), dF_j ($\wedge$),

$$K(\mathbf{r}_i, \mathbf{r}_j) = \frac{d\varphi(dF_i - dF_j)}{dF_j},$$

where $\dfrac{d\varphi(dF_i - dF_j)}{dF_j}$ is a function that depends on the position and orientation of the sites dF_i, dF_j; $d\varphi$ is the angular coefficient which

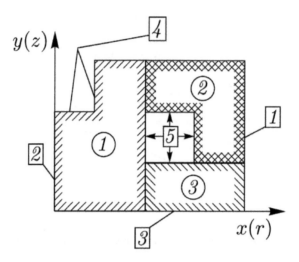

Figure 1.25. Division of the study area in which temperatures are determined; values in the circles are number of sections in which temperature are calculated, and the values in rectangles are numbers of the boundaries of the sections in which the boundary conditions are specified.

determines the part of the energy falling from surface dF_i on surface dF_j.

The equation must be supplemented by initial and boundary conditions of the (1–3rd) kind, and the boundary conditions of the 4th kind are satisfied at the interface between the contacting portions. The heat capacity and thermal conductivity coefficients of the sections are assumed to be given values and not dependent on the temperature in the expected range of temperature change.

Problems of heat conduction in complex structures of arbitrary shape are solved by the method of equivalent thermal circuits [75, 76], which is a variant of the finite difference method.

In this method, the study area (where temperatures are calculated) must be presented as a figure whose sides are parallel to the coordinate axes, for example, as shown in Fig. 1.25. If the boundaries of the study area where the boundary conditions are specified, or the interfaces between parts (materials or environments) are tilted with respect to the coordinate axes, they are recorded as stepped surfaces (areas).

The essence of the method of equivalent thermal circuits is as follows.

The investigated two-dimensional domain, for example, such as shown in Fig. 1.25, is divided into small blocks (cells) (Fig. 1.26), followed by a transition from a medium with continuously distributed properties to a circuit consisting of a thermal resistance R or conductivities G connecting the centres of 'i-th' blocks (Fig. 1.27)

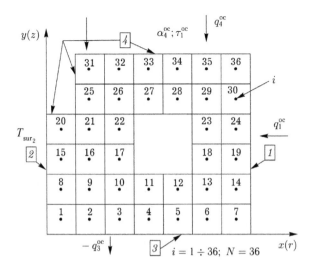

Figure 1.26. Diagram showing the division of the two-dimensional study area into cells; T_{sur} – the boundary conditions of the first kind; α^{oc} – the boundary conditions of the second kind; T^{oc} – the boundary conditions of the third kind.

and having sources, corresponding to $\hat{Q}_i$ and boundary conditions. At the same time, the numbering of nodes (centres of blocks) is assumed to be strictly consistent either in the direction of the axis $y(z)$ or in the direction of the x-axis (r).

In the nodes of the circuit we can find temperatures (potentials), which give a picture of the distribution of temperature field $T_i = T_i(x, y)$ or $T_i(r, z)$. From the determined values of temperatures in the centres of the border blocks, the values $\hat{Q}_i$ in these blocks (if $\hat{Q}_i \neq 0$) and the given heat exchange conditions at the boundaries it is possible to determine the surface temperature, if required for the problem to be solved.

For the stationary problem, the system of equations for all the blocks can be solved by the simple Gauss Seidel method of successive approximations or by the exact method of successive elimination by the Gauss scheme, which takes into account the special features of the thermal conductivity matrix.

Significant reductions (5–10 times) in computation time (due to the strong acceleration of iterations) in solving any variants of problems by simple modification of the given method to the so-called upper relaxation method [74].

The matrix conductivities in the problems of calculating the temperature field has a symmetrical band structure. In this case, any i-th row of the matrix consists of a maximum of five non-zero

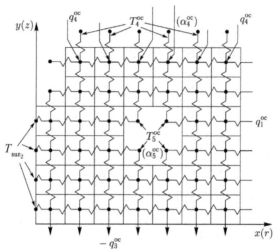

Figure 1.27. Equivalent electrical circuit consisting of thermal resistances or conductivities connecting the centres of the block, for calculating temperatures in the two-dimensional region shown in Fig. 1.26.

elements located on the width of the strip. These non-zero elements reflect the relationship of each i-th node of the equivalent circuit with the surrounding sites and include: the diagonal element, two elements close to each other and two distant elements. As already noted, in the presented programs the numbering of nodes (blocks) of the product is successive in the direction of the x-axis (r) (Fig. 1.26). Therefore, the elements that are directly next to the diagonal ones (one of them above the diagonal, the other – below the diagonal) reflect the relationship with the neighbouring blocks ($i + 1$) and ($i - 1$), located on the horizon, and the elements distant from the diagonal and positioned above and below the diagonal, reflect the relationship of the i-th node with the neighbouring nodes, located along the vertical. The maximum distance of the element above (below) the diagonal element corresponds to the half-width of the strip of the matrix.

Accounting for changes over time of the temperatures in the nodes of the equivalent thermal circuit of the analyzed region of thermal conductivity is based on the application of the implicit scheme proposed and physically justified by D. Liebmann [75, 76].

The system of equations, similar to those used for the steady-state calculations, has a matrix of the symmetric band structure and the exact solution for all T_i (t_m), representing the temperature distribution after the passage of an elementary segment (step) of time.

The essence of the calculation algorithm here is the sequential change of the stationary states, i.e. to determine at each time step the temperature distribution obtained from the exact solution of the system of equations, in which the matrix [**A**] and the array of the right sides, corresponding to the stationary field $\partial T / \partial t = 0$, are supplemented by the terms which take into account the presence of the value $CV_i\,(\partial T / \partial t)S$ in the finite difference representation. In this case, the negative values of $CV_{lk}S/\,\Delta t_m$ are added to the diagonal matrix elements of the steady-state temperature field (SSTF), and the values $CV_{lk}ST_{lk}$ $(t_m-1)/\Delta t_m$ are subtracted from the right hand sides of equations of the SSTF. At $t_m-1 = 0$, the values $T_{lk}(t_m-1)$ are taken from the primary (original) distribution.

In the programs [76], the initial temperature distribution $T_i(t = 0)$ for all $i = 1 - N$ points of the study can be set arbitrarily or fixed by the law of the stationary distribution (through the calculation of steady-state T_i with the values of the original function $Q\,(t)$, $T_{sur}\,(t)$, $T^{oc}(t)$, q^{oc} (t), corresponding time to $t = 0$). The formation of the initial values of $T_i\,(t = 0)$ is defined by specifying specially organized parameters of the set of the initial data.

Calculations of the unsteady temperature field (UTF) with uniform steps Δt_m are often not justified. So, setting Δt_m = const may cause some inconvenience for solving many thermal problems of laser technology, such as pulsed-periodic action $q^{oc}(t)$ or $Q\,(t)$. Therefore, simpler and more convenient programs have been developed and are available for mass use that allow to also consider additional arbitrary radiative heat transfer between the surfaces of products included in the study area, as well as between its external borders and the environment. Information about these programs will be published in a future study [74].

References

1.   Fowles G.R., Silfvast W.T., Appl. Phys. Lett. 1965. Vol. 6. No. 12. P. 236–237.
2.   Piltch M., Walter W.T., Solimene N., Gould G., Bennett W.R., Appl. Phys. Lett. 1965. Vol. 7. No. 11. P. 309–310.
3.   Walter W.T., Piltch M., Solimene N., Gould G., Bull. Amer. Phys. Society. 1966. Vol. 11. No. 1. P.113.
4.   Walter W.T, Solimene N., Piltch M., Gould G., IEEE J. Quantum Electronics. 1966. Vol. QE-2. No. 9. P.474–479.
5.   Piltch M., Gould G., Review Scientific Instrument. 1966. Vol. 37. No.7. P. 925–927.
6.   Budnikov P.P. et al. Chemical technology of ceramics and refractories. Moscow: Izd. Liter. po Stroit. 1972.
7.   Isaev A.A, Kazaryan M.A., Petrash G.G., Pribory Tekh, Eksper. 1973. No. 1. P. 188–189.
8.   Kazaryan M.A., Research of pulsed metal vapor lasers. Thesis Cand. Phys. and Mathematic Sciences, FIAN. Moscow: 1974, P. 151 .
9.   Isaev A.A., Kazaryan M.A., Petrash G.G., Copyrights evidence of USSR. No. 446239 (14.07.1971).

10. Burmakin V.A., Bylkin V.I., Doroshkin A.A., Isaev A.A., Kazaryan M.A., Petrash G.G., Authors Cert, USSR. No. 555776 (11.03.1974).

11. Isaev A.A., Kazaryan M.A., Petrash G.G., Pisma v ZhETF. 1972. Vol. 16.No.1. P. 40–42.

12. Alger T.V., Bennett W.J., Rev. Sci. Instrum. 1982. Vol. 53. No. 6. P.762–764.

13. Direktor L.B., Malikov M.M, Skovorod'ko S.N., et al., Teplofizika vysokikh temperatur. 1983. Vol. 21. No. 1. P. 161–166.

14. Direktor L.B., V. Katchalov, Malikov M.M,et al., Teplofizika vysokikh temperatur. 1985. V.23. No.1. P.193-195.

15. Direktor L.B., Fomin V.A., Malikov M.M., Teplofizika vysokikh temperatur. 1990. V.28. No.3. P.427-432.

16. Babeyko Yu.A., Vasiliev L.A., Orlov V.K., et al., Kvantovaya elektronika. 1976 V.3. No.10. P.2303-2304.

17. Isaev A.A., Kazaryan M.A, Petrash G.G., Kvantovaya elektronika. 1973. V.6 (18). P.112-115.

18. Batenin V.M., Klimovskii I.I., Selezneva L.A., Teplofizika vysokikh temperatur. 1980. V.18. No.4. P.707-712.

19. Asmus J.F., Moncur N.K. , Appl. Phys. Lett. 1968.No.13.No.11. P.384-385.

20. Fedorov A..I, Sergeenko V.P., Tarasenko V.F., Kvantovaya elektronika. 1977. V.4. No.9. P.2036-2037.

21. Fedorov A.I., Sergeenko V.P., VF Tarasenko, et al., Izv. VUZ. Fizika. 1977. No.2. P.135-136.

22. Isakov I.M., Leonov A.G., Petrushevich Yu, Starostin A.N., Zhurnal tekhnicheskoi fiziki. 1981. V.51. No.3. P.525-532.

23. Shuhtin A.M.,Mishakov V.G., Fedotov G.A., Pisma Zh. Tekh. Fiz. 1977 V.3. P.750-752.

24. Shuhtin A.M., Fedotov G.A., Mishakov V.G., Kvantovaya elektronika. 1978 V.5. No.7. P.1592-1595.

25. Vasiliev L.A., Hertz V.E., Direktor L.B., et al., Teplofizika vysokikh temperatur. 1982. V.20, No.5. P.995-997.

26. Malikov M.M,, Fomin V.L., Shevchenko A.L., Shpil'rain E.E., Teplofizika vysokikh temperatur. 1985. V.23. No.5. P.966-971.

27. Ferrar C.M. , IEEE J. Quantum Electronics. 1973. No.QE-9. No.8. P.856-857.

28. Bashilov V.A., Gerasimov L.I., Smilga V.I., Abstracts. Moscow:VNTIC GKNT.1978. P.376-377.

29. Russel G.R., Nerheim N.M., Piviritto T.J., Appl. Phys. Lett. 1973. No.21. No.2. P.565-567.

30. Gudkov E.B., Egorov V.G., Pavlenko V.S., Kvantovaya elektronika. 1979 V.6 No.12. P.2633-2636.

31. Granqvist C.G., Buhrman R.A. , Appl. Phys. Lett. 1976. No.47. No.5. P.2200-2219.

32. Voronyuk L.V., Research of the effect of cesium impurity on the lasing characteristics of metal-vapor lasers. Thesis, UGU. Uzhgorod. 1988, P.190.

33. Belokrinitsky N.S., Voronyuk L.V., Glushchenko O.A., et al. Investigation of the kinetics and energy parameters of CuCs-laser. Preprint No.6, Institute of Physics, Academy of Sciences of the USSR. Kiev. 1988.

34. Bohan P.A., Silant'ev V.I., Solomonov V.I., Kvantovaya elektronika. 1980. V.7. No.6. P.1264-1269.

35. Vetter A.A., Nerheim N.M. , Appl. Phys. Lett. 1977. No.30. No.8. P.405-407.

36. Lesnoi M.A., Kvantovaya elektronika. 1984. V.11. No.1.P.205-208.

37. Kushner M.J., Culick F.E.C. , IEEE J. Quantum Electronics. 1979. No.QE-15.

No.9. P.835.
38. Cross L.A., Jenkins R.P., Gokay M.P. , Appl. Phys. Lett. 1978. No.49. No.1. P.453-454.
39. Alexandrov I.S., Babeyko U.A., Babayev A.A., et al., Kvantovaya elektronika. 1975. V.2. No.9. P.2077-2079.
40. Buzhinskiy O.I., Krysanov S.I., Slivitsky A.A., Pribory Tekh. Eksper., 1979. No.4. P.274.
41. Sokolov A.V., SviridovA.V., Kvantovaya elektronika. 1981 V.8. No.8. P.1686-1696.
42. Fahlen T.P. , J. Appl. Phys. 1974. No.45. No.9. P.4132-4133.
43. Anders A.K., Tobin R.C., Appl. Phys. Lett. 1988. No.64. No.9. P.4285-4292.
44. Anders A.K., Tobin R.C., J. Appl. Phys. 1989. No.66, No.7. P.2794-2799.
45. Smilansky I., Kerman A., Levin LA, Erez G., IEEE J. Quantum Electronics 1977. No.QE-13. No.1. P.24-26.
46. Isaev A.A., Petrash G.G., Pisma v ZhETF. 1968. V.7. P.204-207.
47. Karabut E.K., Mikhalevsky V.S., Papakin V.F.,Sam M.F., Zhurnal tekhnicheskoi fiziki. 39, 1969. No.10. P.1923-1924.
48. Kotelnikov R.B., Bashlykov S.N., Galiakbarov E.G., Chestnut A.K., Particularly refractory elements and compounds. Moscow: Metallurgiya. 1969. P.372.
49. Physical and chemical properties of oxides. Ed. Samsonov G.V., Moscow: Metallurgiya. 1978. P.471.
50. Bohan P.A., Metal vapor lasers with collisional de-excitation of the lower working states. Doctor Thesis. Sci. Sciences. IOA. Novosibirsk. 1988.P. 418 .
51. Bohan P.A., Nikolaev V.N., Solomonov V.I., Kvantovaya elektronika. 1975. V.2. No.1. P.159-162.
52. Bohan P.A., Klimkin V.M., ZhPS. 1973. V.9. No.3. P.414-418.
53. Frisch S.E., Determination of the concentration of normal and excited atoms and oscillator forces and methods of emission and absorption of light. In: Spectroscopy of gas-discharge plasma. Leningrad. Nauka, 1970. P.7-62.
54. Bohan P.A., Vlasov G.Y., Gorokhov A.M., et al., Kvantovaya elektronika. 1977. V.4. No.6. P.1395.
55. Bohan P.A., Gerasimov V.A., Author's certificate, USSR No.755136. Byull. No. 41. Publ.7.11.1984.
56. Bokhan P.A., Gerasimov V.A. Invention de la Republique Francaise 2529401. BOPI Brevets - 1985. No.6. P.8354.
57. Bohan P.A., Shcheglov V.B. , Kvantovaya elektronika. 1978 V.5. No.2. P.381-387.
58. Kim J.J., Convey J.F., Rev. Sci. Instrum. 1982. No.53. No.10. P.1623-1682.
19   Bohan P.A., Sorokln A.R., Zhurnal tekhnicheskoi fiziki. 1985. 1.55. No.1. P.88-95.
60. Bohan P.A.,Sorokin A.R., Pisma v ZhETF. 1982 V.8. No.15. P.947-950.
61. Bohan P.A., Sorokin A.R., Pisma v ZhETF. 1984. No.10. No.10. P.620-623.
62. Kirkpatrick C. The Theory and properties of disordered materials. Moscow: Mir. 1977. P.249-292.
63. Tables of physical quantities. A handbook, Ed. I.K. Kikoin. Moscow: Atomizdat, 1976. P.1006.
64. Kazaryan M.A., Trofimov A.N, Kvantovaya elektronika. 1978 V.5. No.11. P.2471-2472.
65. Piper J.A., Optics communications. 1975. No.14. No.3. P.296-300.
66. Piper J.A., IEEE J. Quantum Electronics. 1978. No.QE-14. No.6. P.405-408.
67. Pulsed Metal Vapour Lasers. Proceedings of the NATO Advanced Research

Workshop on Pulsed Metal Vapor Lasers – Physics and Emerging Applications in Industry, Medicine and Science. St. Andrews, U.K., 6-10, 1995; edited by Chris E. Little and Nikola V. Savotinov, NATO ASI Series, Kluwer Academic Publishers. 1996. P.125-138.

68.   Jones Q.R., Little C.E., IEEE J. Quantum Electronics. 1992. No.QE-28. No.3. P.590-593.

69.   Jones Q.R., Little C.E., Optics communications. 1992. No.91. No.3, 4. P.223-228.

70.   Jones Q.R., Little C.E., Opt. Quantum Electronics. 1992. No. 24. No.1. P.67-72.

71.   Jones Q.R., Little C.E. , Optics Communications. 1992. No.89. No.1. P.80-87.

72.   Jones Q.R., Maitland A., Little C.E., IEEE J. Quantum Electronics. 1994.No. QE-30. No.10. P.2385-2390.

73.   Sabotinov N.V., Akerboom F., Jones Q.R., Maitland A., Little C.E., IEEE J. Quantum Electronics. 1995. No.QE-31. No.4. P.747-753.

74.   Tykhotsky V.V., Calculations of temperature in laser technology. Tutorial for high school students. Moscow: MFTI. 1991. P. 150 .

75.   Liebman G., British Journal of Applied Physics. 1955. No.6. No.4. P.129-135.

76.   Liebman G., Transaction of the American Society of Mechanical Engineers. 1956. No.78. No.3. P.655-665.

Excitation circuit and its effect on the lasing characteristics of self-heated copper vapour lasers

2.1. Electrical characteristics of the discharge

The electrical characteristics of the discharge (see, e.g. [1–6]) mean in the first place characteristics such as the voltage in the discharge (at the electrodes of the gas discharge tube (GDT)) $U_d(t)$, the current through the discharge $I_d(t)$, active resistance $R_d(t)$ and inductance L_d of the discharge. For the formation of general ideas about the characteristics of these discharges in self-contained pulsed and repetitively pulsed (pulsed-periodic) lasers we take a look at the results of [2, 3], in which the studies were conducted with self-heated copper vapour lasers with an unsealed tube (Fig. 2.1) similar to those described in [7,8], with a GDT with a diameter of 1.2 cm and 70 cm long and with the excitation circuit (Fig. 2.2 a), which is one of the variants of the

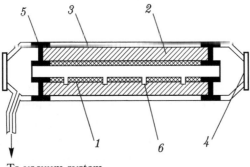

To vacuum system

Figure 2.1. Schematic of the emitter of a self-heated copper vapour laser tube. 1 – alundum GDT, 2 – thermal insulation, 3 – vacuum-tight shell, 4 – windows for discharge of laser radiation, 5 – current leads with the electrodes, 6 – copper vapour generators.

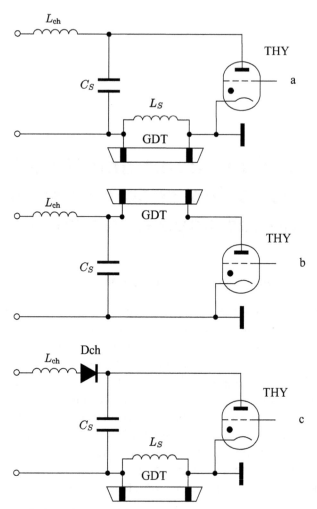

Figure 2.2. Excitation of lasers with direct discharge of the storage condenser on the discharge gap and the resonant recharge of the storage capacitor. L_{ch} – charging inductance (choke), L_S – the inductance shunting the discharge gap during the recharge time of the storage capacitor; C_S – storage capacitor; THY – thyratron; D_{ch} – charging diode.

circuit with direct discharge of the storage capacitor to the discharge gap and with resonant recharging of the storage capacitor (see Fig. 2.2). These circuits are sometimes called circuits with a choke charge of the storage capacitor, a diode in the presence of the charging circuit (Fig. 2.2 c) – circuits with a diode–choke charge storage capacitor.

Buffer gases [2,3] were He, Ne and Ar at pressures from 2 to 14 kPa. Voltage U_r at the rectifier output ranged from 3 to 6 kV, and the pulse repetition rate from 5 to 15 kHz. The thyratron switch was TGI1-

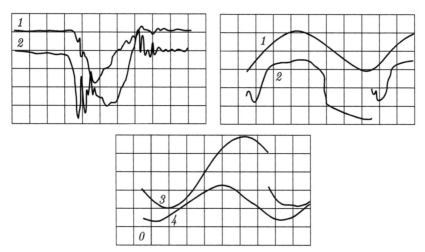

Figure 2.3. Oscillograms of the currents (1.4) and voltage in the circuit shown in Fig. 3.1, a) $L_{ch} = 0.8$ H, $C_S = 3300$ pF, $L_S = 7$ mH, diameter of the GDT 12 mm, the length of GDT 70 cm, the pressure of neon 13.3 kPa; a) during the the excitation pulse: 1 – the current through the discharge (sensitivity 60 A/div), 2 – the voltage across the discharge gap (sensitivity 3.3 kV/div), sweep duration 100 ns/div, b) for the relaxation time of the plasma: 1 – discharge current (1.5 A/div), 2 – the voltage drop across the discharge gap (132 V/div), 3 – the voltage on the storage capacitor C_S (72 kV/div), 4 – the current through the inductance L_S (1 A/div). Duration of the sweep 20 µs/div, the straight lines correspond to zero values of the signals.

2000/35. The charging inductance L_{ch} and inductance L_S, shunting the discharge, were respectively 0.8 H and 7 µH. The electrical capacitance of the storage capacitor C_S was 3.3 nF.

In [2,3] the excitation circuit of the copper vapour laser is not optimized to achieve the maximum efficiency or to obtain the maximum output power. Perhaps for this reason the studies [2,3] have identified a number of factors affecting the characteristics of pulsed self-contained lasers. Typical oscillograms of the current and voltage pulses at various power circuit elements during excitation and relaxation are shown in Fig. 2.3. Oscillograms of current pulses through the discharge and voltage across the electrodes of the GDT indicate the inductive nature of the discharge. Calculation of the inductance of the discharge L_d, carried out in [2,3] by the equation

$$U_d = U_R + U_{L_d} = I_d R_d + L_d \frac{dI_d}{dt} \,, \qquad (2.1)$$

where, besides the well-known notations, U_R and U_L are respectively the voltage across the active resistance and inductance of the discharge, for the point in time when the current through the discharge is equal to

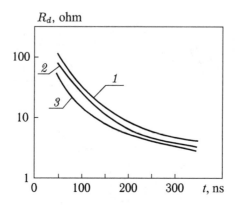

Figure 2.4. The change of active resistance of the discharge during the excitation pulse at different neon buffer gas pressures: 1 – 13.3 kPa, 2 – 6.7 kPa, 3 – 3.3 kPa, l_p = 70 cm, d_p = 12 mm, Q_r = 1.8 kW, f = 7 kHz.

zero, showed that the value of L_d is not dependent on the experimental conditions (type and pressure of the buffer gas, the voltage at the rectifier, etc.) and is L_d = 1.6 ± 0.1 µH. For comparison, the inductance of the straight conductor whose dimensions match the dimensions of the discharge gap of the GDT, calculated by the formula [9]

$$L_p = \frac{\mu_0}{2\pi} l_p \left(\ln \frac{4l_p}{d_p} - 1 \right),$$
(2.2)

where l_p and d_p are the length and diameter of the conductor, coinciding with the length and diameter of the GDT, is about 0.7 µH, i.e. approximately two times less than the inductance, measured in the experiment.

Without going into detailed analysis of all the reasons for the increase of inductance in the discharge [2,3], let us mention at least three of them: inductance of the current leads of the laser tube, the inductance of the shunt, designed to measure the discharge current pulses and secured in the experiments [2,3] directly to cathodic current leads, the mutual induction of the discharge with the other parts of the discharge circuit. Therefore, strictly speaking, the inductance, measured in [2,3], should be viewed as an overall inductance of the discharge, the shunt and current leads of the laser tube.

Figure 2.4 shows the calculated resistance of the discharge R_d [3] using the equation (2.1) from the oscillograms of current pulses I_d and voltage U_d with the inductance of the discharge taken into account. According to these results, first, during the excitation pulse, the discharge resistance is reduced from a few hundred ohms to several ohms, and, secondly, the value of R_d at the end of the excitation pulse depends only weakly on the buffer gas pressure, which apparently is a consequence of the termination of energy input into the discharge (at $R_d \approx 3$ ohms). Termination of the energy supply can occur with a decrease in the resistance level to values that are much smaller than the resistance of the thyratron R_T, or the wave impedance of the discharge circuit $\rho_C = (L_C/C_S)^{1/2}$, where L_C is the inductance of the discharge circuit.

2.2. Distribution of electrical energy consumed by a rectifier in different elements of the charger and discharge circuits

To answer the question of how the energy, required from the rectifier, is distributed between the discharge, the thyratron and charging circuit elements, measurements were taken in various papers (see, e.g., [1–3,5,6]) of the electrical power input to the discharge and the power dissipated by the thyratron. Consider the results of [2, 3] and compare with the results of [1,6]. The energies G_d and G_{rel}, respectively, supplied to the discharge during the time of the excitation pulse τ_{ex} and in interpulse interval τ_{rel}, were calculated in [2,3] for the corresponding oscillograms of discharge current and voltage at the electrodes of the GDT using formulas

$$G_d = \int_0^{\tau_{ex}} I_d U_d dt, \quad G_{rel} = \int_{\tau_{ex}}^{\tau_{ex}+\tau_{rel}} I_d U_d dt. \tag{2.3}$$

The time-averaged power supplied to the discharge at different time intervals was defined as

$$\bar{Q}_d = G_d f, \quad \bar{Q}_{rel} = G_{rel} f. \tag{2.4}$$

The time-averaged power $\bar{Q}_T$ released in the thyratron was measured by the colorimetric method of heating water, cooling the thyratron. The time-averaged power $\bar{Q}_c$ dissipated by the elements of the charging circuit, is determined from the equation:

$$\bar{Q}_c = Q_r - \left(\bar{Q}_d + \bar{Q}_{rel} + \bar{Q}_T\right),$$

where Q_r is the power consumed from the rectifier. The results of measurements of $\bar{Q}_d$, $\bar{Q}_{rel}$, $\bar{Q}_T$ and calculations of $\bar{Q}_c$ are shown in Table 2.1 and indicate that in the considered experiments in the thyratron and the elements of the charging circuit the power loss is respectively 27–30% and 13–20% of the power consumed from the rectifier. These

Table 2.1.

No.	p_{Ne}, kPa	U_r, kV	I_r, A	Q_r, W	$\overline{Q_d}$, W	$\overline{Q_{rel}}$, W	$\overline{Q_T}$, W	$\overline{Q_c}$, W
1	13.3	4.6	0.39	1800	820	130	480	370
2	6.7	4.6	0.39	1800	780	140	520	360
3	3.3	4.6	0.39	1800	870	150	550	230

data agree well with the results of [1], in which, apparently, the authors measured for the first time the loss of the power consumed from the rectifier, in the various elements of the excitation circuit of the self-heated copper vapour lasers with different GDT, whose diameter was varied from 0 6 to 2.4 cm, and the length from 30 to 100 cm. According to [1] up to 30% of the power consumed by a rectifier, dissipated in the thyratron, and 25–35% in the charging circuit. In later studies [6] of self-heated copper vapour laser (l_p = 95 cm; d_p = 1.5 cm; p_{Ne} = 3.3–13.3 kPa; f = 5–20 kHz) it was found that the efficiency of supply of energy to the active zone of GDT with the thyratron excitation circuit in the commuted energy range 0.03–0.3 J/pulse is in the range 50–70%. These data agree well with the results of [1–3].

2.3. Effect of the charging circuit on the lasing characteristics of self-contained lasers

As soon as the reservation that in this case the lasing characteristics of the laser primarily involve the lasing power $\overline{P}_1$ and the efficiency of conversion to laser radiation of the power consumed from the rectifier (the practical efficiency η_r).

When analyzing the data in Table 2.1 attention should be paid to a significant energy input ($\overline{Q}_{rel} \approx$ 150 W) in a discharge between excitation pulses, which can significantly affect the relaxation processes of the plasma and, consequently, the lasing characteristics. For example, the current flow in the active medium before the next excitation pulse can lead to an additional population of the metastable levels and the consequent reduction due to this population of the lasing energy in the pulse.

To verify the existence of the negative influence of the current recharge of the storage capacitor on the lasing characteristics of pulse-periodic metal vapour lasers measurements were taken in [2,3] of the lasing power of two self-heated copper vapour lasers, which differ only in the location of the discharge tube in the excitation circuit (circuits in Fig. 2.2 a and 2.2 b). Comparison of the measurement

Table 2.2.

p_{Ne}, kPa; f, kHz		13.3	6.6	3.3
		6.8	6.6	6.52
P_{rz}, W	Circuit I (Fig. 2.2 a)	0.5	0.06	0
	Circuit II (Fig. 2.2 b)	1	0.68	0.4
$U_r = 4.6$ kV, $I_r = 0.39$ A				

results, shown in Table 2.2, clearly shows the negative impact of the recharge current on the lasing characteristics. This is also indicated by the results of studies [10] of self-heated copper vapour lasers with two different excitation circuits depicted in Fig. 2.2 a (circuit I) and Fig. 2.2 b (circuit II), respectively. As follows from the results [10] presented in Fig. 2.5, the exclusion of the flow of the recharge current of the storage capacitor through the discharge gap (through the active medium of the laser) results in the conditions used in [10] not only in an increase in power generation, but also in an increase of the maximum achievable lasing power.

Confirmation of the possibility of the recharge current flowing through the active medium of the laser when it is excited by the circuit shown in Fig. 2.2 a, was obtained in [11] in investigating the excitation circuit with resonant recharge of the storage capacitor, but with the resistance instead of shunting inductance L_b.

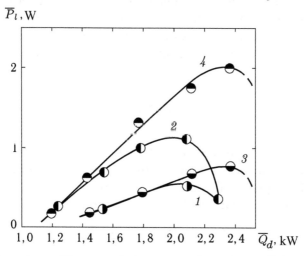

Figure 2.5. Dependence of the average lasing power on the power supplied to the discharge, $f = 7$ kHz, $l_p = 100$ cm, $d_p = 1.2$ cm, $p_{Ne} = 13.3$ kPa. Scheme I (Fig. 2.2 a): 1 – lasing power on the yellow line, 2 – total lasing power. Scheme II (Fig. 2.2 b): 3 – lasing power on the yellow line, 4 – the total lasing power.

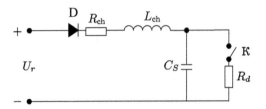

Figure 2.6. The equivalent circuit diagram of the charge of the storage capacitor.

Strictly speaking, by choosing the value of L_S we can minimize part of the recharge current that flows through the discharge gap in the excitation of the discharge by the circuit shown in Fig. 2.2 a. However, we can assume that this minimization is effective only for relatively low pulse repetition frequencies and small GDT radii. Confirmation to this assumption are the results of [12, 13], in which a copper vapour laser with a pulse repetition frequency of 235 kHz could work only with the scheme shown in Fig. 2.2 b, and additionally equipped with a diode as indicated in the diagram shown in Fig. 2.2 c.

As is well known, charging a capacitor from a DC voltage source (rectifier) via a resistor (resistance) is carried out with an efficiency of 50%. The use of resonant charging of the storage capacitor with a suitable choice of the capacitance of the storage capacitor C_S, resistance R_{ch} and inductance L_{ch} of the charging circuit can significantly reduce energy losses in the charging circuit. In general, the problem of energy losses in the charging circuit elements is solved in [14] for the charging circuit the equivalent circuit of which is shown in Fig. 2.6. For the duration of the discharge the storage capacitor much shorter than the charging time, on the basis of Kirchhoff's equations one can write [14]:

$$U_r - U_c + R_{oh}C_s\frac{dU_c}{dt} + L_{ch}C_s\frac{d^2U_c}{dt^2} , \qquad (2.5)$$

where, besides the well-known notations, U_r is the output voltage of the rectifier, U_C is the voltage on the storage capacitor. The initial conditions for equation (2.5): $t = 0$, $U_C = 0$.

The solution of equation (2.5) has the form:

$$U_c(t) = U_r\left(1 - e^{-\alpha t}\cos\omega t\right) , \qquad (2.6)$$

$$\omega = \sqrt{\omega_0^2 - \alpha^2} , \quad \omega_0 = 1/\sqrt{L_{ch}C_s} , \quad \alpha = R_{ch}/L_{ch}.$$

Charging time T_m of the storage capacitor to the maximum voltage that determines the minimum excitation pulse repetition period is equal to:

$$T_m = \frac{\pi}{\omega} - \frac{1}{\omega}\operatorname{arctg}\frac{\alpha}{\omega}. \tag{2.7}$$

Charging current I_{ch} of the storage capacitor is given by:

$$I_{ch}(t) = C_s \frac{dU_c}{dt} = U_r C_s e^{-\alpha t}(\alpha \cdot \cos\omega t + \omega \cdot \sin\omega t). \tag{2.8}$$

In view of (2.6)–(2.8) the efficiency η_{ch} of the charging circuit is defined as follows:

$$\eta_{ch} = \frac{C_s U_{C\max}^2}{2}\left(\int_0^{T_m} U_r I_{ch}(t)\,dt\right)^{-1} = \frac{1}{2}\left[1+\left(1-\frac{\varphi^2}{4}\right)^{1/2}\right. \times$$

$$\times \exp\left(-\frac{\pi\varphi}{\sqrt{4-\varphi^2}} + \frac{\varphi}{\sqrt{4-\varphi^2}}\operatorname{arct g}\frac{\varphi}{\sqrt{4-\varphi^2}}\right)\right], \tag{2.9}$$

$$U_{C\max} = U_C(T_m), \quad 0 < \varphi = R_{ch}/\sqrt{L_{ch}/C_s} < 2.$$

when $\varphi > 2$, $\eta_3 = 0.5$.

When $\varphi \ll 1$, equation (2.9) is transformed to the known relation:

$$\eta_{ch} = 1/\left(1 + (\pi/4)R_{ch}\sqrt{C_s/L_{ch}}\right).$$

The relationship between the maximum voltage $U_{C\max}$ on the storage capacitor and the efficiency of the charging circuit η_{ch} are given by:

$$U_{C\max} = 2\eta_{ch}U_r. \tag{2.10}$$

Borrowed from [14], the results of calculation of the efficiency of the charging circuit η_{ch} are shown in Fig. 2.7. According to these results, the efficiency of the discharge circuit of 0.9 and more can be achieved only if $\varphi < 0.15$ or $R_{ch} < 0.15\sqrt{L_{ch}/C_S}$.

2.4. Influence of the discharge circuit on the lasing characteristics of self-contained lasers

According to the views available in mid-70s for the mechanism of lasing of self-contained lasers, which, apparently, at that time were

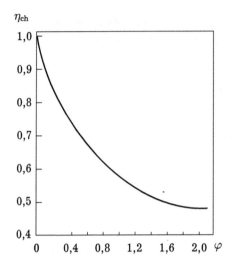

η_{ch}

Figure 2.7. The results of calculation of the efficiency of the charging circuit according to the formula (2.9) [14].

held by most researchers, the specific energy of these lasers is uniquely determined by the steepness of the leading edge of the current pulse or, in other words, the rate of current rise . For this reason, the negative impact of the discharge circuit inductance L_C on the lasing characteristics of various self-contained lasers, observed, for example, in [15, 16], was attributed primarily to the influence of inductance on the steepness of the leading edge of the current pulse.

However, in [17], apparently, it was first reported that experiments carried out by the authors [17] with different copper vapour lasers excited by longitudinal and transverse electrical discharges showed the following: if the steepening of the pulse front of the discharge current dI_d/dt is achieved by increasing the voltage at the output of the rectifier, then an increase of the steepness results in saturation of the energy output which at low copper vapour pressures (1–10 Pa) is gradually transformed into a decrease. In [17] it is suggested that at low copper vapour pressures and high-power pumping most of the copper atoms are ionised before lasing occurs, with the lasing time being finite.

Quantitative understanding of the values dI_d/dt, at which the energy output in the copper vapour laser is saturated can be formed on the basis of the data presented in [10] where experimental investigation was carried out of the influence of the rate of current rise dI_d/dt on the average lasing power of the self-heated copper vapour laser at a time when at a change dI_d/dt the concentration of copper atoms in the discharge remains constant. The experiment were carried out using the excitation scheme shown in Fig. 2.2 b, which excluded the impact of current charge of the storage capacitor on the experimental

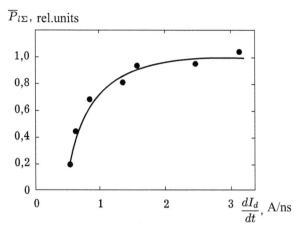

Figure 2.8. Dependence of the average lasing power on the steepness of the current pulse at the time of the lasing pulse. p_{Ne} = 3.3 kPa, f = 7 kHz, l_p = 100 cm, dp = 12 mm. Circuit II (Fig. 2.2 b).

results. Changing the rate of current rise was carried out in [10] by the short-term change (one to two seconds) of the voltage at the rectifier output U_r. At short-term changes in U_r and, consequently, the power level supplied to the discharge, the temperature of the GDT and, consequently, the concentration of copper atoms in the discharge volume remained unchanged since the time of heating the thermal insulation from 300°C to operating temperature in the experimental conditions [10] was about an hour.

The results of measurements [10] of the dependence of $\overline{P}_{l\Sigma}$ on dI_d/dt, shown in Fig. 2.8 suggest that starting with certain values dI_d/dt, the rate of current rise is not a parameter that determines the energy and average lasing power in pulsed and repetitively pulsed metal vapour lasers, which confirms the findings of [17].

The findings of [10, 17] are in good agreement with the results of [18, 19]. The first of them shows that in the self-heated copper vapour laser (l_p = 30 cm; d_p = 3.4 cm; f = 4 kHz) the lasing energy increases with increasing energy input G_d into the discharge from 50 to 300 mJ and then in the range of the G_d values from 300 to 500 mJ remains constant. In [19] a a pulse-periodic copper vapour laser (active medium length 55 cm; d_p = 3 cm; f = 4 kHz) was used to measure the dependence of the maximum concentration of copper atoms $n_{r\,max}$ at $^2P^0_{1/2}$ during the excitation pulse on the amplitude of the current pulse. According to these measurements, the value $n_{r\,max}$ first increases with increasing amplitude of the current pulse, and then saturates. Since the specific energy output and the maximum concentration of atoms in the upper laser level are interconnected, the

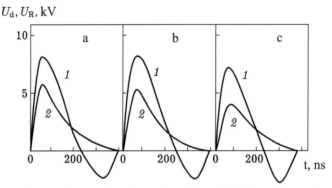

Figure 2.9. Voltage pulses on the electrodes of the GDT (1) and the active resistance of the discharge (2). The pressure of neon: a – 13.3 kPa, b – 6.7, c – 3.3; l_p = 70 cm, d_p = 1.2 cm.

results of [19] also suggest that the amplitude of the pulse of discharge current and, consequently, its slope, are not parameters that uniquely determine the lasing energy of metal vapour lasers.

The nature of the influence of the discharge circuit inductance on the lasing characteristics can be judged on the basis of comparison of the voltage pulses $U_d(t)$ and $U_R(t)$ borrowed from [3] and presented in Fig. 2.9 respectively on the electrodes of the GDT of the copper vapour laser and the active resistance of the discharge. The pulses U_R (t) were calculated in [3] using the equation (2.1) from the pulses of the discharge current and voltage pulses on the discharge, similar to those shown in Fig. 2.3. As can be seen, the presence of the inductance of the discharge leads to a substantial decrease in the voltage across the active resistance of the discharge compared to the voltage across the electrodes of the GDT. For example, at a neon pressure of 3.3 kPa the voltage on the active resistance of the discharge is almost two times less than the voltage at the discharge.

Comparison of the pulses U_d (t) and U_R (t) shows that the presence of the discharge circuit inductance reduces the strength of the electric field in the discharge and, consequently, reduces the average energy (temperature) of the electrons, which in turn leads to a decrease in the rate of population of the upper laser level and an increase of the rate of population of the lower laser level and, ultimately, to a decrease in lasing energy.

As shown in the introduction, the lifetime of the population inversion of two levels with the probability of spontaneous transition between them A_{21} is close to the value A_{21}^{-1}. This means that the characteristic time τ_E of establishment of such strength of the electric field in the discharge, which corresponds to the electron temperature, providing efficient excitation of the upper laser level, must satisfy the condition:

$$\tau_E \ll A_{21}^{-1}. \tag{2.11}$$

Actually, the condition (2.11) is a condition on the value of the inductance of the discharge circuit, which determines to a large extent the rate of increase of the discharge current at the beginning of the discharge and, consequently, the rate of increase of the electric field strength in the discharge. Failure to comply with the conditions (2.11) for any particular self-contained laser leads, at least, to a decrease of the energy output compared with the maximum possible output.

In addition to the already discussed impact on the field strength in the discharge, the presence of the inductance of the discharge circuit can lead to a marked reduction in the physical and practical efficiency of the laser, since the energy stored in the inductor at the end of the lasing pulse

$$G_L = \frac{L_c I_d^2}{2} \tag{2.12}$$

(here I_d is the discharge current at the end of the lasing pulse), is released in a discharge after lasing.

We briefly consider the main mechanisms of the effect of the discharge circuit inductance on the lasing characteristics of self-contained lasers in the direct discharge of the storage capacitor to the discharge gap, and we note that those wishing to get acquainted with theoretical and experimental studies of these mechanisms may consult [20, 21].

The presence of the negative impact of the discharge circuit inductance on the lasing characteristics of the self-contained lasers led to attempts to use the coaxial arrangement of the GDT and a return conductor, since it is known that the inductance of a straight wire with a coaxial current lead with an inside diameter d_b, defined by the formula [9]:

$$L_p = \frac{\mu_0}{2\pi} l_p \left(\ln \frac{d_b}{d_p} \right), \tag{2.13}$$

is significantly less than the inductance of a straight wire which is calculated by the formula (2.2).

For example, in [22] a manganese vapour laser was studied using the design shown in Fig. 2.10. A distinctive feature of this emitter is the presence of a reverse conductor located coaxially with the GDT,

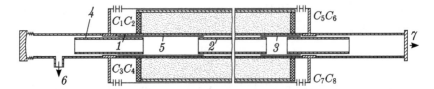

Figure 2.10. The design of the manganese vapour laser [22]. 1 – quartz tube with a diameter of 2.8 cm, 2 – BeO discharge tube, 2 cm in diameter and the length of the discharge 46 cm, 3 – anode, 4 – quartz liner, 5 – cathode, 6 – evacuation, 7 – radiation yield, 8 – reverse conductor, diameter 10.5 cm; C1–C8 – storage capacitors.

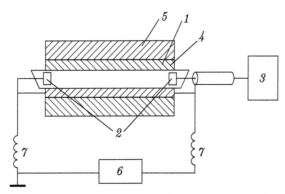

Figure 2.11. Schematic representation of a metal vapour laser [23].

which reduced the discharge circuit inductance to 0.2 µH. The switch is [22] was a TGI1-1000/25 thyratron. The capacitors C1–C8 were in form of low-inductance capacitors KVI-3-1000. Discharge of the capacitors to the discharge gap was carried out in the Blumlein circuit.

The transmitter shown in Fig. 2.10 c [22] provides the following parameters of excitation pulses: amplitude of the voltage pulse at the electrodes of the GDT 35 kV, pulse amplitude of the discharge current 1.5 kV, pulse excitation duration at the base 50–70 ns; excitation pulse repetition frequency 20 kHz.

In [23] the authors proposed a construction of the emitter of a repetitively pulsed metal vapour laser tube (Fig. 2.11), which includes the GDT (1) with electrodes (2) connected to the excitation unit (3) and the heater (4), placed in a thermal insulation (5) and connected to a voltage source (6) of industrial frequency. The heater is designed as a conductor made of high-conductive ceramics with the resistance lower than the resistance of the discharge at the end of the excitation pulse, and is located coaxially with the GDT. One of the GDT electrodes is connected to the excitation unit by a conductor–heater, so that one end of the conductor–heater is connected to this electrode, and the other – to the excitation unit. In [23] it is proposed to produce the conductor–heater either from a material in the CiS–MoSi$_2$ system or calcium-doped chromate III of lanthanum or yttrium. The source of the industrial frequency voltage is protected against high-voltage pulses using choke coils (7). It is possible that the implementation in practice of the emitter structure, proposed in [23], will simultaneously reduce time required by the metal vapour lasers to reach the operating conditions and to increase the lasing power and efficiency of such lasers.

Thus, the reduction of the discharge circuit inductance is one way to improve the output characteristics of pulsed lasers at self-contained transitions. However, as shown in [24], reducing the inductance of the discharge circuit by reducing the diameter of the reverse conductor may lead under certain conditions to an undesirable increase in distributed capacitance and, consequently, to an additional electrical mismatch of the discharge with the external electrical circuit.

As is known, one way of increasing the efficiency of self-contained lasers is to use partial discharge of the storage capacitor, or, in other words, the forced termination (break) of current and voltage pulses. The implementation of this method in [25] using a copper vapour laser with a GDT of 6 mm diameter and 30 cm long allowed to reach the values of efficiency of 9%[1] The neon pressure in the GDT was 6.7 kPa, and the current was interrupted at 20 A. Judging by the specified value of efficiency, the negative influence of the discharge circuit inductance in [25] is reduced, apparently, to a minimum. However, it remains unclear how and under what GDT diameters and neon pressures the negative impact of the discharge circuit inductance on the characteristics of self-contained lasers becomes evident.

The answer to the question can be given to some extent on the basis of the results of [26], which analyzed the impact of the discharge circuit inductance on the characteristics of self-contained lasers with forced interruption of current in the discharge circuit. The discharge circuit (see Fig. 2.12 a) includes a storage capacitor with electrical capacitance C_S and charged to voltage U_0; an ideal switch, which provides closure and opening of the discharge circuit; the inductance of the discharge circuit L_C and the discharge resistance R_d changing with time.

Under the assumption that the value of electrical capacitance of the capacitor C_S is so great that the voltage on it does not change during

---

[1] In [25] it is not mentioned what type of efficiency is meant (physical or practical). According-ing to information published in [25] which is not discussed here, [25] refers to the physical efficiency at the end of the excitation pulse.

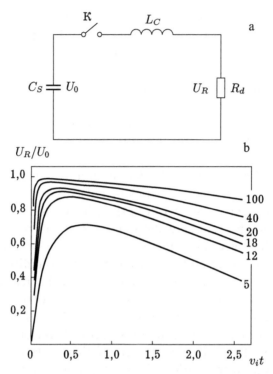

Figure 2.12. Dependences of relative voltage on the active resistance of the discharge on the value $v_i t$ [26]. The numbers on the curves are the value of the parameter $R_0 / v_i L_C$.

the excitation pulse, and that the plasma parameters are distributed uniformly over the cross section of the GDT and the ionization frequency v_i of atoms of the working metal in the discharge does not depend on the average electron energy (of the strength of the electric field E), the voltage U_R at the active resistance of the discharge was defined in [26] by solving a system of equations:

$$L_C \frac{dI_d}{dt} + U_R - U_0, \quad U_R - I_d R_d,$$

$$R_d = \frac{1}{\sigma} \cdot \frac{l_p}{\pi r_p^2}, \quad \sigma = \frac{n_e e^2}{mv}, \quad (2.14)$$

$$\frac{dn_e}{dt} = v_i n_e,$$

where, in addition to the standard notations, e, m is the charge and mass of the electron, n_e is the electron concentration, l_p and r_p is the length and radius of the GDT, v is the effective frequency of elastic collisions of electrons with heavy particles.

The initial conditions for the system of equations (2.14): at $t = 0$, $n_e = n_{e0}$, $I_d = 0$.

In [26] it was assumed that the frequency v and voltage U_0 are constant during the excitation pulse. The condition of constancy of U_0 actually means that the discharge circuit considered in [26] is equivalent to a discharge circuit without the storage capacitor directly connected to a DC voltage source with the voltage at the output of this source being $U_r = U_0$.

The solution of the system of equations (2.14) has the form:

$$\frac{U_R}{U_0} = \frac{R_0}{L_c v_i} \cdot e^{-v_i t} \cdot e^{\frac{R_0}{L_c v_i} e^{-v_i t}} \int e^{-\frac{R_0}{L_c v_i} e^{-v_i t}} v_i dt, \qquad (2.15)$$

where R_0 is the resistance of the discharge at $t = 0$.

Figure 2.12 b shows the calculated [26] (using formula (2.15) dependences of the ratio U_R/U_0 on the value $v_i t$ for different values of $R_0/v_i L_c$. According to these dependences, for a given value $v_i \tau_{ex}$ we can find a value $R_0/L_c v_i$, indicated in [26] as B_{cr} that provided

$$R_0 / L_c v_i \geq B_{cr} \qquad (2.16)$$

the reduction of the voltage on the active resistance of the discharge with respect to the voltage on the storage capacitor during the excitation pulse ($v_i t \leq v_i \tau_{ex}$), except for its leading edge, does not go beyond the prescribed range.

Given the previously discussed formula for the resistance of the discharge (2.14), the condition (2.16) takes the form:

$$v / n_{e0} r_p^2 \geq B_{cr} e^2 \pi v_i L_c / m l_p. \qquad (2.17)$$

In [26] equation (2.17) was used to evaluate the diameter of GDT, at which the voltage drop on the active resistance of the discharge during

Table 2.3.

Type of thyratron	Maximal voltage across the anode	Maximal anode current	Maximum rate of rise of anode current	Pulse duration	Limiting reverse voltage on the anode
	$U_{a\,lim}$, kV	$I_{a\,lim}$, A	$(dI/dt)_{a\,lim}$, A/μs	t_p, μs	$U_{rev\,lim}$, kV
TGI1-500/16	16	500	2000	0.5-10	3.2
TGI1-1000/25	25	1000	4000	50	5

the excitation time τ_{ex} is no more than 20% of the pre-pulse voltage on the storage capacitor, which according to [26] corresponds to the value B_{cr} = 14. The evaluation was conducted for a coaxial conductor with the ration of the conductor radius to the radius of GDT equal to 5 and the following parameters: pressure of neon p_{Ne} = 2 kPa; τ_{ex} = 35 ns; pre-pulse concentration of copper atoms n_{Cu0} = 1.4 · 10^{15} cm^{-3}; v_i = 4 · 10^7 s^{-1}, the temperatures of the gas and electrons, averaged over the cross section of the GDT, are equal to T_g = 2530 K and T_e = 4.2 eV, respectively. According to the estimates [26], the diameter of GDT at which the voltage reduction at discharge in relation to the voltage on the storage capacitor does not exceed 20%, should not exceed a certain critical diameter $d_{p_{cr}}$, roughly equal to 4 mm. As follows from (2.17), the critical diameter of GDT may be increased by increasing the frequency v of collisions between electrons and heavy particles, i.e. by increasing the buffer gas pressure. For example, when the neon pressure was increased to 6.7 kPa the maximum diameter of the GDT increases to 7 mm. This value agrees well with the diameter of GDT (d_p = 6 mm) used in the previously mentioned study [25] in the implementation of the efficiency of the copper vapour laser at 9%. This means that the efficiency achieved in [25] was close to the limit, and that without any special measures that reduce the negative impact of the discharge circuit inductance on the lasing characteristics, achieving the efficiency of 9% for the GDT with the diameters significantly exceeding 6 mm, it is hardly possible.

Despite a number of simplifying assumptions made in [26] in the derivation of (2.17), it allows, for example, to conclude that the increase in the pre-pulse electron concentration, which occurs with increasing frequency of excitation pulses, should lead to a reduction of the strength of the field in the discharge and, consequently, to a deterioration of the lasing characteristics. This conclusion confirms the results of [27], indicating a decrease in the amplitude of the voltage across the electrodes of the GDT of the copper vapour laser with an increase in the excitation pulse repetition frequency. Furthermore, from (2.17) it follows that the coordination of the discharge with the excitation source, with other things being equal, can be substantially improved by increasing the parameter $R_0/L_C v_i$ only by reducing the pre-pulse electron concentration. This conclusion is consistent with the results of [21] according to which the increase of the lasing power of the copper vapour lasers with the addition of hydrogen to their active medium is due to the increase of the rate of triple recombination of electrons due to their more rapid cooling in collisions with hydrogen atoms and molecules. Obviously, the increase in the rate of triple recombination of electrons leads to a decrease in the pre-pulse electron concentration.

It is possible that the decrease in the pre-pulse electron concentration is also associated with the improved lasing performance of a hybrid (CuBr/HBr) laser [28] compared with the known copper halide vapour lasers. The reduction of the pre-pulse concentration in [28] can occur, firstly, by the mechanism discussed above due to the presence in the active medium of hydrogen atoms resulting from dissociation of the molecules HBr and, secondly, due to dissociative electron attachment to molecules of Br_2 and HBr (see, e.g. [29]). As shown in [30], the presence in the active medium of the copper bromide vapour laser of the Br_2 molecules with a concentration of $3 \cdot 10^{15}$ cm^{-3} can lead to a significant increase in the relaxation rate of the electron concentration in comparison with the relaxation rate in triple recombination.

Concluding the consideration of the effect of the discharge circuit inductance on the lasing characteristics, we discuss briefly the results of studies of the effect of the parameters of the discharge circuit on the lasing characteristics of the copper vapour laser, made in [31, 32]. According to these studies, conducted with a copper vapour laser with a GDT with a diameter of 11 mm and a length of 30 cm, increasing the capacity of the storage capacitor increases the lasing pulse duration and its energy and reduces the energy input into the discharge by the end of the laser pulse, i.e. increases the physical efficiency of the laser, calculated for the end of the laser pulse.

Based on the analysis of the experimental results, in [31, 32] it is concluded that for efficient pumping of the upper laser level it is necessary to form on the electrodes of the GDT a voltage pulse with a steep leading edge. Therefore, the discharge of the storage capacitor must have an aperiodic character. This condition can be achieved by increasing the capacitance of the storage capacitor. Thus, to achieve high efficiency of the conversion of energy, supplied to the discharge, to the energy of stimulated radiation, the excitation pulses (voltage and current pulses) should be interrupted at the end of the laser pulse.

The strength of influence of the discharge circuit inductance on the lasing characteristics of self-contained lasers in the terms recommended in [31, 32], can be estimated on the basis of the results of the previously considered study [26], since, as noted earlier, this paper analyzed the operation modes of lasers in which, firstly, by forced termination of the excitation pulse and, secondly, on the storage capacitor voltage during the excitation pulse does not change, which corresponds to the aperiodic nature of the discharge of the storage capacitor.

2.5. Features of the thyratron in repetitively pulsed copper vapour lasers

Pulsed hydrogen thyratrons are used most widely used in the excitation circuits of the repetitively pulsed self-contained lasers [33]. The characteristics of two of them are shown as an example in Table 2.3. As shown in section 2.2, the energy losses in the thyratrons, working in the excitation circuits of the repetitively pulsed metal vapour lasers, can reach 30–40% and, therefore, significantly affect the practical efficiency of these lasers. According to [33], the energy losses in the thyratron due to the passage of a pulse current through it, are divided into three categories: start-up losses G_{cm}, losses during the conduction period G_{np} and post-pulse losses G_{nu}. In [5] the authors analyzed the relationship between these losses and the parameters of the charge and discharge circuits of the excitation scheme with resonant charging of the storage capacitor.

The magnitude of the start-up loss is determined by the rate of rise of the current with the limiting value $(dI/dt)_{lim}$ of this rate specified in the certified data for the thyratron (Table 2.3) and can be used to evaluate the applicability of various types of thyratrons in the excitation circuits in metal vapour lasers by comparing $(dI/dt)_{lim}$ with the value of $(dI/dt)_C$, determined by the parameters of the discharge circuit.

In one extreme case, when the inductance L_C is high and the rate of current rise is determined solely by its values, we can write:

$$\left(dI / dt\right)_c = \left(dI / dt\right)_{L_c} = U_0 / L_c. \tag{2.18}$$

(2.18) defines, in essence, the maximum possible rate of current rise in the discharge circuit with inductance L_C at the pre-pulse voltage on the storage capacitor U_0.

Using typical values for self-heated lasers $L_C \approx 1.5$ μH, $U_0 \approx 12$ kV (2.18), we find that $(dI/dt) = 8 \cdot 10^3$ A/μs and exceeds twice the limiting slope of the front of the current pulse for the TGI1-100/25 thyratron, and for the TGI1-500/16 thyratron 4 times. This fact, apparently, is one of the reasons that the TGI1-1000/25 thyratrons are more widely used compared to TGI1-500/16 thyratrons even in lasers, in which the amplitudes of the discharge current do not exceed a few hundred amperes .In addition, the above differences between

$(dI/dt)_{L_C}$ and $(dI/dt)_{\text{lim}}$ for the TGI1-1000/25 thyratron mean that in the low-pressure range of the buffer gas the rate of current rise in the discharge circuit may exceed the permissible value, thus leading to a sharp increase of the start-up losses in the thyratron.

At the other extreme, when the discharge circuit inductance is minimized (e.g., for coaxial laser cells) the rate of current rise is determined by the rate of increase in electron density in the discharge and can not exceed the value

$$\left(dI/dt\right)_c = \left(dI/dt\right)_{R_d} \approx \frac{U_0 \pi r_d^2 \langle n_e \rangle_0 \alpha_i n_{Cu}}{l_p m v}, \qquad (2.19)$$

where in addition to the known symbols $\langle n_e \rangle_0$ is the average pre-pulse electron concentration in the cross-section of the GDT, α_i is the rate constant of ionization of copper atoms, n_{Cu} is the pre-pulse concentration of copper atoms, v is the effective frequency of electron collisions with heavy particles.

To estimate the quantity $(dI/dt)_C$ in the case where $(dI/dt)_{L_C} \approx (dI/dt)_{R_d}$ we can use the relation

$$\left(dI/dt\right)_C = \frac{\left(dI/dt\right)_{L_c} \left(dI/dt\right)_{R_d}}{\left(dI/dt\right)_{L_c} + \left(dI/dt\right)_{R_d}}, \qquad (2.20)$$

which gives the correct values of $(dI/dt)_C$ in the two limiting cases considered.

In view of (2.20) the condition of applicability of the thyratron in the excitation circuit of the laser at self-contained transitions in respect of the rate of rise of current can be written as

$$\left(dI/dt\right)_{\text{lim}} \geq \left(dI/dt\right)_c. \qquad (2.21)$$

The conditions (2.21) were verified in [5], based on data from [34], in which the repetitively pulsed copper vapour laser with a coaxial discharge chamber achieved the maximum practical efficiency of 2.9%. Calculations using (2.20) for the conditions [34] ($p_{Ne} = 2.7$ kPa; $\langle n_e \rangle_0 \approx 3 \cdot 10^{13}$ cm^{-3}; $n_{Cu} \approx 10^{15}$ cm^{-3}; $r_p = 1$ cm; $l_p = 26.5$ cm; $U_0 \approx 6$ kV; $q_i = 10^{-8}$ cm^3/s; $L_C = 70$ nH, $T_g \approx 2500$ K) in [5] resulted in the value of $(dI/dt)_C \approx 4500$ A/μs, is almost identical to the maximum slope of the front edge of the pulse current for the TGI1-1000/25 thyratron used in [34].

Thus, despite the approximate nature of the relations (2.18)–(2.20), they can be used to assess the applicability of particular thyratrons in the excitation circuits of specific lasers at self-contained transitions.

The energy dissipated by the thyratron during the conduction period, i.e. when the discharge in the thyratron is fully developed and passed into the stationary combustion mode, can be estimated as

$$G_{cond} \approx \overline{I}_d U_{ac} \tau_p \approx \tau_p \frac{U_{ac}^2}{R_T}, \tag{2.22}$$

where $\overline{I}_d$ is the mean value of discharge current during the pulse, U_{ac} is the voltage on the thyratron during the conduction period, R_T is the resistance of the thyratron in the conduction period.

According to [33] the value of U_{ac} for high-power thyratrons with maximum currents and pulse duration of 10–30 μs is 150–200 V. When the pulse duration is of the order of fractions of a microsecond, the value U_{ac} is 3–4 times higher.

Given the smallness of the conduction losses G_{cond} compared to the energy G_0, accumulated by the beginning of the excitation pulse in the storage capacitor,

$$G_{cond} << G_0 = C_s U_0^2 / 2 \tag{2.23}$$

and the assumption that the average time for the discharge current pulse is approximately equal to

$$\overline{I}_d \approx \frac{U}{2\overline{R}_d} = \frac{U_0 \langle \overline{n}_e \rangle \cdot e^2}{2mv} \cdot \frac{\pi r_p^2}{l_p}, \tag{2.24}$$

where $\langle n_e \rangle$ is the electron concentration averaged over the time of the excitation pulse and the cross section of the GDT, it is easy to determined the condition for the voltage on the thyratron U_{ac}, under which the conduction losses are small compared with energy G:

$$U_{ac} << \frac{C_s U_0^2}{2} \cdot \frac{mv}{\langle \overline{n}_e \rangle \cdot e^2} \cdot \frac{l_p}{\pi r_p^2}. \tag{2.25}$$

It is easy to show that inequality (2.25) corresponds to the inequality

$$R_T << \overline{R}_d.$$

For typical conditions of self-heated copper vapour lasers with small GDT: $U_0 = 12$ kV, $C_s \approx 2.2$ pF, $\tau_p = \tau_{ex} \approx 200$ ns, $\overline{I}_d \approx 300$ A and in accordance with the above data [33], $U_{ac} \approx 800$ V. Substituting these values into (2.22) and (2.23), we find that $G_{np} = 0.024$ J, $G_0 = 0.16$ J and, consequently, in lasers of this type in the typical conditions of their work, the conduction losses in the thyratron should be about 15% of energy stored in the storage capacitor.

Due to the mismatch of the discharge resistance R_d with the characteristic impedance ρ_C of the discharge circuit immediately after the current pulse the storage capacitor is charged to a voltage, called reverse U_{rev} that is less than the original and is compared with the opposite sign, so that the negative voltage is applied to the anode of the thyratron. At the same time the total current flows through a decaying plasma in the discharge gap of the thyratron and its value in powerful thyratrons can reach tens of amperes for the duration of its passage of 0.1–0.2 μs [33]. At the admissible values of U_{rev} the post-pulse energy loss per pulse G are less than 0.01 J. However, increasing the reverse voltage can lead to a substantial increase of the post-pulse losses, to reverse discharge ignition and, consequently, ot an unstable thyratron. According to [33], the permissible value of reverse voltage for high-power thyratrons does not exceed 5 kV.

Confining ourselves to the analysis of the effects of the scheme with resonant charging of the storage capacitor, shown in Fig. 2.2 c, the value of pre-pulse U_0 and reverse voltage U_{rev} on the thyratron and assuming for simplicity that the thyratron is an ideal switch with zero internal resistance, and the resistance of the discharge is constant in time and equal to its pre-pulse value R_{d0}, in [5] with [35] taken into account the authors derived an expression for the magnitude of reverse voltage

$$U_{rev} = -U_0 \exp(-\alpha\pi / \omega),$$

$$\omega = \left(\omega_0^2 - \alpha^2\right)^{1/2}, \quad \alpha = R_{d0} / 2L_c, \quad \omega_0^2 = 1 / L_c C_s.$$

$$(2.26)$$

It should be noted that the relation (2.26) is valid if $\alpha < \omega_0$.

Given the existence of the reverse voltage and the fact that the circuit shown in Fig. 2.2 c, provides recharging of the storage capacitor to the maximum possible voltage

$$U_0 = 2U_r + |U_{rev}| .$$

$$(2.27)$$

From (2.26) and (2.27) we can get relations determining the values of U_0 and U_{rev} in steady-state operation of a repetitively pulsed laser

$$U_0 = \frac{2U_r}{1 - \exp(-\alpha\pi / \omega)},$$

$$(2.28)$$

$$U_{rev} = \frac{2U_r \cdot \exp(-\alpha\pi/\omega)}{1 - \exp(-\alpha\pi/\omega)} . \tag{2.29}$$

Obviously, for stable operation of the thyratrons it is necessary to satisfy the conditions:

$$U_0 \leq U_{a\,lim}, \quad U_{rev} \leq U_{rev\,lim}, \tag{2.30}$$

where U_a and $U_{rev\,lim}$ are the limiting values of the anodic and reverse voltage on the thyratron (Table 2.3). Otherwise there is either a substantial increase of the energy losses in the thyratron or a spontaneous breakdown of the thyratron takes place.

Equations (2.28) and (2.29) show that the decrease in the resistance of the discharge, i.e. reduction of the buffer gas pressure or increase of the diameter of GDT and pre-pulse electron density (for example, by increasing the pulse repetition frequency) at a given voltage in the rectifier lead to the situation in which one of the voltages U_0 or U_{rev} or may exceed the permissible value. For this to not happen, it is necessary to fulfil the conditions (2.28)–(2.30) which, on the one hand, determine through the frequency of elastic collisions between electrons and buffer gas atoms the minimum permissible (limiting) buffer gas pressure at the given radius and length of GDT and, on the other hand, the radius and length of GDT at a given pressure:

$$\frac{mv}{\langle n_e \rangle_0 \, e^2} \cdot \frac{l_p}{r_p^2} > 2\left(L_c/C_s\right)^{1/2} \ln\left(\frac{U_{lim}}{U_{a\,lim} - 2U_r}\right), \tag{2.31}$$

$$\frac{mv}{\langle n_e \rangle_0 \, e^2} \cdot \frac{l_p}{r_p^2} > 2\left(L_c/C_s\right)^{1/2} \ln\left(\frac{2U_r + U_{rev\,lim}}{U_{rev\,lim}}\right). \tag{2.32}$$

Figure 2.13 shows the results of calculations [5], using the relations (2.31) and (2.32), the critical pressures of neon, below which the values of U_0 and U_{rev} exceed the maximum allowable values for the TGI1-1000/25 and TGI1-500/16 thyratrons. The calculations were performed for $\langle n_e \rangle_0 = 3 \cdot 10^{13}$ cm^{-3}, $(L_C/C_S)^{1/2} = 15$ ohms, and $r_p = 1$ cm. The gas temperature averaged over the cross section of the GDT was assumed to be 2500 K. The frequency v is calculated only taking into account collisions of the electrons with the neon atoms according to [36] for the value of the electron temperature $T_e = 3$ eV. Analysis of these results suggests the following conclusions. Firstly, the critical pressures corresponding to the condition $U_0 = U_{a\,lim}$ and calculated for the GDT 100 cm long, are in good agreement

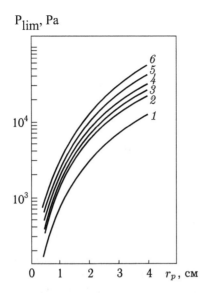

P_{lim}, Pa

Figure 2.13. Dependence of the limiting (critical) pressures of neon, corresponding to the condition $U_0 = U_{a\;lim}$ (curves 1, 2, 5) and condition $U_{rev} = U_{rev\;lim}$ (curves 3, 4, 6) on the radius of the GRD. 1, 3, 5, 6 – TGI1-1000/25 thyratron; 2, 4 – thyratron TGI1-500/16, 1, 2, 3, 4, 5 – l_p = 100 cm, 6 – l_p = 50 cm. 1, 2, 3, 4, 6 – U_r = 5 kV, 5 – U_r = 10 kV; d_p = 2 cm.

with the pressures under which the work of the self-heated lasers, which use TGI1-1000/25 thyratrons in the excitation units, becomes unstable. Second, shortening the length of GDT n times leads to an n-fold increase in the critical pressure p_{cr}, corresponding to the condition $U_0 = U_{a\;lim}$. At the same time, doubling the voltage at the rectifier output leads to an increase in p_{cr} three times (lines 1 and 5 in Fig. 2.13). Third, the critical pressure p_{cr} corresponding to the condition $U_{rev} = U_{rev\;lim}$ is higher than the p_{cr} values corresponding to the condition $U_0 = U_{rev\;lim}$ (curves 1 and 3 and curves 3 and 4 in Fig. 2.13). Fourth, the critical pressure of neon for the TGI1-500/16 thyratron is significantly higher than the critical pressures for the TGI1-1000/25 thyratron. Moreover, an analysis of the dependences shown in Fig. 2.13 shows that the TGI1-500/16 thyratron is not suitable for use with lasers with a diameter of GDT greater than 3 cm at neon pressures $p_{Ne} \leq 4$ kPa. The latter conclusion is consistent with the fact that for excitation of lasers with the GDT diameter of about 3 cm or more in carried out in Russian research studies using only TGI1-1000/25 thyratrons.

Figure 2.14 shows the dependence of the critical pressure of neon for the thyratrons TGI1-1000/25 TGI1-500/16 on the voltage at the

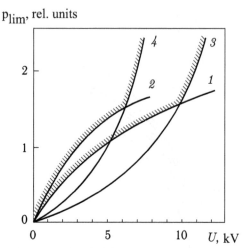

Figure 2.14. Dependence of the limiting neon pressure on voltage at the output of the rectifier for thyratrons TGI1-1000/25 (curves 1, 3) and TGT1-500/16 (curves 2, 4); d_p = 2 cm, curves 1 and 2 correspond to the condition $U_{rev} = U_{rev\ lim}$; curves 3, 4 – condition $U_0 = U_{a\ lim}$.

rectifier output U_r. It is seen that in practically the whole range of values U_r the condition for the reverse voltage on the storage capacitor is more stringent than the condition on the pre-pulse voltage U_0.

Along with the theoretical analysis of the operating conditions of the thyratron in the excitation circuit of pulse-periodic lasers at self-contained transitions, in [5] an experimental study of the operation of the TGI1-1000/25 thyratron in the excitation circuit of the self-heated copper vapour laser was carried out. The experimental conditions in which the UL-101 industrial emitter with a GDT 2 cm in diameter and 50 cm in length was used were consistent with conditions of the above estimates and calculations. Excitation of the laser was carried out using a circuit with resonant charge of the storage capacitor shown in Fig. 2.2 a. The capacitance of the storage capacitor was 2.2 nF. The buffer gas pressure (neon) ranged from 2 to 24 kPa, the voltage at the rectifier output from 4 to 5.5 kV, pulse repetition frequency from 7 to 13 kHz.

Pre-pulse U_0 and reverse U_{rev} voltage in the storage capacitor, the rate of current rise $(dI/dt)_C$, the energy input into the discharge G_d, and energy losses in the thyratron were determined from the oscillograms of pulses of the discharge current and voltage at the anode of the thyratron and GDT electrodes.

The results of measurement of U_0, U_{rev} and $(dI/dt)_C$ are shown in Fig. 2.15. The observed increase in all three parameters with decreasing neon pressure qualitatively agrees with (2.20), (2.28)

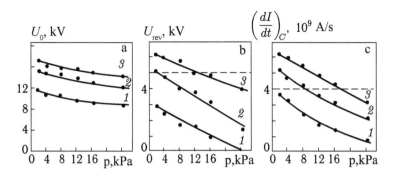

Figure 2.15. Dependence of pre-pulse voltage U_0 (a), the amplitude of the reverse voltage on the anode of the thyratron U_{rev} (b) and the rate of current rise $(dI/dt)_C$ (c) on the neon pressure in the GDT at different voltages at the output of the rectifier: 1–4 kV, 2–5 kV, 3–5.5 kV. Dashed lines show the limiting values of the parameters for the TGI1-1000/25 thyratron. Pulse repetition frequency f = 12.5 kHz.

and (2.29). Within the experimental error, the measured value of U_0 coincides with the calculations using (2.27) and in the whole range of experimental conditions is less than the maximum permissible values for the TGI1-1000/25 thyratron. At the same time, the reverse voltage on the thyratron U_{rev} exceeds its maximum allowable value in a wide range of experimental conditions. Comparison of the results of the experiment (Fig. 2.15, b) and calculations (Fig. 2.13, curve 6) shows that the critical pressure of neon, corresponding to the condition $U_{rev} = U_{rev\ lim}$ and determined in the experiment at a voltage at the rectifier output U_r = 5 kV, equals 2.7 kPa, and agrees well with the critical pressure, calculated by relation (3.32) for U_r = 5 kV, l_p = 50 cm.

As can be seen from Fig. 2.15 c, at voltages on the output of the rectifier of 5 and 5.5 kV the rate of current rise in a wide range of neon pressures exceeds the maximum permissible value for the TGI1-1000/25 thyratron $4 \cdot 10^3$ A/μn, without, however, an unacceptable increase in the energy losses in the thyratron. The reason is, apparently, the fact that in the considered experimental conditions [5] the pre-pulse voltage U_0, along with the rate of current rise determines the magnitude of the start-up losses that are much smaller than its limiting value of 25 kV.

Figure 2.16 shows typical oscillograms of pulses of the discharge current and voltage at the anode of the thyratron and the calculated (from the oscillograms) time dependence of the instantaneous power Q_T, dissipated by the thyratron. As can be seen from this dependence the energy losses in the thyratron clearly include the start-up losses, the losses during the conduction period and the post-pulse losses. The

Table 2.4.

Q_r, W	f, kHz	p_{Ne}, kPa	$\bar{Q}_d$, W	$\bar{Q}_{cm}$, W	$\bar{Q}_{np}$, W	$\bar{Q}_{nu}$, W	$\bar{Q}_T$, W	$\bar{Q}_T/Q_r$	$\bar{Q}_d/Q_r$	R_d, ohm	R_T, ohm
1875	12.5	24	1077	170	226	124	520	0.28	0.57	55	13
1875	12.5	16	1051	159	247	218	624	0.33	0.56	51	11
1875	12.5	12	984	154	271	241	666	0.35	0.52	40	9
1875	12.5	8	786	175	257	360	792	0.42	0.42	30	9
1875	12.5	4	718	252	267	416	935	0.49	0.38	23	8

Table 2.5.

Q_r, W	f, kHz	p_{Ne}, kPa	$\bar{Q}_d$, W	$\bar{Q}_{cm}$, W	$\bar{Q}_{np}$, W	$\bar{Q}_{nu}$, W	$\bar{Q}_T$, W	$\bar{Q}_T/Q_r$	$\bar{Q}_d/Q_r$
2525	12.5	8	1228	196	527	427	1148	0.46	0.49
1989	12.5	8	1085	108	318	251	677	0.34	0.55
1610	12.5	8	880	60	290	183	533	0.33	0.55
1770	14.5	8	1060	125	222	240	587	0.33	0.6
1770	10	8	968	99	346	254	699	0.39	0.55
1770	7.7	8	769	82	291	377	750	0.42	0.43

energies dissipated by the thyratron in each type of loss (G_{cm}, G_{cond} and G_{nu}), were determined from the time dependence on the power Q_T, released in the thyratron, by integrating this function in the corresponding time intervals (Fig. 2.16). The time-averaged power losses in the thyratron $\bar{Q}_{cm}$, $\bar{Q}_{cond}$, $\bar{Q}_{nu}$ were calculated in [5]

$$\bar{Q}_j = fG_j, \quad \bar{Q}_T = \sum_i \bar{Q}_j, \qquad (2.33)$$

where j denotes the kind of loss.

The results of calculation of the time-averaged powers, put into the discharge $\bar{Q}_d$ of (I.10) and scattered by the thyratron $\bar{Q}_{cm}$, $\bar{Q}_{cond}$, $\bar{Q}_{nu}$ calculated from by (2.33) are presented in Tables 2.4 and 2.5. In addition, Table 2.4 shows the values of resistances of the discharge and the thyratron corresponding to the maximum electrical power supplier to the discharge.

Based on the analysis of data in the Tables 2.4 and 2.5, we can draw the following conclusions. The start-up energy losses increase with decreasing neon pressure (Table 2.4), with an increase in the power consumed by the rectifier, and an increase in the pulse repetition frequency (Table 2.5). In all of the experimental conditions [5], the

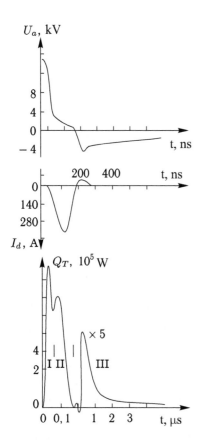

Figure 2.16. Oscillograms of voltage pulses at the anode of the thyratron and discharge current, and the time dependence of the power dissipated in the thyratron calculated from the oscillograms. I – start-up loses, II – conductivity losses, III – post-pulse losses. The power consumed by the rectifier $Q_r = 2{,}5$ kW, neon pressure 8 kPa. Time-averaged start-up losses $Q_{cm} = 196$ W, the conduction losses $Q_{cond} = 527$ W, post-pulse losses $Q_{nu} = 527$ W.

power loss during the conduction period are 15–20% of the power consumed by the rectifier, which agrees well with earlier results of the evaluation. These losses constitute the bulk of the energy losses in the thyratron at high neon pressures (Table 2.4) and at high power levels Q_r, consumed by the rectifier (Table 2.5). Despite the fact that at a constant level of Q_r a reduction of the neon pressure leads to an increase of energy losses during the conduction period, their share in the overall balance of the energy losses in the thyratron decreases. The reason is that with decreasing neon pressure neon the discharge resistance decreases much more than the resistance of the thyratron during the conduction period (Table 2.4). However, the decrease in the resistance of the discharge is accompanied not only by direct redistribution of energy between the discharge and the thyratron during the conduction period, but also by an increases of the post-pulse as well as start-up energy losses. The post-pulse energy losses are sufficiently large in most of the investigated operating modes of the thyratron. In particular, their fraction increases at low neon pressures. For example, at a pressure of $p_{Ne} = 4$ kPa (Table 2.4), they

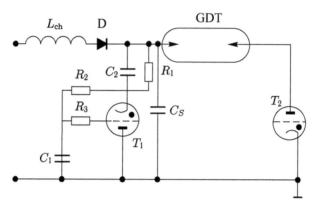

Figure 2.17. Schematic diagram of the stabilization of the reverse voltage at a given level.

represent more than 40% of the power dissipated by the thyratron, which, in turn, with other things being equal, depends strongly on the neon pressure (Table 2.4) and the reduction of pressure from 24 up to 4 kPa increased by 1.75 times, while achieving 40% of the power consumed from the rectifier.

Thus, the experimental studies [5] gave results consistent with the results of previous analysis of the work of the thyratron in the excitation circuit of the pulse-periodic lasers. In addition, these studies have shown that the in transition to low neon pressures, optimum for obtaining high lasing power and efficiency of the copper vapour lasers, the energy losses in the thyratron rise sharply and primarily due to growth of the post-pulse losses. As a consequence, the allowable power, commutated by the thyratron and its service life, decreases.

As is well known [33], to reduce the post-pulse losses the thyratron uses a circuit of removing the reverse voltage from the storage capacitor through the discharge of the latter and ballast resistance. In [5] a circuit is proposed for stabilizing the reverse voltage U_{rev} (Fig. 2.17) with the main distinguishing feature being the parallel connection of the ballast resistance R_1 of the capacitor C_2. The circuit works as follows. When the voltage on the storage capacitor C_S is reduced, thyratron T_1 is unlocked by the discharge current of the capacitor C_1 flowing through R_2, the grid-cathode of the thyratron T_1 gap and ballast resistor R_1. When the polarity of the voltage across the thyratron T_2 is changed the capacitors C_S and C_2 are connected in parallel and charged to a reverse voltage smaller than that to which the storage capacitor without the stabilization circuit of this voltage is charged. The value R_1, which determines the time constant of the discharge of the capacitor C_2 and C_S $\tau_C \approx 2.3 \ R_1 \times (C_2 + C_S)$, is chosen so that the time during which the anode of the switching thyratron T_2 receives negative voltage is sufficient to restore its insulating properties.

Completing the description of the circuit shown in Fig. 2.17, we note first that, instead of the thyratron T_1 we can use a high-voltage vacuum diode, and the GDT may be located not in the anode but in the cathode circuit of the thyratron T_2 as shown in Fig. 2.2 a. Secondly, the use in [5] of the stabilization circuit of U_{rev} with the TGI1-1000/25 thyratron as the thyratron T_1 to excite the self-heated copper vapour laser with a UL-101 transmitter led to about a four-cold reduction in post-pulse losses in the switching thyratron TGI1-1000/25. In the most severe operating conditions (at low neon pressures of neon) its service life has increased from tens to hundreds of hours.

References

1. Kirilov A.E., Polunin Y.P., Soldatov A.N., Fedorov V.F., Lasers metal vapor for atmospheric research. Measuring instruments for study, the surface layers of the atmosphere. – Tomsk, Institute Atmospheric Optics SB RAS. 1977. P.59-79.
2. Klimovskii I.I., Selezneva L.A., Teplofizika vysokikh temperatur. 1979. V.17. No.1. P.27-30.
3. Selezneva L.A., Influence of discharge parameters on the lasing characteristics of the self-heated copper vapor laser. PhD Thesis. Sci. Sciences. Moscow:1980.
4. Isaev A.A., Knaipp, H., Rentsch M., Kvantovaya elektronika. 1983. Vol.10. No.6. P.1183-1189.
5. Kelman V.A., Klimovskii I.I., Fuchko V.Yu., Zapesochnyi I.P. The study of features of the work of the thyratron in the excitation circuit of the copper vapour laser. Preprint KIYAI-85-16. Kiev. 1985. (Kel'man V.A., Klimovskii I.I., Fuchko V.Yu., Zapesochnyi I.P. Kvantovaya elektronika. National Inter-departmental collection of scientific papers. Naukova Dumka, 1988. No.34. S.17-23).
6. Isaev A.A., Lemmerman G.Yu. Power circuit of the pulsed metal vapour Metal vapour metal halide vapour lasers. (Proceedings of the Lebedev Physics Institute. No.181). Moscow: Nauka. 1987. P.164-179.
7. Burmakin V.A., Evtyunin A.N., Forest M.A., Bylkin V.I., Kvantovaya elektronika. 1978 V.5. No.5. P.1000-1004.
8. Delyaev V.P., Tikhov V.V., Lemnyi M.A., et al., Elektronnaya Promut. 1981. Vol. 5-6. P.101-102.
9. Ginkin G.G., A handbook of radioelectronics. Moscow–Leningrad.: Gosenergoizdat.1948.
10. Batenin V.M., Klimovskii I.I., Selezneva L.A., Teplofizika vysokikh temperatur. 1980. V.18. No.4. P.707-712.
11. Karras T.W. Variation of pulsed width in copper lasers. Proc. Intern. Conf. Lasers'80. VA: STS Press McLean. 1981. P.139-147.
12. Soldatov A.N., Fedorov V.F., Investigation of the parameters of the active medium of Cu vapor lasers at high repetition rates. X Siberian Conference on Spectroscopy: Abstracts. Tomsk: Publishing House of TSU. 1981. p.85.
13. Soldatov A.N., Fedorov V.F., Izv. VUZ Fizika. 1983. No.9. P.80-84.
14. Opachko I.I., Fenchak V.A., The effectiveness of repetitively pulsed charging

the storage capacitor. Metrological support of production and test engineering. Scientific and technical collection. Uzhgorod. 1988. P.167-170.

15.   Vetter A.A. , IEEE Journal Quantum Electronics. 1977. Vol.QE-3. No.11. P.889-891.

16.   Kirilov A.E., Kukharev V.I., Soldatov A.N., Tarasenko V.F. Izv. VUZ Fizika. 1977. No.10. P.146-150.

17.   Bohan P.A., Gerasimov V.A., Solomonov V.I., Shcheglov V.B., Kvantovaya elektronika. 1978. V.5. No.10. P.2162-2173.

18.   Smilanski I., Kerman A., Levin LA, Erez G., Optics Communications. 1978. Vol.25. No.1. P.79-82.

19.   Tenenbaum J., Smilanski I., Lavi S., et al., Optics Communications. 1981. Vol.36. No.5. P.391-394.

20.   Maltsev A.N., Kinetics repetitively pulsed copper vapour lasers. Preprint No.1 IOA, Tomsk branch of the Academy of Sciences of the USSR. 1982.

21.   Bohan P.A., Metal vapour lasers with collisional de-excitation of the lower working states. PhD Thesis. Tomsk. 1988.

22.   Bohan P.A., Burlakov V.D., Gerasimov V.A., Solomonov V.I., Kvantovaya elektronika. 1976 V.3. No.6. P.1239-1244.

23.   Klimovskii I.I., Pashchenko V., Pakhomov E.P., Repetitively pulsed metal vapor lasers. Patent No.1804675. The priority of invention since 05.27.1991.

24.   Blau P. , IEEE J. Quantum Electronics. 1994. Vol.QE-30. No.3. P.763-769.

25.   Soldatov A.N., Fedorov V.F., Yudin N.A., Kvantovaya elektronika. 1994. V.21. No.8. P.733-734.

26.   Klimovskii I.I., Teplofizika vysokikh temperatur. V.27. 1989. No.6. P.1190-1198.

27.   Bohan P.A., Silant'ev V.I., Solomonov V.I., Kvantovaya elektronika,1980. V.7. No.6. P.1264-1269.

28.   Jones D.R., Maitland A., Little C.E. , IEEE J. Quantum Electronics. 1994. Vol.30. No.10. P.2385-2390.

29.   Smirnov B.M., Negative ions. Moscow: Atomizdat. 1978.

30.   Zayakin A.V., Klimovskii I.I., Kvantovaya elektronika. 1983. Vol.10, No.6. P.1092-1097.

31.   Demkin V.P., Soldatov A.N., Yudin N.A., Optika atmosfery i okeana. V.16 1993, No.6. P.659-665.

32.   Yudin N.A., The efficiency of excitation of the active medium and control of the energy characteristics of a copper vapour laser. Dissertation. Tomsk: 1996.

33.   Fogel'son T.B., Breusova L.N., Vagin L.I. Pulsed hydrogen thyratrons. Moscow: Sovetskoe Radio, 1974.

34.   Bohan P.A., Gerasimov V.A. Kvantovaya elektronika. 1979. V.6. No.3. P.431-455.

35.   Aghakhanyan T.M., Gavrilov L.E., Mishchenko V.G., Basics of nanosecond pulse technology, Ed. Aghakhanyan T.M. Moscow: Atomizdat. 1976.

36.   Baille P., Chang J.-S., Claude A. et al., J. Phys. B: At. Mol. Physics. 1981. Vol.14. No.9. P.1485-1495.

Excitation circuits of self-contained lasers

3.1. Discharge circuits that increase the steepness of the leading edge of the voltage pulse on electrodes of gas discharge tubes

Corrections of the voltage pulse shape (steepening of the leading edge) with corrective (peaking) capacitors C_c, connected in parallel to the discharge gap is widely used in high-power nanosecond pulse generators used as switches, spark gaps of different types [1]. In a number of works on copper vapour laser (see, e.g., [2–4]) correction was also applied to the excitation pulse shape by connecting the corrective (peaking) capacitor in parallel with the discharge gap. It should be noted that, as a rule, record results in terms of efficiency and output power of the copper vapour lasers were obtained using corrective capacitances. In each case the electric capacitance C_c of the peaking capacitor was chosen empirically. For example, the maximum practical efficiency for a copper vapour laser (l_p = 26.5 cm, d_p = 2 cm) [3], operating in the frequency mode with f = 8 kHz, was obtained at C_c = C_S. At the same time, in [4] where a copper vapour laser (l_p = 80 cm, d_p = 2.8 cm) briefly obtained the average output power of 43.5 W at a practical efficiency of 1%, the capacitance of the correction capacitor was 130 pF at the capacitance of the storage capacitor of 2000 pF.

Besides the traditional discharge circuits (direct discharge circuit of the storage capacitor and the circuit with the peaking capacitor) the copper halide vapour lasers [5–8] were excited and investigated using two innovative discharge circuits (see Fig. 3.1): the circuit with two peaking capacitors and the circuit with interacting discharge circuits. The circuit with two peaking capacitors [5] (Fig. 3.1, a) comprises a storage capacitor C_S, two metal plates P_1 and P_2, located between the electrodes along the gas discharge tube (GDT), and two peaking capacitors C_1 and C_2, connecting the metal plates with the GDT electrodes. Furthermore, this circuit includes the inductance L_s, shunting the discharge during recharging of the storage capacitor, and inductance L_c, reducing (correcting) the rate of current rise in the discharge circuit. In contrast to the scheme with two peaking capacitors, the circuit with interacting discharge circuits [6] (Fig. 3.1 b) consists of two discharge circuits formed by metal plates (P_1, P_2) and two capacitors (respectively C_1, C_2 and C_3, C_4), connecting these plates with the GDT electrodes. The electrical capacitance of the capacitors C_2 and C_3 is much smaller than that of the smaller capacitors C_1 and C_4. According to the data [8], the maximum efficiency of the CuBr vapour laser with a GDT with a diameter of 2 cm and 50 cm long is obtained at the following the capacitance values of the capacitors: C_1 = 950 pF, C_2 = C_3 = 215 pF, C_4 = 3140 pF. An important feature of the circuit with the interacting discharge circuits is that the GDT electrodes are not grounded and are under some potential with respect to the ground.

In conventional circuits with the peaking capacitors or without them, as well as in the circuit with two peaking capacitors the GDT cathode significantly warmer than the grounded electrode (anode). In the scheme with the interacting discharge circuits the electrodes of the laser tube are heated to almost the same temperature, so the use of this scheme leads to a considerable increase of the average laser power and

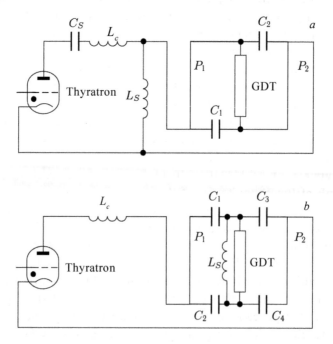

Figure 3.1. Schematic diagrams of the discharge circuits [8]: with two peaking capacitors (*a*) and interacting discharge circuits (*b*).

efficiency compared to conventional circuits, or the scheme with two peaking capacitors.

In [7,8], the results of comparative studies of the CuBr vapour lasers with different discharge circuits are presented. The length and diameter of the discharge tube were 2 and 50 cm, respectively. The buffer gas was a mixture of neon (2 kPa) and hydrogen (40 Pa). The excitation pulse repetition frequency was equal to 21 kHz, the power absorbed by the rectifier was 1.2 kW. The pre-pulse voltage on the storage capacitor (in the scheme of interacting discharge circuits – on the block of the storage capacitors) was 10.3 kV. In the transition from the excitation circuit with the peaking capacitor to the excitation circuit with two peaking capacitors and with interacting discharge circuits the lasing power increased, respectively, from 10.25 to 14.50 W and 16.12 W, and the physical efficiency (the ratio of lasing energy to electric power inputted in the discharge by the end of the laser pulse) increased from 1.43% to 1.92% and 2.07% respectively. Some variation in the capacitance of a storage capacitor and an increase in pre-pulse voltage up to 11.3 kV increased the average lasing power and the physical efficiency of the laser excited by the scheme with interacting discharge circuits, respectively to 17.75% and 2.17 W. As stated in [8] the cause of the increase of the average lasing power and physical efficiency of the laser at the transition from the circuit with the peaking capacitor to the circuit with two capacitors and interacting discharge circuits is that this increases the discharge current and the amount of energy supplied to the discharge until the end of the lasing pulse, and in addition, increases the ratio of the energy supplied to the discharge, to the energy stored in the storage capacitor (block of the storage capacitors).

It should be noted that the scheme with the interacting discharge circuits produced several record results. In [8] a GDT 50 cm long and 2 cm in diameter produced a lasing power of 24 W at the practical efficiency of 1.8%. For the same laser circuitry with a peaking condenser the efficiency was 1.1%.

In [9,10] experiments were carried with a CuBr vapour laser and a CuBr/HBr-laser with a GDT with apertures 4.5 mm in diameter, located near the electrodes with the distance between the electrodes being 30 cm. The active area volume of the GDT was 4.77 cm^3. In [9] the maximum lasing power was 6.7 W at an average power of 1.4 W/cm^3. In [10], a CuBr/HBr-laser in which the CuBr compound formed by interacting of pure copper, placed in GDT, with hydrogen bromide, produced an average lasing power of 9.5 W, which corresponds to the specific lasing power of 2 W/cm^3.

B [11, 12] in tests with a CuBr/HBr-laser with a GDT of 6 cm diameter and a length of 200 cm the circuit with interacting discharge circuits yielded a lasing power of 201 W at a practical efficiency of 1.9%. At the output laser power of 120 W the practical efficiency was 3.2%.

In concluding discussion of the various discharge circuits, we present a system of differential equations that allow to calculate the discharge current and the voltage on the active resistance of the discharge for each of these circuits [7] with the equivalent circuits shown in Fig. 3.2.

For the scheme of direct discharge of the storage capacitor (Fig. 3.2 a):

$$dI_d / dt = \left(U_{C_1} - R_d I_d - \varphi \right) / L_c,$$
$$dU_{C_1} / dt = I_d / C_1. \tag{3.1}$$

For the scheme with the peaking capacitor (Fig. 3.2 b):

$$\frac{dI_{C_2}}{dt} = \left(U_{C_1} - U_{C_2} - L_c \frac{dI_d}{dt} - \varphi \right) / L_c,$$

$$\frac{dI_d}{dt}\left(U_{C_2} - R_d I_d \right) / L_d,$$

$$\frac{dU_{C_1}}{dt} = \frac{\left(I_d + I_{C_2} \right)}{C_1}, \tag{3.2}$$

$$\frac{dU_{C_2}}{dt} = \frac{I_{C_2}}{C_2}.$$

For the scheme with two peaking capacitors (Fig. 3.2 c)

$$\frac{dI_{C_3}}{dt} = \left(U_{C_1} - U_{C_2} - \left(L_c + L_2 \right) \frac{dI_{P_2}}{dt} - \varphi \right) / L_c,$$

$$\frac{dI_{P_2}}{dt}\left(U_{C_3} - R_d I_d - L_d \frac{dI_R}{dt} \right) / L_2, \tag{3.3}$$

$$\frac{dI_d}{dt} = \left(U_{C_2} - R_d I_d - L_1 \frac{dI_{C_3}}{dt} \right) / \left(L_1 + L_d \right),$$

$$\frac{dU_{C_1}}{dt} = -\frac{\left(I_{P_2} + I_{C_3} \right)}{C_1},$$

For the scheme with interacting discharge circuits (Fig. 3.2 d):

$$\frac{dU_{C_2}}{dt} = \frac{\left(I_{P_2} - I_d\right)}{C_1},$$

$$\frac{dU_{C_3}}{dt} = \frac{I_{C_3}}{C_3}.$$

$$\frac{dI_{C_3}}{dt} = \left(U_{C_2} + U_{C_4} - \left(L_c + L_2\right)\frac{dI_{C_4}}{dt} - \varphi\right)/L_c,$$

$$\frac{dI_{C_4}}{dt} = \left(U_{C_4} - U_{C_3} - R_d I_d - L_d \frac{dI_d}{dt}\right)/L_2,$$

$$\frac{dI_d}{dt} = \left(U_{C_1} - U_{C_2} - R_d I_d - L_1 \frac{dI_{C_3}}{dt}\right)/\left(L_d + L_1\right),$$

$$\frac{dU_{C_1}}{dt} = -\frac{\left(I_{C_3} + I_d\right)}{C_1}, \tag{3.4}$$

$$\frac{dU_{C_2}}{dt} = \frac{\left(I_{C_4} - I_d\right)}{C_2},$$

$$\frac{dU_{C_3}}{dt} = -\frac{I_{C_3}}{C_3}.$$

$$\frac{dU_{C_4}}{dt} = -\frac{I_{C_4}}{C_4}.$$

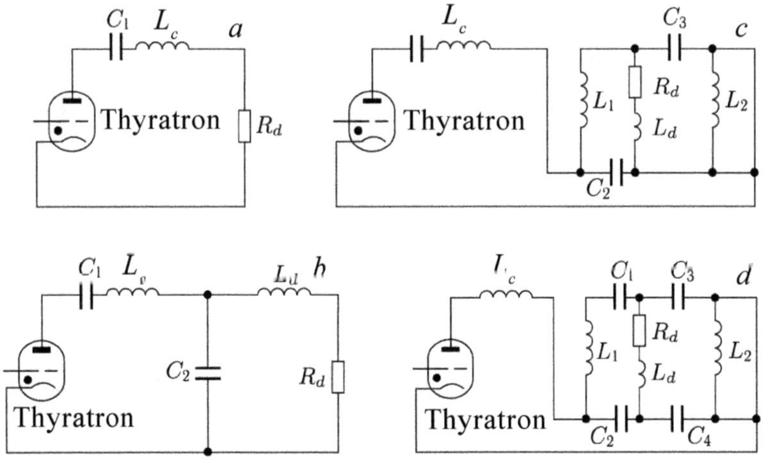

Figure 3.2. Equivalent circuits of various discharge circuits [7]: *a* – discharge circuit for direct discharge of the storage capacitor (inductance L_c^0 includes a correction inductance, discharge inductance and the parasitic inductance of the connecting wires); *b* – discharge circuit with the peaking capacitor (hereinafter L_c and L_d include parasitic inductances of the respective connecting wires); *c* – diagram of the discharge circuit with two peaking capacitors; *d* – discharge circuit with interacting discharge circuits.

3.2. Excitation circuits with pulse transformers

Pulse transformers are used in excitation circuits of self-contained lasers, either to increase the voltage on the electrodes of the GDT (see, e.g., [13–15] or in combination with adjusting elements of the discharge circuit to correct the shape of the voltage pulse on the GDT electrodes (see for example [16]), or to match the impedance of the discharge with the output impedance of the excitation unit (see, e.g., [17–19]).

Figure 3.3 shows the diagram of the excitation unit described in [13]. This unit uses the step-up pulse cable transformer with the windings made of coaxial cable RK–106. Twelve turns of the cable are wound on 10 ferrite F-1000 rings with an inner diameter of 80 mm and the outer diameter of 120 mm. The cable core served as a secondary winding, and its sheath was divided into three segments in parallel with 3.5 turns each and formed the primary winding of the transformer. Thus, the ideal transformation ratio for the transformer under consideration is $n = 3.4$. At the voltage across the primary winding of 16 kV, the voltage on the secondary winding could be up to 50 kV, i.e. the actual transformation ratio reached $n = 3$ and was very close the ideal value,

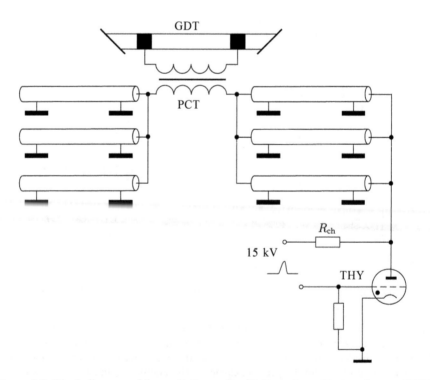

Figure 3.3. Block diagram of the excitation unit with the pulse cable transformer (PCT) and the symmetric charge (storage) line [13] (THY – thyratron).

indicating that relatively small energy losses in the transformer. The symmetric storage line was produced from five parallel 5 m segments of the RK-49 cable. In [13] it is shown that capacitors can be used instead of the storage line. The excitation circuit presented in Fig. 3.3 produced current pulses with a rise time of 15–20 ns and a total duration of 100–150 ns, and the current pulse had the form of a damped oscillation with a maximum amplitude of 200–300 A.

In [14, 15] the authors used the excitation circuits with a pulse step-up transformer (transformer ratio 3:1) which differed from each other by the fact that, in the first case, the cable line, used as the storage device, was charged through the load resistor, and the second – through a charging inductance and resistance. The scheme of the power supply, which was used in [15], is shown in Fig. 3.4. In [14] a CuCl vapour laser with a GDT 1 cm in diameter and 30 cm in length with the excitation pulse repetition frequency of 20 kHz produced record (up to now) values of the specific lasing energy of 35 $\mu J/cm^3$ (for all lasers on copper halide vapours) and the specific average lasing power of 0.7 W/cm^3 (for lasers on copper halide vapours with the GDT with a diameter of about 1 cm.) The practical laser efficiency of the laser, described in [14], was 1%. It should be noted that the above value of specific energy was close to the limit of the achievable energy given by the theory [20], at a concentration of Cu atoms of 10^{15} cm^{-3}, and hence the value of the specific average lasing power is evidently also close to the limiting achievable values for the GDT with a diameter of

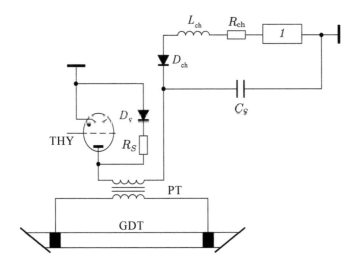

Figure 3.4. The excitation unit of a laser on copper bromide vapours [15]. 1 – rectifier, PT – pulse transformer; D_S – shunt diode; R_S – shunt resistor.

about 1 cm. In [15] using a copper bromide vapours laser with a GDT with a diameter of 2.5 cm and a length 200 cm with a pulse repetition frequency of 16.7 kHz the total average lasing power on both lines was 19.5 W. The practical and overall efficiencies were respectively equal to 1.1% and 0.7%.

Completing discussion of the studies, in which the excitation of the lasers at self-contained transitions of metal atoms was carried out using step-up pulse transformers, it should be noted that, despite the fact that the excitation units with step-up pulse transformers have produced some record results, such transformers have not found any widespread use in excitation units of lasers at self-contained transitions of metal atoms. The main reason for this apparently is that, on the one hand, the step-up pulse transformers increase the voltage on the electrodes of the GDT and, consequently, the slope of the leading edge of the voltage pulse, and the other, which is more important – degrade matching of the discharge with the output impedance of the excitation unit, which affects the efficiency of lasers.

Figure 3.5 is a schematic diagram of the excitation unit with a step-up pulse cable autotransformer and a magnetic chopper [16] which in fact is a correction circuit of the shape of the voltage pulse on the electrodes of the GDT. The pulse cable autotransformer contains 23 turns of the PVTFE-5 cable, wound on six 400 NN 125 × 80 × 12 ferrite rings. The braid and the central strand of the cable are connected in series. The transformer has forced air cooling. To improve cooling a 1 mm gap is made between the ferrite rings. The insulation, preventing the breakdown of turns between each other and the ferrite core is formed by fluoropolymer rings with grooves.

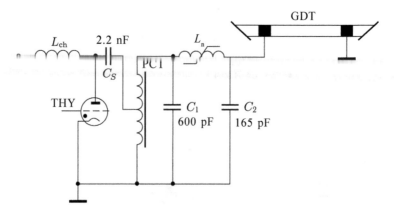

Figure 3.5. Schematic diagram of the excitation circuit of the pulse cable autotransformer and a magnetic chopper [16]. PCT – pulse cable autotransformer; L_n – nonlinear inductance.

The non-linear choke L_n is a conductor with 70 ferrite rings of the 3000 NMA K 16 × 18 × 6. The choke is placed in a glass tube and water cooled. The application of the non-linear choke coil led to a decrease in the duration of the leading edge of the current pulse from 100 to 40 ns.

The excitation circuit, shown in Fig. 3.5, was used [16] in experiments with self-heating copper vapour lasers with a GDT with a diameter of 12 mm and 20 mm and the heated section 860 mm long. The maximum average lasing output (10 and 18 W) was obtained at a neon pressure of 44 kPa and the excitation pulse repetition frequency of 8 ± 0.2 kHz. The overall efficiency for the GDT with a diameter of 20 mm with the whole power consumption taken into account was 0.58%. When the laser is connected to the excitation circuit without the transformer the average lasing power for the GDT with a diameter of 12 mm was reduced to 5 W, and for the GDT with 20 mm diameter – to 10.5 W. The total efficiency of the laser with the GDT with a diameter of 20 mm was reduced to 0.42%.

Thus, the use in [16] of the pulse transformer and the correction of the shape of the excitation pulse increased the average lasing power by 80–100%, indicating the feasibility of applying the so-called compression schemes to increase the average laser power obtained from a single GDT. However, since the shortening of the leading edge of the excitation pulse in such schemes is followed by a loss of energy in the non-linear inductors, their application may not ensure the creation of high-performance lasers with 10% efficiency (see section 3.12).

In those cases where the discharge resistance is sufficiently small as is the case in copper vapour lasers with a transverse discharge, it is recommended to apply matching transformers. For example, when using a coaxial laser with transverse discharge [17], experiments were conducted with a nanosecond pulse generator, assembled according to the Blumlein scheme, with a matching transformer on long lines. The storage device was in the form of two pieces of cable with a total capacitance of $C_S = 5$ nF. Figure 3.6 is a diagram of the final stage of the excitation unit with a transformer on long lines assembled from segments of the coaxial cable connected in series on the generator side and by the discharge cell in parallel. The length of the individual segments l_c is determined by the inequality $l_c \geq v_{ex}\tau_{ex}$, where v_{ex} is the velocity of propagation of the electromagnetic wave along the cable. The impedance of the transformer at a transformer ratio of 8 is 0.37 ohms. As noted in [17], at an excitation pulse duration of 150 ns the pulse shape of the voltage on electrodes of the laser cell

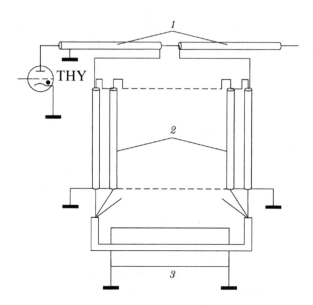

Figure 3.6. The scheme of the final stage of the excitation unit with a transformer at long lines [17]. *1* – cable storage; *2* – transformer on long lines, *3* – coaxial laser cuvette.

is practically a linear function of the voltage on the storage unit with a coefficient of proportionality, equal to the transformation coefficient.

In [18] a copper vapour laser with a radially transverse discharge was excited using the same generator with the cable transformer as in [17]. The output impedance of the pumping generator depending on the number of cables in the transformer was 0.15 or 0.4 ohms, and the transformation ratio was 10. The results of [17, 18] suggest that it is possible to create excitation units with very low output impedance and, therefore, it is possible to match the output impedance of the excitation unit with the discharge resistor in a wide range of conditions. However, the cable transformers have the disadvantage that part of the excitation pulse, reflected from the discharge, dissipates for some time in the discharge, thereby leading to a decrease in the physical and practical efficiency.

There is, apparently, the only work [19] which studied the feasibility of using step-down transformers in the excitation units of self-heating copper vapour lasers. According to [19], the use of a step-down transformer in the excitation unit of the copper vapour laser with a GDT diameter of 2 cm and 50 cm long leads at neon pressures of about 10 kPa to a significant improvement of the discharge impedance of the matching circuit and the discharge resistance and to a reduction of the energy loss of a thyratron to 10% of the energy stored in the storage capacitor.

3.3. Thyratron excitation units with higher excitation pulse repetition frequency and increased switching capacity

Increasing the power of the excitation units of the self-contained lasers can be done either through increasing the excitation pulse repetition frequency f at the same energy level W_r, switched by the thyratron in a single pulse, or by increasing W_T at a constant value of f, either by increasing both simultaneously. The limiting pulse repetition frequency is defined by recovery time τ_T of the electric strength of the thyratron [21]. During this time, the current flowing through the thyratron should not exceed a permissible value $I_{T\ max}$. For a given frequency f the limiting value of W_T is determined by the power admissible for the thyratron and dissipated at its electrodes, above which the thyratron overheats and ultimately fails.

As discussed in [22], in the conditions used in [23] for the TGI1-1000/25 thyratron in excitation blocks of the copper vapour laser at 1% lasing efficiency it was possible to switch only 1.5–1.6 kW. Apparently, the limiting switching power for the thyratron TGI1-1000/25 is about 2 kW. Such power level is required, for example, when working with self-heating copper vapour lasers with a GDT with a diameter of 2 cm and about 100 cm long. Increase of the switching power level in excess of 2 kW requires use of several TGI1-1000/25 thyratrons in the excitation units.

One way to increase the excitation pulse repetition frequency and the level of switching power is the inclusion of several similar excitation units (modules) per single GDT. Each module has its own charge storage capacitor and the induction coil. The main advantage of this method is that when using thyratrons and capacitors with low allowable pulse repetition frequency values, this arrangement allows to obtain virtually any delay between excitation pulses and as a result, very high excitation pulsed repetition frequencies. This, for example, is shown by the results of [24], in which the method of double excitation pulses was used for investigations. Each of the two storage capacitors was charged through its charging circuit consisting of a choke coil and a diode and discharged through its thyratron. This scheme allowed operation with a repetition frequency of double pulses of 2 kHz when the time delay between pulses was changed from 0 to 130 μs.

In [25] a copper chloride vapour laser with a GDT diameter of 3 cm and a length of 70 cm was excited by two blocks with the diode–choke charge of the storage capacitor and its subsequent discharge to the load through the TGI1-1000/25 thyratron. The excitation units worked alternately with the total pulse repetition frequency of 20 kHz.

The capacitance of the storage capacitor in each unit was 1100 pF, the voltage on the anode of the thyratrons 17 kV. The average power, switched by both thyratrons, was 3.2 kW.

The main disadvantage of the modular excitation units is the large number of charging circuits which in the circuits with the resonant charge of the storage capacitor results in a significant increase in overall weight and dimensions of the excitation unit. The excitation units of the metal vapour lasers were developed using various methods of increasing the excitation pulse repetition frequency and the switched (consumed by the rectifier) power ensuring the maximum possible use of the thyratrons with respect to both frequencies f and the power switched by them.

Strictly speaking, the value of the charging inductance L_{ch} should be greater than a certain value defined by the characteristics of the thyratron L_{chT}, at which over time τ_T of restoration of the electric strength of the thyratron the current strength flowing through it does not exceed the admissible values $I_{T\,max}$ [26]

$$L_{ch} \geq L_{chT} = \frac{U_r}{I_{T\,max}} \tau_T. \qquad (3.5)$$

According to (3.5), the maximum excitation pulse repetition frequency f_T and maximum switching (consumed by the rectifier) power Q_r, determined by the characteristics of the thyratron, are respectively [26]

$$f_T = \frac{1}{\pi\sqrt{L_{chT}C_S}}, \qquad (3.6)$$

$$Q_r = f_T \frac{C_S U_0^2}{2}, \qquad (3.7)$$

$$U_0 \approx 2U_r.$$

In [26] it was proposed to increase the excitation pulse repetition frequency f by using a saturable inductance as the charging inductance, which allows, providing the time is sufficient for recovering the electric strength of the thyratron, to drastically reduce the recharging time of the storage capacitor, thereby realizing the frequency f much greater than frequency f_T, defined by equation (3.6).

The authors of [26] developed a pulse generator for excitation of repetitively pulsed high-power lasers. The thyratron, used in this generator, had the following characteristics: $U_{a\,max} \approx 40$ kV, $\tau_T \approx 20$ ms, $I_{T\,max} \approx 1$ A. The storage capacitor with 10 nF capacitance was charged up to 30 kV through the diode and the charging inductance, which was a system of 15 toroidal coils wound on ferrite cores with an inner diameter of 34 mm, thickness of 9 mm and a height of 27 mm, with an air gap of 2 mm. The whole system was immersed in transformer oil. The inductance of the unsaturated charging inductance was 150 mH.

The generator developed in [26] has the following characteristics: excitation pulse repetition frequency $f \approx 15$ kHz pre-pulse voltage on the storage capacitor $U_0 = 30$ kV, switching (absorbed by the rectifier) power $Q_r = 67.5$ kW.

In the case where the recovery time of the electric strength of the thyratron provides the required excitation pulse repetition frequency and only an increase the power of the excitation pulse is required, it is efficient to apply the schemes with a charging circuit and the parallel connection of several thyratrons. In this case, the main difficulty is to provide simultaneous operation of all thyratrons. The fact is that there is some time lag of the anode current pulse in relation to the trigger pulse of the thyratron. The time delay is determined by the time of development of discharge in the thyratron and depends on several factors: hydrogen pressure, filament voltage, grid pulse parameters, etc. [21]. Moreover, the time delay varies for different thyratrons of the same type. When heating the thyratron the hydrogen density is redistributed in the discharge volume and, consequently, the time delay changes. The time delay increases during long-term operation of the thyratron.

These features of the thyratrons were the reason why in the development of high-power modulators on thyratrons the researchers also developed a number of techniques that provide simultaneous operation of thyratrons.

Figure 3.7 is a diagram of the modulator on two parallel thyratrons, illustrating various methods ensuring synchronous operation of

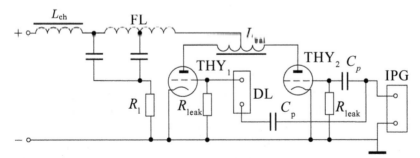

Figure 3.7. Schematic diagram of the parallel connection of pulsed thyratrons THY_1 and THY_2 [21]. DL – adjustable delay line of the grid pulse; L_{bal} – balanced choke; IPG – igniting pulse generator, FL – forming line; R_l – load resistance; R_{leak} – leakage resistance; C_p – coupling capacitor.

thyratrons. The scheme with some simplifications was taken from [21]. Simultaneous operation of the thyratrons and uniform current distribution between them are ensured by powering two grids from a general powerful generator of trigger pulses. Furthermore, in the grid of each or several thyratrons there are adjustable delay lines which compensate the difference in the delay of the anode current in relation to the voltage pulse in the grid formed during heating of thyratrons is or eventually during their work.

Finally, to balance the currents in the thyratrons special transformers are connected between their anodes or balanced chokes L_{bal} connected in the counterphase whose windings have an equal number of turns. The presence of the balanced choke causes that when one of the thyratron is unlocked the voltage at its anode is reduced by the voltage drop across the balanced choke and the speed of discharge development in the open thyratron decreases. On the other hand, the over-voltage on the anode of the second thyratron arising from the coupling of counter windings of the balanced choke contributes to its unlocking.

It should be made clear that the use of balanced chokes in the excitation units of the copper vapour lasers is hardly advisable, as it should lead to an increase in the inductance of the discharge circuit and, as a consequence, to all those previously discussed negative effects of the influence on the inductance of the discharge circuit on the lasing parameters.

Direct discharge of the storage capacitor to the GDT through four thyratrons at the same time was used in [27] for the excitation of a copper vapour laser with a GDT with a diameter 3.8 cm and a length of 125 cm. As indicated in [27], there is a correlation between the synchronicity of triggering of thyratrons and the lasing power of the laser: the scatter of up to 10 ns in triggering of the thyratron has a significant impact on lasing power. Accurate short-term adjustment of the thyratrons for a single mode allows to receive single results, significant greater than those that have been achieved in continuous operation of the thyratrons.

It is possible that in [27] the disruption of synchronicity of triggering (ignition) of the thyratrons is connected in some way to the instability of tension of the thyratrons, caused, in turn, by fluctuations the voltage in the electrical network. The stability of tension of the thyratrons can be significantly improved by the device used to stabilize the tension of the thyratrons proposed in [28]. The device described in [28] differs from the previously known devices by: the lack of a power transformer operating at a frequency of 50 Hz; a high degree of stabilization at

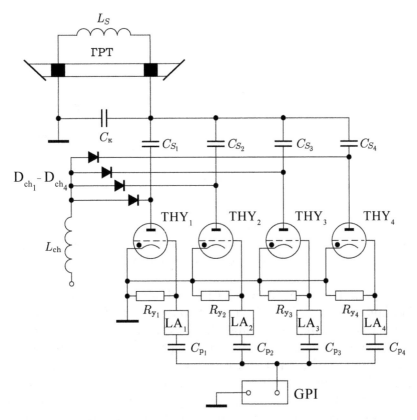

Figure 3.8. Schematic diagram of the parallel connection of pulsed thyratrons TGI1-1000/25 to the excitation unit of the copper vapour laser [29].

the minimum number of circuit elements; the lack of an auxiliary transformer for powering the device.

The device described in [28] provides stabilization of the filament voltage of the thyratrons to within 1.5% with fluctuations of the mains voltage of 220 V/50 Hz ± 10%.

Apparently, because of the difficulty of synchronizing the start of several thyratrons in the discharge through them of one storage capacitor, high-power excitation units of copper vapour lasers use widely the circuits with simultaneous direct discharge of several storage capacitors through an appropriate number of thyratrons [22,29]. Figure 3.8 is a diagram of parallel connection of the thyratrons in the excitation block of lasers on different metals vapours [29]. Synchronous operation of thyratrons was started in [29] by ignition pulses from a generator and by the inclusion of delay lines in the grid circuit of the thyratrons. Automatic bias on the grid of the thyratrons was ensured by the capacitors $C_{P_1} - C_{P_4}$ and $R_{y_1} - R_{y_4}$ resistors.

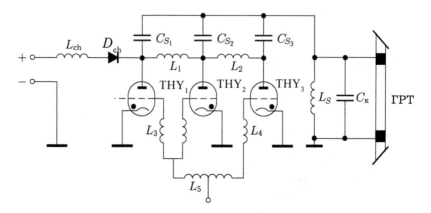

Figure 3.9. Schematic diagram of the parallel connection of three thyratrons with one GDT [22]. $C_{S_1} = C_{S_2} = C_{S_3} = 0.73$ nF; $L_{ch} = 0.8$ H; D_{ch} – type D1008A diodes; $L_S = L_1 = L_2 = 100$ µH; $THY_1 - THY_2$ – thyratrons TGI1-1000/25.

Voltage pulses of positive polarity with an amplitude of 1.5 kV and a duration of 300 ns formed on the thyratron grids. Even if is assumed that the duration of the leading edge of the voltage pulse in the grids of the thyratrons is only half of its length, i.e. 150 ns, the steepness of the voltage pulse on the grid in terms of [29] is not less than 10^4 V/µs. This is more than 4 times higher than the maximum allowable value for the TGI1-1000/25 thyratron [21]. It is obvious that in such a mode the service life of thyratrons will be significantly lower than their rated durability.

The diagram of the excitation unit used to pump copper vapour lasers described in [22] is shown in Fig. 3.9. Distinctive features of this scheme are, first, the tension of all the thyratrons is created heated from one source of charging voltage. Secondly, the charge of the three storage capacitors passes through a single charging diode and these capacitors are uncoupled by the inclusion between the anodes of the thyratrons of decoupling inductors L_1 and L_2. Third, the grid ignition pulse is supplied from a source through the balanced inductors L_3, L_4 and L_5. The power supply (the circuit is shown in Fig. 3.9) provides the excitation pulse repetition frequency of 5–20 kHz and switching (consumed by the rectifier) power of 4.5–5.0 kW.

Gradual starting of the TGI1-1000/25 thyratrons connected in parallel (hereinafter the parallel connection of the thyratrons with sequential start-up in time will be called parallel-series connection), with a single charge inductor and a storage capacitor (Fig. 3.10) was used [30] when using a self-heating lead vapour laser with a GDT diameter of 5 cm, length 100 cm. The thyristors were started up with a time lag, with

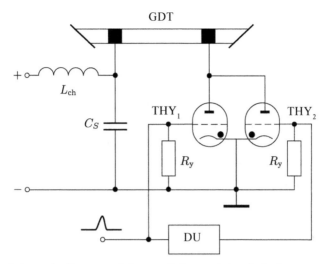

Figure 3.10. Schematic diagram of the parallel-to-serial switch thyratrons [30]. DU – delay unit.

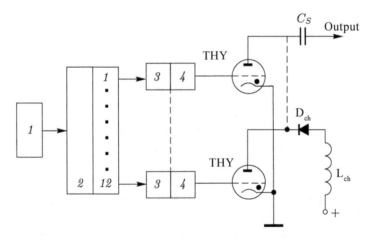

Figure 3.11. Block diagram of the excitation with a parallel-series connection of 12 thyratrons TGI1 1000/25 [31]. *1* – master oscillator, *2* – ring counter, *3*, *4* – Strengthening-forming stage.

the start-up of the second thyratron relative to the first adjusted with the delay unit in the range from a few to hundreds of microseconds. The total excitation pulse repetition frequency varied from 1.2 to 3.2 kHz. The capacitance of the storage capacitor was 10 nF, switching power reached 3 kW.

For switching the mean power of 24–30 kW, with a frequency of 8–9 kHz and 8–20 kW with a frequency of 20 kHz [31] the authors used a parallel-series connection of twelve TGI1-1000/25 thyratrons. The flow chart of the 12-thyratron generator is shown in Fig. 3.11.

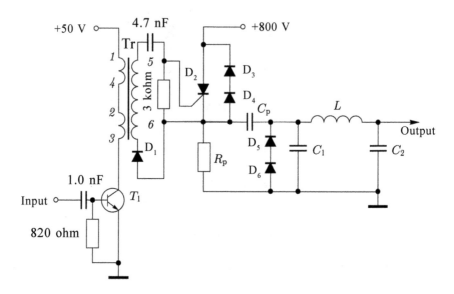

Figure 3.12. Schematic diagram of the amplifying–shaping cascade. [31] T_1 – KT205R; D_1, D_3 – D_6 – KD205Zh; D_2 – TCh50-9; C_1, C_2 – KVI3 – 100 × 12 kV; C_p – K75-15-0.05 × 3 kV; R_p – PEV100-3-5 kohm; T_p – MIT-4V.

The repetition frequency of the excitation pulses f is determined by the master oscillator *1*. Ring counter *2* divides f by 12 and provides alternate switching of the thyratrons by the amplifying and shaping cascades *3* and *4*. The choice in [31] of the circuit with a ring counter, assembled from avalanche transistors P416 [32], is due to the fact that such a scheme is not susceptible to interference created by the thyratrons during operation.

The signal from the output of the ring counter travels to the input of the amplifying–shaping cascade (Fig. 3.12), is amplified by the KP805A transistor, is then applied to the control electrode of the thyristor D_2 and opens it. Through the open thyristor and the 'grid–cathode' gap of the thyratron the capacitor C_p is rapidly charged. The charging of the capacitor C_p is accompanied by the formation of a triggering pulse. After charging C_p the thyristor is closed and the capacitor C_p is discharged through a resistor R_p and diodes D_5, D_6. During the discharge of the capacitor C_p a negative bias is created on the grid of the thyratron. The minimum value of the resistance R_p is determined by the closing current of the thyristor, the maximum value – by the triggering speed of the thyristor $f/12$, because the capacitor C_p must be discharged during the time between pulses. The P-filter protects the amplifying–shaping cascade from high-voltage pulses appearing on the grid of the thyratron at the time of its breakdown. Inductor coil L is wound with a wire with a cross section of 0.8 mm². Its diameter is 20 mm, the number of turns – 15.

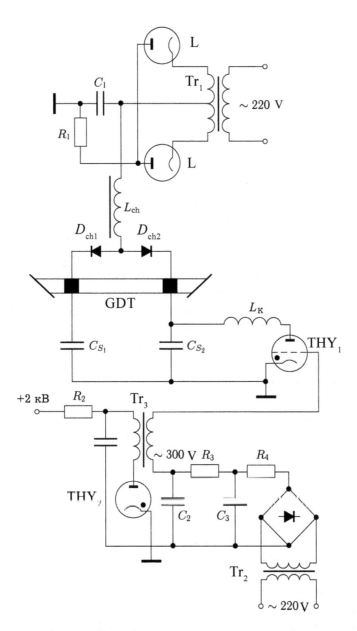

Figure 3.13. Schematic diagram of the excitation unit with the Blumlein circuit and a diode–choke charge of storage capacitors [33]. THY_1 – thyratron TGI1-500/16. THY_2 – TGI1-35/3, L – vacuum diode V1-01/40, R_1 = 750 ohms, R_2 = 30 kohms, $R_{3,4}$ = 100 ohms, C_1 = 5 uF, C_{S_1} = C_{S_2} = 500 pF, C_2 = 0.5 μF, C_3 = 20 μF. L_{ch} = 0.8 H, L_c = 3 μH.

In concluding the discussion of the thyratron excitation units with direct discharge of the storage capacitor, it should be noted that the circuit with parallel-series connection of the thyratrons provide in powerful excitation units longer service life than the schemes with parallel connection of the thyratrons.

3.4. Excitation units with Blumlein circuits

Excitation units of lasers on self-contained transitions of metal atoms use quite widely discharge circuits based on the principle of doubling Blumlein voltage, for brevity, simply called Blumlein circuits. Figure 3.13 is a diagram of the excitation unit of a copper vapour laser [33], which combines the principle of doubling Blumlein voltage and resonant charge of the storage capacitor. Capacitors C_{S_1} and C_{S_2} are charged via choke L_{ch} and diodes D_{ch_1} and D_{ch_2} to a voltage of about 12 kV. After starting the thyratron the capacitor C_{S_1} is recharged, so that between the GDT electrodes there are two series-connected capacitors with a total capacity of 250 pF and a voltage of about 24 kV. The circuit shown in Fig. 3.13 uses a negative bias of 300 V on the grid of the thyratron. The thyratron is started up by the pulse of positive polarity with an amplitude of 2 kV and 300 ns duration, i.e., the slope of the leading edge of the grid voltage pulse was at least $7 \cdot 10^3$ V/µs and was much higher than the permissible value [21]

Correction inductance L_c, ensuring matching of the switching pulse with the rate of discharge development, was 3 µH, so that taking into account the entire circuit inductance, which included the thyratron, the steepness of the current pulse through the thyratron is about $4 \cdot 10^3$ A/µs, i.e. close to the maximum permissible value.

The circuit shown in Fig. 3.13 was used to form current pulses with an amplitude of 500 A and duration at half-height of 40 ns, which could follow at a frequency of 15 kHz.

A certain advantage of the Blumlein circuit is that at the same voltage at the rectifier the voltage produced by the circuit between the GDT electrodes is about twice the voltage produced by the direct discharge circuit of the storage capacitor in the discharge gap. However, one of the significant disadvantage of the Blumlein circuit is the fact that all the energy stored in the capacitors C_{S_1} and C_{S_2} (see Fig. 3.13) (with the exception of the energy dissipated in the thyratron) is released in a discharge. Capacitors C_{S_1}, C_{S_2}, the GDT and the connecting wires form an oscillatory circuit and electrical fluctuations in the circuit in the operating conditions of [33] continue for more than 180 ns, i.e., the lifetime of the discharge is much higher than the previously mentioned

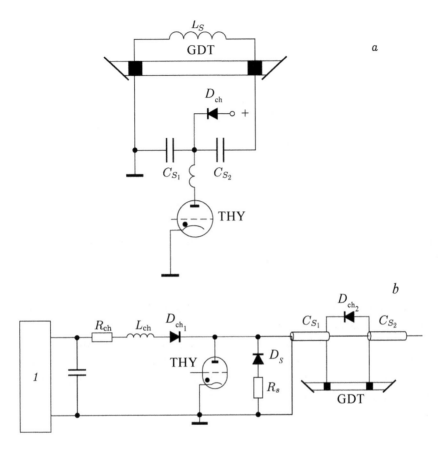

Figure 3.14. Schematic diagram of the excitation laser on lead vapours (*a*) [34] and CuCl vapours (*b*) [35]. *1* – rectifier.

current pulse duration at half-height that actually determines the length of the first half-wave of the current pulse.

In the excitation circuit shown in Fig. 3.13 both GDT electrodes are under a high potential, which creates certain problems in the operation of the laser. In [34, 35], the excitation of lasers on lead and copper chloride vapours was carried out using the excitation units with the Blumlein circuit (see Fig. 3.14) where one of the electrodes of the GDT is grounded. The discharge circuit in the excitation unit schematically shown in Fig. 3.14 b is formed by the same cable lines as those used as the storage devices, with a capacitance of 0.101 nF/m and the signal passage time of 5 ns/m The characteristic impedance changes from 50 ohms for a single cable to 6.25 ohms for eight cables connected in parallel. The cable length is selected to ensure that the total capacitance is equal to 1.5, 3 or 6 nF. As indicated in [35], in

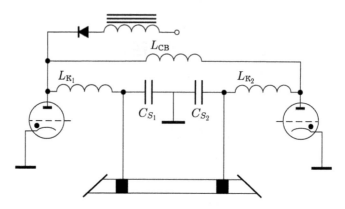

Figure 3.15. Schematic diagram of the generator of excitation pulses on lasers on vapours of copper halides [36]. L_{co} – coupling choke.

reality, in the circuit shown in Fig. 3.14 b the voltage on the GDT electrodes was not doubled.

In [35], together with the experimental results there are also expressions obtained by theoretical analysis for the voltage on the active resistance and for the current through it in the discharge on this resistance of cable lines connected in the Blumlein circuit.

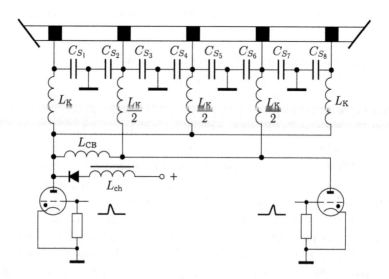

Figure 3.16. Schematic diagram of the generator of excitation pulses of a multiple-section laser on copper halides vapours [36].

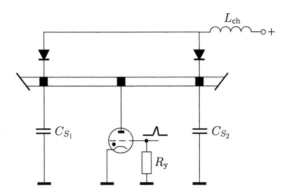

Figure 3.17. Two-circuit diagram of the excitation circuit of a lead vapour laser [37]..

3.5. Excitation units reducing the effect of cataphoresis on the loss of working metal from GDT

In lasers at self-contained transitions longitudinal cataphoresis can play a significant role in the removal of metal into the cold zones of GDT and therefore lead to a reduced service life of lasers. Work [36] describes the excitation circuit (Fig. 3.15) of lasers producing pulses of different polarity on the electrodes of the GDT. Storage capacitors C_{S_1} and C_{S_2} are charged through the choke L_{ch} and diode D_{ch} to twice the voltage at the rectifier output. The values of capacitances C_{S_1} and C_{S_2} were varied in [36] from 0.5 to 5 nF, the inductances were equal to respectively L_{ch} = 0.1 H, L_{co} = 1 mH. The influence of cataphoresis in the circuit shown in Fig. 3.15, is eliminated by the fact that in sequential operation of the thyratrons the discharge current flows in opposite directions. Corrective inductance L_{c_1} and L_{c_2} were chosen in such a way that the change of the polarity of the voltage on the capacitances occurred during about 200 ns and in the conditions of [36] were about 3 μH.

In operation with a GDT filled with helium at a pressure of 1.3 kPa, the circuit shown in Fig. 3.15, produced the pulse repetition frequency of 20 kHz at the level of the power consumed by the rectifier of 2 kW.

The length of the laser at a predetermined voltage on the storage capacitors may be increased by partitioning the laser tube, the various parts which are connected in series optically and by pulsed supply – in parallel, as shown in Fig. 3.16 [36]. To ensure that the excitation pulse times in all sections of the GDT were identical, the corrective inductances in the middle section should be less than half the inductance

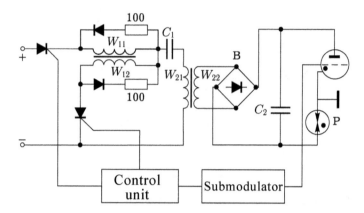

Figure 3.18. Schematic diagram of the thyratron–thyristor excitation pulse generator [38]. D_1, D_2 – TCh-100-8-364; D_3, D_4 – VL-10-10, B – VL-25-12 (8 pcs. in each arm), THY – TGI1-1000/25 thyristor; C_1 – FGTI-0,1 × 2 (20 pcs.); C_2 – KVI-3; TR_1: NN300 core K180 × 110 × 20, w_{11}–w_{12} = 4 coils, wire BPVL-6; TR_2: NN300 core K180 2 × 110 × 20, w_{21} = 20 coils (wire BVPL-6), w_{22} = 700 turns (MGSHV wire-0.12, five layers).

in the ends of the tube. The pulse repetition frequency at the value of capacitances of the storage capacitors of 0.5 nF and L_c = 4.5 µH [36] with the circuit shown in Fig. 3.16 was 15 kHz. The power released in a discharge was 5 kW.

Figure 3.17 shows the excitation diagram of a two-section laser on lead vapours [37], which is one of the variants of the circuit with the direct discharge of the storage capacitor. In this circuit, the central electrode is the cathode, so cataphoresis occurs in the direction of the central electrode and does not lead to removal of the working metal to cold portions of the GDT. Moreover, in the considered circuit cataphoresis and diffusion of atoms take place in different directions, so that cataphoresis can lead to partial compensation of the consumption of working metal caused by diffusion of its atoms to the cold ends of the GDT. The two-section GDT with a diameter of 10 mm and a total length of 60 cm with a capacitance of $C_{S_1} - C_{S_2}$ = 1000 pF, a neon pressure of 4.7 kPa, and the voltage at the rectifier output of 2.9 kV produced an excitation pulse repetition frequency of 23 kHz.

3.6. Thyratron–thyristor excitation drive units

One of the design flaws of the above excitation circuits of copper vapour lasers is the presence of a high-frequency transformer with industrial frequency, essentially determining the weight and dimension characteristics of the laser as a whole. A significant reduction in the size of the power supply may be achieved using the proposed [38] excitation pulse generator built on the basis of thyristor converters powered from DC sources.

The schematic diagram of the thyratron–thyristor generator is shown in Fig. 3.18. The generator operates as follows. The start pulse with a current rise rate of 2.5 A/µs comes from the control unit to the control electrode of the thyristor D_1 and opens it. The capacitor C_1 is charged by the rectifier through the primary winding of the transformers TR_1 and TR_2. When charging the capacitor C_1 a voltage pulse forms in the secondary winding of the transformer TR_2 and the storage capacitor C_2 is charged through one of the rectifier arms B. Some time after capacitor C_2 is charged, the thyratron grid receives an ignition pulse and the capacitor C_2 is discharged to the discharge gap P. Subsequently the next control pulse opens thyristor D_2 and the capacitor C_2 is discharged through the other arm of the rectifier. After that, the grid of the thyratron receives the next pulse and the cycle of the generator is repeated.

Transformer TR_1 to counter-connnected windings promotes stable operation of the generator, providing at triggering of one of the thyristors a short pulse of reverse voltage to the second thyristor electrode, thereby preventing false triggering of the second thyristor.

The parasitic capacitance of the windings of the pulse transformer TR_2 and its inductances are chosen so that after the end of the current flow through the thyristors they have reverse voltage of 220 V, thereby reducing the recovery time of the thyristors and increasing the reliability of their connection on the upper boundary of the operating frequency of the generator.

For the capacitances of the capacitors C_1 and C_2, the relation $C_1 < n^2 C_2$, where $n = w_{22}/w_{21}$, is fulfilled. By fulfilling the above ratio each thyristor receives the reverse voltage prior to application of direct voltage thereby increasing the allowable rate of rise of the direct voltage [39].

The considered thyratron–thyristor generator provides the maximum excitation pulse repetition rate of 12 kHz at the voltage on the thyratron anode of 10 kV and an output power of 1.6 kW. In excitation of a copper vapour laser with a discharge gap with a diameter of 1.6 cm and a length of 63.5 cm, this generator produced an average lasing power of 10 W at an efficiency of 0.6%. As stated in [38] the switching power (excitation pulse repetition frequency and the output voltage) may increase through cascade (parallel–series) activation of the thyristors,

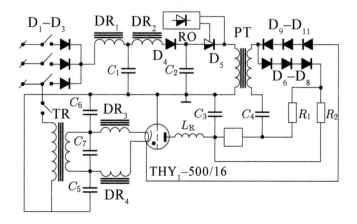

Figure 3.19. Schematic diagram of the excitation pulse generator of a nitrogen laser [40]. $D_1 - D_4$ – D203G diodes, D_5 – thyristor T-15, $D_6 - D_{11}$ – rectifier stacks KTs201G; C_1 – capacitors K75-18 (2×100 μF $\times 1000$ V); C_2 – MBG4 capacitors (14×100 μF $\times 1000$ V); C_3 – capacitors KVI-3 (11×470 pF $\times 16$ kV); C_4 – capacitors KVI-3 (7×470 pF $\times 20$ kV) DR$_1$ – choke D43-0.08-0-2.2; DR$_2$ – choke D52-0.01-12.5, DP$_3$, DP$_4$ – chokes (20 turns on a ferrite ring with a diameter of 120 mm); PT – pulse transformer (coiled wire PEL-0.8 on a cylinder with a diameter of 40 mm, length 150 mm, glued from ferrite rings (primary winding – three parallel connected wires – contains 25 turns, secondary winding – 1500); Tr – filament transformer; R_1, R_2 – WEC 50 resistors (30 ohm); RO – relaxation oscillator, L_k – laser cuvette.

increasing the capacitance of the capacitor C_1 and the corresponding increase in the transformation ratio of the pulse transformer.

It was mentioned earlier that in the Blumlein circuit, both GDT electrodes are usually under high potential (see, e.g., Fig. 3.13), which is not always convenient, for example, in the development of the excitation units with low-inductance discharge circuits. As stated in [40], the desire to reduce the inductance of the discharge circuit requires a compact design of the laser and the excitation unit, which reduces the electrical strength of the elements which are under a voltage of more than 10 kV during operation of the laser. Therefore, in the development of low-inductance excitation units with the Blumlein circuit it is necessary to try to ensure that both electrodes are under the potential for the minimum possible time.

In [40] an excitation unit was developed for a nitrogen laser with the Blumlein circuit, in which the time during which the elements are under high voltage is reduced to a minimum. The scheme of this excitation unit used in principle also in excitation units of metal vapour lasers is shown in Fig. 3.19. One of the distinguishing features of this scheme is that the thyratron grid is used as the cathode. Choke DR$_2$ and the diode

D$_4$ in the scheme under consideration provide a diode–choke charge of capacitor C$_2$, whose capacitance is much smaller than the capacitance C$_1$, so that at the end of the charging period, the voltage on capacitor C$_2$ is nearly doubled compared with the voltage on the capacitor C$_1$. Capacitor C$_2$ is discharged through the primary winding of the pulse transformer and thyristor D$_5$, which is connected by relaxation oscillator RO assembled on the dinistor. During discharge of the capacitor C$_2$ the secondary winding of the PT produces damped sinusoidal current, the first positive half-wave of which charges, through the diodes D$_6$ – D$_8$ and resistance R$_1$, R$_2$ during 25 μs the storage capacitors C$_3$, C$_4$ to a voltage of 22 kV. The thyratron is started up through the diodes D$_9$ – D$_{11}$ by the first negative half-wave of current in the secondary winding of the PT. The time during which the electrodes of the laser are under high voltage is about 30 μm. Chokes DP$_3$, DP$_4$ and capacitors C$_5$ – C$_7$ protect the filament circuit of the thyratron from the effect of the triggering pulse. It should be noted that in this circuit the regulation (correction) inductance L$_c$ has been reduced to a minimum determined by the possibility of the maximum non-inductive connection of the thyratron to the capacitor C$_3$.

In [40] it is indicated that the nitrogen laser excited by the circuit, presented in Fig. 3.19, worked about 250 hours at a pulse repetition frequency of 50–100 Hz.

Noting the advantages of the scheme shown in Fig. 3.19, it is necessary to specify that as a result of using the thyratron grid as a cathode the slope of the fronts of the current pulse, switched by the

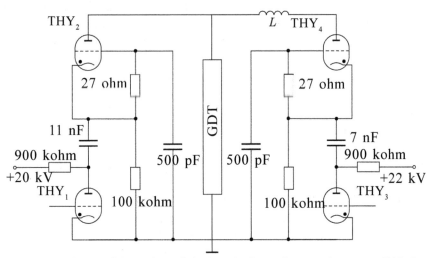

Figure 3.20. Scheme of formation of dual excitation pulses on thyratrons [43]. *L* – additional inductance.

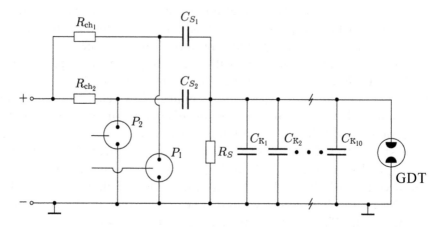

Figure 3.21. The generator of dual excitation pulses at controlled dischargers [42]. $R_{ch_1} = R_{ch_2} = 250$ kOhms; $C_{S_1} = C_{S_2} = 0.01$ μF; $R_S = 1$ kOhm; $C_{c_1} = C_{c_2} = ... = C_{c_{10}} = 1000$ pF.

thyratron, is considerably steeper than the certified value. Assuming that the inductance of the discharge circuit with the TGI1-500/16 thyratron is about 1 μH and considering that the voltage across the capacitor C_3 is 22 kV, we find that steepness of the current pulse front switched by the thyratron in Fig. 3.15 is about 2.2×10^4 A/μs, i.e., an order of magnitude greater than the steepness of the anodic current pulse admissible for the TGI1-500/16 thyratron.

3.7. Generators of dual pulses and excitation pulse trains

Lasing in of copper halide vapours was carried out for the first time upon excitation of the copper chloride vapours by dual excitation pulses [41], the first one of which was then called 'dissociation pulse', and the second one – 'excitation pulse'. The first pulse provides the dissociation of molecules of copper halides, while the second one – excitation of copper atoms generated by dissociation of copper atoms and lasing. Dual pulse generators were made both on the basis of controlled gas dischargers and thyratrons. Figure 3.20 shows a typical dual pulse generator based on controlled dischargers used in [42] for the excitation of the active medium of the laser on copper iodide vapours using the transverse discharge.

Figure 3.21 is a diagram of the dual-pulse generator used in [43] to investigate the effect of the rate of current rise controlled by an additional inductance, in the excitation pulse on the characteristics of the laser on copper chloride vapours excited by dual pulses. A unique features of this scheme is that it avoids leakage of the charge current

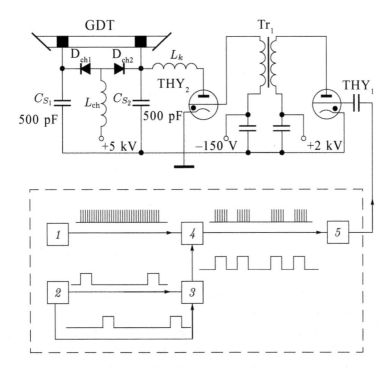

Figure 3.22. The generator circuit of trains of excitation pulses of a laser on CuCl vapours [45]. *1* – pulse generator G5-28A, *2* – pulse generator G5-30A; *3* – mixer of signals of direct and delay channels, *4* – thyristor switch, *5* – signal generator G5-27A; THY_1 – TGI1-10/1; THY_2 – TGI1-1000/25.

of the storage capacitors through the discharge gap and, apparently, it is therefore possible to work with the thyratrons at anode voltages close to the maximum allowable value.

In the study of lasers on metal halide vapours special attention has been paid to the method of excitation of lasers by pulse trains [44,45]. The advantages of this method of excitation of lasers include, first, the ability to analyze processes occurring in the active medium of the self-contained laser, on the basis of the nature of the changes in the lasing in a train from pulse to pulse. Second, the train mode of excitation of the lasers can maintain a constant power level added to the discharge, when the repetition frequency of the excitation pulses is changed by changing the period of repetition of the trains. The condition of constant power level inputted to the discharge, and hence the temperature of the working medium, can be written in the form [44,45]

$$\frac{f \cdot T_{tr}}{T} = \text{const}, \tag{3.8}$$

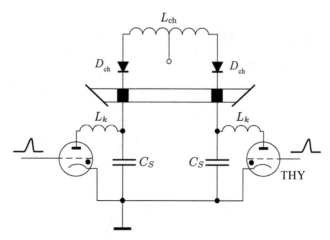

Figure 3.23. The scheme of the tacitron generator of excitation pulses of a copper vapour laser [47].

where f is the excitation pulse repetition frequency in the train, T and and T_{tr} is the repetition period and the duration of the trains.

Figure 3.22 is a diagram of the circuit of formation of trains of pulses [44, 45] which produces both periodically repetitive trains of excitation pulses and pairs of trains with adjustable duration, the delay between the trains and their repetition frequency. The excitation pulse repetition frequency is set by the G5-28A generator. The relative position of pairs of trains, their duration and repetition frequency are set by the G5-30A generator. The pulses from the generator arrive at the input of the switch produced on the MP-39A transistor and carry out strobing of the sequence of pulses arriving at the switch from the G5-28A generator through the mixer of signals OM. Signal generator G5-27 converts the pulses received at its input to start-up pulses of a submodulator formed on the TGI1-10/1 thyratron. The circuit shown in Fig. 3.22 produces trains of pulses with a duration of 10^{-6}–1 s with a repetition period of 10^{-4}–1 s and pairs of pulse trains with a delay between them of zero and one second at the level of switching power about 1 kW.

3.8. Excitation pulse generators on tacitrons and vacuum triodes

A distinctive feature of tacitrons intended like pulse thyratrons for switching current and voltage pulses, is the ability to interrupt the flow of current through the tacitron and significantly higher pulse repetition frequencies [46]. The disadvantages of the tacitrons include power losses greater than in the thyratrons.

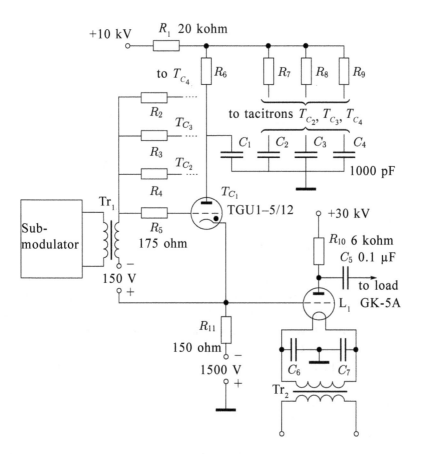

Figure 3.24. Schematic diagram of the tube generator of excitation pulses of copper halide vapour lasers [25].

Despite the fact that the tacitrons are not widely used in the generator of excitation pulses, in the few cases where they were used they produced record lasing pulse repetition frequencies. Figure 3.23 is a diagram of the excitation circuit of a laser on copper vapours [47], collected on two alternately triggered tacitrons with water cooling. When using a GDT with a diameter of 7 mm and 50 cm long, the circuit shown in Fig. 3.23, provided the following parameters: the excitation pulse repetition frequency of 100 kHz, the voltage of 12 kV on the capacitors with the capacitance of 670 pF, the current pulse duration 20 ns.

In [48], the application of tacitrons allowed to raise the excitation pulse repetition frequency in a copper vapour laser up to 235 kHz, and the duration of the excitation pulse was varied in the range of 20 to 200 ns, depending on the parameters of the excitation circuit and the active environment.

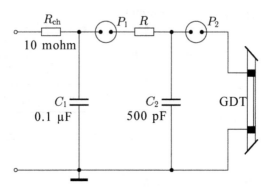

Figure 3.25. Schematic diagram of the excitation circuit of a laser on copper iodide vapours [57]. C_1 – charging capacitor; C_2 – storage capacitor; P_1 – uncontrolled air gap; P_2 – uncontrolled SF_6-discharger.

In [49], the application of a TGU1-5/12 tacitron produced the excitation pulse repetition frequency of 100 kHz for a copper vapour laser with a GDT with a diameter of 4.5 mm and the average specific output power of 1.3 W/cm³ close to the maximum achievable specific lasing power for the GDT of this diameter [50].

Figure 3.24 shows a diagram of the generator of excitation pulses with a powerful GC-5A triode [25]; the driving pulse on the grid is formed by four TGU-5/12 tacitrons running parallel. When using copper halide vapour lasers [25] with a GDT 70 cm long and 3 cm in diameter a vacuum tube with an anode voltage of 20 kV produced current pulses with a total duration of 100 ns and an amplitude of 300 A. When using a vacuum tube to excite a laser on CuCl vapour with the buffer gas neon (neon pressure 0.25–1.3 kPa) [25] at the repetition frequency of 10 kHz the average lasing power was 10 W and the practical efficiency 0.6%.

3.9. Excitation circuits of pulse-periodic lasers with dischargers as switches

Pulse dischargers have been used widely in high-voltage pulse technology from the early days of their development (see, e.g., [51–53]). The advantages of the dischargers include high values of voltage, amplitudes and rates of increase of switching current, small size and, consequently, low self-inductance, the absence of filament circuits, and virtually instant readiness to work; ease of assembly and operation, and finally a low price. The main disadvantages of the dischargers, limiting their use in the excitation circuits of lasers at self-contained transitions include low switching frequency and greater erosion of the

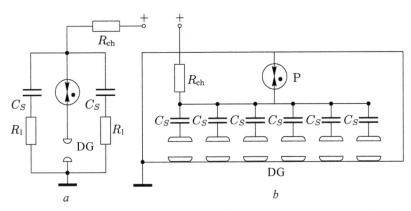

Figure 3.26. Schematic diagram of the discharge circuit of single-section (*a*) and multi-section (*b*) copper vapour lasers with a transverse discharge [59]. P – RU-62 discharger, DG – discharge gap; R_i – shunt to measure pulse currents. The diagram of the multi-section laser does not show the inductances required for charging the storage capacitors shunting each discharge gap.

electrodes, resulting in a limited number of operations of dischargers which for the best of them does not exceed apparently 10^7.

However, dischargers with movable electrodes were recently developed and are being studied (see, for example [54–56]). These dischargers consist of a system of fixed electrodes mounted on the fixed dielectric disks and the systems of movable electrodes fixed to the dielectric disks, rotating between the stationary disks. At the time of convergence of the movable and fixed electrodes a breakdown of the discharge gap is initiated and the discharger starts up. Such dischargers have (in addition to the advantages of the dischargers with the fixed electrodes) new advantages: an increase in switching frequency and reduced electrode erosion, resulting in an increase in the number of operations of the dischargers. Now, of course, it is difficult to estimate how the dischargers with the movable electrodes are suitable for use in the excitation circuits of self-contained lasers. The capabilities of these switches have not been sufficiently studied and it is quite possible that in the future they will find a particular application in these excitation circuits.

Spark dischargers were used in a number of studies [57–59] of lasers at self-contained transitions of metal atoms. For example, in [57] a laser on copper iodide vapours with an excitation circuit, which included two spark dischargers (see Fig. 3.25), produced for the first time pulse-periodic lasing in copper halides vapours. In the experiment described in [57] the authors used a fully heated GDT with a discharge gap length of 5 cm and hollow electric electrodes with a diameter of 5 mm. The storage capacitor C_1 with a capacitance of 0.1 μF was charged to about 45 kV and discharged through an uncontrolled air gap P_1 in the RC-circuit consisting of a resistor R and a capacitor C_2 with a capacitance of 500 pF. When charging the capacitor C_2 to a voltage of about 10 kV the uncontrollable SF_6-discharger P_2 spontaneously broke down and the capacitor C_2 was discharged to the GDT. The instability in the breakdown voltage of the C_1 led to unstable energy input to the discharge and also unstable excitation pulse repetition frequency, which ranged from about 5.5 to 8 kHz. In [57] the implementation of the excitation pulse repetition frequency at 18 kHz was also reported. In addition, it is stated that the specific lasing energy in individual pulses reached 35 $\mu J/cm^3$. The use of dischargers for the excitation of copper vapour lasers with a small volume GDT at frequencies up to 18 kHz was reported in [58].

To determine the characteristics of the dischargers as switching elements in the pulse-periodic lasers at self-contained transitions of metal atoms, the results of the studies in [59] of a copper vapour laser with a transverse discharge and the excitation unit (see Fig. 3.26), produced using the type RU-62 discharger will be discussed. To reduce the inductance of the discharge circuit all elements of the excitation circuit are mounted directly on the high-voltage input of the laser cell.

When using a one-section laser cell with a discharge gap 15 cm long, 4 cm wide, and the inter-electrode distance of 5.5 cm, the electrode temperature was varied from 1300 to 2000 K, the pressure of the buffer gas (neon) – from 5.3 to 60 kPa, the capacitance of the storage capacitor – from 10 to 40 nF, and the level of charging voltage – from 4 to 10 kV. The excitation pulse repetition frequency varied from 10 to 60 Hz.

Throughout the range of these parameters the voltage and current pulses were oscillatory in nature, and for the first current half-wave, depending on the experimental conditions, between 70 and 95% of the energy stored in the storage capacitors was released in the discharge The amplitude of the current reached 11 kA, and the slope of the current wavefront 2.5×10^5 A/μs, i.e. almost two orders of magnitude greater than the permissible steepness of the current pulse for the TGI1-1000/25 thyratron. With the one-section laser cell the lasing pulse energy in the optimum conditions was 4.5 mJ.

In [59] a multi-section cell [59] was excited using both a single RU-62 discharger located, as shown in Fig. 3.26, b, in the centre of the longitudinal axis of the laser cell, and three RU-62 dischargers, each of which operated on two parallel connected sections of the laser.

In the first case, the service life of the discharger was 10^5 operations at the amplitude of the total current of 30 kA and a pulse repetition frequency of current pulses of 40 Hz. It was found that the front of the discharge current in extreme sections was extended by 30–40 ns relative to the front of the discharge current in the central sections, due to the large value of the inductance of the discharge circuits of the end sections caused by the large length of the return conductor.

In parallel operation of all three dischargers the inductances of all discharge circuits were the same. At the charge voltage of 10 kV the instability of formation of the current wavefront in all six sections did not exceed ±10 ns, and increased to ±40 ns with decreasing charge voltage up to 7 kV. When using three dischargers and a storage capacitance of 20 nF the amplitude and steepness of the front of the current pulse in each section were 5.5–6.0 kA and $6.3 \cdot 10^4$ A/μs respectively. Thus the lasing energy of the pulse was 15 mJ.

As stated in [59], the characteristics of the excitation circuit and the lasing parameters of a copper vapour laser, excited by a transverse discharge, can be significantly improved compared with the value realized in [59], if instability of the switches operating in parallel does not exceed 0.1–1 ns.

3.10. Excitation circuits with semiconductor switches

Gas discharge switches, employed in the excitation units of self-contained lasers, have fundamental flaws caused by the very nature of the processes occurring during in discharges in gases. It is firstly triggering instability, complicating the synchronization of complex systems, and a short service life, depending on the rapid destruction of the electrodes. Furthermore, thyratrons require special circuits for

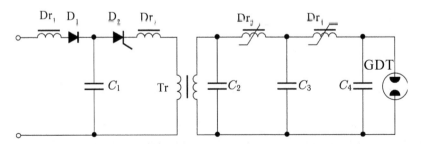

Figure 3.27. Schematic diagram of the magneto-thyristor generator [60]. C_1 = 2 μF (FGTI-0.1 × 2). The magnetic circuit $Dr_1 - Dr_4$ and Tr – ferrite NN300 K180 × 110 × 20; winding data: $Dr_1 - Dr_4 - w$ = 90, 11, 90, 30 turns respectively, Tr – w_1 = 8 turns. w_2 = 400 turns; D_1 – VCh2-160-8. D_2 – TChI100-10-468.

heating cathodes consuming quite a lot of power, which leads to an even greater decrease in the overall efficiency of lasers.

In order to create safe, economical and compact excitation units of metal vapour lasers, studies [60,61] described the development and investigations magneto-thyristor generators of excitation pulses. The circuit of the magneto-thyristor generator, described in [60], is shown in Fig. 3.27. The main features of this circuit are as follows. Due to the presence in the circuit of the diode–choke charge of the capacitor C_1 there is no direct voltage at the anode of thyristor D_2 during thyristor turn-off time. The duration of voltage rise on the capacitor C_1 is about 150 μm. The first stage of compression (approximately 30 times) is implemented by switching thyristor D_2. Due to this the rise time of the voltage across the capacitor C_2 is 5 μm. In the same stage, the voltage is increased by the pulse transformer Tr. Choke DP_2 is connected in the circuit to reduce the switching losses of the thyristor.

The second stage is characterised by the magnetic compression of the pulse by saturable choke coils Dr_3 and Dr_4 on ferrite elements. The duration of the voltage rise on the capacitors C_3 and C_4 respectively 900 and 300 ns. The presence of the peaking capacitor C_4 at the output of the second unit is due to the specific nature of the load – the discharge gap of the copper vapour laser. As stated in [60], the direct connection of the discharge gap to the output of the second unit of magnetic compression degrades the output characteristics of the laser, since in this case inductance of the choke Dr_4 is added to the intrinsic inductance of the laser. In addition to the first, without the peaking capacitor C_4 the working conditions of the output unit of the generator are impaired, as the relatively high resistance of the discharge at the beginning of the excitation pulse hampers the flow of the magnetizing current of the choke Dr_4.

When working with a copper vapour laser with a GDT with a diameter of 14 mm and 635 mm long, the excitation pulse generator, described in [60], provides the following parameters of the excitation pulse: amplitude of voltage pulse – 10 kV, amplitude of the discharge current pulse – 250 A, the current pulse duration at half height – about 100 ns, excitation pulse repetition frequency – from 0 to 5 kHz. At frequency $f = 5$ kHz the power introduced into the discharge was 1 kW, average lasing power – 3 W.

In contrast to [60], in [61] the magneto-thyristor excitation circuit was developed using the circuit of the magneto-thyristor generator operating in the symmetric mode and having the smallest energy losses in the magnetic reversal of the cores and the highest efficiency. However, unlike [60], it [61] it was not possible to achieve a satisfactory matching of the magneto-thyristor generator with the discharge. It is possible that this is due to the fact that the generator, developed in [61], did not have a peaking capacitor, significantly improving, according to the data in [60], matching of the generator with the discharge.

The magneto-thyristor generator, described in [61], was used for excitation of a copper vapour laser with an UL-101 tube. The excitation pulse repetition frequency was 10 kHz, discharge current pulse amplitude – 300 A, the duration of the leading edge of current – 120 ns, the voltage on the capacitor in the last compression unit – 15 kV. At the mains power of 3.6 kW the average lasing power did not exceed 1.5 W.

Significant progress in the development of excitation units self-contained lasers should be expected in connection with the development a new switching principle – via the controlling plasma layer [62], when the switching of high powers by semiconductor devices is carried out by a sharp increase in the conductivity of the area which in the initial state has a very high resistance and blocks the external voltage applied to the device. This area is usually represented by the completely exhausted (by the strong field) volume charge region (VCR) of the reverse p–n-junction. The sharp increase in the conductivity of this region is due to filling of the region with well-conducting electron–hole plasma.

Implementation of the new principle of switching increased the power switched by semiconductor devices by almost an order of magnitude. This principle was used in the development of plasma-controlled analogues of the thyristor and the transistor (microsecond range). New principles underlying the work of powerful semiconductor switches required the development of new circuit design [62], often very different from the conventional; the main feature of this new circuitry is that almost all new devices are of the two-electrode type.

As stated in [62], the rejection of the gate electrode and replacement of this electrode by a plasma layer homogeneous over the area allows, in principle, to form a conductive plasma channel with an area equal to the area of the Si wafer. However, it is quite difficult to produce such a layer. A variety of ways of its formation was studied – a pulsed avalanche breakdown of the collector p–n-junction, impact ionization in a strong microwave field, the ionization by powerful pulses of coherent and incoherent light.

Study [63] describes a powerful nanosecond thyristor switch (semiconductor switch with light control – SSL) switched by a light pulse. The SSL, described in [63], is capable of switching currents with an amplitude of about 10 A. It is based on a powerful high-

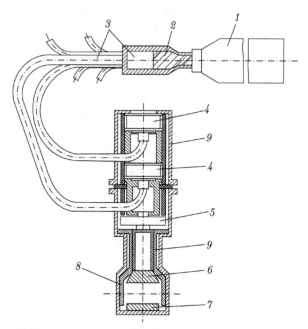

Figure 3.28. The scheme of the working cell for excitation of a laser using SSL [64]. *1* – LTIPCh-6 laser; *2* – focon; *3* – fiber optics collector with taps; *4* – SSL; *5* – block of storage capacitors; *6, 7* – the electrodes of the discharge gap; *8* – its current conductor; *9* – isolators.

voltage *p–n–p–n* structure, capable of blocking a voltage of 7.5 kV. When illuminating the structure with pulsed laser radiation through photodetector windows, electron–hole pairs with high concentration form in one of the electrodes and, consequently, SSL becomes conducting.

The use of the SSL allows to generate the excitation pulses, similar in characteristics to the excitation pulses formed by the dischargers. Figure 3.28 shows the diagram of the working cell for pumping lasers, including two series-connected SSLs designed for switching a voltage of 10 kV [64]. The source of pulsed laser radiation was a LTIPCh-6 laser. Waveguides are used to transport light to the photodetector windows of the SSL.

In [64] a low-inductance circuit with a storage capacitor with a capacitance of 0.5 μF, a capacitor voltage at 10 kV and parasitic inductance of the discharge circuit ≲17 nH, the switch on two series-connected SSLs in the mode close to the short circuit mode provides switching of current with an amplitude of 12 kA. The pulse repetition rate did not exceed 100 kHz. Unfortunately, [63] does not provide

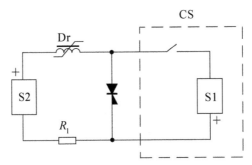

Figure 3.29. The basic layout of pulsed RAD generators [62]. F_1, F_2 – low-voltage and high-voltage power shapers; Dr – choke, SW – switch, CS – control system.

specific examples of the use of the excitation units with SSL for the pumping of copper vapour lasers.

As noted in [64], the benefits of SSLs compared with dischargers should include a very short time scatter in the activation of various SSLs, which is estimated not to exceed several tens of picoseconds.

Apparently, the shortcomings of SSLs include the fact that after transition to the conducting state they conduct in both directions, so that sufficiently sustained oscillations of the current can occur in the discharge circuit.

Based on the results of [64], it can be assumed that the major obstacle in the use of SSLs in the excitation circuits of lasers of self-contained transitions metal atoms is their complicated use (a pulse laser, optical fibers, etc. are required). A much easier and energetically favourable way to control the plasma layer is the so-called reverse-injection control, at which the controlled plasma layer is created by short-term changes in the polarity of the voltage applied to the semiconductor device [62]. This method was used to construct several new classes of devices, including two classes of switching devices (plasma-driven counterparts of the thyristor and the transistor) which were named to the reversibly-activated dinistor (RAD) and reverse-controlled transistor (RCT); these devices work in micro- and submicrosecond ranges. To date, all the circuitry was designed primarily for RAD [62].

The basic scheme of pulse RAD generators is shown in Fig. 3.29. When the switch SW is turned, the shaper S_1 of the control system CS is discharged through RAD to form a pump current pulse. The saturable core choke Dr which is in the unsaturated state, delays the growth of the current flowing from the power storage S_2 to S_1. At this point, all the voltage of the power storage S_2 is applied to the choke. This choke must be designed so that its core is saturated after the end of

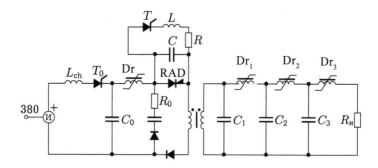

Figure 3.30. Schematic diagram of the high-frequency RAD generator for pumping a copper vapour laser [62]. RAD – reversely-activated dinistor; D – cutoff diode, T – thyristor pump; T_0 – thyristor of the charger; $f = 8$ kHz $Q_r = 3$ kW.

the pumping process of RAD, i.e. 1–2 μs after closing the switch SW. After saturation, the RAD, in which the controlling plasma layer has also formed, receives the voltage from the shaper S_2; RAD switches a powerful current pulse I to the load R_1. Thus, the principal basic feature of the circuit of the RAD generator is the saturated choke Dr connected in series with the RAD and the relatively low-current but high-voltage switch SW connected in series with shaper S_1 which blocks the total voltage of high-voltage power shaper S_2 and low-voltage shaper of F_1. If the energy of the shaper S_2 is not completely dissipated in the load and the shaper is recharged, and re-pumping of RAD is undesirable, a powerful diode in connected in series with the load and prevents discharge through the RAD of the recharged shaper S_2. The design and appearance of a powerful periodic RAD-switch, constructed according to the scheme shown in Fig. 3.29, are shown in [62].

Figure 3.30 shows a diagram of a high-frequency RAD generator for pumping a copper vapour laser [62]. The basis of the circuit is a power cell on the generator RAD (operating frequency – 15–20 kHz, diameter of the semiconductor element 40 mm) producing a pulse with an amplitude of 1 kA and a half-width of 1 μs at a frequency of 8–10 kHz; pump current has an amplitude of 50 A and a half-width of 0.25 μm. After transformation and three stages of magnetic compression (chokes $Dr_{1,2,3}$ with ferrite or metglas cores) the half-width of this pulse on the load is 0.1 μs and voltage 10 kV. The average power of the generator is 3 kW at a frequency of 8–10 kHz with air cooling. At double-sided water cooling the semiconductor unit may provide an average power an order of magnitude greater.

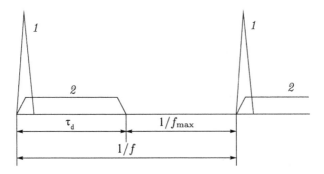

Figure 3.31. Diagram of the method of excitation of a laser with stabilization of lasing energy at changes of the excitation pulse repetition frequency [67]. *1* – excitation pulses, *2* – additional pulses.

Unfortunately, study [62] does not provide any specific examples of the use of the generator of excitation pulses with the RAD for pumping of copper vapour lasers.

3.11. Methods for controlling the characteristics of metal vapour lasers

The main problem of controlling the characteristics of repetitively pulsed metal vapour lasers is when regulating one of the characteristics (e.g., pulse repetition frequency) it is necessary to ensure that all other characteristics (e.g., energy and lasing pulse duration) remain unchanged. The problems of controlling the lasing characteristics of the repetitively pulsed lasers at self-contained transitions of metal atoms are discussed in [65–67].

The basis of various methods for stabilizing the lasing pulse characteristics of self-heating lasers [65, 67] is that when the frequency f is varied it is necessary to maintain constant pre-pulse plasma parameters, especially the concentration and temperature of the electrons and the concentration of atoms of the working metal. One of the methods of stabilization for self-heating lasers is to use two repetitively pulsed discharges [65, 67] (Fig. 3.31). The pulses of the first of them – the excitation pulses (EP) – carry out excitation of the active medium. The pulses of the second discharge – additional pulses (AP) – stabilize the power dissipated in the discharge and, as a consequence, stabilize the temperatures of the gas and the GDT wall. In this method of stabilizing the AP amplitude is lower than the amplitude of the EP and the energy that is put into the discharge during the additional pulse is K times greater than the energy supplied to the

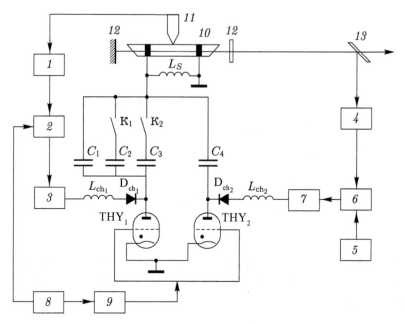

Figure 3.32. Circuitry of the laser with the stabilization of the output parameters [65]. *1* – temperature meter, *2, 6* – electronic regulators, *3, 7* – high-voltage rectifiers, *4* – meter of stabilized output parameter of laser radiation, *5* – device setting the stabilized output parameter, *8* – master oscillator, *9* – modulator, *10* – gas discharge tube *11* – thermocouple, *12* – cavity mirrors, *13* – beam splitter; K_1, K_2 – high-voltage vacuum relay; C_1–C_4 – storage capacitors.

discharge during the excitation pulse. The values of the duration of AP τ_d and K must satisfy the relations [65, 67]:

$$\tau_d = (f_{max} - f)/f \cdot f_{max}, \quad K = (f_{max} - f)/f, \qquad (3.9)$$

where f_{max} is the upper limit of the excitation pulse repetition frequency.

If necessary, the characteristics of the lasing pulse can also be stabilized by changing the amplitude of the additional pulse and the parameters of the excitation pulse.

Figure 3.32 is a block diagram of the excitation circuit of the self-heating copper vapour laser with the stabilization of the output characteristics [65]. In this circuit, the excitation pulse repetition frequency is determined by the master oscillator *8*. From the oscillator *8* the master pulses travel to the modulator *9*. Thyratron THY_2 and capacitor C_4 form excitation pulses and thyratron THY_1 and capacitors C_1–C_3 generate additional pulses.

The laser has two feedback channels: stabilization of the output parameters of laser radiation (lasing energy per pulse, mean or pulsed lasing power) and temperature stabilization of GDT. The laser radiation parameters are stabilized as follows. From light splitter *13* laser radiation travels to the device for measuring the stabilized output parameter and the signal from the device is fed to one of the inputs of an electronic regulator *6*. The second input of the regulator receives reference voltage from the specifier of the values of the output parameter. If there is a potential difference between the reference voltage and the voltage of the meter of the output parameter the electronic controller changes charging voltage of the capacitor C_4, resulting in the required variation of the output parameter. The GDT temperature is stabilized in the same manner as stabilization of the lasing parameters. In order to reduce overloading of the rectifier at a rapid increase in the excitation pulse repetition frequency f the voltage regulator *2* and the master oscillator *8* are connected by a link providing a reduction in the voltage of the rectifier *3* at increasing f.

The excitation circuit, shown in Fig. 3.32, stabilizes the lasing energy (0.25 mJ) of the self-heating copper vapour laser within 1.5% when the excitation pulse repetition frequency changes from 1 kHz to 6 kHz at rates of 1–2 kHz/min.

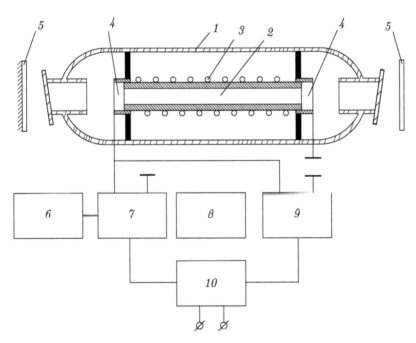

Figure 3.33. Scheme of the stabilized laser with an integrated ohmic heater [67]. *1* – hermetic casing, *2* – gas discharge channel; *3* – heater, *4* – electrodes, *5* – mirrors of the resonator, *6* – master pulse generator *7* – heating source, *8* – synchronization unit, *9* – generator of excitation pulses, *10* – regulation unit.

A method of controlling the laser pulse duration, proposed in [66, 67], is based on the change in the pre-pulse parameters of plasma due to the superposition of an additional pulse on the discharge gap having the same pulse repetition frequency as the excitation pulse, but with an adjustable time delay between AP and EP. In this case, the additional impulse heats up recombining plasma and thereby changes the pre-pulse parameters of plasma before the next excitation pulse. The thermal balance in the active element is preserved by fulfilling the conditions $(E_{ex} + E_{ad}) f$ = const, where E_{ex} and E_{ad} is the energy inputted to the discharge in respectively the excitation pulse and the additional pulse. This method also works as a way of controlling the colourity of the lasing, when it is carried out simultaneously in several spectral lines (for example, in a copper vapour laser at λ = 510.5 nm and λ = 578.2 nm).

As stated in [67], an advantage of lasers in which the lasing characteristics are governed by the formation of additional pulses is the simple design of self-heating GDTs. However, for any given GDT the additional pulse reduces the maximum attainable lasing power values, pulse repetition frequency and efficiency. Therefore, in [67] it is proposed to use combined heating of GDT using the circuit shown in Fig. 3.33 with a built-in ohmic heater manufactured in the form of a spiral. In this circuit, the GDT is heated simultaneously GDT by heating and excitation sources. After reaching the operating temperature of the GDT the synchronization unit activates the excitation source with a repetition frequency determined by the master oscillator. The regulation unit decreases the voltage amplitude of the heating source to the value providing the steady thermal mode of the GDT so that the following conditions are met:

$$I_0 U_0 = P - \Delta E f, \ U_0 < U_a, \tag{3.10}$$

where U_0 and I_0 is the amplitude of the sinusoidal voltage and the current of the heater; U_a is the burning voltage of the arc discharge in the active laser medium; P is the power needed to heat the laser to the operating temperature; ΔE is the fraction of the energy of the excitation pulse, dissipated as heat.

As stated in [67], the combined method of heating the GDT has the following advantages:

- optimizes the excitation pulses at a maximum lasing power or efficiency;
- reduces the power dissipated by the thyratron, which leads to an increase in the service lifetime of the laser as a whole;

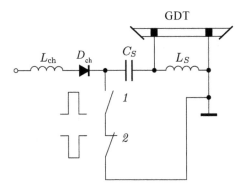

Figure 3.34. The circuit of a copper vapour laser [71]. *1* – thyratron *2* – tacitron.

- allows the laser to operate in the standby mode, when its readiness for operation is only supported by the heater power;
- allows to adjust the pulse repetition rate of zero to f_{max}.

3.12. Prospects for the development of excitation pulse generators of self-contained lasers

Obviously, future improvements will be made in switches of various types and their corresponding elemental base, and therefore to improve the excitation circuits created on the basis of these switches. However, completing the consideration of the various excitation pulses of self-contained lasers, it should be noted that practically all the above generators have one significant drawback: the inability to stop the excitation pulse at the end of the laser pulse. That is, the application of the above circuits of formation of excitation pulses does not allow without any additional means to expect to receive the maximum achievable efficiency values of the copper vapour lasers defined by the theory (see for example [68, 69]).

The problem of termination of the excitation pulse at the end of the laser pulse has been discussed to some extent by many authors. For example, in [70], based on studies of copper vapour lasers with a GDT with a diameter of 1 cm and 40 cm long it was concluded that at forced interruption of the excitation pulse at the end of the lasing pulse the efficiency may reach 4–6%.

Forced termination of the excitation pulse is implemented in [71] and [72,73] for copper vapour lasers with a GDT 6 mm in diameter and 300 and 170 mm long. In both cases, the pressure of the buffer gas (neon) was 6.7 kPa. The excitation circuit used in [71–73] in shown in Fig. 3.34.

The integral switch is formed by series-connected thyratron TGI1-270/12 (1) and tacitron TGU1-27/7 (2) and allows the discharge current to be terminated at 20 A. In the initial state the grid of the tacitron has a positive bias and it is in an open state. After starting the thyratron the grid of the tacitron receives a negative pulse, locking the tacitron. Consequently, the duration of the excitation pulse can be regulated and the energy input into the GDT can be terminated by completion of the lasing pulse. The efficiency of a copper vapour laser in the mode of complete termination of energy input after completion of the lasing pulse was 9%.

As stated in [71], in the regime of complete termination of the energy input there was no characteristic glow of the buffer gas in the cold parts of the GDT. A more detailed study of the discharge plasma has shown that this mode is characterized by a significant reduction of the degree of excitation of the levels of the copper atom positioned above resonance levels, and also of the degree of ionization of copper, with the character of the glow of spectral lines, beginning with the first resonance levels of the copper atom, remaining unchanged.

Given the results of the studies described in [71–73] it is apparent that one of the most promising ways of development of excitation circuits of self-contained lasers is the development of units providing forced termination of the excitation pulse.

In [70], the requirements for the parameters of excitation pulses are described and the conditions for their implementation are formulated, which apparently determined to a large extent the future path of the development of excitation units of self-contained lasers:

1) for the efficient pumping of the active transitions of metal vapour lasers it is necessary to form the excitation pulse with a steep front of the voltage pulse on the active element of the laser, which is terminated at the end of the lasing pulse, subject to the conditions of the aperiodic nature of discharge for the duration of the excitation pulse;

2) such an excitation pulse can be formed only by partial discharge of the storage capacitor, so for efficient pumping of metal vapour lasers it is preferred to use vacuum tubes or transistor switches as switches.

References

1.   Mesiats G.A., Generation of powerful nanosecond pulses. Moscow: Soviet radio. 1974.
2.   Bohan P.A., Gerasimov V.A., Solomonov V.I., et al. On the mechanism of lasing of copper vapor lasers. Effective discharge metal vapor lasers (Digest of

articles). Tomsk: Publishing Laboratory IAO SB RAS,1978, P. 133-159.

3. Bohan P.A.,Gerasimov VA , Kvantovaya elektronika, 1979 V.6. No.3. P.451-455.
4. Isaev A.A., Lemmerman G.Y., Kvantovaya elektronika, 1977, Volume 4, No.7. P.1413-1417.
5. Vuchkov N.K., Astadjov D.N., Sabotinov N.V., Optical and Quantum Electronics. 1988. Vol.20. No.2. P.433-438.
6. Vuchkov N.K., Astadjov D.N., Sabotinov N.V., Opt. Quantum Electronics. 1991. Vol.23. No.2. P.S549-S553.
7. Vuchkov N.K., Astadjov D.N., Sabotinov N.V. , IEEE J. Quantum Electronics. 1994. Vol.30. No.3. P.750-758.
8. Vuchkov N.K. Novel Circuits for Excitation of Metal Vapour Lasers , Pulsed Metal Vapor Lasers. Edited by Chris E. Little and Nikola V.Sabotinov. Nato ASI Series (1. Disarmament Technologies - Vol.5). Dordrecht / Boston / London: Kluwer Academik Publishers. 1996. P.183-188.
9. Astadjov D.N., Dimitrov K.D., Little C.E. at al. , IEEE J. Quantum Electronics. 1994. Vol.30. No.8. P.1358-1360.
10. Sabotinov N.V., Akerboom A.F., Jones D.R, Maitland A., Little C.E. , IEEE J. Quantum Electronics. 1995. Vol.31. No.4. P.747-753.
11. Jones D.R., Maitland A., Little C.E. , IEEE J. Quantum Electronics. 1994. Vol.30. No.10. P.2385-2390.
12. Little C.E., Jones D.R., Fairlie P.A., Whyte C.G. Metal HyBrID Lasers, Pulsed Metal Vapour Lasers. Edited by Chris E. Little and Nikola V.Sabotinov. Nato ASI Series (1. Disarmament Technologies - Vol.5). -Dordrecht / Boston / London: Kluwer Academic Publishers. 1996. P.125-136.
13. Isaev A.A., GG Petrash Research pulsed gas lasers on atomic transitions. Pulsed gas discharge lasers on transitions atoms and molecules (Proceedings of the Lebedev Physics Institute. V.81). Moscow: Nauka. 1975. P.3-87.
14. Chen C.J., Russel G.R., Appl. Phys. Lett. 1975. Vol.26, No.9. P.504-505.
15. Chen C.J., Bhanji A.M., Russel G.R., Appl. Phys. Lett. 1978. Vol.33. No.2. P.146-148.
16. Zubov V.V., Lyabin N.A., Mishin V.I., et al., Kvantovaya elektronika. 1983. Vol.10. No.9. P.1908-1910.
17. Babeyko Yu., Vasiliev L.A., Sviridov A.V., et al., Kvantovaya elektronika. 1979 V.6. No.5. P.1102-1105.
18. Sokolov A.V., Sviridov A.V., Kvantovaya elektronika. 1981 V.8. No.8. P.1686-1696.
19. Galkin A.F., Klimovskii I.I., On the issue of improving energy characteristics of self-heating copper vapor lasers. Labor and Employment at Parvanatsionalna conference on optics and laser technology - "Optika'82." Panagyurishte. 22-25 septembri.1982. P.127-129.
20. Vokhmin P.A., Klimovskii I.I., Teplofizika vysokikh temperatur, 1978. V.16, No.5. P.1080-1085.
21. Fogelson T.B., Breusova L.N., Vaginitis L., Pulse hydrogen thyratrons. Moscow: Sovetskoe Radio. 1974.
22. Isaev A.A., Lemmerman G.Y., System power pulsed lasers copper vapor. Metal vapor lasers and their halides (Proceedings of the Lebedev Physics Institute.V.181). Moscow: Nauka. 1987. P.164-179.
23. Isaev A.A.,Kazaryan M.A., Petrash G.G., Zh. Prikl. Spektroskopii. 1973. V.18. Vol.3. P.483-484.
24. IsaevA.A., Kazaryan M.A., Petrash G.G., Brief Communications in Physics. 1973. No.2. P.27-29.

25. Abas oglu Y.R., Aboyan P.A., Abrosimov G.V., et al., Kvantovaya elektronika. 1981, V.8. No.3. P.648-650.

26. Karny Z., Rotstien A., Chuchem D., Milanski T.S., Rev. Sci. Instrum.1980. Vol.51. No.10. P.1426-1428.

27. Bohan P.A., Zhurnal tekhnicheskoi fiziki. 1981. V.51. Issue 1. P.206-209.

28. Lizogub V.G., Topuchkanov V.P., Optika atmosfery i okeana. 1995 V.8, No.11. P.1687-1689.

29. Kirilov A.E., Polunin Y.P., Soldatov A.N., Fedorov V.F., Lasers metal vapor for atmospheric research. Measuring instruments for study, the surface layers of the atmosphere. Tomsk, Institute of Atmospheric Optics, SB RAS. 1977. P.59-79.

30. Batenin V.M., Galkin A.F., Klimovskii I.I., Kvantovaya elektronika, 1981 V.8. No.5. P.1098-1100.

31. Voronov V.I., Kirilo A.E., Soldatov A.N., et al., Pribory Tekh. Eksper. 1982. No.1. P.151-152.

32. Deacons V.P., Avalanche transistors and their use in pulse devices. Moscow: Sovetskoe Radio, 1973.

33. Bohan P.A., Nikolaev V.N., Solomonov V.I. , Kvantovaya elektronika. 1975. V.2. No.1. P.159-162.

34. Kirilov A.E., Kuharev V.N., Soldatov A.N., Tarasenko V.F., Izv. VUZ Fizika. 1977. No.10. P.146-149.

35. Nerheim N.M., Bhanji A.M., Russel G.R., IEEE J. Quantum Electronics. 1978. Vol.QE-14. No.9. P.686-693.

36. Pack J.L., Liu C.P., Feldman D.W., Weaver L.A ., Rev. Sci. Instrum. 1977. Vol.48. No.8. P.1047-1049.

37. Kirilov A.E, Kuharev V.N., Soldatov A.N. , Kvantovaya elektronika. 1979 V.6. No.3. P.473-477.

38. Dashuk P.P., Potapov P.E. , Pribory tekh. Eksper. 1982 No.6. P.145-147.

39. Magetto G. Thyristor electronics / Translation from the French. Zheglov V.A. Moscow. Energiya. 1977.

40. Bonch-V.A., Long Y.u., Timokhin A.A., Optiko-Mekh.Prom.. 1980. No.4. P.22-24.

41. Chen C.J., Nerheim N.M., Russel G.R., Appl. Phys. Lett. 1973. Vol.23. No.9. P.514-515.

42. Piper J.A. , Optics Communicationa. 1975. Vol.14. No.3. P.296-300.

43. Vetter A.A., IEEE J. Kvantovaya elektronika. 1977. Vol.QE-13. No.11. P.889-891.

44. Gordon E.B., Egorov V.G., Pavlenko V.P., Kvantovaya elektronika. 1978.V.5. No.9. P.162-164.

45. Egorov M.G., Finding ways to improve the working concentrations of atoms in self-contained metal atoms lasers. PhD Thesis. Chernogolovka. 1982.

46. Kachanov I.L., Ionic devices. Moscow: Energiya. 1972.

47. Alaev M.A., Baranov A.I., Vereshchagin N.M., et al., Kvantovaya elektronika. 1976 V.3. No.5. P.1134-1136.

48. Soldatov A.N., Fedorov V.F., Investigation of the parameters of the active medium of the Cu vapor lasers at high repetition rate. X Siberian Conference on Spectroscopy (population inversion and lasing at transitions in atoms and molecules), Abstracts. Tomsk: Publishing House of TSU. 1981. p.85.

49. Vorob'ov V.B., Kalinin P.V., Klimovskii I.I., et al., Kvantovaya elektronika. 1991. V.18. No.10. P.1178-1180.

50. Galkin A., Klimovskii I., Computed model of copper-vapor laser with the average specific output power above 1 W/cm², Metal Vapor Lasers and Their

Applications: CIS Selected Papers, G.G. Petrash, Editor, Proc. SPIE 2110. 1993. P.90-99.

51. Kiselev Yu.V., Cherepanov V.P., Spark arresters. Moscow: Sovetskoe radio. 1976.
52. Koval'chuk V.M., Kremnev V.V., Potalitsyn Y.F., High-current nanosecond generators, Ed. Mesyats G.A. Novosibirsk: Nauka. Siberian Branch. 1979.
53. High power switching systems. Translated from English. Moscow: Mir. 1981.
54. Pavlov V.A., Pichugin Yu.P., Romanenko I.N., Pribory Tekh. Eksper. 1980. No.6. P.205-206.
55. Pavlov V.A., Pichugin Yu.P., Romanenko I.N., Elektrichestvo. 1986. No.4. P.56-58.
56. Pavlov V.A., Romanenko I.N., Romanov Yu., Shilin N.V., Elektrichestvo. 1988. No.9. P.51-55.
57. Liberman I., Babcock R.V., Liu C.P. et al., Appl. Phys. Lett. 1974. Vol.25. No.6. P.334-335.
58. Anderson R.P., Springer L., Bricks B.G., Karras T.W., IEEE J. Quantum Electronics. 1975. Vol.QE-11. No.4. P.172-174.
59. Buzhinskiy O.I, Efimov A.V., Slivitsky A.A., Kvantovaya elektronika. 1982. v.9. No.9. P.1854-1856.
60. Dashuk P.P., Potapov P.E., Pribory Tekh. Eksper. 1983. No.1. P.155-156.
61. Okunev R.I., Pakhomov L.N., Petrunkin V.Y., Stepanyants A.L., Pisma Zh.Tekh. Fiz. 1983. V.9. No.11. P.670-673.
62. Turkevich V.M., Grekhov I.V., New principles of switching high power semiconductor devices. Leningrad: Nauka. 1988.
63. Volle V.M, Voronkov V.B., Grekhov I.V. et al., Zhurnal Tekhnicheskoi Fiziki. 1981. T.51. Issue 2. P.373-379.
64. Alexandrov V.M., Buzhinskiy O.I., Grekhov I.V. et al., Kvantovaya elektronika. 1981 V.8. No.1. P.191-193.
65. Soldatov A.N., Fedorov V.F., Kvantovaya elektronika. 1983. Vol.10. No.5. P.974-980.
66. Yevtushenko G.S., Kirilov A.E., Kruglyakov V.L., et al. , Zh. Prikl. Spektroskopii. 1988. T.49. No.5. P.745-751.
67. Soldatov A.N., Optika atmosfery i okeana. 1993 V.6. No.6. P.659-665.
68. Vokhmin P.A., Klimovskii I.I., Teplofizika vysokikh temperatur. 1978. V.16. No.5. P.1080-1085.
69. Batenin V.M., Vokhmin P.A., Klimovskii I.I., Selezneva L.A., Teplofizika vysokikh temperatur. 1982. V.20. No.1. P.177-180.
70. Denikin V.P., Soldatov A.N., Yudin N.A., Optika atmosfery i okeana. 1993 V.6 No.6. P.659-665.
71. Soldatov A.N., Fedorov V.F., Yudin N.A., Kvantovaya elektronika. 1994. V.21 (8). P.733-734.
72. Soldatov A.N., Sukhanov V.B., Fedorov V.F., Yudin N.A., Optika atmosfery i okeana. 1995 V.8. No.11. P.1626-1636.
73. Soldatov A.N. MVL parameter management through electron plasma components, Pulsed Metal Vapour Lasers. Edited by Chris E. Little and Nikola V. Sabotinov. NATO ASI Series (1. Disarmament Technologies – Vol.5). Dordrecht/Boston/London: Kluwer Academic Publishers. 1996. P.175-182.

Repetitively pulsed self-contained lasers

4.1. The history of research of repetitively pulsed metal vapour lasers

The initial history of studies of different lasers on self-contained transitions of metal atoms is reflected in some detail in [1,2]. Therefore, in this chapter attention is given to the history of development of only the repetitively pulsed lasers of this type which is of greatest interest for practical applications.

The repetitively pulsed mode of the laser at self-contained metal atoms was realized for the first time in [3] with a manganese vapour laser with gas discharge tube (GDT) with a diameter d_p = 1 cm and a length l_p = 100 cm at an excitation pulse repetition frequency of f = 360 Hz. The total average lasing power in the green and infrared spectral regions was respectively 2.1 and 1.2 mW. In [4] a copper vapour laser with a GDT 1 cm in diameter and the heated length of l – 80 cm at a frequency f = 660 Hz produced the average lasing power of 15 mW and 5 mW on the green and yellow lines, respectively. In [5] experiments were carried out with repetitively pulsed operation of a copper vapour laser (d_p = 5 cm, l_a = 80 cm) at a frequency of f = 1.2 kHz with an average lasing power of 0.5 W on the green line with the practical efficiency slightly exceeding 1%.

In [6] by using a thyratron TGI1-500/16 as the switch the excitation pulse repetition frequency in lasers on copper vapours (d_p = 3 mm, l_a = 25 cm), gold vapours (d_p = 3 mm, l_a = 25 cm) and lead vapours (d_p = 4.5 mm, l_a = 40 cm) was raised to 2.5 kHz. The average lasing power on both lines of the copper atom is 25 mW, for the red line of the gold atom 10 mW, and the two lines of a lead atom (λ = 406.2; 722.9 nm) 35 mW, including the red line 25 mW. With the manganese vapour laser (d_p = 4.5 mm, l_a = 20 cm, f = 1 kHz) in [6] the average lasing power in the green spectral region was 5 mW. The increase in the lead vapour laser of the diameter and length of the heated part of the GDT respectively to 1.5 and 60 cm and the transition to the excitation of the discharge with a pulse transformer [7] enabled at the same excitation pulse repetition frequency of 2.5 kHz to increase the average lasing power on the 722.9 nm line to 0.3 W.

Substantial progress in improving the lasing power and overall efficiency of repetitively pulsed metal vapour lasers has been achieved through the implementation of the self-heating mode repetitively pulsed metal vapour lasers [8,9]. In [8] the authors reported the results of studies of two repetitively pulsed copper vapour lasers with an alundum GDT with active volumes of 35 cm³ (d_p = 0.8 cm, l_a = 70 cm) and 125 cm³ (d_p = 1.5 cm, l_a = 70 cm). respectively. A distinctive feature of these lasers was that their GDTs, filled with inert gases at a pressure of several kilopascals, were warmed up to an operating temperature of about 1500°C without a special furnace, due to the heat generated in the discharge. The highest total average lasing power of the laser with a GDT of smaller diameter was obtained at a frequency of f = 18 kHz and equalled $\bar{P}_{l\Sigma}$ = 6 W at the practical efficiency of $\eta_{r\Sigma}$= 0.35%. With a laser with a GDT of larger diameter at a repetition frequency of excitation pulses of 18 and 20 kHz the laser power was $\bar{P}_{l\Sigma}$= 15 W and practical efficiency $\eta_{r\Sigma}$= 1%. In [9] it was reported that a self-heating lead vapour laser operating at a frequency f = 12 kHz, produced an average lasing power of 1.3 W at the 722.9 nm line at the practical efficiency of 0.2%.

Starting around 1975–1976, after the implementation of the self-heating mode of the repetitively pulsed copper and lead vapour lasers, research and development of lasers based on self-contained transitions of metal atoms were developed mainly in three areas:

- Research and development of copper vapour lasers with longitudinal discharge;
- Lasing and improvement of the performance of pulsed-periodic lasers at self-contained transitions of atoms and ions of other metals;
- Research and development of copper vapour lasers with a transverse discharge.

The comparative lasing characteristics obtained by laser on vapours of gold, manganese, lead, barium, strontium and europium until the middle of 1983, are given in [2]. The progress, such as that which occurred with the output characteristics of copper vapour lasers, was not achieved with the characteristics of these lasers. Therefore, in this place it seems appropriate to limit considerations to the indication of several of the most recent publications on some of the above as well as some other types of lasers (except copper vapour lasers), which apart from the latest results of research contain more or less complete lists of publications containing the results of studies of the lasers. These publications, in our opinion, are the following: [10, 11] – the gold vapour laser, [10, 12, 13] – bismuth vapour lasers, [14–17] – laser at self-contained transitions of the europium ion , [17] – europium vapour laser, [18–20] – barium vapour laser, [21] – lead vapour laser, [22] – manganese vapour laser.

Copper vapour lasers with transverse discharge were rather intensively studied and developed in the USSR (see [23–45]). Judging from the published works, the best results were obtained in [34, 35, 43]. In [34] the authors reported the results of studies of copper vapour lasers with a transverse discharge with two designs of gas-discharge chamber (GDC): with a rectangular and or a coaxial chamber with a useful volume to 4 l. Both GDCs obtained similar results: the optimal excitation pulse repetition frequency lies in the range of 3-4 kHz, the neon buffer gas pressure - 0.05–0.15 MPa, the concentration of copper atoms 10^{15} cm^{-3}. ×With a 4-modular design, using a GDC, with a cross-sectional dimension of 1.5 × 12 cm and a length of 120 cm the average lasing power was 180 W at a frequency of $f = 4$ kHz, pulse energy of more than 0.04 mJ, peak power $1.5 \cdot 10^6$ W. With the tri-modal structure consisting of coaxial GDCs, with the inner and outer diameters of respectively 5 and 8 cm and a length of 120 cm, the average lasing power was 120 W at a frequency of $f = 3$ kHz. The discharge circuit in the excitation block of these lasers was produced using the Blumlein circuit with a storage line and several parallel working TGI1-2500/50 thyratrons. When the excitation pulse energy of 1020 J the thyratron worked in the microsecond regime, and the subsequent magnetic compression pulse to 150 ns was carried out with the help of several non-linear circuits.

In [35] the authors reported on reaching the average lasing power of about 100 W at an excitation pulse repetition frequency of 3 kHz using a copper vapour laser with a gas-discharge chamber (GDC), with a cross-section of 5 × 5 cm. The optimum temperature of the GDC wall was 1720 K, and the neon pressure 0.07 MPa. The lasing energy in the pulse was 33 mJ and the peak power exceeded 10^8 W. These characteristics more or less coincide with the characteristics of repetitively pulsed copper vapour lasers with a longitudinal discharge with a GDT (gas discharge tube)diameter of 5–6 cm.

Finally, [43] presents the data for a copper vapour laser with a transverse discharge with a GDC with the dimensions of $6 \times 10 \times 160$ cm. With an unstable resonator with a frequency of $f = 2$ kHz and at a neon pressure of 10^5 Pa the average power of this laser was 104 W.

Despite the fact that with certain successes have been achieved with the copper vapour lasers with transverse discharge, such lasers still have not found practical application. The main reason, apparently, is that to maintain in these lasers the GDC wall temperature at the working level it is necessary to use special heaters [28, 43], which leads both to technological difficulties in the manufacture of the GDC, and to a significant reduction in the overall efficiency of the copper vapour lasers with a transverse discharge. However, it is possible that the GDC of the laser with transverse discharge will be used in future for some plasma-chemical technologies that do not require heating of the walls of the GDC to high temperatures.

The greatest advances has been made in the first of the above directions. The maximum temperature of the GDT and the corresponding concentration of copper atoms were achieved in [46, 47]. In [46] experiments were carried out using GDTs with a diameter of 3 mm and 4 mm and a length of 7 and 8.5 mm. Buffer gases were helium, neon and argon at pressures 1.3–4 kPa. With a GDT of a smaller diameter at an excitation pulse repetition of $f = 4.2$ kHz the specific lasing energy was $W = 33$ µJ/cm³. For a GDT with a larger diameter the frequency was $f = 6.8$ kHz, $W = 39$ µJ/cm³. The wall temperature T_w of the GDT was 1800–1850°C, which corresponds to the concentration of copper atoms $3 \cdot 10^{16}$ cm⁻³. In [47] with the GDT of the same design at $T_w = 1800$°C and a frequency $f = 6$ kHz the average lasing power was 2 W at a specific energy of 40 µJ/cm³. In [47] it is stated that when the concentration of copper atoms is greater than 10^{16} cm the lasing pulse consisted of four separate peaks with a duration of 5–6 ns. Interferometric measurements showed that each peak has its own lasing frequency. When the concentration of copper atoms is reduced to $3 \cdot 10^{15}$ cm⁻³ the lasing pulse acquired the usual form.

In the first 10–12 years after the publication of [8], the average lasing power of repetitively pulsed copper vapour lasers was increased mainly by increasing the diameter of GDT.

In [48], the results of research of self-heating copper vapour lasers with different GDTs were published. The maximum lasing power of 38 W was obtained in [48] with the GDT diameter of 20 mm and a length of 100 cm ($f = 16.5$ kHz; $p_{Ne} \approx 2$ kPa). The practical efficiency was 0.43%. Reducing the excitation pulse repetition frequency to 10 kHz led to a reduction of the lasing power up to 20.5 W. The practical efficiency increased to 0.45%. Unfortunately, in [48] it was not mentioned whether the values of power relate to the stationary regime of lasers or were obtained in a short time in the transient mode of operation. With a GDT diameter of 2.4 cm and a length of 53 cm at an excitation pulse repetition frequency of 13 kHz [48] the average lasing power was 16 W.

In [49] in experiments with self-heating copper vapour lasers with a GDT diameter of 28 mm and a length of 80 cm (p_{Ne} = 4 kPa, f = 16.7 kHz) the average lasing power in a short transient regime of the laser was 43.5 W at a practical efficiency of 1%. Judging from the experimental results presented in [49], in the steady-state operation the lasing power was 20 W.

In [50] a copper vapour laser with a GDT diameter of 3.8 cm and a length of 125 cm was studied. The neon buffer gas pressure was about 2 kPa, the rate of pumping of neon through the GDT was approximately $0.5 \cdot 10^{-2}$ l/s. The maximum average lasing power was 36 W at an excitation pulse repetition frequency of 5 kHz. In this case the practical efficiency was 0.58%.

In [51] by increasing the diameter of the GDT (d_p = 6 cm), the average lasing power of a copper vapour laser was reduced to 55 W, in [52] to 85 W (d_p = 7 cm), and finally in [53] to 110 W (d_p = 8 cm; f = 5 kHz; η_r = 1%).

According to the experimental data presented in the review [54], the lasing power of the copper vapour lasers with longitudinal discharge was proportional to the cross sectional area of the GDT. However, as pointed out in [54], it was difficult to expect a further increase in the average lasing power of the copper vapour lasers, since the increase of the cross section of the GDT results in a large drop of the discharge resistance, which creates significant challenges for the energy input into the discharge.

It is worth noting that the average lasing powers achieved with a copper vapour laser with a longitudinal discharge up to 1982 were comparable to maximum lasing power, determined theoretically at that time. For example, in [55] the authors calculated the parameters of a repetitively pulsed copper vapour laser and it was shown that the GDT with a diameter of 8 cm and an active medium length of 122 cm can produce the average lasing power of about 130 W and with a GDT 12 cm in diameter and, apparently, with the same length of the active medium – 210 W at an efficiency of 1.5–2%.

Thus, by 1983 the possibilities of increasing the average lasing power of copper vapour lasers with longitudinal discharge by increasing the diameter of GDT were mostly exhausted. Still searching for ways to further improve the average lasing power of copper vapour lasers with longitudinal discharge did not cease. With varying degrees of activity these searches were conducted in three areas:

- the use of additives of molecular gas to neon;
- increase in the steepness of the leading edge of the voltage pulse to reduce the population of the lower laser levels in the initial period of the discharge;
- The development of laser systems, consisting of a low-power oscillator and a powerful amplifier.

The effect of additives of molecular gas (hydrogen) on the lasing characteristics of the repetitively pulsed copper vapour laser with a GDT diameter of 2.7 cm and a length of 25 cm was studied in [56]. According to the results of these studies, at a neon pressure of 2.4 kPa the discharge circuit inductance $L_c = 2$ μH, the addition of hydrogen at a pressure of 47 Pa leads to an increase in the optimal excitation pulse repetition frequency from about 3.2 to 5 kHz. The addition of hydrogen at a pressure of 665 Pa increases the optimal frequency f to more than 9 kHz. In addition, in [56] it is indicated that the addition of hydrogen leads to an increase in the average lasing power of the laser. In [56] it is concluded that the positive effect of hydrogen on the output characteristics of a copper vapour laser is associated with an increase in the rate of cooling of electrons due to collisions with light hydrogen atoms, which in turn increases the rate of triple recombination of electrons and, consequently, reduces the prepulse electron concentration and leads to a good matching of the discharge impedance with the wave impedance of the discharge circuit.

In [57] the authors reported on the implementation of a copper vapour laser at an average lasing power of 60 W. Judging by the results, an important role in obtaining record lasing power was played by adding hydrogen to the buffer gas.

In [58] it is reported that the addition of hydrogen at a pressure of 13 Pa to the buffer gas in a gold vapour laser with a GDT diameter of 2.7 cm and 50 cm long provided a fivefold increase in the average lasing power on the line 312.2 nm, so that the power reached about 100 mW.

Despite the apparent positive effect of small amounts of hydrogen on the lasing characteristics of the copper vapour lasers, such additions are not currently used in the creation of powerful copper vapour lasers. One reason for this is that in the non-pumped lasers the hydrogen impurity disappears with time due to some obscure unexplained chemical reactions disappear, therefore, it is quite difficult to ensure the stationary operating mode of the laser in respect of power. However, we can assume that the addition of hydrogen to the buffer gas can be used in pumped copper vapour lasers.

Moreover, the very possibility of increasing the power of repetitively pulsed copper vapour lasers by use of various molecular impurities remained, nevertheless, very attractive. In a recent paper [59] the authors reported the doubling of the average lasing power of two copper vapour lasers by adding a halide in their active media but its name was not given. By use of this halide the lasing power of the copper vapour laser with a GDT with a diameter of 2.5 cm and 100 cm long increased from 25 to 50 W, and the lasing power with a GDT with diameter of 4 cm and 150 cm long increased from 50–55 W to 100 W. As indicated in [59], the lasing power of 100 W coincides with the lasing power of a conventional copper vapour laser with the volume of the active medium 4 times larger. According to the results in [59], improving the output characteristics of lasers is associated with improved spatial and temporal characteristics of laser radiation. It is quite possible that the improvement of these characteristics is a consequence of reducing the radial inhomogeneity of the prepulse plasma parameters.

There are at least two principal possibilities of shortening the front edge of the voltage pulse

– use as switches of vacuum tubes,

– the use of non-linear inductances.

In the literature on metal vapour lasers in recent years the first of these methods seems to be generally disregarded. On the second method there are at least two publications – [60, 61]. The first one proposes a mathematical model for calculating the parameters of excitation pulses generated by compression circuits. In [61] the authors describe a generator excitation pulses based on the Blumlein circuit with a non linear inductor with a saturable core, and the results of investigations of copper vapour lasers with a GDT 2 cm in diameter and a length of the electrode gap of 93 cm at a neon pressure of 20 kPa and an excitation pulse repetition frequency of 10 kHz are presented. According to the results in [61], the application of the voltage doubling circuit with a non-linear inductance increases the average lasing power from 20 (excitation scheme with a direct discharge of the storage capacitor) to 37 W and increases the practical efficiency from 0.8 to 1%.

The oscillator–amplifier system was used for the first time in copper vapours in the aforementioned previous work [47], which studied self-heating copper vapour lasers with a small GDT. The comparison in [47] of the output characteristics of a copper vapour laser, operating either as a generator or an amplifier, showed that in the laser generator mode average lasing power $\bar{P}_l$ = 250 mW at a pulse duration τ_g = 20 ns. In operation of the laser as an amplifier the $\bar{P}_l$ value for the same parameters of the excitation pulse increased to 450 mW, and the duration τ_g was reduced to 5 ns. The observed increase in lasing power was explained by the authors of [47] bu suggesting that in operation of the laser in the generator mode a significant proportion of the energy is lost as a result of the output of induced radiation output to the GDT wall.

Omitting the consideration of the results of numerous studies of laser systems, consisting of the generator and one or more amps, we will dwell on the results of two recent papers on this subject [62,63]. The first of these papers describes a laser system consisting of a generator and three amplifiers. The excitation pulse repetition frequency was 4.4 kHz. The generator power was 20 W at a pulse duration of 50 ns. The power of each amplifier (the dimensions of the GDT and buffer gas pressure are not specified in the paper) was 250 W at a total output power of the laser system over 750 W. The laser system described in [63] consists of a generator and four amplifiers but the size of the GDT is also not listed. The maximum average lasing power, taken from one amplifier, is 560 W at a total power of the laser system of 1902.

In addition to studies by increasing the lasing power of lasers at self-contained transitions of metal atoms (primarily copper vapour laser) studies have been performed to improve the efficiency of such lasers, as a rule without regard to an increase in lasing power. Already in 1967, in one of the first works with a copper vapour laser (d_n = 5 cm, l_p = 80 cm) [64] the practical efficiency reached 1.2%. Study [64] clearly demonstrated the ability to create high-performance lasers in the visible wavelength range of lasers based on self-contained transitions of metal atoms.

The next significant progress in increasing the practical efficiency of the copper vapour lasers achieved after eleven years in 1978, in the study of copper vapour lasers with a GDT diameter of 2 cm and 15 cm long with an external heater [65]. The excitation pulse repetition frequency was 50 Hz, the neon buffer gas pressure 3 kPa. The growth of the practical efficiency in [65] from 0.6% to 2.1% was achieved by reducing the discharge circuit inductance by using coaxial current leads.

Further increase of the practical efficiency of the copper vapour lasers is realized in [66] with the GDT of the same design as in [65]. The studies were conducted in the single and double excitation pulse modes with a repetition frequency of 50 Hz, as well as in the regular mode of excitation pulses with a frequency of 8 kHz. In [66] it was reported that the optimization of the diameter of the GDT, the vapour pressure of copper, type and buffer gas pressure and the excitation mode enabled to achieve the maximum efficiency of 3% with the GDT of diameter 2 cm and a length of 26.5 cm with a specific energy of 20.8 µJ/cm^3 and a pulse repetition frequency of 50 Hz. The same GDT was used in the repetitively pulsed mode of laser operation at a frequency f = 8 kHz with a practical efficiency in relation to the power extracted from the rectifier of 2.6% and with respect to the energy stored in the storage capacitor of 2.9% .

And finally, as a result of the implementation in [67, 68] of the aperiodic discharge of a storage capacitor and the forced termination of the excitation pulse in two copper vapour lasers with a GDT 6 mm diameter and a length of respectively 300 and 170 mm the physical efficiency reached 9%. More detailed results of [67, 68] are described in section 3.12.

Discussing the work aimed at improving the efficiency of copper vapour lasers, it is important to mention the study [69], which attempted to implement the theoretically [70–74] steady-state lasing of a copper vapour laser with a GDT of a small diameter with an expected practical efficiency at the level of 5–8%. In the experiment [69] with the GDT 4.5 mm in diameter and 30 cm long (p_{Ne} = 600 Pa, f = 70 kHz, C_S = 330 pF) the average specific lasing power was 1.3 W/cm^3, in good agreement with theoretical results [70–74]. However, the physical efficiency obtained in [69] was about 1%. The authors of [69] believe that such a low value of efficiency is due to the presence in the discharge of cold zones with the length comparable to the length of the laser active medium.

4.2. The radial inhomogeneity of the plasma parameters in repetitively pulsed self-contained lasers

As shown by one of the first experimental studies of the properties of discharges in the self-heated copper vapour lasers [75–79], these discharges have a significant radial inhomogeneity of the plasma, which is expressed in the radial distributions of the lasing rate [75,77,79], the spontaneous radiation of the plasma during the excitation pulse [76] and plasma radiation in the interpulse period [76,78]. The question of the causes of the radial plasma inhomogeneities and peculiarities of their effect on the lasing characteristics was formulated for the first time in 1977–1978 in [75,80]. However, to date the problem of the formation of prepulse distributions of plasma parameters in the repetitively pulsed lasers at self-contained transitions, apparently, has not been solved in general form. This solution is extremely difficult because it requires consideration in the energy and particle balance equations of not only volume processes but also the transport processes in the radial direction. For this reason, in most numerical calculations of the energy characteristics of self-contained lasers the radial distribution of the plasma parameters is assumed to be homogeneous, and the transfer of energy and particles to the GDT wall is taken into account in the so-called zero-dimensional approximation. Therefore, it seems appropriate to assess the conditions of formation of inhomogeneities in the distribution of plasma parameters that determine the output characteristics of these lasers.

4.2.1. The heterogeneity of the temperature distribution of gas

Based on the most general concepts, we can say that the energy injected into the discharge during the excitation pulse is almost all expended on heating of electrons, excitation of the atoms of the working metal and the inert gas and the ionization of especially the atoms of the working metal. Part of this energy is transferred during the excitation pulse from the active medium by the induced (through the end of the GDT) and spontaneous (to the GDT wall) radiation. Another part is transferred in the interpulse interval to the GDT wall diffusing charged particles and radiation. And finally, a small or large part of this energy is passed through the electron gas to heavy particles (atoms and ions), which leads to gas heating in the post-pulse period. Obviously, the fraction of energy used for heating the gas, depends on several factors and, above all, on the type and pressure of the buffer gas and the GDT diameter or transverse dimensions of the GDT. The heating period of heavy particles in collisions with electrons is replaced by cooling due the thermal conductivity outflow of heat from the gas to the wall of the GDT. In the steady state operation of a repetitively pulsed laser heating of the gas due to the introduction of energy in the discharge during the excitation pulse is compensated by its cooling so that the prepulse radial temperature distribution of gas T_{gst} (r) and its average value $\langle T_g \rangle_{st}$ over the cross section of the GDT remains constant from pulse to pulse.

In the conditions typical for the majority of repetitively pulsed metal vapour lasers, according to which the characteristic time of heating of heavy particles in collisions with electrons

$$\tau_{ge} \approx \left(\frac{2m}{M_a} v_{ea} + \frac{2m}{M_i} v_{ei} \right)^{-1} \tag{4.1}$$

is considerably shorter than the repetition period of excitation pulses $T = f{-}1$, the level of heating gas in such lasers will to a large extent depend on the ratio of the pulse repetition frequency f and the characteristic gas cooling time

$$\tau_{g\chi} \approx \frac{r_p^2}{6\chi_a}, \tag{4.2}$$

where χ_a is the thermal diffusivity of the buffer gas.

Provided $\tau_{g\chi} \ll f^{-1}$ the gas will be cooled almost to the GDT wall temperature during the interpulse period and the operating mode of a repetitively pulsed laser is close to its mode in excitation by single pulses.

Provided $\tau_{g\chi} \ll f^{-1}$ the gas in the GDT is heated to such a temperature that the magnitude of its fluctuations is much smaller than its average value over time. In this case, with good accuracy the temperature of the gas can be regarded as constant in time, i.e. $T_g(t, r) \approx T_g(r)$. The distribution $T_g(t)$ depends on the distribution of the time-averaged specific energy (energy input) $G(r)$ in the discharge. If the buffer gas pressure and the GDT radius are reduced, most of the charged particles will go to the wall of the GDT, taking with them a large amount of energy supplied to the discharge thus leading to a decrease in the gas heating rate. However, it follows from the results of the measurements described in section 4.3, that under the conditions where the convective heat loss from the GDT is not important, at neon pressures of $p > 3$ kPa, and the radii of GDT $r_p > 0.5$ cm practically all the energy injected into the discharge is expended in heating the gas.

At $\tau_{g\chi} \sim f^{-1}$ there are considerable changes in the gas temperature in the interpulse interval, and to determine the prepulse temperature distribution it is necessary to solve the problem similar to that solved in [80] in calculating the prepulse radial distribution of gas temperature in the pulsed-periodic discharges.

The relationship between the thermal conductivity of gas λ_g and its temperature T_g can be represented as [81]:

$$\lambda_g = AT_g^B, \tag{4.3}$$

where A and B are some constants. Their values, determined from the dependence $\lambda_g = f\,(T_g)$, given in [82], are:

for helium $A = 1.35 \cdot 10^{-5}$ W·cm^{-1}·K$^{-1.787}$, $B = 0.787$;

for neon $A = 8.96 \cdot 10^{-6}$ W·cm^{-1}·K$^{-1.683}$, $B = 0.683$;

for argon $A = 2.66 \cdot 10^{-6}$ W·cm^{-1}·K$^{-1.728}$, $B = 0.728$;

From the known dependence (4.3) we can determine the temperature dependences of the coefficients of thermal conductivity of inert gases

$$\chi_a = A_\chi \frac{T_g^{B_\chi}}{p},\qquad (4.4)$$

where p is the pressure of inert gas in Pa.

for helium: $A_\chi = 5.3$ cm^2 s^{-1} ·K$^{-1.787}$ Pa, $B_\chi = 1.787$;

for neon: $A_\chi = 3.4$ cm^2 s^{-1} ·K$^{-1.683}$ Pa, $B_\chi = 1.683$;

for argon: $A_\chi = 1.1$ cm^2 s^{-1} ·K$^{-1.728}$ Pa, $B_\chi = 1.728$;

Figure 4.1 shows the calculated (using (4.2), (4.4)) dependences $\tau_{g\chi}$ of the neon buffer gas pressure, allowing to make a visual representation of the neon pressure range, the GDT radii and excitation pulse repetition frequencies of the nearly stationary or non-stationary heating of the gas in the GDT. For example, it is clear that when the radius of the GDT is $r_p \approx 1$ cm, the pressures of neon $p_{Ne} > 2.7$ kPa and the frequency $f = 10$ kHz, the condition $\tau_{g\chi} \gg f^{-1}$, therefore, in this case the oscillations of the gas temperature in the interpulse period will be insignificant relative to its average value over time and the temperature distribution of the gas can be considered practically constant in time.

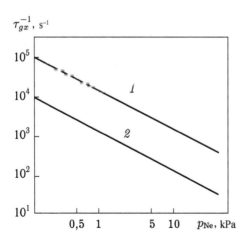

Figure 4.1. Dependence of the characteristics cooling time of neon on pressure $T_g = 2000$ K. 1) $r_p = 0.3$ cm; 2) $r_p = 1$ cm.

4.2.2. Heterogeneity of prepulse density distributions of the concentration of the atoms of inert gas and working metal in the ground state

In cases where the concentration of the buffer gas atoms significantly exceeds the concentration of all other particles (electrons, ions and atoms of the working metal), the characteristic time t of equalization of the pressure in the GDT, which is increased by heating of the gas, can be estimated as

$$\tau_p \sim r_p / v_{so},\qquad(4.5)$$

where v_{so} is the speed of sound in the buffer gas.

For the neon buffer gas, the GDT 2 cm in diameter and the gas temperature of 2000 K the estimate using (4.5) yields $\tau_p \approx 8$ μs, which is much less than the typical interpulse interval durations and the characteristic times of buffer gas cooling. This means that at small fluctuations in the temperature of gas relative to its average value the pressure in the GDT will remain practically constant and, therefore, the relationship of the radial distributions of gas temperature $T_g(r)$ and the concentration of buffer gas atoms $n_a(r)$ will be determined by the relation:

$$p_g = kT_g(r)\cdot n_a(r) = \text{const.}\qquad(4.6)$$

In contrast to the distribution of the concentration of buffer gas atoms, which can be considered independent of time, the concentration distribution of the atoms of the working metal $n_M(t, r)$ evolves over time. The atoms of the working metal disappear as a result of ionization during the excitation pulse and then their concentration is restored due to volume recombination and diffusion from the walls of the GDT. In the case when the prepulse concentration of the metal atoms n_M is restored to its maximum possible value, the relationship between the distributions $T_g(r)$ and the prepulse concentration $n_{Mst}(r)$ is determined by the same ratio (4.6):

$$p_M = kT_g(r)\cdot n_{M\ st}(r),\qquad(4.7)$$

where p_M is the partial vapour pressure of the working metal.

In real lasers operating with high excitation pulse repetition frequencies one should expect larger or smaller increase in the heterogeneity in the radial distribution of the prepulse concentration of atoms of the working metal, compared with the heterogeneity defined by (4.7).

4.2.3. The heterogeneity of prepulse electron concentration distribution

The prepulse distribution of the concentration of electrons n_{est} (r) is formed in the interpulse period of time due to diffusion to the wall of the GDT and bulk recombination of the electrons produced both during the excitation pulse and at the beginning of the interpulse period. A qualitative understanding of the nature and magnitude of the radial heterogeneity of the plasma parameters can be formed as in [83] by comparing the characteristic times of bulk processes and energy and particle transfer processes. The characteristic time τ_a of ambipolar diffusion of electrons on the GDT wall

$$\tau_a \sim \frac{L_n^2}{D_a}, \tag{4.8}$$

where L_n is the characteristic size of the plasma (the non-uniform distribution of n_e). For a cylindrical GDT and the diffusion distribution n_e (r) $L_n = r_p / 2.4$.

The characteristic time of the annihilation of the electrons as a result of three-body recombination is equal to

$$\tau_r \sim \left(n_e^2 \beta_e \right)^{-1}. \tag{4.9}$$

In the case of the predominance of the diffusion annihilation of electrons $(\tau_a \ll \tau_p)$ the prepulse distribution n_{est} will be close to the diffusion one [84, 85]. At the dominant role of volume recombination $(\tau_a \gg \tau_p)$ the distribution n_{est} (r) is almost flat in the axial zone of the GDT and falls off sharply at the walls of the GDT [84]. The characteristic length L_n at which there is a sharp decrease in the concentration of electrons due to ambipolar diffusion can be estimated from the ratio

$$L_n \sim \left(\frac{D_a}{\beta_e n_e^2} \right)^{1/2}. \tag{4.10}$$

That is, if $\tau_a \gg \tau_p$ the distribution of the electron concentration can be assumed with a good degree of accuracy to be uniform at $r < r_p - L_n$.

Figure 4.2 shows the dependence on the radius of the GDT on the neon pressure, calculated from (4.8) and (4.9) for the case of equality of the characteristic times $\tau_r = \tau_a$. The calculation was performed for the values of $n_e = 5 \cdot 10^{13}$ cm^{-3} and $T_e = 3300$ K. These values n_e and T_e are the typical prepulse values for the self-heating copper vapour

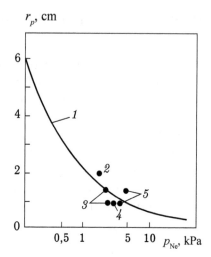

Figure 4.2. Dependence (1) of the GDT radius at which the conditions $\tau_a = \tau_r$ is fulfilled for the axial zone of GDT, on neon pressure. 2 – [50], 3 – [56], 4 – [66], 5 – [49].

lasers. The gas temperature is assumed to be constant over the cross section and equal to 2500 K. In the region of the GDT radii and neon pressures that lie above and below the dependence shown in Fig. 4.2, the axial zone of the GDT is characterized by the dominance of either the bulk recombination of electrons or their diffusion to the wall of the GDT.

In addition to this dependence, the dots in Fig. 4.2 denote the radii of the GDT and the working pressure of neon of several copper vapour lasers mentioned in section 4.1, which produced relatively high output characteristics of lasers. It is seen that for these lasers the p_{Ne} and r_p values lie mainly in the area where the rate of annihilation of the electrons due to diffusion and volume recombination are comparable to each other. In this case, the radius of the zone of homogeneous distribution $n_{est}(r)$ can be estimated as $\Lambda_n \approx r_p - L_n$. For the considered lasers Λ_n does not exceed 0.65 r_p and, therefore, all these lasers operate in the conditions of the heterogeneous prepulse distribution of the electron concentration. It should be noted that the decrease in prepulse concentration on the axis of the GDT compared with the value used in evaluating τ_r will lead to an increase in the role of diffusion annihilation of electrons and, consequently, the growth of the non-uniformity of the prepulse electron distribution.

4.2.4. The heterogeneity of prepulse electron temperature distribution and concentration of metastable atoms

The heterogeneity of the distribution of electrons and heavy particles leads to both non-uniform heating on the electrons in the GDT cross section in the three-body recombination and to their non-uniform cooling in collisions with heavy particles.

The characteristic time of recombination of heating of electrons is equal

$$\tau_{er} \sim T_e \left(I_M n_e^2 \beta_e \right)^{-1}, \qquad (4.11)$$

where I_M is the ionization potential of the atom of the working metal.

The characteristic time of electron cooling τ_{eg} in elastic collisions with buffer gas atoms and metal ions is equal to τ_{ge} and is defined by (4.1).

The heterogeneity of the prepulse distribution of electron temperature T_{est} (r) in the GDT radius, formed as a result of non-uniform heating and cooling of electrons, is compensated by the electronic thermal conductivity, facilitating alignment of the electron temperature in the cross section of the GDT. The characteristic time of spatial relaxation (alignment) of the electron temperature in the GDT can be estimated from the ratio

$$\tau_{eT} \sim \frac{L_T^2}{\chi_e}, \quad \chi_e \approx D_e, \qquad (4.12)$$

where L_T is the characteristic size of the plasma (region of the non-uniform distribution of electron temperature), in the case of cylindrical geometry is equal to $L_T = r_p/2.4$; χ_e is the coefficient of electronic thermal diffusivity, D_e is the electron diffusion coefficient calculated for the inert gases in [86]. For given values of τ_{er} and τ_{eg} the formula (4.11) can be used to estimate the characteristic size Λ_T of equalization of the electron temperature. Provided $\Lambda_T \geq L_T$ the distribution T_e (r) can be assumed to be uniform, and at $\Lambda_T < L_T$ the distribution T_e (r) is inhomogeneous.

Figure 4.3 shows the results of calculation using the ratio (4.11) of the characteristic size of the equalization of the electron temperature Λ_T, corresponding to some characteristic time τ, equal to the minimum of the characteristic times τ_{er} and τ_{eg}. The calculation of these relationships was carried out for $T_e = 3300$ K, $T_g = 2500$ K and two values of n_e: $5 \cdot 10^{13}$ cm^{-3} and 10^{14} cm^{-3}. The same figure gives the radii of the GDT and the neon pressure for the same lasers as

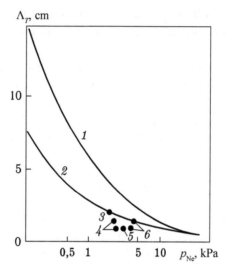

Figure 4.3. Dependence of the characteristic size of the equalization of the electron temperature of the neon pressure. $T_e = 3300$ K, $T_g = 2500$ K. 1) $n_e = 5 \cdot 10^{13}$ cm^{-3}, 2) $n_e = 10^{14}$ cm^{-3}, 3) [50], 4) [56], 5 [66], 6 [49].

in Fig. 4.2. It is seen that for these lasers we can be expect a fairly uniform prepulse distribution of T_{est} (r).

Thus, the above analysis shows that, depending on the radius of the GDT, the pressure and kind of the buffer gas, the pulse repetition frequency in the repetitively pulsed metal vapour lasers, different combinations of the distributions of plasma parameters can form prior to the next pulse: from substantially inhomogeneous distributions of plasma parameters to more or less homogeneous (when the characteristic times of all the above discussed processes are much smaller than the excitation pulse repetition period $T = f^{-1}$).

As the prepulse plasma parameters essentially determine the lasing energy and the laser pulse duration, we should expect a significant impact of the heterogeneity of the prepulse distributions of plasma parameters on the lasing characteristics and primarily on the efficiency of repetitively pulsed metal vapour lasers.

4.2.5. The heterogeneity of the distribution of plasma parameters during the excitation pulse

During the excitation pulse, the initial radial distributions of the plasma parameters evolve in time. Moreover, if the characteristic times of all the previously considered processes are much larger than the pulse duration τ_{ex}, then for any given value of the current radius r the plasma parameters (concentration of metal atoms in the main, metastable and resonance states, electron concentration) can be calculated independently. Estimates show that for copper vapour lasers the characteristic times of all the previously considered processes of energy transfer and particles are much longer than τ_{ex}, except for the characteristic time of equalization of the electron temperature τ_{eT} (4.11), which at small radii of GDT and low neon pressures can be not only less than τ_{ex} but also less than the characteristic cooling time of electrons in inelastic collisions with heavy particles τ_{eH}. For the values of secondary electron energies $\bar{\varepsilon}$ comparable with the excitation energies of the resonance levels, the characteristics time τ_{eH} can be defined as

$$\tau_{eH} \sim T_e\left(E_r\alpha_g n_M\right)^{-1}, \qquad (4.13)$$

where α_g is the total constant of the rate of failure of the ground state of the working metal by the electrons, E_r is the the excitation energy of the resonant level of the ground state.

Figure 4.4 shows the dependence on the length of equalization of the electron temperature Λ_T during the excitation pulse, corresponding to the condition $\tau_{eH} = \tau_{eT}$ with the parameters typical of a copper vapour laser during the excitation pulse: $n_M = 10^{15}\ 10^{-8}\ \text{cm}^3$, $T_e = 4\ \text{eV}$, $\alpha_g \approx 8\cdot10^{-8}\ \text{cm}^3\ \text{s}^{-1}$. It is seen that for r_p and r_{Ne} typical of the existing copper vapour lasers ($r_p \geq 0.5\ \text{cm}$, $p_{Ne} > 2\ \text{kPa}$), the characteristic length of equalization Λ_T of T_e during the excitation pulse is smaller and even much smaller than the radius of the GDT used in [49, 56, 66].

This fact means that during the excitation pulse in the conditions typical of copper vapour lasers, the electron concentration and temperature in each small section of the GDT evolves independently

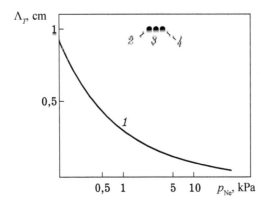

Figure 4.4. The dependence of the characteristic size of the equalization of the electron temperature during the excitation pulse on neon pressure (1). 2) [56]; 3) [66]; 4 [49].

of n_e and T_e in other areas. In practice, in most cases it determines the inhomogeneous distribution of electron concentration and temperature, as well as the lasing intensity in the cross section of the GDT.

4.3. The gas temperature in repetitively pulsed copper vapour lasers

The gas temperature in the repetitively pulsed copper vapour lasers excited by longitudinal discharge was measured using different methods in a number of studies [81,87–94] and in a transverse discharge in [33, 37]. However, before proceeding to the presentation of measurement results, let us consider the results of certain estimates and analytical solutions to the problem of heating of gas in the repetitively pulsed lasers at self-contained transitions of metal atoms.

Assessment of maximum possible fluctuations in the gas temperature in the conditions [81,87,88] was carried out in [87,88] based on the assumption that all the energy injected into the discharge for a single excitation pulse is spent on heating of the gas:

$$G_d = \frac{5}{2} k \Delta T_g n_{Ne} V_p, \tag{4.14}$$

where, besides the well-known notations, V_p is the volume of the GDT.

Calculations in [87, 88] using (4.13) for $T_g = 3000$ K at $p_{Ne} = 3.3$ kPa, $G_d = 0.13$ J gives $\Delta_g \approx 500$ K, and at $p_{Ne} = 13.3$ kPa and $G_d = 0.13$ J $\Delta_g \approx 130$ K. In fact, the real gas temperature fluctuations will be much lower because of the heat sink on the wall of the GDT and the finiteness of the duration of energy transfer from electrons to the gas. That is, the gas temperature in the self-heated metal vapour lasers can be characterized with a reasonable degree of accuracy by time-averaged value of temperature T_g which is not much different from both its minimum and maximum values.

Calculation of the temperature of the gas, taking into account all the factors leading to the inhomogeneous distribution of temperature in the cross section of the GDT in repetitively pulsed discharge conditions is a complex problem whose solution can be found only by numerical methods. However, given the smallness of the fluctuations of the gas temperature in the self-heated copper vapour laser it is expedient to solve the problem of heating the gas in repetitively pulsed lasers at self-contained transitions in the stationary approximation for different distributions of specific energy $\bar{G}(r)$ [W/cm³] in the radius of the GDT.

For the first time the problem of heating the gas in the pulsed-periodic lasers at self-contained transitions of metal atoms was solved in [95] under the assumption that the distribution of energy input in the GDT cross section is uniform and that only half the power input to the discharge is expended on heating the gas.

In [92,93], the problem of heating the gas was solved under the assumption that all the energy, injected into the discharge, is expended in heating the gas, and taking into account the fact that the radius of the discharge r_d may be less than the radius of GDT ($r_d \le r_p$). Models of the specific energy distributions $\bar{G}$ in the GDT cross section were represented by three distributions corresponding to the same level of heat input power $\bar{Q}_{dl}$ supplied to the discharge:

$$\text{homogeneous } \bar{G}_{\text{hom}} = \bar{Q}_{dl} / \pi r_d^2, \tag{4.15}$$

$$\text{parabolic } \bar{G}_{\text{par}} = \frac{2\bar{Q}_{dl}}{\pi r_d^2}\left(1-\frac{r^2}{r_d^2}\right), \tag{4.16}$$

$$\text{triangular } \bar{G}_{\text{tri}} = \frac{3\bar{Q}_{dl}}{\pi r_d^2}\left(1-\frac{r}{r_d}\right), \tag{4.17}$$

$$\bar{Q}_{dl} = 2\pi \int_0^{r_d} \bar{G}(r)\, r dr.$$

It was also assumed that the relationship between thermal conductivity λ_g and temperature of the gas T_g is defined by (4.3).

Solving the heat conductivity equation in the energy generation range ($0 \le r \le r_d$), the corresponding distributions of specific energy input (4.14)–(4.16) have the form:

$$T_g^{B+1}(r) = T_w^{B+1} + \frac{\bar{Q}_{dl}(B+1)}{2\pi A}\left[\frac{1}{2} + \ln\frac{r_p}{r_d} - \frac{r^2}{2r_d^2}\right], \tag{4.18}$$

$$T_g^{B+1}(r) = T_w^{B+1} + \frac{\bar{Q}_{dl}(B+1)}{2\pi A}\left[\frac{1}{2} + \ln\frac{r_p}{r_d} - \frac{r^2}{r_d^2} + \frac{r^4}{4r_d^4}\right], \tag{4.19}$$

$$T_g^{B+1}(r) = T_w^{B+1} + \frac{\bar{Q}_{dl}(B+1)}{2\pi A}\left[\frac{5}{6} + \ln\frac{r_p}{r_d} - \frac{6r^2}{4r_d^2} + \frac{2r^3}{3r_d^3}\right]. \tag{4.20}$$

For the region $r_d \le r \le r_p$ the solution for all three cases is the same:

$$T_g^{B+1}(r) = T_w^{B+1} + \frac{\bar{Q}_{dl}(B+1)}{2\pi A}\ln\frac{r_p}{r} . \tag{4.21}$$

The average gas temperature in the cross section of the GDT is given by:

$$\langle T_g \rangle = \frac{\displaystyle\int_0^{r_p} n_{Ne}(r)T_g(r)\,rdr}{\displaystyle\int_0^{r_p} n_{Ne}(r)\,rdr} = \frac{r_p^2}{2\displaystyle\int_0^{r_p}\left[r/T_g(r)\right]dr} . \tag{4.22}$$

In experiments [81,87,88] the gas temperature in the self-heating copper vapour laser was measured using an emitter. Its schematic representation is given in Fig. 2.1; the GDT length was 70 cm and the diameter 1.2 cm The excitation of the laser was carried out with the help of circuits shown in Figures 2.2 a (scheme I) and 2.2 b (scheme II). The buffer gas was neon at pressures from 3.3 to 13.3 kPa. Its temperature T_g (0) in the axial zone of the GDT was determined from the Doppler broadening of eight neon lines using the well-known method [96] by detecting the radiation emitted through the end of the GDT. The radiation was separated from the axial zone of the GDT by means of two diaphragms 1 mm in diameter.

To determine the point in time to which the measurements of gas temperature are related, a monochromator and a photomultiplier were used to study the temporal variation of the intensity of these neon lines. It was found that radiation pulses in these lines are either the same in duration as the duration of the current pulse, or have some afterglow. However, the corresponding analysis of the results of measurements of the temperature T_g (0) showed that the presence of the afterglow has no effect on these results and, therefore, the gas temperature measured from the Doppler broadening of the neon lines corresponds to the excitation pulse.

The results of measurements [81,87,88] of the gas temperature in the axial zone of the GDT of the self-heating copper vapour laser are shown in Fig. 4.5. Comparison of these results shows that when the laser is excited by circuit I the gas temperature $T_g(0)$ is significantly higher than when it was excited by circuit II. A marked difference in gas temperatures is associated with a greater heterogeneity of the discharge at its initiation through the circuit I where high currents of the storage capacitor flow through the discharge. This is shown by the special investigation [78] of the intensity distribution of discharge glow in the GDT cross section in the interpulse interval in the laser

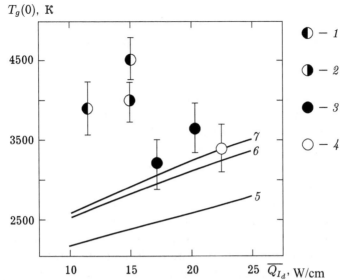

Figure 4.5. Dependence of gas temperature on the axis of GDT on the level of heat input to the discharge l_p = 70 cm, d_p = 12 mm. Scheme I: 1) the pressure of neon p_{Ne} = 13.3 kPa, f = 7.2 kHz; 2) p_{Ne} = 3.3 kPa, f = 7 kHz. Scheme II: 3) p_{Ne} = 13.3 kPa, f = 7 kHz; 4) p_{Ne} = 3.3 kPa, f = 6.7 kHz. 5) calculation in [94], T_w = 1800 K, 6) calculation by (4.19), 7) by (4.20), T_w = 1880 K, r_d = r_p.

excited by scheme I, carried out under the same conditions as those in which the gas temperature was measured. According to the results [78], immediately after the excitation pulse the discharge occupies the entire cross section of the GDT. Its diameter then gradually decreases and the discharge is a distinct luminous channel, separated from the GDT walls by a dark space. Over time, the diameter of the luminous channel continues to decline and before the next excitation pulse is equal to 0.4–0.5 of the GDT diameter. It should be noted that the results of [78] agree well with the results of earlier work [76].

In order to answer the question to what extent the measured gas temperatures $T_g(0)$ correspond to the minimum prepulse temperature $T_{gst}(0)$, in [87, 88] the relation (4.1) was used to estimate the characteristic time τ_{ge} of gas heating in collisions with electrons. In this case, the effective frequencies of elastic collisions of the electron with the neon atoms v_{ea} and with copper ions v_{ei}, respectively, were determined according to the date presented in [86] and [97]. The calculation using (4.1) showed that for T_g = 3000 K, the maximum neon pressure of 13.3 kPa for the experimental conditions and the concentration of copper ions $5 \cdot 10^{14}$ cm^{-3}, the characteristic time τ_{ge} is around 3 μs and significantly exceeds the duration of the excitation pulse. This means that the gas temperature values shown in Fig. 4.5 are close to their prepulse, i.e. minimum values.

The results of calculation of the gas temperature $T_g(0)$ in the axial zone of the GDT [81, 88], conducted using (4.19) and (4.20), assuming that the discharge occupies the entire section of the GDT, are shown in Fig. 4.5. Also given are the results of the calculation of $T_g(0)$ carried out in [81, 88] using the formula [95], obtained under the assumption of uniform distribution of specific energy $\bar{G}(r)$ along the radius of the GDT and the independence of the thermal conductivity of neon λ_{Ne} on its temperature. In the calculations in [81, 88] it was assumed that λ_{Ne} = $2 \cdot 10^{-3}$ W/cm. However, unlike in [94], in calculating the temperature $T_g(0)$ in [81, 87] it was assumed that all the energy injected into the discharge is expended in heating the gas.

The large difference between the results of measurements of $T_g(0)$ and calculations using the formula [95] is due to the fact that in the case of the self-heating metal vapour lasers the assumption of the uniform distribution of energy input $\bar{G}(r)$ in the GDT cross section is not realistic. It is known that the average electron energy and, consequently, the energy released in the discharge depend on the parameter E/N (N is the concentration of heavy particles), whose distribution over the GDT cross section is determined in this case by the distribution of the gas temperature $T_g(r)$. An inhomogeneous radial distribution of the parameter E/N actually determines the inhomogeneous distribution of energy input. Factors contributing to the equalization of energy input in the GDT cross section are the electronic thermal conductivity, which plays an important role in the GDT of small diameter and low buffer gas pressures, the skin effect, which manifests itself in the large diameter GDT [55], and the heterogeneity of the prepulse radial distribution $n_{Cu}(r)$ which has a minimum at the GDT axis.

A satisfactory agreement between the results of measurements of $T_g(0)$ in excitation of the laser by scheme II, which excludes the flow of charge current of the storage capacitor through the discharge and the results of calculation of $T_g(0)$ using (4.18) and (4.19) indicates a non-uniform distribution of specific energy $G(r)$ through the GDT section in repetitively pulsed metal vapour lasers.

Along with the works [81,87,88], measurements of gas temperature in the self-heating copper vapour laser on the basis of Doppler broadening of the neon line 540.0 nm were taken in [89]. The experiments in [89] were conducted using a GDT 1 cm in diameter and 70 cm long, pulse repetition frequency was 10 kHz, neon pressure 2 kPa. Neon radiation was recorded simultaneously throughout the whole section of the GDT so that the gas temperatures, measured in [89], are averaged over the GDT cross section in a certain way. At the power consumed from the rectifier of $Q_r = 2.5$ kW the GDT wall temperature was $T_w = 1800$ K, and the gas temperature $\langle T_g \rangle$ measured in [89] was 2500 K. Given that only about half the power Q_r is supplied to the discharge, the quoted temperature corresponds to the heat input capacity of $\overline{Q}_{dl} \approx 18$ W / cm.

The analysis [98] of the agreement of the results of measurements of gas temperature in the axial zone of the GDT $T_g(0)$ [81, 88] and the temperature averaged over the cross section of GDT [89] shows their good agreement, indicating that in the repetitively pulsed copper vapour laser with the GDT diameter of about 1 cm and a length of about 70 cm under typical conditions of operation the specific energy G distribution over the cross section of the GDT is highly non-uniform and that almost all the energy injected into the discharge is expended in heating the gas

The experiments [90,92,93] with a GDT 2 cm in diameter and 50 cm long were carried out using industrial emitter UL-101. The buffer gases were helium, neon and argon at a pressure of 5 to 50 kPa. The excitation pulse repetition frequency was 12.5 kHz. The average (in the volume) gas temperature of the GDT was measured using the method first proposed in [90]. The essence of this method consists in measuring the pressure change Δp in the active element when disconnecting the discharge and in converting this change to the average temperature of the gas in the GDT as follows:

$$\langle T_g \rangle = \frac{\left[T_w \left(1 + \Delta p / p \right) \right]}{\left[1 - \left(\Delta p / p \right) \cdot \left(n_\Sigma / n_p - 1 \right) \right]},$$ (4.23)

$$n_p = p / kT_w,$$

where, besides the well-known notations, n_Σ and n_p are respectively the total number of atoms of inert gas in the emitter and in the GDT with the discharge switched off, p is the buffer gas pressure in the emitter with the discharge switched off.

The value n_Σ was determined in [92,93] by supplying a known amount of the buffer gas in the active element. Δp was measured using manotrons with short-term (for about one second) switching off the discharge. Indications of the manotrons were stabilized during about 0.01 s which is caused, apparently, by the inertial properties of the manotrons whose intrinsic resonant frequency is 150-200 Hz. The temperature of the wall of the GDT T_w was determined as follows. After turning off the discharge the time dependence of T_w was measured with a pyrometer EOP-66 with the interval between successive measurements of

$\Delta t = 10$ s. The approximation this time dependence at the moment of switching off the discharge made it possible to determine the value of T_w in the stationary operating mode of the laser.

The error in calculating the absolute value of $\langle T_g \rangle$ by (4.22) is determined by the error of measurement of each of the quantities in this formula, and is estimated to approximately equal ± 6%. The relative measurement error of the mean gas temperature in the GDT volume is determined by the measurement error of Δp and is about ±3%.

The results of measurements of $\langle T_g \rangle$ by the method described above are shown in Fig. 4.6. As expected, the gas temperature increases with increasing atomic weight in the sequence He, Ne, Ar, which is

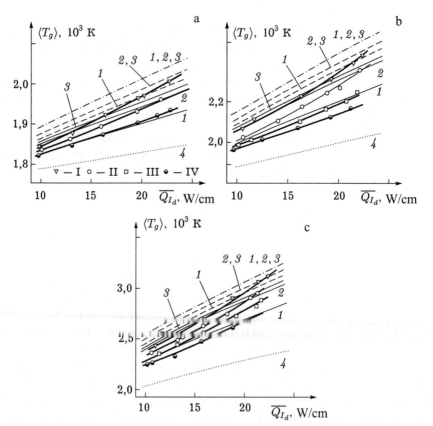

Figure 4.6. Dependence of the mean gas temperature in the GDT of a self-heating copper vapour laser on the level of heat input supplied to the discharge. a) helium, b) neon, c) argon. Experiment: I) inert gas pressure of 50 kPa; II) 30; III) 10; IV) 5. Calculation: solid curves – $r_d / r_p = 1$; dotted line – 0.7, dot-dash – 0.2. The radial distribution of the specific energy input: 1) homogeneous; 2) parabolic; 3) triangular; 4) the calculation for a homogeneous distribution of $\bar{G}$ (r), $r_d / r_p = 1$ under the assumption that the heating of the gas uses half the energy input to the discharge.

associated with a corresponding decrease in thermal conductivity in this sequence of the inert gases. For example, at a pressure of these gases of 50 kPa and at the heat input of $\overline{Q}_{dl} \approx 23$ W / cm the average gas temperature exceeds the temperature of the wall GDT in helium at 250, 600 in neon and argon at 1100 K. It is also worth noting that with increasing the buffer gas pressure, its temperature $\langle T_g \rangle$ increases markedly that shows the growth of non-uniformity of energy input.

Comparison of results of measurements of $\langle T_g \rangle$ and the calculation results obtained using (4.17) – (4.19), also shown in Fig. 4.5, shows that at neon pressures of 5 and 10 kPa measured temperature $\langle T_g \rangle$ is lower than that which would have occurred at the homogeneous distribution of the specific energy $\overline{G}(r)$ in the cross section of the GDT. According to the authors of [92,93] this fact testifies to the removal of energy from the discharge in the given experimental conditions.

Analysis [92,93] of the removal of energy by the charged particles and radiation from the discharge zone to the wall GDT and convective gas flows through the ends of the GDT for the buffer gas (neon) showed firstly that at the neon pressure of 5 kPa and the level of heat input power $\overline{Q}_{dl} = 10$ W/cm, the charged particles carry to the wall of the GDT about 1.3 W/cm, i.e. about 13% of the value $\overline{Q}_{dl}$. As the neon pressure and heat input $\overline{Q}_{dl}$ increase the power transferred by the charged particles to the GDT wall decreases. Secondly, at the levels of heat input power supplied to the discharge of 10 and 20 W/ cm the fraction of the power transferred by induced and spontaneous radiation respectively through the ends and the wall of GDT does not exceed 5%. And finally, thirdly, in [92,93] the transfer of energy from the discharge zone through the end of the GDT by forced convection of the gas [99] was taken into account. Confirmation of the existence of such convection are visually observable (due to the scattering of laser radiation) flows of small particles from the GDT in the direction of the output windows of the emitter. The corresponding estimate [92,93] shows that in the given experiment conditions approximately 30% of the power supplied to the discharge is removed form the GDT by gas convection.

Accounting for the fraction of energy carried out from the discharge by diffusion of charged particles and by convection leads to a shift of the experimental curves in Fig. 4.6 b to the right, so that the results of measurements of $\langle T_g \rangle$ at pressures of 5 and 10 kPa are combined with the results of calculations for the inhomogeneous energy input.

The radial distributions of the gas temperature $T_g(r)$ in the self-heating copper vapour laser ($l_p = 50$ cm, $d_p = 2$ cm) were measured in [91] using a thermal lens and these distributions were used to calculate

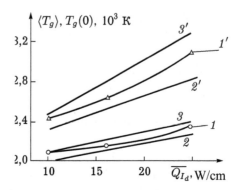

Figure 4.7. The results of measurements of gas temperature in the self-heating copper vapour laser [91]. 1) the gas temperature averaged over the cross section of the GDT $T_g\langle 0\rangle$, 1') the gas temperature $T_g(0)$ in the axial zone of GDT. 2, 2') the results of calculation $\langle T_g\rangle$ and $T_g(0)$ respectively for a uniform radial distribution $\bar{G}$ (r) for $r_d = r_p$; 3, 3') the results of the calculation of $\langle T_g\rangle$ and $T_g(0)$ for a parabolic distribution G (r) for $r_d = r_p$.

the corresponding values of the average temperatures $\langle T_g\rangle$. The results of [91] are shown in Fig. 4.7. This figure also shows the calculated gas temperature $T_g(0)$ in the axial zone of the GDT and the average temperature $\langle T_g\rangle$ for the homogeneous and parabolic distributions of energy input along the radius of the GDT. At first glance, the comparison of the calculated and experimental results indicates the heterogeneity of energy generation in the cross section of the GDT but this heterogeneity is small. However, if we consider that the experiments in [91] were performed with the same laser as in [92,93], then we should expect that in the conditions in [91] as well as in [92, 93], a significant fraction of energy input to the discharge was removed by the convective gas flows through the ends of the GDT. With this in mind, the experimental curves in Fig. 4.6 will shift to the right and will be more consistent with the results of calculations for the parabolic distribution of energy generation over the radius of the GDT.

Study [94] shows the radial distribution of the gas temperature $T_g(r)$ in the pulsed-periodic copper vapour laser ($d_p = 4.2$ cm, $l_p = 150$ cm, $f = 6.5$ kHz, $\bar{P}_{I\Sigma} = 40$ W). The distribution of $T_g(r)$ was calculated from the radial distribution of the copper atoms in the ground state measured by the hook method. At the wall temperature of the GDT of 1700 K the gas temperature on the axis of the GDT is about 2600 K. Unfortunately, [94] does not give any information about the power required from the rectifier or the power input to the discharge. However, if we proceed from the assumption that the practical efficiency of the laser was about 1% and that 50% of the power required from the rectifier was lost in the switch, the power input to the discharge in the conditions of [94] was about 2 kW and the unit power input $\overline{Q}_{dl} \approx 13$ W/cm. With a parabolic distribution of energy input this level of $\overline{Q}_{dl}$ corresponds to the gas temperature in the axial zone of the GDT of $T_g (0) \approx 2700$ K (see Fig. 4.6). The good agreement of this value with that measured in [94] indicates that the laser investigated in [94] has an inhomogeneous (near parabolic) radial distribution of gas temperature.

So, if we consider that for the GDT with $l_p/d_p \leq 25$ a large part of the energy supplied to the discharge shall be transferred through the ends of the GDT by convective flows of gas, all the results of measurements [81,87–94] of the gas temperature in the repetitively pulsed lasers copper vapour indicate an inhomogeneous distribution of energy over the cross section of the GDT. In this case, as indicated by the results of calculations [92,93] presented at Fig. 4.8, for GDTs with small diameters and corresponding levels of power input of $\overline{Q}_{dl} \approx 15$ W/cm at different radial distributions of specific energy inputs $\overline{G}(r)$ we obtain fairly similar values of the gas temperature $T_g (0)$ in the axial zone of the GDT. However, with increasing diameter of the GDT and the corresponding increase of $\overline{Q}_{dl}$ to 25 and 50 W/cm the difference between the values of the temperature $T_g(0)$, corresponding to uniform and triangular distributions of specific energy, significantly

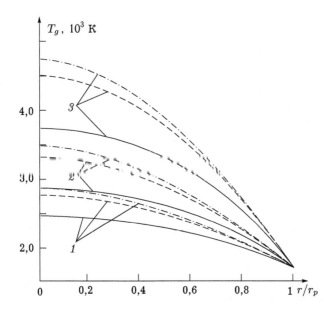

Figure 4.8. The temperature distribution in the self-heating of gas lasers pairs of copper for the neon buffer gas [92, 93]. The distribution of energy input at $r_d = r_p$: solid curves – uniform, dotted line – parabolic; dash-dotted line – triangular. Power per unit length, introduced into the discharge $\overline{Q}_{dl}$: 1 – 15 W/cm; 2 – 25; 3 – 50.

increases and reaches for the specified values of $\overline{Q}_{dl}$ respectively 600 and 1000 K. This means that in the metal vapour lasers at values $\overline{Q}_{dl}$ of 25–50 W/cm the heterogeneity of the energy input to the discharge is a factor that leads to a substantial increase in the gas temperature in the axial zone of the GDT, that is to an increase of the population of lower laser levels, and as a consequence to the deterioration of the lasing characteristics.

It should be noted that at present the problem of overheating of gas in the active medium of copper vapour lasers with longitudinal discharge is apparently recognized. This is evidenced, for example, in the work [100], which discusses the possibility of reducing the gas temperature in the GDTs of larger diameters due to the use of longitudinal heat-conducting walls.

In concluding the measurements of gas temperature T_g in the repetitively pulsed copper vapour lasers, we discuss briefly the results of measurements of T_g in a copper vapour laser excited by transverse discharge [33, 37]. In [33] the integral absorption in the resonance line of the copper atom was used to measure the concentration distribution of copper atoms in the cross section of a GDT 6 × 6 cm in size (length of the GDT is not indicated) and determine the temperature profile in the gas discharge. The excitation pulse repetition frequency in the conditions of [33] was 2.8 kHz. According to the data [33], at the power supplied to the discharge of 90 W/cm the gas temperature at the axis of the GDT was 2770 K, and the coefficient of conversion of electrical energy into heat (the heating of heavy particles) was about 30%. In [33] it is assumed that the rest of the energy input to the discharge appears to be re-emitted and transferred to the walls of the GDT by radiant heat.

When comparing the results of [33] with the measurements of gas temperature in a copper vapour laser with longitudinal discharge it should be noted that the temperatures T_g (0) ≈ 2800 K are realized in terms of [91] for values of Q_{dl} ≈ 20 W / cm (see Fig. 4.7) 4.5 times smaller than in [33]. This difference is difficult to explain like the assumption of the role of radiation in the removal of 70% of the energy input to the discharge to the GDT wall. Recall that the estimates made in [92, 93] showed that in the copper vapour laser with longitudinal discharge with a GDT with a diameter of 2 cm and 50 cm long radiation transfers to the GDT wall less than 5% of the energy input to the discharge.

Results similar to [33] were obtained in [37] when measuring the gas temperature at the axis of the GDT with a cross section of 5 × 5 cm and a length of 40 cm. The temperature was measured using two methods: spectral (Doppler width of the emission lines of the neon buffer gas) and interferometric (the shift of the bands in the Mach–Zehnder interferometer in moments of sudden switching on and off the discharge). Both methods gave values of $T_g(0)$ close to 2200 K at a power input to the discharge of 1.1 kW, i.e $Q_{dl} = 27.5$ W/cm. Figure 4.7 shows that for a laser with longitudinal discharge [91] such a value of $T_g(0)$, if obtained, is recorded only for values of $Q_{dl} < 10$ W/cm. Some of the reasons that could explain the low values of $T_g(0)$, measured in [37], are discussed in [97]. One reason is the configuration of the discharge which leads to the fact that the specific energy release in the relatively cold near-electrode regions of the plasma should exceed the specific energy release at the GDT axis. In addition, it should be noted that the above results of the measurements the temperature of gas in the repetitively pulsed copper vapour laser indicate a clear downward trend in these temperatures with decreasing ratio lp/dp, indirectly indicating the growing importance of convective transport of energy from the GDT and GDC through their ends with a decrease in their length and increasing the transverse dimensions. Therefore, it is likely that the low values of gas temperatures measured in [33,37] must first of all be due to the transfer of energy from the discharge through the ends of the GSC, with the ratio of the length to its transverse dimension of the cross section in [37] is only 8. It should be noted that in addition to the experimental results, the study [37] presented the results of calculation of the non-steady temperature fields in the discharge gap of the rectangular cross section, conducted on the basis of the numerical solution of the non-steady two-dimensional heat equation.

4.4. The results of measurements of concentrations of metastable and resonance-excited atoms of working metal in lasers at self-contained transitions of metal atoms

There are quite a number of publications containing the results of measurements of concentrations of resonantly excited n_r and metastable n_m copper atoms in a copper vapour laser [55,92,100–127]. In addition, there is a relatively small number of works describing the measurements of the concentration n_m in the interpulse interval of the bismuth vapour [128–130], lead vapour [131], barium vapour [132] and strontium vapour [133] lasers. As regards the methods of measuring all of these works can be divided into three groups. The first group includes the papers (see, e.g. [102–104]), in which the concentrations of resonance and metastable atoms of copper were measured by the complicated Rozhdestvensky hook method [134.135]. This method is used primarily to measure the concentrations of copper atoms at different levels during the excitation pulse. Fairly widely used is the method of resonant absorption of radiation by the same laser used for the first time apparently in [128]. This method was used to measure the concentrations of metastable atoms in the interpulse interval. Finally, the third method is the method of resonant absorption of radiation from a tunable dye laser. This method was first applied in [101]. Its features in the measurement of the concentration of copper atoms in the ground state and at the excited levels in the copper vapour laser are discussed in [111,120]. It seems that with this method the most interesting results were obtained when measuring the concentration of copper atoms in the metastable and resonance levels during the excitation pulse in the self-heating copper vapour laser with a GDT 2 cm in diameter and a length of the active zone of 40 cm. The excitation pulse repetition frequency in this laser was 10 kHz, the neon buffer gas pressure 40 kPa. The discharge switch was a generator lamp GMI-29, working in the mode of incomplete discharge of the storage capacitor. The power required from the rectifier was fixed and was about 2 kW. With this power, the wall temperature T_w of the GDT passed during heating the entire temperature range lasing of the laser with respect to the copper vapour density. The maximum average lasing power on both lines was up to 6 watts (practical efficiency 0.3%) at T_w = 1570–1590°C.

Without describing all the measurements, we consider only those which visibly identify the features of the evolution of the concentration of resonance n_r and metastable n_m atoms during the excitation pulse and the interpulse period. Figure 4.9, taken from [122], shows, together with measurements of concentrations of copper atoms at different levels during the excitation pulse, the results of measurements of the discharge current pulses I_d and voltage U_d and the calculated (from these voltage pulses) voltage in the active resistance of the discharge capacity U_R and power Q_{in} injected into the discharge. Figure 4.9 shows clearly that the prepulse concentration of metastable atoms exceeds the concentration of resonant atoms and that the equalization of the concentrations n_r and n_m, corresponding to cessation of the laser pulse, occurs at the trailing edge of the voltage pulse U_R.

Changes of the radial distributions of concentrations of the resonance and metastable atoms in a copper vapour laser, occurring as a result of the excitation pulse, can be evaluated by the distributions shown in Fig. 4.10 [122, 123]. When searching for causes of these changes, it should be taken into account that they are based on changes of the radial distributions of temperature and electron concentration and the

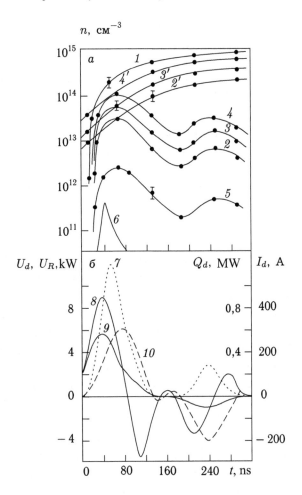

Figure 4.9.   The results of measurements of concentrations of copper atoms in different levels during the excitation pulse (a) and the discharge current pulses and voltage on the discharge (b) in the repetitively pulsed copper vapour laser [122]. d_p = 2 cm, l_a = 40 cm, p_{Ne} = 40 kPa, f = 10 kHz, Q_{in} = 2 kW, a) the ground state $\Delta n_g \equiv n_{gst} - n_g$ (*t*) (1); level $^2P_{3/2}$ (2 4), level $^2D_{5/2}$ (2'–4'), level of $^2D_{1/2}$ with an excitation energy of 44544 cm^{-1} (5) at T_w = 1500°C (2, 2', 5) 1590 (1, 3, 3') 1625 (4, 4'); 6) temporal behaviour amplified probe radiation at T_w = 1590°C. b: 7) Q_d, 8) U_d, 9) U_R, 10) I_d, T_w = 1590°C.

concentration of copper atoms in the ground state taking place during the excitation pulse. With greater or lesser degree of certainty we can assert that the presence of a maximum in the radial distribution of resonance atoms at T_w > 1700°C indicates a greater degree of efficiency of excitation and ionization of copper atoms in the axial zone of the GDT than in its wall regions. The minimum in these distributions at T_w ≤ 1525°C indicates a significant shortage of copper atoms on the GDT axis. Finally, the maximum in the radial prepulse distributions of the metastable atoms is most likely associated with the heterogeneity of the prepulse electron temperature distribution, which has a maximum on the GDT axis.

Joint analysis of measurement results presented in Figs. 4.9 and 4.10 reveals that at T_w = 1590 K the concentration n_r in the first maximum (Fig. 4.9) reaches approximately $n_{r\,max}$ of (5–6) $\cdot 10^{13}$ cm^{-3}, which is about (2.5–3)% of the prepulse concentration of copper atoms n_{gst}, which according to [114] is approximately equal to $2 \cdot 10^{15}$ cm^{-3}. Given these values $n_{r\,max}$ and the time dependence n_m (t), shown in Fig. 4.9, we can estimate the gain factor in the line center from the relation:

$$\kappa_0 = \sigma_{p_0}\left(n_r - \frac{g_r}{g_m}n_m\right). \qquad (4.24)$$

Substituting into (4.23) $n_r \approx$ (5–6)$\cdot 10^{13}$ cm^{-3}, $n_m \approx 4.6 \cdot 10^{13}$ cm^{-2}, $g_r =$ 4, g_m = 6, $\sigma_{p0} \approx 1.8 \cdot 10^{-14}$ cm^2, we obtain $\kappa_0 \approx 0.34$–0.52 cm^{-1}, which corresponds to the value $\kappa_0 l_a \approx 13.6$–20.8.

After the end of the excitation pulse, the plasma parameters and with them the populations of metastable levels relax to their prepulse values. Figure 4.11 shows the results of measurements of the concentration of metastable atoms n_m at the level $^2D_{5/2}$ in the interpulse interval of time [92, 107]. The experiments were performed with a copper vapour laser (d_p = 2 cm, l_a = 40 cm, f = 8 kHz) at a low GDT wall temperature T_w = 1673 K, which ensures the possibility of measuring the concentration n_m immediately after the excitation pulse. The qualitative feature of the dependence $n_m = f(t)$, presented at Fig. 4.11 is the non-monotonic form of three of them. It is seen that at neon pressures greater than 10 kPa the concentration of metastable atoms after the excitation pulse decreases rapidly (faster with increasing pressure), then within 5–10 µs increases, followed by a monotonic decrease before the next excitation pulse.

Figure 4.12 illustrates the effect of the type of buffer gas on the nature of the relaxation of the concentration of metastable atoms in a copper vapour laser [92,116]. As noted in these studies, the qualitative features of the relaxation of the concentration of metastable atoms in mixtures of copper vapour with He, Ne, Ar are similar. At a low buffer gas pressure a drop in the concentration n_m with time has a monotonic character. At buffer gas pressures greater than 10 kPa the form of the dependence $n_m = f(t)$ is as a rule non-monotonic. After an initial rapid decline, which lasts in helium less than 1 µs, 1.2 µs in neon and 2–3 µs in argon, the concentration of metastable atoms increases reaching the maximum at 0.8 µs for helium, 5–6 µs for neon and 14–20 µs for argon.

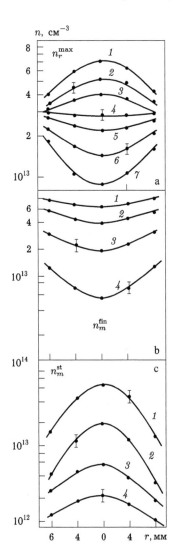

Figure 4.10. The radial concentration distribution of resonance and metastable atoms in a repetitively pulsed copper vapour laser [122]. The laser parameters are same as for Fig. 4.9. a) maximum concentration of the atoms n_r in the excitation pulse at the level $^2P^0_{3/2}$: 1) $T_w = 1600°$C; 2) 1585; 3) 1570, 4) 1550, 5) 1525; 6) 1500, 7) 1465. b) postpulse (after the end of the excitation pulse), the concentration atoms at the level $^2D_{5/2}$: 1) $T_w = 1620°$C; 2) 1590; 3) 1550; 4) 1470. c) prepulse concentration of atoms at the level $^2D_{5/2}$: 1) $T_w = 1630°$C; 2) 1550; 3) 1540, 4) 1485.

Figure 4.13 shows the results of measurements of the concentrations of metastable levels in interpulse interval in a copper vapour laser (d_p = 2 cm, l_a = 40 cm, f = 10 kHz) [112]. The dependences $n_m = f(t)$ in this figure have the form typical for all the lasers at self-contained transitions of metal atoms, and are usually described by two characteristic times of fast τ_1 and slow τ_2 population relaxation of the metastable level, corresponding to a rapid (at the beginning of the interpulse interval) and a slow decrease in the concentration of metastable atoms. Attention is drawn (Fig. 4.13 d) to the result showing that the characteristic times τ_1 and τ_2 depend on the buffer gas

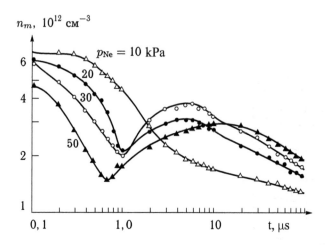

Figure 4.11. The dependence of the concentration of metastable atoms (level $^2D_{5/2}$) on the axis of GDT in the copper vapour laser on time at various neon pressures (d_p = 2 cm, l_a = 40 cm, f = 8 kHz, U_r = 4.5 kV, C_S = 3.3 nF).

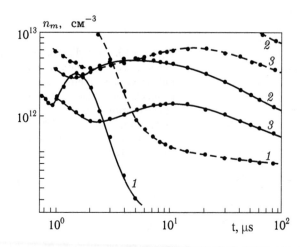

Figure 4.12. Time dependence of the concentration of metastable atoms in the interpulse interval in the copper vapour laser (d_p = 2 cm, l_a = 40 cm, f = 8 kHz, p = 50 kPa). Solid lines – the level $^2D_{3/2}$, dashed line – $^2D_{5/2}$. 1) helium buffer gas, T_w = 1803 K, Q_r = 2.09 kW, 2) neon, T_w = 1838 K, Q_r = 1.82 kW, 3) argon, T_w = 1793 K, Q_r = 1.63 kW.

pressure, and τ_1 increases with decreasing pressure and τ_2 decreases. In addition, attention is drawn to (Fig. 4.13 b) that the value of τ_1 for the GDT axis is greater than τ_1 for the GDT wall, and the corresponding values of τ_2 have an inverse relationship.

The dependences $n_m = f(t)$ for the axial and wall regions of the GDT showed an inhomogeneous distribution of the concentration of

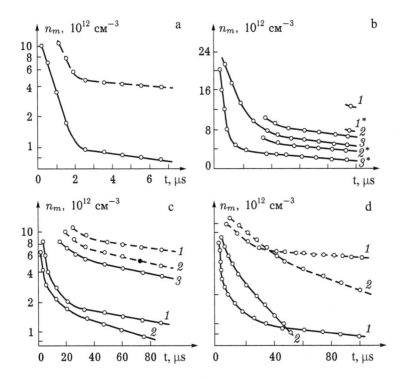

Figure 4.13. Time dependence of the concentration of metastable atoms in the interpulse interval on the axis and the wall (*) of the GDT of a copper vapour laser (d_p = 2 cm, l_a = 40 cm, f = 10 kHz). Solid line – the level $^2D_{3/2}$, dotted line – the level $^2D_{5/2}$, the neon pressure of 40 kPa, in Figure d curves 2 relate to neon pressure 4 kPa. Q_r = 1.3 (a), 2.2 (b: 1), 2.1 (b: 2), 1.97 (b: 3), 1.9 (c: 1.1) 1.65 (c: 2.2) 1.93 (c: 3.3), 1.7 (d: 1) and 1.5 kW (d: 2); T_w = 1450 (a), 1714 (b: 1), 1675 (b: 2), 1600 (b: 3 and c:3.3); 1520 (c: 1.1); 1470 (c: 2.2 and d: 2) and 1500°C (d: 1).

metastable atoms in the interpulse interval. Study [112] presents the results of measurements of the distribution $n_m(r)$ in the copper vapour laser in the interpulse interval at a pressure p_{Ne} = 40 kPa and different levels of power input to the discharge. A visual representation of the distribution $n_m(r)$, their evolution over time and impact on the evolution of the buffer gas pressure are given by the results of measurements taken from [92,116] and presented at Fig. 4.14.

Before proceeding to discuss the processes that determine the relaxation of the concentration of metastable atoms in the interpulse interval in the copper vapour laser (Figs. 4.11–4.14), we discuss the question of to what extent the concentration of metastable atoms n_m are in equilibrium with the electron temperature T_e. In [92], the answer to this question is given on the assumption that the equilibrium between n_m and T_e occurs in the case where the characteristic time of

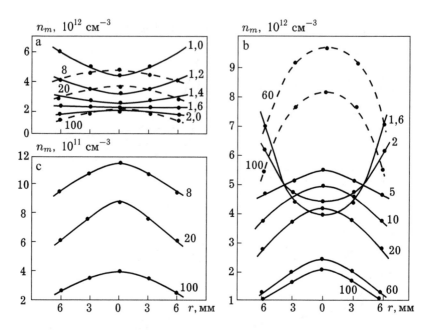

Figure 4.14. The radial distribution of the concentration of metastable atoms in the interpulse interval in the copper vapour laser (d_p = 2 cm, l_a = 40 cm, f = 8 kHz, C_S = 3.3 nF). Solid lines – the level $^2D_{3/2}$; dashed lines – $^2D_{5/2}$. a: p_{Ne} = 5 kPa, Q_r = 2.03 kW, Q_{dl} = 19 W/cm, T_w = 1753 K; b: p_{Ne} = 50 kPa, Q_r = 1.82 kW, Q_{dl} = 28 W/cm, T_w = 1838 K.

de-excitation of the metastable levels by the electrons is $\tau_m^e \approx (n_e \alpha_{mg})$, where α_{mg} is the quenching rate constant of the metastable level to the ground state, less than the characteristic time τ_m of changes in the concentration of metastable atoms. Estimates made in [92] using the approximation formula for α_{mg} from [136] showed that in all laser modes in [92,116] the condition $\tau_m^e \leq \tau_m$ is satisfied for the entire prepulse interval. In [92] it is concluded that in the interpulse interval of time the population of metastable levels at different times and at different points in the GDT cross section is given by the Boltzmann relationship, that is,

$$n_m(r,t) = \frac{g_m}{g_g} n_g(r,t) \exp\left[-\Delta E_{gm} / T_e(t)\right]. \tag{4.25}$$

A more rigorous estimate of the deviation of concentration of metastable atoms from its equilibrium value can be made taking into account the fact that the transitions between the energy levels corresponding to the excited states of the outer *s*-electron (unbiased system of levels) and *d*-electron (biased system of levels) is unlikely as compared with transitions between the levels of the same system (see, e.g. [137]), and the recombination flux of electrons J_e can be assumed to be equal to the sum of the non-interacting recombination fluxes J_{e1} and J_{e2} for unbiased and biased systems levels

$$J_e = J_{e1} + J_{e2} = \beta_e n_e^2 n_{i1} + \beta_e n_e^2 n_{i2}, \tag{4.26}$$

$$n_e = n_{i1} + n_{i2},$$

where, besides the well-known notations, n_{i1} and n_{i2} are the concentrations of copper ions in the $3d^1_{10}S_0$ state and in the states $3d^9 4s^3 D_{3,\,2,1}$ and $3d^9 4s^1 D_2$, are related by the Boltzmann equation.

Under the assumption that in each of the resonant levels ($^2P^0_{1/2}$ and $^2P^0_{3/2}$) is half of the recombination flux J_{e1}, and through each of the metastable ($^2D_{3/2}$ and $^2D_{5/2}$) – half of J_{e2}, we can get the relationship to calculate the concentration of metastable atoms at the levels $^2D_{3/2}$ and $^2D_{5/2}$ for given values of n_g, n_e and T_e:

$$n_m = \frac{\left(\dfrac{\beta_e n_e n_{i2}}{2} + n_g \alpha_{gm}\right)\left(\alpha_{rg} + \alpha_{rm} + \dfrac{A_{rm}}{n_e}\right)}{\left(\alpha_{rg} + \alpha_{rm} + \dfrac{A_{rm}}{n_e}\right)\left(\alpha_{mg} + \alpha_{mr}\right) - \alpha_{mr}\left(\alpha_{rm} + \dfrac{A_{rm}}{n_e}\right)} +$$

$$+ \frac{\left(\alpha_{rm} + \dfrac{A_{rm}}{n_e}\right)\left(n_g \alpha_{gr} + \dfrac{\beta_e n_e n_{i1}}{2}\right)}{\left(\alpha_{rg} + \alpha_{rm} + \dfrac{A_{rm}}{n_e}\right)\left(\alpha_{mg} + \alpha_{mr}\right) - \alpha_{mr}\left(\alpha_{rm} + \dfrac{A_{rm}}{n_e}\right)}. \tag{4.27}$$

Estimates using (4.27) of the deviations of the concentrations of metastable atoms at these levels from their values equilibrium with T_e, were obtained for two sets of plasma parameters – postpulse ($n_e = 5 \cdot 10^{14}$ cm^{-3}, $n_g = 10^{14}$ cm^{-3}, $T_e = 0.5$ eV) and prepulse ($n_e = 3 \cdot 10^{13}$ cm^{-2}, $n_g = 10^{15}$ cm, $T_e = 0.3$ eV) showed that in both cases the deviation is less than one percent.

It should be noted that in evaluation the energies of the above states of copper ions were taken from [138]. In addition, the evaluation used the values of α_{gr} and α_{gm} calculated using the data on excitation cross sections of resonance and metastable levels in the near-threshold energy regions listed in [139]. The rate constants for excitation in laser transitions were calculated by the semi-empirical formula of Van Regemorter [140]. The oscillator strengths for transitions $^2D_{3/2} \rightarrow {}^2P^0_{1/2}$ and $^2D_{5/2} \rightarrow {}^2P^0_{3/2}$ required for the calculations were taken from [141].

Thus, the various estimates show that under typical operating conditions of repetitively pulsed copper vapour lasers the population of metastable levels in the interpulse interval is in equilibrium with the electron temperature. Having established this fact, we turn to the analysis of the experimental results, based on the physical model of relaxation of the plasma parameters formulated in [83]. According to this model, after completion of the excitation pulse ionization of copper atoms is due to the stored kinetic energy of the electrons. The electron temperature T_e during the time

$$\tau_i \sim T_e \left(n_g \alpha_i I_{Cu} \right)^{-1} \tag{4.28}$$

falls to the ionization temperature T_i, corresponding to equilibrium between ionization and recombination. It should be noted that if during time τ_i the processes of radial transport of energy and particles do not manage to advance to any significant extent, then after this specified time there will be established some radial distribution Ti (r) defining in terms of the Saha equation [142] the local correlation between the concentrations of electrons and copper atoms in the ground state. If these processes are essentially manifested during time τ_i, the equilibrium between ionization and recombination is to be understood in the integral form in the GDT section – recombination in the centre of the GDT is balanced by the ionization near its walls.

Then, for some characteristic time τ_{cool} electrons lose energy in elastic collisions with heavy particles and by diffusion to the wall of the GDT and cool down to a temperature at which these energy losses are balanced by recombination heating. This is followed by the establishment of the quasi-stationary regime of slow cooling of electrons determined the rate of decrease n_e with time. In this case, due to the exponential dependence the ionization rate constants α_i on T_e become small. The characteristic time τ_{cool} in cylindrical geometry is given by

$$\tau_{cool} \sim \left[\tau_{eg}^{-1} + \left(\tau_q l_e \right)^{-1} \left(v + \Delta v \right) \right]^{-1}, \tag{4.29}$$

$$v = T_e \ln \left(r_p / 2.4 \lambda \right), \quad \lambda \approx D_a \sqrt{M_i / T_g} ,$$

$$\Delta v = \frac{1}{2} T_e \ln \left(M_i T_e / m T_g \right)^{1/2}.$$

Here, the value $\tau_{eg} = \tau_{ge}$ and is defined by (4.1); Δv and v determine the energy loss of the diffusing electron in overcoming the potential drop in the boundary layer and in work against the forces of the ambipolar field in the plasma; λ is the thickness of the wall layer.

We estimate the characteristic time τ_i for copper vapour lasers, assuming that at the end of the excitation pulse $T_e \approx 1.5$ eV, $n_{Cu} \approx 3 \cdot 10^{14}$ cm^{-3}. Given the importance of the ionization potential of the copper atom $I_{Cu} \approx 7.76$ eV and using as constant α_i the total rate constants of excitation of resonance levels $\alpha_{gr\Sigma} \approx 2 \cdot 10^{-8}$ cm^3 s^{-1}, we obtain from (4.27) $\tau_i \approx 30$ ns. Even if we assume that this value τ_i is significantly understated, it still turns out that in terms of the operation of repetitively pulsed copper vapour lasers, the electron temperature decreases to a value of T_i in a time not exceeding a few hundred nanoseconds.

As will be shown in the following section, in a copper vapour laser with prepulse concentrations $n_{Cu} \approx 10^{15}$ cm^{-3} after the excitation pulse the degree of ionization of copper atoms δ_i reaches about 40% or more. The corresponding calculations by the Saha equation shows that the value of $\delta_i \approx 10\%$ corresponds to the $T_i \approx 0.4$ eV; $\delta_i \approx 50\%$ – Ti ≈ 0.5; $\delta_i \approx 90\%$ – Ti ≈ 0.6 eV and $\delta_i \approx 98\%$ – Ti ≈ 0.7 eV.

Evaluations have shown that under typical conditions of pulsed-periodic copper vapour lasers the electron temperature decreases in no more than hundreds of nanoseconds to a temperature determined by the degree of ionization of copper atoms and lying in the range from about 0.5 to 0.7 eV.

Knowing the electron temperature at the beginning of the interpulse interval, we estimate the characteristic time τ_{cool} for experimental conditions, the results of which are shown in Fig. 4.11. Given the cooling of electrons only in collisions with neon atoms, $\tau_{cool} = \tau_{eg}$, assuming $T_e = 0.5$ eV, $T_g = 2500$ K and defining v_{ea} in accordance with [86], we find that for $p_{Ne} = 10$ kPa $\tau_{eg} \approx 6$ μs; for to 20 kPa 3 μs; for 30 kPa 2 μs and for 50 kPa 1.2 μs. It is seen that these values of τ_{eg} are in good agreement with the time of the rapid decrease in the concentration of metastable atoms at various neon pressures. This means that the initial rapid cooling in the concentration of metastable atoms in the first place due to the relaxation of the electron temperature from T_i to the values of T_e set in quasi-stationary electron cooling. It should be noted that the relationship of the rapid relaxation of the populations of metastable levels in lasers based on self-contained transitions of metal atoms with the rapid cooling of electrons was apparently first stated in [133].

Given that the electron temperature decreases with time, the growth of the concentration of metastable atoms, in Fig. 4.11 at neon pressures of 20, 30 and 50 kPa can be caused only by increasing concentration of copper atoms in the axial zone of the GDT as a result of three-body recombination of electrons. Note that the strong effect of the recovery of the concentrations of atoms of the working metal in the axial zone of the GDT on the nature of the dependence of the concentration of metastable atoms on time was shown in [92,107] for a copper vapour laser and in [143] for a bismuth vapour laser.

The increase the characteristic time τ_{eg} in the sequence He, Ne, Ar explains the shift observed in Fig. 4.12 into the region of long minimum times in the corresponding dependence $n_m = f(t)$ for the level $^2D_{3/2}$. This once again confirms that the nature of fast relaxation process n_m defines a set of electron cooling processes and recovery of the working atoms in the ground state, which can occur in various ways not only in different modes of operation of lasers, but also in the same mode but in different zones of the section of the GDT. Confirmation of what was said earlier is a marked difference in the characteristic times of fast relaxation n_m in the axial and wall areas of the GDT (see Fig. 4.13 b). This difference is easily explained by the fact that the axial zone of the GDT is characterized by a more significant ionization of copper atoms and, consequently, the high rate of their recovery as a result of triple recombination of electrons reflected in the increase of the characteristic time τ_1 of fast relaxation in the axial zone of the GDT.

After discussing the reasons determining the special features of the relaxation of concentrations of the metastable atoms in the interpulse interval in a copper vapour laser, we consider the reasons for the heterogeneity of prepulse distribution n_m (r) (see Fig. 4.14). Assessment of the feasibility of an inhomogeneous prepulse distribution n_m (r), observed in Fig. 4.14 b, was carried out in [92,116] by comparing the characteristic times of the equalization of the electron temperature along the radius of the GDT τ_{eT} (4.11) and recombination electron heating

$$\tau_{er} \sim T_e \left(I_{Cu} \beta_e n_e^2 \right)^{-1} \tag{4.30}$$

and cooling of the electrons in collisions with heavy particles τ_{eg} (4.1). The evaluation was conducted for two sets of parameters close to those realized in the conditions under which the results of measurements, presented in Fig. 4.14, were obtained:

1) $p_{Ne} = 50$ kPa, $T_e = 3500$ K, $n_e = 5 \cdot 10^{13}$ cm^{-3}, $\langle T_g \rangle = 2600$ K;
2) $p_{Ne} = 5$ kPa, $T_e = 2500$ K, $n_e = 1 \cdot 10^{13}$ cm^{-3}, $\langle T_g \rangle = 2200$ K.

According to the estimates, at $p_{Ne} = 50$ kPa $\tau_{eT} \approx 2.5 \cdot 10^{-6}$ s, $\tau_{er} \approx 1.3 \cdot 10^{-5}$ s and $\tau_{eg} \approx 1.6 \cdot 10^{-6}$ s ($\tau_{eg} < \tau_{eT} < \tau_{er}$). These values of the characteristic times imply that the electrons are cooled and that the electronic thermal conductivity can not compensate fully the loss of energy in collisions with heavy particles. That is at the neon pressure of 50 kPa the prepulse distribution T_e (r) and the corresponding distribution n_m (r) must be heterogeneous, as is observed experimentally (see Fig. 4.14 b). At a neon pressure of 5 kPa $\tau_{eT} \approx$ $3 \cdot 10^{-7}$ s, $\tau_{er} \approx 5.2 \cdot 10^{-5}$ s and $\tau_{eg} \approx 1.5 \cdot 10^{-5}$ s $(\tau_{eT} < \tau_{eg} < \tau_{er})$, so the prepulse distribution T_e (r) should be more or less homogeneous, as is observed experimentally (see Fig. 4.14 a). Since the start of the interpulse interval is realized at significantly higher concentrations of electrons, the heterogeneity in the distribution of $T(r)$ can also take place at a neon pressure of 5 kPa.

In concluding the qualitative interpretation of the measurement results of the concentration of resonance and metastable atoms in a copper vapour laser, it should be noted that at present there is a self-consistent numerical model of a copper vapour laser [124,144], which takes into account not only the volume processes in the plasma of a repetitively pulsed gas discharge but also the processes of radial transport of particles. The kinetic part of the model takes into account 11 atomic and ionic levels of the copper atom, 4 levels of neon, 70 collisional and radiative processes. The results of calculations according to this model of the time evolution of the concentration of resonance and metastable atoms are in good agreement with experimental results.

4.5. The concentration and temperature of electrons in self-contained lasers

4.5.1. The concentration and temperature of the electrons during the excitation pulse

For the first time, the electron concentration in the self-heating copper vapour laser during the excitation pulse was defined in [88,145] on the basis of the resistance of the discharge. Figure 4.15 shows the results of calculation of the average electron concentration in the cross section of the GDT $\langle n_e \rangle$ [88] during the excitation pulse in the copper vapour laser with a GDT diameter of 1.2 cm and a length of 70 cm, calculated using the formula

$$\langle n_e \rangle = \frac{\langle \sigma \rangle m v_{ea}}{e^2}, \quad \langle \sigma \rangle = \frac{l_p}{\pi r_p^2 R_d}, \quad (4.31)$$

where v_{ea} is the effective frequency of collisions with the neon atoms, calculated according to [146] for a Maxwellian electron energy distribution. The calculation of $\langle n_e \rangle$ was performed in the following order. From the measured discharge current pulses and the voltage across the GDT electrodes we determined the voltage on the

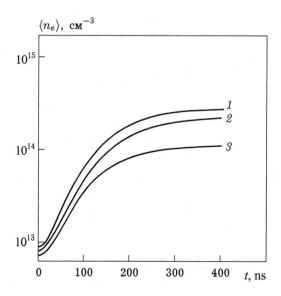

Figure 4.15. The time variation of the electron concentration during the excitation pulse in the self-heating copper vapour laser. Scheme of excitation – in Fig. 2.2 a, l_p = 70 cm, d_p = 1.2 cm, Q_r = 1.8 kW, $f \approx$ 7 kHz. p_{Ne}: 1 – 13.3 kPa, 2 – 6.7, 3 – 3.3.

active resistance of the discharge U_R, according to which, under the assumption that the average gas temperature in the GDT cross section is equal to $\langle T_g \rangle$ = 3000 K, we calculated the parameter E/p_0, where p_0 is the pressure of neon reduced to normal conditions. According to this parameter and the ratio n_{Cu}/n_{Ne}, in which the concentration of copper atoms was assumed to be 10^{15} cm^{-3}, using data from [147] calculations were carried out of the average energy $\bar{\varepsilon}$ and the corresponding electron temperature $T = 2\bar{\varepsilon}/3$. Then using (4.20) the concentration $\langle n_e \rangle$ was determined. The calculation results, presented in Fig. 4.15, shows that during the excitation pulse, the average electron concentration in the GDT cross section increases by at least an order of magnitude.

Calculation of the discharge current pulse, performed in [88] on the basis of the measured voltage pulse at the GDT electrodes, yielded the results that agree well with experiment, and showed that the main role in the ionization of the active medium of copper vapour lasers is played by their ionization through resonant levels $^2P^0_{3/2}$ and $^2P^0_{1/2}$. At not too high pressures of the buffer gas neon ($p_{Ne} \leq$ 20 kPa) for a correct description of the kinetics of ionization it is sufficient to consider the ionization of copper atoms in these levels and the direct ionization.

Apparently, the only paper in which results of measurements of electron temperature in the copper vapour laser are presented is

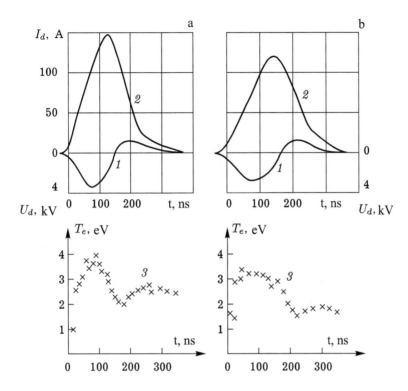

Figure 4.16. The time dependence of voltage (1), current (2) and electron temperature (3) in the discharge of a copper vapour laser (l_p = 50 cm, d_p = 2 cm, p_{Ne} = 3.3 kPa, C_S = 2.2 nF, $Q_r \approx$ 1 kW). a: U_r = 4.2 kV, f = 8 kHz, $\bar{P}_{l\Sigma}$ = 3 W. b: U_r = 3.6 kV, f = 12 kHz, $\bar{P}_{l\Sigma}$ = 2 W.

[148]. The measurements were performed with a laser with a GDT diameter of 2 cm and a GDT length of 50 cm (f = 8–12 kHz, $Q_r \approx$ 1 kW, p_{Ne} = 3.3 kPa, the average output power reached 3.5 W). The electron temperature T_e was determined from the distribution of copper atoms in the excited levels lying near or above the ionization limit. Populations of these levels were obtained by measuring the relative and absolute intensities of spectral lines of the copper atom. Methods of measurements and experimental setup are described in [149,150].

The results of measurements [148] are presented in Fig. 4.16. As noted in [148], the lasing pulse coincided with the peak current. Its duration at half maximum was 15 ns, and at the base more than 50 ns. It is noteworthy that the dependence $T_e = f\,(t)$ in Fig. 4.16 is almost identical with the type of dependence $n_m = f(t)$ in Fig. 4.9. This agreement once again indicates that the electron temperature in a pulsed discharge in lasers based on self-contained transitions of metal atoms tracks the electric field in the discharge.

This conclusion is consistent with the results of theoretical work [151] in which in the approximation of an infinitely powerful flow at the excitation threshold E_{ex} and assuming that the electron concentration is constant in time $n_e(t) = n_e 0$ and that all electrons available at the moment when the electric field is applied have the same energy

$$n_e(\varepsilon, \, t=0) = n_{e0}\delta(\varepsilon - \varepsilon_0), \quad \varepsilon_0 < E_{ex},$$

the authors obtained a relation for the characteristic time of excitation τ_{ex} determining the rate of excitation in the steady state ($t \rightarrow \infty$),

$$\tau_{ex} = \frac{E_{ex}m(v_a^2 + \omega^2)}{e^2 E^2 v_a}. \tag{4.32}$$

Here v_a is the frequency of pulse transfer in the scattering of electrons on atoms, ω is the frequency of the electric field.

The characteristic time delay of excitation $\tau_{ex.d}$ is introduced in [151] by the relation determining the dependence of the concentration of excited atoms on the time

$$n_{ex}(t) = \frac{n_{e0}}{\tau_{ex}}(t - \tau_{ex.d}).$$

According to [151], the maximum delay time, corresponding to the condition $\varepsilon_0 \ll E$, is 0.29 τ_{ex}.

In the same study, in the approximation of instantaneous ionization of excited atoms the authors obtained relations for the characteristic time of ionization τ_i

$$\tau_i = \frac{1.26 E_{ex}m(v_a^2 + \omega^2)}{e^2 E^2 v_a} \tag{4.33}$$

and the delay time of ionization $\tau_{i.d}$

$$\tau_{i.d} = \tau_i \ln\left(1.24\frac{2.18r}{sh(2.18r)}\right), \tag{4.34}$$

$$r = (\varepsilon_0 / E_{ex})^{1/2}$$

that determines the development of ionization by the relation

$$n_e(t) = n_{e0}\exp\left[(t - \tau_{i.d})/\tau_i\right].$$

According to [151] maximum $\tau_{i.d}$ is 0.21 τ_{ex}.

The estimate obtained from (4.32) for the experimental conditions, whose results are shown in Fig. 4.16 a, gives the value of the characteristic time $\tau_{ex} \approx 4$ ns and the corresponding values $\tau_{ex.d} \approx 1.2$ ns and $\tau_{i.d} \approx 0.8$. These values of the characteristic times, confirm that in the repetitively pulsed copper vapour lasers the electron temperature during the excitation pulse tracks the electric field strength and T_e can be used by the results of calculation of stationary EEDF[1].

In this connection, it should be noted that the results of [148] not only give an idea about the level of electron temperatures realized in the typical conditions of copper vapour lasers, but also to confirm the results of calculations of the stationary EEDF in mixtures of copper vapour with neon [147]. The estimate obtained in [147] of the electron temperature at the time corresponding to the maximum discharge current (Fig. 4.16 a), yields a value of $T_e \approx 3$ eV, in good agreement with measurements.

Along with the electron temperature, in [148] the distribution of copper atoms on closely spaced levels with the energy gap $\Delta E \sim 0.01$ eV and the excitation energy of 9 eV were used to measure the gas temperature $\langle T_g \rangle$ averaged over the cross section of the GDT equal to 2400 ± 150 K.

4.5.2. Concentration and temperature of electrons in the interpulse interval in lasers with a GDT of small diameters ($d_p \leq 2$ cm)

The electron concentration, realized in the interpulse interval of time, in discharges of self-heating copper vapour lasers with GDTs of small diameters was measured in [88,145,152–154]. In experiments [88,145,153] with self-heating copper vapour lasers with GDT 70 cm long and with a diameter of 1.2 cm, excited by the circuit shown in Fig. 2.2 a, measurements of electron concentration in the axial zone of GDT $n_e(0)$ were carried out by the known method [155] with a two beam interferometer (at a wavelength of CO_2 laser of 10.6 μm) with the temporal and spatial resolution of 2 μs and 2–3 mm.

According to the results of measurements of the electron concentration in the axial zone of the GDT, given in Fig. 4.17, in the early interpulse interval there is a strong decrease of the electron concentration $n_e(0)$ in the axial zone of the GDT. The rate of this decrease depends on the type and pressure of buffer gas. Estimates of the characteristic cooling time τ_{eg} of the electrons in collisions with buffer gas atoms, similar to those in the previous section, show that when the neon pressure is varied from 13.3 kPa to 1.8 kPa τ_{eg} varies

---

[1]This conclusion does not apply to the full extent to the case of reduction of the strength of the electric field in the discharge – Editor's note.

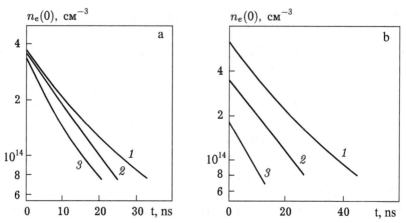

Figure 4.17. Time dependence of the electron concentration on the axis of HRT in the interpulse interval in the copper vapour laser (lp = 70 cm, dp = 1, 2 cm, f ≈ 7 kHz, Qr ≈ 1, 8 kW). a: 1 - argon, ð = 2, 9 kPa, 2 - neon, 3.3 kPa, 3 - helium, 3.3 kPa. b: neon: 1 - ð = 13 3 kPa, 2 - 3.3 3 - 1.8 kPa.

from about 5 to 33 μs. In the sequence of buffer gases He, Ne, Ar at a pressure of about 3 kPa the characteristic time τ_{eg} increases from 5 μs to 50 μs. The measured values of τ_{eg} show the extent to which dependences $n_e(0) = f(t)$, shown in Fig. 4.17, correspond to the time interval of rapid cooling of the electrons, and the extent to which they correspond to the time interval of the quasi-stationary slow cooling.

In addition to [88,145,153], the time dependences of the electron concentration in the interpulse period of self-heating copper vapour laser were measured in [152] with GDTs 40 cm long and 1 cm in diameter (p_{Ne} = 3.3–26.7 kPa; f = 7–12 kHz; Q_r = 420–1400 W) and in [154] with GDTs 30 cm long and 0.7 cm in diameter (p_{Ne} = 2.6 kPa, Q_r = 1760 W, the frequency f is not specified). In [152], the electron concentration, averaged over the cross section of the GDT, was measured from the Stark broadening of hydrogen lines H_β, present in the discharge as an impurity. Overall, the results [152] are consistent with the results of [88,145,153]. The main difference is the presence of increasing concentrations of electrons at the beginning of the interpulse interval. At low neon pressures this increase is very large, 2–2.5 times. However, in later work [154], in which electron concentration was calculated from the measured electron temperature and the absolute populations of highly excited atoms of copper, the presence of the growth of the electron concentration in the early interpulse interval has not been confirmed. In [154] in the first few microseconds, the electron concentration is practically constant, and then for about 15 μs is reduced by five times.

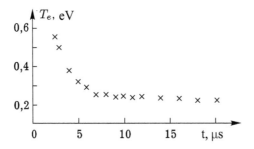

Figure 4.18. Time variation of electron temperature in the interpulse interval in the copper vapour laser (l_p = 50 cm, d_p = 2 cm, p_{Ne} = 3.3 kPa, f = 8 kHz, $Q_r \approx$ 1 kW, P_{IZ} = 3.5 W).

The electron temperature, averaged over the cross section of the GDT and realized in the interpulse interval in the copper vapour laser, was measured in [156] in the conditions matching the conditions of measurement of T_e during the excitation pulse [148]. In [156], by measuring the absolute intensities of spectral lines, the authors obtained the distribution of copper atoms in the excited states in time, and these distributions were used to determine the electron temperature in the interpulse interval. The results of measurements of T_e, presented in Fig. 4.18, are generally consistent with the previously discussed model of the relaxation of the plasma parameters in the interpulse interval. According to the results of measurements in the early interpulse interval, the electron temperature is about 0.55 eV, which according to earlier estimates of the ionization temperature shows an almost complete ionization of copper atoms during the excitation pulse. The duration of the rapid decrease in the electron temperature is about 7 µs and about two and a half times smaller than the characteristic cooling time of electrons in collisions with buffer gas atoms, which seems to indicate a significant role in cooling of the electrons by their collisions with copper ions or diffusion transport of the energy to the GDT wall. The rapid decrease in the electron temperature is over when T_e is close to the gas temperature $\langle T_g \rangle$, measured in [148] and equal to about 2400 K. The results of measurements of T_e in the interpulse interval of time in the self-heating copper vapour laser (l_p = 30 cm, d_p = 7 mm, p_{Ne} = 2.6 kPa) are generally consistent with the results shown in Fig. 4.18, and differ from them only by the higher rate of reduction of T_e in the first few microseconds.

In [88,153], the electron temperature T_e (0) in the axial zone of the GDT was calculated by analogy with [157] from the measured dependence $n_e(0) = f(t)$ (see Fig. 4.17) using the equation

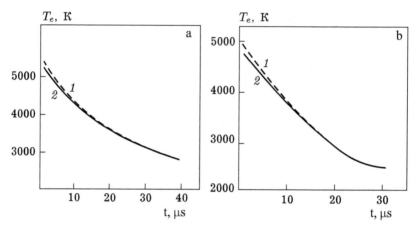

Figure 4.19. Time dependence of the electron temperature in the interpulse period of time in self-heating copper vapour laser. a – neon, $p = 13.3$ kPa, b – neon, $p = 3.3$ kPa. Conditions are the same as for Fig. 4.17.

$$\frac{dn_e(0)}{dt} = -\beta_e n_e^3(0) + \alpha_i n_g(0) n_e(0) - \frac{6D_a}{r_p^2} n_e(0), \tag{4.35}$$

where D_a – the ambipolar diffusion coefficient, calculated taking into account data on the mobility of ions of various metals in inert gases [158], α_i and β_i – coefficients of stepwise ionization and three-body recombination, calculated from [159].

The results of calculation of $T_e(0)$ to (4.33) are presented in Fig. 4.19. The calculations assumed that the average gas temperature in the cross section of the GDT is 3000 K. In addition, to assess the effect of the concentration of copper atoms in the ground state n_{Cu} on the accuracy of reconstruction of the time dependences of the electron temperature $T_e(0)$, the calculation of these curves was performed for two values of n_{Cu}: 10^{15} cm^{-3} and 0. As can be seen, the electron temperature $T_e(0)$ is weakly dependent on the concentration n_{Cu} and, therefore, the ambiguity in the meaning of n_{Cu} which occurs when restoring $T_e(0)$, can not have any significant impact on the accuracy of the reconstruction.

In analyzing the results presented in Fig. 4.19, attention is drawn to the fact that in the early interpulse interval the electron temperature is 5000 K, indicating a high degree of ionization of copper atoms. The rate of decrease in T_e is much smaller than in [156] (cf. Fig. 4.19 b and Fig. 4.18). It is possible that this difference is due to storage capacitor charge currents flowing in [153] during the interpulse interval of time after discharge. Finally, in [153], the electron temperature, as in [156], relaxes to the gas temperature average in the cross section of the GDT.

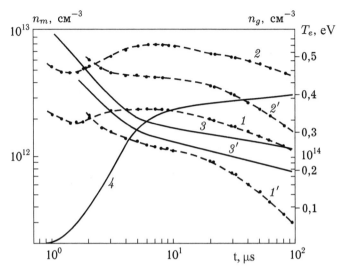

Figure 4.20. The results of measurements of the concentration of metastable atoms (1, 1 '– the level $^2D_{3/2}$, 2, 2' – the level $^2D_{5/2}$) in the interpulse interval at the GDT axis of a copper vapour laser (d_p = 2 cm; l_p = 50 cm; $f \approx 8$ kHz) and the results of calculations from these concentrations of electron temperature (3, 3') and the concentration of copper atoms in the ground state (4). 1, 2, 3, 4 – p_{Ne} = 50 kPa, T_w = 1793 K, $\bar{Q}_{dl}$ = 23 5 W/cm. 1', 2', 3 '– p_{Ne} = 10 kPa, T_w = 1753 K, $\bar{Q}_{dl}$ = 19 W/cm.

In [92, 116] to a self-heating copper vapour laser, the electron temperature in the interpulse time range was calculated from the measured concentrations of metastable atoms at the levels $^2D_{3/2}$ and $^2D_{5/2}$ under the assumption that the populations of these levels are in equilibrium with T_e. The results of measurements of the time dependence of the concentration of metastable atoms in the axial zone of GDT and the derived time dependence of the electron temperature and the concentration of copper atoms in the ground state are shown in Fig. 4.20. We see that the dependences $T_e = f(t)$ have all previously established features within 5–8 μs they fall off from $T_e \approx 0.5$ eV to temperatures comparable with the gas temperature, and then slowly decrease during the interpulse interval. In all this interval, the electron temperature at a pressure of 50 kPa of neon is higher than at a pressure of 10 kPa. When p_{Ne} = 10 kPa ($\bar{Q}dl$ = 19 W/cm) T_e strives to a temperature 2230 K, while the average temperature of the gas $\langle T_g \rangle$, determined from the data presented in Fig. 4.6 b, is about 2200 K. When p_{Ne} = 50 kPa ($\bar{Q}dl$ = 28 W/cm) T_e tends to 2900 K, while the value $\langle T_g \rangle$, defined by extrapolating the corresponding dependence on Fig. 4.6 b is about 2600 K. That is, in the latter case, the temperature T_e relaxes not to the gas temperature average in the cross section of the GDT, as at low pressures, but to a

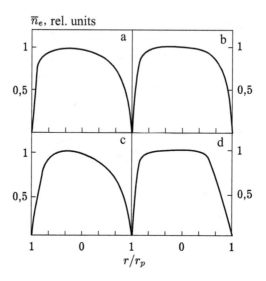

Figure 4.21. The distribution of electron concentration along the radius of HRT in the first 10-20 ms interpulse interval for Ne, 13.3 (a) and 3.3 kPa (b); Ar, 3.3 kPa (a) No, 3.3 kPa (g). dp = 1, 2 cm; lp = 70 cm; f ≈ 7 kHz; Qr ≈ 1, 8 kW.

higher temperature. This indicates the heterogeneity of the prepulse distribution $T_e (r)$, which is in good agreement with the estimate made in the previous section and indicating the heterogeneity of the distribution at p_{Ne} = 50 kPa.

In order to assess the nature of the radial distributions of temperature and concentration of electrons in the interpulse interval of time in self-heating copper vapour lasers, in [88, 153] a standard method [160] was used to measure the time-averaged post-pulse intensity of the photorecombination continuum at $^2P^0_{3/2}$ and $^2P^0_{1/2}$ levels of copper atoms in the wavelength range from 260 to 320 nm for various values of the current radius of the GDT. The results of these measurements using the technique [161] were used to calculate the time-averaged distribution of the concentration $n_e(r)$ and electron temperature $T_e(r)$ along the radius of GDT in the interpulse interval.

Figure 4.21 shows the distributions of the relative electron concentration $n_e(r)/ n_e(0)$ along the radius of the GDT, calculated from the intensity distribution of the recombination continuum at a fixed frequency v_{rad}. The corresponding analysis in [88, 153], taking into account the dependences of the electron concentration in the axial zone of GDT on time (Fig. 4.17), showed that the measured intensity of the recombination continuum and the electron temperature and concentration are averaged over time for the first 10–20 μs after the end of the excitation pulse. Therefore, the observed shape of the

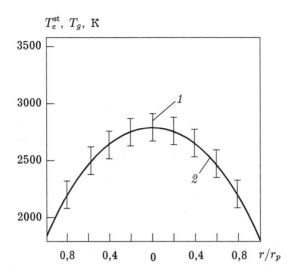

Figure 4.22. Prepulse radial distribution of electron temperature in a copper vapour laser (1) and radial distribution of gas temperature (2) calculated in the approximation of uniform energy generation in the cross section of the GDT, l_a = 40 cm; d_p = 2 cm, f = 10 kHz, p_{Ne} = 40 kPa; T_w = 1530°C; Q_r = 2 kW.

distribution $n_e(r)$ is connected, apparently, to the fact that for these 10–20 μs the distributions $n_e(r)$ due to more rapid recombination of electrons in the axial zone of the GDT.

Measurements of the time-averaged distributions of the electron temperature $\overline{T}_e(r)$ corresponding to the distributions $\overline{n}_e(r)$, shown in Fig. 4.21 for different values of the current radius of GDT for decreasing intensity of the recombination continuum, showed that in all the buffer gases the temperature $\overline{T}_e(r)$ is constant in the cross section of GDT and equal to: for p_{Ne} = 13.3 kPa $\overline{T}_e(r)$ = 4250 ± 300 K; for p_{Ne} = 33 kPa $\overline{T}_e(r)$ = 4900 ± 350 K; for p_{He} = 3.3 kPa $\overline{T}_e(r)$ = 4200 ± 200 K; for p_{Ar} = 2.9 kPa $\overline{T}_e(r)$ = 4300 ± 350 K These data indicate that in the conditions of [88, 153] (d_p = 1.2 cm; p ≤ 13.3 kPa) in spite of the heterogeneity of heating of electrons in the three-body recombination and cooling in collisions with heavy particles, the electron thermal conductivity equalizes the electron temperature across the cross section of the GDT . Comparison for the buffer gas of neon of the above values of $\overline{T}_e$ with the time dependences of the electron temperature in the axial zone of the GDT (see Fig. 4.19) shows that there is satisfactory agreement between the average values of $T_e(0)$ for the first 10–20 μm reduced from the dependence $n_e(0) = f(t)$ and measured from the decay of the recombination continuum.

The heterogeneity in the distribution of $T_e(r)$ is clearly manifested in the self-heating copper vapour laser with a GDT 2 cm in diameter at pressures of buffer gas neon $p_{Ne} \approx 50$ kPa [92,115,116]. In [92, 116], this heterogeneity is found by calculating the prepulse distribution $T_e(r)$ from the Boltzmann ratio based on measured radial distributions of the populations of the levels $^2D_{3/2}$ and $^2D_{5/2}$ atoms of copper. In [115] the distribution of $T_e(r)$ is also calculated from the Boltzmann ratio, but calculations are based on measurements of the populations of the metastable level $^2D_{5/2}$ and the ground state of copper, which gives more accurate values of T_e. The results of calculation of $T_e(r)$ [115], presented in Fig. 4.22, firstly, show significant heterogeneity of the prepulse distribution $T_e(r)$ in the GDT of the self-heating copper vapour laser with a neon pressure of 50 kPa. Second, they are remarkably the same as those presented in the same figure obtained in the calculated distribution of the gas temperature $T_g(r)$, performed in [115] provided that $T_w = 1530°C$ and $\bar{Q}_{dl} = 25$ W / cm.

Summarizing the results of the above measurements of electron temperature in the interpulse interval of time in self-heating copper vapour lasers, it can be concluded: at the values of the parameter $p_{Ne} \cdot d_p^2 \leq 20$ kPa $\cdot$ cm^2 the electron temperature distribution is uniform over the cross section of the GDT and during the interpulse interval T_e relaxes to the average gas temperature in the cross-sectional of the GDT which in typical operating conditions of these lasers is about 2500 K or less and only weakly dependent on the degree of heterogeneity of the radial distribution of specific energy. This means that at the indicated values of the parameter $p_{Ne} \cdot d_p^2$ the prepulse population of the metastable levels of the copper atom can not be a factor limiting the repetition frequency f of excitation pulses. The main reason for restriction of f in this case is the increase in the prepulse electron concentration, leading to a deterioration the of coordination of the discharge with the discharge circuit. This conclusion confirms the results of [162, 163] on the effect of prepulse electron concentration on the characteristics of repetitively pulsed copper vapour lasers and other metals. However, when $p_{Ne} \cdot d_p^2 \geq 100$ kPa $\cdot$ cm^2 the radial inhomogeneity in the distribution of $T_g(r)$ is manifested in the distribution of $T_e(r)$, which can lead to an unacceptable prepulse population of the metastable levels and, consequently, to a significant decline in both the lasing power and efficiency of the lasers. These negative effects should be clearly manifested in lasers with GDT diameters of about 4 cm or more.

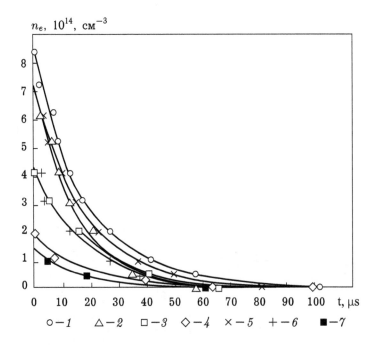

Figure 4.23. Time dependence the time of the electron concentration for different values of the current radius r: 0) 1 cm; 2) –0.5; 3) –1.0, 4) –1.5; 5) –0.5; 6) –1.0; 7) –1.5. Discharge parameters: Q_r = 2.9 kW; U_r = 5 kV; f = 2.5 kHz, p_{Ne} = 13.3 kPa.

4.5.3. Concentration and temperature of electrons in large-diameter GDTs

In order to determine the relaxation characteristics of the plasma in repetitively pulsed metal-vapour lasers with large GDT diameters in [70,164,165] studies were made of repetitively pulsed discharges in mixtures of vapours of bismuth and lead with helium and neon, excited in a GDT with a diameter of 5 cm and a length of 100 cm using the circuit shown in Fig. 3.10. The diameter of the electrodes located coaxially with the GDT, was 4 cm. Specific metals Bi and Pb in these papers were selected for the following reasons. First, the lead vapour laser is a typical representative of the lasers with self-contained transitions metal atoms. Second, the concentration of working atoms at the level of 10^{15} cm^{-3}, characteristic of self-heating copper vapour lasers, is obtained in lead and bismuth vapours at temperatures of the GDT wall of 1000° C. Third, the excitation energy of the resonance levels E_r and of ionization I_M of Bi and Pb atoms are close to the E_r and I_M of the

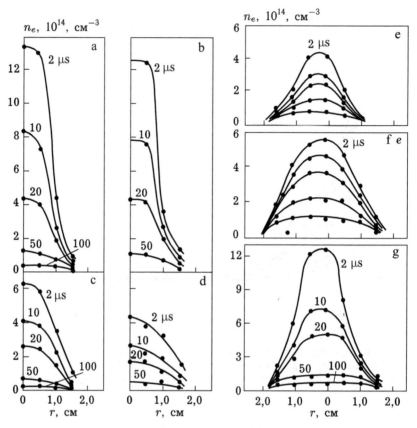

Figure 4.24. The distribution of electron concentration along the radius of GDT at different times after the excitation pulse in the mixtures of Bi and Pb vapours with helium and neon. The numbers on the curves – the time interval after the excitation pulse. a: Bi–He, Q_r = 2.9 kW, U_r = 5 kV, p_{He} = 6.7 kPa, f = 3.0 kHz, b: Bi–He, Q_r = 2, 9 kW, U_r = 5 kV, p_{He} = 3.3 kPa, f = 2.7 kHz, c: Bi–He, Q_r = 2, 9 kW, U_r = 5 kV, p_{Ne} = 6.7 kPa, f = 2.5 kHz; d: Bi–Ne, Q_r = 2.9 kW, U_r = 5 kV, p_{Ne} = 3.3 kPa, f = 2.2 kHz, e: Pb–Ne, Q_r = 2.4 kW, U_r = 3.9 kV, p_{Ne} = 6.7 kPa, f = 3.9 kHz, f: Pb–Ne, Q_r = 2.4 kW, U_r = 4.6 kV, p_{Ne} = 6.7 kPa, f = 2, 2 kHz, g: Pd–He, Q_r = 2.9 kW, U_r = 5 kV, p_{He} = 6.7 kPa, f = 2.1 kHz.

copper atoms, which allows a sufficiently correct comparison of results of investigations of the repetitively pulsed discharges in mixtures of Bi and Pb with Ne and He with the results of previous studies of the self-heating copper vapour lasers with a GDT diameter of 1.2 cm. Here it is appropriate to note that in experiments with pulsed-periodic discharge in mixtures of lead vapour with neon lasing was observed at the 722.9 nm line, and in mixtures with neon and helium in the afterglow of the discharge infrared lasing was detected at the lines 3.15 and 7.15 μm [166].

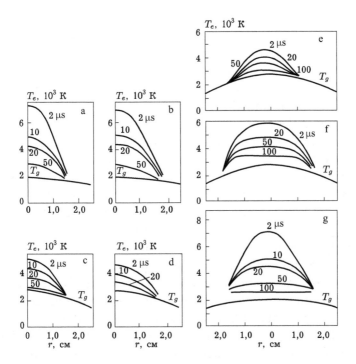

Figure 4.25. The distribution of electron temperature along the radius of GDT at different times after the excitation pulse. Conditions are the same as for Fig. 4.24. The radial distribution of T_g is calculated in the approximation of a parabolic distribution of energy distribution along the GDT radius.

Time dependences of the electron concentration in the interpulse time interval of repetitively pulsed discharges in mixtures of vapours of Bi and Pb with He and Ne were measured by the interferometric method for fixed values of the current radius of the GDT r: 0, ± 5, ± 10 and ± 15 mm. The results are shown in Figs. 4.23 and 4.24. The first, as an example, the time dependence of the electron concentration, directly measured in the experiment. In the second graph there are radial distributions of the electron concentration $n_e(r)$ at different times after the excitation pulse and plotted using such relationships. In the whole investigated range of parameters (type and buffer gas pressure, pulse repetition frequency, etc.), these distributions are significantly heterogeneous. Moreover, their appearance, which is realized within a few microseconds after the excitation pulse, is strongly dependent on the voltage across the storage capacitor, the pulse repetition frequency, the wall temperature of the GDT, and the type and pressure of buffer gas.

To estimate the proportion of energy from the energy inputted to the discharge in the excitation pulse, which is consumed in the ionization of atoms of the working metal, in [70,165] the authors compared the values of the average (in the cross-section of the GDT) concentration of electrons $\langle n_e \rangle_{max}^{exp}$ in the early interspike interval, calculated from experimental data, and the concentration $\langle n_e \rangle_{max}^{cal}$ calculated on the assumption that the entire energy supplied to the discharge is expended on the ionization of atoms of the metal. Naturally, this assumption is valid only when the postpulse electron concentration in the axial zone of GDT is lower than the prepulse concentration of atoms in the same area. Based on this comparison, in [70,165] it is concluded that almost all the energy injected into the discharge during the excitation pulse is in the long run (at the time of establishment of ionization equilibrium in the early interspike interval) used for the ionization of atoms of the working metal. This conclusion is qualitatively confirmed by the synthesis of several studies of copper vapour lasers [127]. According to this generalization, the postpulse electron concentration, measured in a variety of copper vapour lasers, is proportional to the specific energy.

Figure 4.25 shows the results of calculations [70, 165] of the radial distributions of electron temperature in the interpulse time interval from the experimentally measured dependence $n_e(r, t)$ using the equation of the local balance of the electron concentration

$$\frac{\partial n_e(r,t)}{\partial t} = -\beta_e(r,t)n_e^3(r,t) + \alpha_i(r,t)n_M(r)n_e(r) -$$
$$-D_a(r)\frac{1}{r}\cdot\frac{\partial}{\partial r}\left[r\frac{\partial n_e(r,t)}{\partial r}\right],$$

(4.36)

where D_a is the ambipolar diffusion coefficient calculated from the ion mobility of bismuth and lead in Ne and He [158].

These figures also show the distribution of gas temperature, calculated in the approximation of a parabolic distribution of specific energy on the radius of the GDT. As follows from the form of the distribution of $T_e(r)$, shown in Fig. 4.25, unlike the GDT diameter of about 1 cm, in the GDTs 5 cm in diameter in the interpulse interval there are significantly non-uniform distributions of electron temperature along the radius of GDT. That is, in large diameter GDTs there is no equalization of the electron temperature over the cross section due to the electron thermal conductivity, as is the case with GDTs of small diameters.

In [70, 165] the authors estimated the unit volume electron energy losses q_e due to the electronic heat conductivity and compared them with the energy losses by the electrons in elastic collisions q_{ea} with heavy particles. As a result of this comparison it was revealed that for the pressures of neon p_{Ne} = 3.3–13.3 kPa and n_e = 10^{14}–10^{15} cm^{-3} the ratio q_e/q_{ea} does not exceed a few percent. This means that in the GDT 5 cm in diameter the GDT wall heat loss due to electronic heat conductivity does not play a significant role in the thermal balance.

4.6. The concentration of atoms of the working metal in self-contained lasers

For the first time the change in the concentration of atoms of the working metal in consequence of the excitation pulse was measured in [167] with a self-heating barium vapour laser with a GDT diameter of 3 cm and length 55 cm. The excitation pulse repetition frequency was 1.5 kHz, the neon buffer gas pressure 4 kPa. Measurement of the concentration of barium atoms in the ground state was carried by the hook method at 1 µs after the discharge. The measurements revealed that in the whole GDT temperature range studied (815–970 K) the decrease in the concentration of atoms due to the excitation pulse was about 70%.

The first measurements of the spatial distributions of the concentration of copper atoms in the repetitively pulsed copper vapour laser were given in [33] for a copper vapour laser with transverse discharge. In [33] the integral absorption in the resonance line of 324.7 nm was used to measured the distribution of copper atoms in the cross section of a GDC 6 × 6 cm in size. Measurements were performed in the interval between excitation pulses, following with a frequency of 2.8 kHz. The measurements in [33] showed that the concentration of copper atoms n_{Cu} (0) in the axial zone of the GDC is much lower than the concentration $n_{Cu\ w}$ in its near-wall region. In [33] this decrease in the concentration is attributed to heating of the gas and the profile n_{Cu} (x) was used to calculated profile T_g (x). According to the results [33], under optimal lasing power conditions ($\overline{Q}_{dl}$ ≈ 90 W/cm) the gas in the centre of the GDC is heated to about 1000 K with respect to the wall temperature of the GDC.

The first measurements of spatial and temporal distributions of the concentrations of copper atoms in the ground state in the afterglow of the discharge of the copper vapour laser with external heating were made in [168, 169] by the hook method. The experiments were performed with two lasers excited by single pulses, with a diameter of GDT of 2 and 1 cm and 50 cm long. The buffer gases were helium and neon at a pressure of 665 Pa. As shown by the measurements of

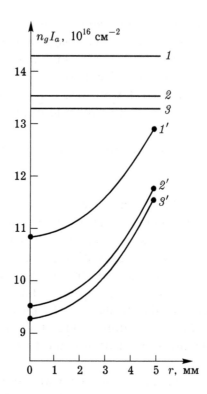

Figure 4.26. The distribution of the concentration of copper atoms in the cross section of GDT in a copper vapour laser excited by single pulses p_{Ne} = 665 Pa, 1, 2, 3 – no discharge, T_w = 1570°C (1), 1540 (2), 1530 (3), 1', 2', 3'– with discharge, $C_s = C_p$ = 4.7 nF, U_r = 10 kV, T_w = 1570°C (1'), 1540 (2'), 1530 (3').

the concentration of copper atoms in the ground state n_{Cu} in a GDT 2 cm in diameter taken at different GDT wall temperatures, the concentration of atoms n_{Cu} ten microseconds after the excitation pulse was 60–60% of the prepulse concentration and reached its equilibrium value after 150–200 μs .

The radial distribution n_{Cu} (r) was studied in [168. 169] with a laser with a GDT 1 cm in diameter. The radial inhomogeneity of the distribution n_{Cu} (r) was observed only at the beginning of the afterglow. For delays over 10 μs and longer no heterogeneity was recorded. As follows from the experimental results presented in Fig. 4.26, in single excitation pulses the axial zone of the GDT is characterized by more effective (excitation and ionization) of the ground state by the discharge than in the wall region of the GDT. Based on the results of measurements, in [168, 169] it is concluded that almost all the energy, injected into the discharge, is expended on the ionization of copper atoms and only a small part of it is converted to stimulated radiation.

In [169] by measuring the concentration of copper atoms in the axial zone n_{Cu} of the GDT (d_p = 1 cm, l_p = 57 cm, f = 10 kHz, p_{Ne} = 2.6 kPa) of a self-heating laser excited with a frequency of 10 kHz, it was found that 5 µs after the excitation pulse the concentration n_{Cu} (0) is equal to $4 \cdot 10^{13}$ cm^{-3}, which is significantly less than the equilibrium concentration value n_{Cu}, equal to about 10^{15} cm^{-3}. Based on this fact, in [169] it is concluded that the concentration of electrons after the excitation pulse can reach values of 10^{15} cm^3. It is also possible that this is the evidence of that in the considered experiment the prepulse concentration of copper atoms $n_{Cu}^{st}(0)$ in the axial zone of the GDT significantly less than its equilibrium value $n_{Cu\,w}$ with the wall temperature T_w of the GDT.

Thus, the results [168,169] agree well with the measurements of electron concentration set forth in the preceding section, and indicate a significant degree of ionization of working atoms during the excitation pulse in pulses and repetitively pulsed copper vapour lasers.

It is worth noting that the first conclusion according to which in the conditions of the predominance of the diffusion annihilation of ions of the working metal the concentration of the working atoms may be less than its value equilibrium with T_w value was made in [79, 88]. In addition, in [79] the authors solved the problem of the difference between the concentration of working atoms in the discharge and its equilibrium value for the case where recovery of the working atoms in the discharge occurs as a result of their diffusion from the wall of the GDT, and the annihilation as a result of ionization during the pulse excitation and subsequent ambipolar diffusion on the wall of the GDT.

In order to identify the main features of the restoration of the concentration of copper atoms in the interpulse interval at high buffer gas pressures, we turn to Fig. 4.20 [92, 116], which shows the time dependence of the concentration of metastable copper atoms in the axial zone of the GDT with a diameter of 2 and 50 cm at a neon pressure of 50 kPa the restored (from this dependence) time dependences of the electron temperature T_e (0) and the concentration n_g (0) of the copper atoms in the ground state. The dependence n_g (0) = $f(t)$ clearly indicates, firstly, the essential ionization of copper atoms. Secondly, the restoration of the concentration of copper atoms from $2 \cdot 10^{13}$ cm^{-3} to $2 \cdot 10^{14}$ cm^{-3} takes within 7 µs, i.e. during the time coinciding with the time of rapid cooling of the electrons. Then, after about 90 µs n_g (0) is restored to about $4 \cdot 10^{14}$ cm^{-3}. And finally, thirdly, the concentration n_g (0) does not have time to recover during

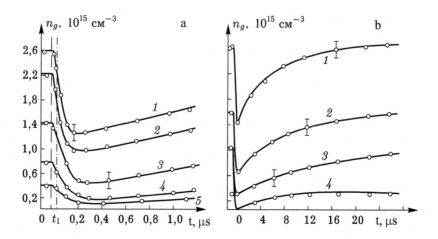

Figure 4.27. Time dependence of the concentration of copper atoms in the ground state on the axis of GDT in the excitation pulse (a) and near afterglow (b) at different temperatures in the GDT wall of the copper vapour laser (d_p = 2 cm, l_a = 40 cm, f = 10 kHz, Q_r = 2 kW). a: 1 – T_w = 1640°C; 2 – 1625; 3 – 1585; 4 – 1550; 5 – 1520; time t_1 corresponds to the maximum amplification in the active medium. b: 1 – T_w = 1650°C; 2 – 1595; 3 – 1565; 4 – 1500.

the interpulse time interval to its value equilibrium with T_w value equal to $1.8 \cdot 10^{15}$ cm^{-3}.

A rather detailed study of the spatio-temporal evolution of the concentration of copper atoms in a self-heating copper laser (d_p = 2 cm, l_a = 40 cm, f = 10 kHz, p_{Ne} = 40 kPa) was carried out in [114, 115]. The results of this study are shown in Figs. 4.27, 4.28. Figure 4.27.b shows that, as in the case of [92, 116], the restoration of the concentration of copper atoms in the axial zone of the GDT to about 80% of the prepulse concentration n_{Cu} occurs in the first 8–12 μs, depending on the temperature of the walls of GDT. From the radial distributions of prepulse concentration of copper atoms, shown in Fig. 4.28 a, it follows that the prepulse concentration $n^{st}_{u}(r)$ of the copper atoms on the axis of the GDT is significantly lower than the concentration $n_{gw}(r)$, equilibrium with T_w. As indicated in [114, 115], the maximum difference between n_{gw} and n^{st} is found at T_w = 1450°C and is fivefold. A similar deficiency of copper atoms on the axis of the GDT was found in [92, 116] in experiments with self-heating copper vapour lasers (d_p = 2 cm, l_a = 40 cm, f = 7.2 kHz, T_w = 1838 K) at a neon pressure of 50 kPa. The prepulse concentration of copper atoms on the axial zone of the GDT, estimated on the basis of measurement results of the concentration of metastable atoms by the Boltzmann ratio of the concentration of the copper atoms in the axial zone of the GDT, is equal to $4 \cdot 10^{14}$ cm^{-3}, while the concentration of

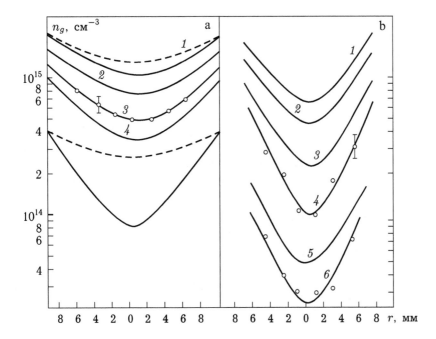

Figure 4.28. The radial distribution of the concentration of copper atoms in the ground state in the prepulse period (a) and at times corresponding to the minimum population of the ground state of copper (b) (characteristics of the laser are the same as for Fig. 4.27) a: 1 – T_w = 1570°C; 2 – 1555; 3 – 1540; 4 – 1500; 5 – 1420. b: 1 – T_w = 1610°C; 2 – 1590; 3 – 1560; 4 – 1520; 5 – 1470; 6 – 1360. The dotted line shows the distribution n_{gw}^{st} $(r) = n_{gw} T_g (r) /T_w$, calculated under the assumption of uniform energy release along the radius of GDT at T_w = 1420 and 1570°C.

the copper atoms in the axial zone of the GDT equilibrium with T_w is $1.8 \cdot 10^{15}$ cm^{-3}. At the same time at a pressure of 10 kPa there is good agreement between the prepulse concentration of copper atoms outside the axis of the GDT $n_{Cu}^{st}(0)$ = $9 \cdot 10^{14}$ cm^{-3} and the equilibrium value $n_{Cu\,w}(0) - 10^{15}$ cm^{-3},

When the recovery of the Cu atoms by bulk recombination predominates in the axial zone of the GDT, the mentioned discrepancy recombination marked discrepancy between the prepulse $n_{Cu}^{st}(0)$ and equilibrium $n_{Cu\,w}(0)$ concentrations of copper atoms is not evident. Therefore, in [92, 116] to verify the possible existence found in [92, 116] of the deficiency of copper atoms in the axial zone of the GDT of the self-heating laser in the conditions of preferential recovery of the concentration of copper atoms as a result of bulk recombination of copper ions the authors numerically solved the problem of reconstructing the concentration of copper atoms in the interpulse interval in the axial zone of the GDT.

The calculation of the concentration of copper atoms in the ground state in the axial zone of GDT in the interpulse interval was performed using the relation

$$n_g(0,t) = n_g(0,t)_\beta + n_g(0,t)_D, \tag{4.37}$$

where the first and second terms on the right side of the equation are the concentrations of copper atoms in the axial zone of the GDT, restored by the bulk recombination of copper ions and electrons and due to diffusion from the wall of the GDT. The time is counted from the start of the interpulse interval.

$n_g(0, t)$ was calculated using a system of equations, written under the assumption that the radial distribution of electron concentration is described by the Bessel function,

$$\frac{\partial n_e(0,t)}{\partial t} = -\frac{6D_a}{r_p^2} n_e(0,t) - \beta_e n_e^3(0,t) - \alpha_i(0,t) n_g(0), \tag{4.38}$$

$$\frac{d\langle T_e \rangle}{dt} = \frac{2}{3} \cdot \frac{6D_a}{r_p^2} \langle T_e \rangle \left[\ln \frac{r_p \langle T_g \rangle^{1/2}}{2,4 D_a M_i^{1/2}} + \ln \left(\frac{M_i \langle T_e \rangle}{m \langle T_g \rangle} \right)^{1/2} \right] -$$

$$- \frac{2m}{M_{Ne}} v_{ea} \left(\langle T_e \rangle - \langle T_g \rangle \right) - \frac{2m}{M_i} v_{ei} \left(\langle T_e \rangle - \langle T_g \rangle \right) + \left(\frac{2}{3} I_{Cu} + \langle T_e \rangle \right) \times \tag{4.39}$$

$$\times \left(\frac{0,22}{0,46} \beta_e n_e^2(0,t) - \alpha_i \langle n_g \rangle \right),$$

$$n_g(0,t)_\beta = \int_0^t \beta_e n_e^3(0,t) \, dt, \tag{4.40}$$

$$n_g(0,t) \leq n_{gw}(0). \tag{4.41}$$

The values v_{ea} and v_{ei} were calculated in [92,116] according to [86] and [97], respectively. n_e was averaged over the cross section of the GDT assuming the Bessel electron concentration distribution along the radius of the GDT: $\langle n_e \rangle = 0.46 \, n_e(0)$, $\langle n_e^3 \rangle = 0.22 \, n_e^3(0)$. The condition (4.39) reflects the fact that when the concentration of copper atoms $n_g(0, t)$ reaches the equilibrium value $n_{gw}(0)$, the growth of $n_g(0, t)$ in the axial zone of the GDT is stopped.

$n_g(0, t)$ was calculated using two relations

$$n_g\left(0,t\right)_D = n_w \cdot \left[1 - \sum_{k=1}^{\alpha} \frac{2}{\mu_k J_1\left(\mu_k\right)} \exp\left(-\mu_k^2 \frac{D_{Cu}t}{r_p^2}\right)\right], \qquad (4.42)$$

$$n_g\left(0,t\right)_D = n_w \left[1 - \exp\left(-\frac{6D_{Cu}}{r_p^2}t\right)\right]. \qquad (4.43)$$

where the relation (4.42) is the exact solution of the non-stationary diffusion equation [170], indicating recovery of Cu atoms due to diffusion from the walls of the GDT on the assumption that during the excitation pulse the copper atoms in the discharge are completely ionized and by ambipolar diffusion travel to the wall of the GDT; J_1 is the Bessel function of the first order; μ_k is the k-th root of the Bessel functions of zeroth order; D_{Cu} is the diffusion coefficient of copper atoms in neon [171]. In the calculation of (4.42) in [92, 116] the first 5 terms in the series were used.

Calculations of the recovery the concentration of copper atoms in the axial zone of the GDT were reported in [92, 116] for a GDT 2 cm in diameter and a neon pressure of 2–50 kPa. The average gas temperature was assumed to be $\langle T_g \rangle$ = 2500 K. The concentration of copper atoms on the wall of the GDT was assumed to be n_w = 2·10^{15} cm^{-3}, the equilibrium concentration of copper atoms on the axis of the GDT n_{gw} (0) = 10^{15} cm^{-3}. At a given excitation pulse repetition frequency f the initial conditions for system (4.36)–(4.39) are given as

$$n_e\left(0,0\right) = n_g\left(0,T\right); \quad \langle T_e \rangle(0) = T_i \approx 0.7$$

$$T = 1/f.$$

The results of the calculation [92,116] of the prepulse concentration of copper atoms in the axial zone of the GDT in the steady state operation of a repetitively pulsed laser, called the stationary prepulse concentration and denoted as n_g(0, T), are shown in Fig. 4.29. According to these results, the value of n_g(0, T) depends on the excitation pulse repetition frequency and with its increase, beginning at some f_{cr} value, hereinafter called critical, the concentration n_g (0, T) is significantly below the equilibrium value. Two variants of accounting the diffusion of copper atoms from the wall of the GDT give qualitatively the same picture of recovery of the concentration of copper atoms in the axial zone of the GDT, but lead to different rates of this recovery, indicating a significant effect of the degree of ionization of working atoms in the excitation pulse on the rate of recovery of the properties of the active laser medium.

Since the calculation of the dependences of the concentration n_g (0, T) on the pulse repetition frequency was performed for conditions close to experimental [92,116], Fig. 4.29 shows the pulse repetition frequency f, at which in [92,116] the authors observed at a neon pressure of 50 kPa a shortage of copper atoms in the axial zone of the GDT of the repetitively pulsed copper vapour laser. It is seen that this value of f is in better agreement with the calculated results corresponding to the case where the calculation of the recovery of the concentration of copper atoms by diffusion from the walls of the GDT was carried out according to (4.41), which in the absence of complete ionization of copper atoms in the entire cross section of the GDT describes more accurately the recovery of the concentration of the copper atoms in the axial zone of the GDT in the early interspike interval. It should be noted here that because the relations (4.40) and (4.41) overestimate the recovery rate of the concentration of copper atoms in the axial zone of the GDT, for values of $n_g(0)$ close to $n_{gw}(0)$ it should be expected that in the real case the dependences n_g (0, T) on $1/f$ will be less steep and the effects of reducing the concentration of copper atoms in the axial zone of the GDT will occur at lower pulse repetition frequencies than would be expected from the curves presented in Fig. 4.29.

Thus, despite the approximate form of (4.37) which gives, strictly speaking, only a qualitative description of the restoration of the concentration of copper atoms $n_g(0)$ in the axial zone of the GDT, this

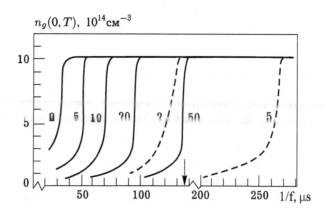

Figure 4.29. Dependence of the stationary prepulse concentration of copper atoms in the axial zone of GDT on the pulse repetition period. The solid curves correspond to taking into account diffusion of copper atoms in (4.43), dotted – by (4.42). The numbers on the curves – the neon pressure in kPa. The arrow marks the experimental value of the pulse repetition frequency at which there is a shortage of copper atoms in the axial zone of GDT at a neon pressure of 50 kPa in [92, 116].

relationship was used to obtain a solution, indicating the possibility of a substantial reduction of $n_g(0)$ with increasing frequency f. Based on the results of calculations presented in Fig. 4.29, in [92, 116] it is concluded that the recovery of copper atoms in the ground state can be one of the factors influencing the formation of the optimum excitation pulse repetition frequencies f.

This factor may play a significant role in cases where the ionization of the working of the metal is achieved in most of the GDT section. This almost always occurs in self-heating lasers with heating of the GDT. Incomplete recovery of the atoms leads to the well-known ring-shape lasing. Lasing takes place in the near-wall region, where a significant contribution to the recovery of atoms makes their diffusion from the wall of the GDT. In heating the width of the lasing ring increases as with the increasing concentration of the working metal atoms full ionization of these atoms is achieved in an increasingly narrow zone near the axis of the GDT and the contribution of the diffusion of atoms in their recovery in the axial zone increases. It is possible that the sharp decrease in the concentration of copper atoms in the axial zone of the GDT is associated with the collapse of lasing (observed in [28] at increasing excitation pulse repetition frequency) in a horizontal strip located in the centre of the discharge gap a copper vapour laser excited by transverse discharge.

In concluding the results of studies of the kinetics of the ground state of the lasers at self-contained transitions of the metal atoms (especially in copper vapour lasers), we should mention studies [172,173], which discuss some of the processes that shape the distribution of copper atoms in the ground state in the cross section of the GDT and leading to a shortage of the concentration of copper atoms in the axial zone of the GDT compared with the concentration equilibrium with the GDT wall temperature.

References

1. Petrash G.G., Usp. Fiz. Nauk. 1971 V.105. Issue 4. P.645-676.
2. Soldatov A.N., Solomonov V.I., Gas-discharge lasers on self-contained in metal vapors. Novosibirsk: Nauka, 1985.
3. Piltch M., Walter W.T., Solimene N., Gould G., Appl.Physics Letters. 1965. Vol.7. No.11. P.309-310.
4. Walter W.T., Solimene N., Piltch M., Gould G., IEEE J. Kvantovaya elektronika. 1966. Vol.QE-2. No.9. P.474-479.
5. Walter W.V., IEEE J. Kvantovaya elektronika. 1968. Vol.QE-4. No.5. P.355-356.
6. Isaev A.A., Kazaryan M.A., Petrash G.G., Zh.Prikl.Spektroskopii. 1973. V.18. Vol.3. P.484-485.

7.   Isaev A.A., Kazaryan M.A., Petrash G.G., Kvantovaya elektronika. 1972. No.5 (11). P.100.

8.   Isaev A.A., Kazaryan M.A., Petrash P.P., Pisma v ZhETF. V. 1972. V.16. Issue 1. P.40-42.

9.   Petrach G.G., Isaev A.A., Kazaryan M.A., IEEE J. Kvantovaya elektronika. 1973. Vol.QE-9. No.6. P.644-645.

10.   Markov S.V., Petrash G.G., through VM Pulse Research gold vapor laser and bismuth. Metal vapor lasers and halides (Proceedings of the Lebedev Physics Institute. V.181). Moscow: Nauka. 1987. P.18-34.

11.   Evtushenko G. Gold-Vapor Laser. Present State of the Art, Design and Application, Metal Vapor Lasers and Their Applications: CIS Selected Papers, G.G. Petrash, Editor. Proc. SPIE 2110. 1993. P.204-214.

12.   Kelman V.A., Klimovski I.I., Shpenik Yu.O., J. Russian Laser Research. 1994. Vol.15. No.1. P.69-73.

13.   Kelman V.A., Klimovski I.I., Shpenik Yu.O. Investigation of Bismuth Vapour Lasers , Pulsed Metal Vapor Lasers. Proceedings of the NATO Advanced Research Workshop on Pulsed Metal Vapor Lasers - Physics and Emerging Applications in Industry, Medicine and Science, SV. Andrews, U.K., Aug.6-10, 1995; C.E. Little and N.V. Sabotinov, EdP. Dordrecht: NATO ASI Series, Kluver Academie Publishers. 1996. P.79-84.

14.   Bokhan P.A., Zakrevsky D.E. Investigations on the collisional He + EuII laser with longitudinal gas discharge excitation at pressures up to 5 atm , Metal Vapor Lasers and Their Applications: CIS Selected Papers, Petrash GG.,, Editor. Proc. SPIE 2110. P.220-235.

15.   Bokhan P.A., Zakrevsky D.E., J. Russian Laser Research. 1995. Vol.16. No.2. P.138-144.

16.   Klimkin V.M., Prokop'ev V.E., Sokovikov V.G., On the problems of inversion population mechanism and the generation thershold for an Eu-ion laser, Atomic and Molecular Pulsed Lasers, V.F. Tarasenko, G.Y. Mayer, Petrash G.G., EditorP. Proc. SPIE 2619. P.104-108.

17.   Bokhan P. High operating pressure metal vapour lasers , Atomic and Molecular Pulsed Lasers, V.F. Tarasenko, G.Y. Mayer, G.G. Petrash, Editors. Proc. SPIE 2619. P.113-122.

18.   Isaev A.A., Lemmerman G.Y., Markova S.V., Petrash G.G., Pulse barium vapor laser. Metal vapor lasers and their halides (Proceedings LPI. V.181). Moscow: Nauka, 1987. P.3-17.

19.   Karabut E.K., Kravchenko V.F., Savranskii V.V., J. Russian Laser Research. 1994, Vol.15. No.1. P.78-80.

20.   Mildren R.R., Brown D.J.W, Piper J.A. An investigation into the factors limiting pulse energies in the barium vapour laser using density measurements obtained by the hook method , International Quantum Electronics Conference, 1996 OSA Technical Digest Series (Optical Society of America, Washington, D.C.). 1996. P.175.

21.   Evtushenko G.P., Filonov A.F., Filonova N.A. Unstable Resonator Lead Vapor Laser , Atomic and Molecular Pulsed Lasers, Tarasenko V.F., G.Y. Mayer, G.G. Petrash, EditorP. Proc. SPIE 2619. P.84-88.

22.   Isakov V.K., Kalugin M.M., Parfyonov E.N., Potapov S.E., Zhurnal tekhnicheskoi fiziki. 1983. V.33. Issue 4. P.704-714.

23.   Alexandrov I.S., Babeyko U.A, Babayev A.A., et al., Kvantovaya elektronika. 1975. V.2. No.9. P.2077-2079.

24.   Babeyko Yuri Vasiliev L.A., Orlov V.K. et al., Kvantovaya elektronika. 1976 V.3. No.10. P.2303-2304.

25. Bohan P.A., Shcheglov V.B., Kvantovaya elektronika. V.5 1978. No.2..381-387.
26. Babeyko Yu., Vasiliev L.A., Sokolov A.V., et al., Kvantovaya elektronika. 1978 V.5. No.9. P.2041-2479.
27. Babeyko Yu., Vasiliev L.A,, Sviridov A.V. et al., Kvantovaya elektronika. 1979 V.6. No.5. P.1102-1105.
28. Artemyev A.J., Babeyko Yu., Bakhtin O.M., et al., Kvantovaya elektronika. 1980. V.7. No.9. P.1948-1954.
29. Arlantsev S.V., Buchanov V.V., Vasiliev L.A., et al., Kvantovaya elektronika. 1980. V.7. No.11. P.2319-2424.
30. Alexandrov V.M., Buzhinskiy O.I., Sin IV et al., Kvantovaya elektronika. 1981 V.8. No.1. P.191-193.
31. Arlantsev S.V., Buchanov V.V., Vasiliev L.A., et al., Reports of the USSR Academy of Science. 1981 V.260. No.4. P.853-857.
32. Sokolov A.V., Sviridov A.V., Kvantovaya elektronika. 1981 V.8. No.8.P.1686-1696.
33. Borowicz B.L., Vasiliev L.A., Hertz V.E., et al., Kvantovaya elektronika. 1981 V.8. No.9. P.1996-2000.
34. Borowicz B.L.Vasiliev L.A, Ryazan V.M., et al. Copper vapour lasers with a transverse discharge. III All-Union Conference "Laser Optics" (4-8 January 1982). Abstracts. L enigrad: GOI. 1981. P.125-126.
35. Artemyev A.J., Borovichi B.L., Vasiliev L.A., et al., Kvantovaya elektronika. 1982. v.9. No.4. P.738-743.
36. Buzhinskiy O.I., Efimov A.V., Slivitsky A.A., Kvantovaya elektronika. 1982. v.9. No.9. P.1854-1856.
37. Borowicz B.L., Grigorian R.A., Kazeko G.P., et al., Kvantovaya elektronika. 1982. V. 9.No.10. P.1983-1992.
38. Buchanov V.V., Young E.I., Tykotsky V.V., Kvantovaya elektronika. 1983. Vol.10. No.3. P.629-631.
39. Artemyev A.J., Borovichi B.L., Vasiliev L.A., et al., Kvantovaya elektronika. 1983. Vol.10. No.7. P.1441-1447.
40. Bukhanov V.V., Young E.I., Yurchenko N., Kvantovaya elektronika. 1983. Vol.10. No.8. P.1553-1560.
41. Borowicz B.L., Nelegach E.P., Rybin V.M., Yurchenko N., Kvantovaya elektronika. 1984. V.11. No.12. P.2471-2479.
42. Borowicz B.L., Yurchenko N.I., Kvantovaya elektronika., 1985 vol.12. No.7. P.1377-1386.
43. Borovich B.L., Transverse-dicharge copper-vapor lasers, Metal Vapor Lasers and Their Applications: CIS Selected Papers, Petrash G.G, Editor.Proc. SPIE 2110. P.16 63.
44. Yurchenko N. Theoretical study of copper-vapor lasers , Metal Vapor Lasers and Their Applications: CIS Selected Papers, G.G. Petrash, Editor. Proc. SPIE 2110. P.64-77.
45. Borovich B.L., Yurchenko N.I. Physics of Transverse-Discharge Copper Vapour Lasers, Pulsed Metal Vapor Lasers. Proceedings of the NATO Advanced Research Workshop on Pulsed Metal Vapor Lasers – Physics and Emerging Applications in Industry, Medicine and Science, SV. Andrews, U.K., Aug. 6-10, 1995; C.E. Little and N.V. Sabotinov, EdP. Dordrecht: NATO ASI Series, Kluver Academie PublisherP. 1996. P.73-7-8.
46. Anderson R.S., Springer L., Bricks B.G., Karras T.W., IEEE J., Kvantovaya elektronika. 1975. Vol.QE-11. No.4. P.172-174.
47. Anderson R.S., Bricks B.G., Springer L.W., Karras T.W., IEEE J.,Kvantovaya

elektronika. 1975. Vol.QE-11. No.9. P.56D-57D.

48. Kirilov A.E., Polunin Y.P., Soldatov A.N., Fedorov V.F., Lasers metal vapor for atmospheric research. Measuring devices for the study, the surface layers of the atmosphere. Tomsk: Institute of Atmospheric Optics SB RAP. 1977. P.59-79.

49. Isaev A.A., Lemmerman G.Y., Kvantovaya elektronika. 1977. V.4. No.7. P.1413-1417.

50. Bohan P.A., Zhurnal tekhnicheskoi fiziki.. 1981. V.51. Issue 1.P.206-209.

51. Warner B.E., Anderson R.P., Grove R.E. CLEOS, 26-27 Feb. 1980, Digest of Technical Paper. Optical Society of America. Wash. D.C. 1980. P.90.

52. Anderson R.S., Warner B.E., Larson C., Grove R.E., CLEO Conf., 10-12 June 1981, Digest of Technical Papers, Optical Society of America, Wash.D.C., 1981.

53. Anderson R.S., Warner B.E., Larson C., Grove C.R., J. Kvantovaya elektronika. 1981. Vol.QE-17. No.12. P.50.

54. Grove R.E., Laser Focus. 1982. Vol.18. No.7. P.45-50.

55. Kushner M.J., Warner B.E., J. Appl. Physics. 1983. Vol.54. No.6. P.2970-2982.

56. Bohan P.A., Silant'ev V.I., Solomonov V.I., Kvantovaya elektronika. 1980. V.7. No.6. P.1264-1269.

57. Bohan P.A., Maltsev A.N., Silant'ev VI., New methods to improve energy characteristics of repetitively pulsed lasers vapor Arts metals. Abstracts of the X All-Union Conference on Coherent and nonlinear optics. (Kiev. October 14-17, 1980). Moscow:1980. Part I. Section I - IV. P.201-202.

58. Bokhan P.A., Dubnishcheva V.Ya., Zakrevskii D.E., Nastaushev Yu.V., J. Russian Laser Research. 1995. Vol.16. No.2. P.164-171.

59. Withford M.L., Brown DJW, Carman R.J., Piper J.A., Kinetic enhancement in copper vapour lasers using halogen donor gas additives. XX International Quantum Electronics Conference, 1996 OSA Technical Digest Series. Optical Society of America, Wash. D.C. 1996. P.238.

60. Molodykh E.I., Pevtsova I.K., Tykotskii V.V., J. Russian Laser Research. 1996. Vol.17. No.4. P.365-367.

61. Lyabin N.A., Zubov V.V., Koroleva M.E., Ugol'nikov S.A., J. Russian Laser Research. 1996. Vol.17. No.4. P.346-355.

62. Hackel R.P., Warner B.E. The copper-pumped dye laser system at Lawrence Livermore National Laboratory. Laser Isotope Separation: CIS Selected Papers, JA Paisner, Editor. Proc. SPIE. 1993. Vol.1859. P.2-13.

63. Konagai C., Aoki N., Ohtani R., Kobayashi N., Kimura H. Copper Vapor Laser System DevelopmenV. Proc. 6th International Symposium on Advanced Nuclear Energy Research. P.637-642.

64. Walter W.W., Bull. American Phys. Society. 1967. Vol.12. No.1. P. 90.

65. Bohan P.A., Gerasimov V.A., Solomonov V.I., Shcheglov V.B., Kvantovaya elektronika. 1978 V.5. No.10. P.2162-2173.

66. Bohan P.A., Gerasimov V.A , Kvantovaya elektronika. 1979 V.6. No.3. P.451-455.

67. Soldatov A.N., Fedorov V.F., Yudin N.A., Kvantovaya elektronika. 1994. V.21 (8). P.733-734.

68. Soldatov A.N., Sukhanov V.B., Fedorov V.F., Yudin N.A., Optika atmosfery i okeana. 1995 V.8. No.11. P.1626-1636.

69. Vorobyov V.B., Kalinin S..V, Klimovskii I.I., et al., Kvantovaya elektronika. 1991. V.18. No.10. P.1178-1180.

70. Galkin A.F.,Radial inhomogeneity of the plasma parameters in the self-heating self-contained lasers with longitudinal discharge and its impact on the performance of the generation. PhD Thesis. Sci. Sciences. Moscow:1985.

71. Galkin A.F.,Klimovskii I.I., Effect of radial inhomogeneity the characteristics of the plasma generating repetitively pulsed lasers copper vapor in a nonuniform distribution of plasma parameters over the section of GDV. Preprint IVTAN No.5-2206. 1987.

72. Galkin A.F.,Klimovskii I.I., Optimal parameters of a repetitively pulsed copper vapor lasers in a nonuniform distribution of plasma parameters over the section of GDV. Preprint IVTAN No.5-228. 1987.

73. Klimovskii I.I., Teplofizika vysokikh temperatur. 1989. V.27. No.6. P.1190-1198.

74. Galkin A., Klimovskii I. Compute,d model of copper-vapor laser with the average specific output power above 1 W/cm^2, Metal Vapor Lasers and Their Applications: CIS Selected Papers, G.G. Petrash, Editor, Proc. SPIE 2110. 1993. P.90-99.

75. Kirilov A.E., Polunin Y.P., Soldatov A.N., Fedorov V.F. Lasers metal vapor for atmospheric research , Measuring devices for the study, the surface layers of the atmosphere. Tomsk Institute of Atmospheric Optics SB RAP. 1977. P.59-79.

76. Elaev V.F. Pozdeev V.V., Soldatov A.N., Radial heterogeneity gas discharge plasma copper , laser measuring devices study, the surface layers of the atmosphere. Tomsk, Institute of Atmospheric Optics SB RAS. 1977. P.94-97.

77. Elaev V.F., Melchenko V.S., Pozdeev V.V., Soldatov A.N., Effect of radial inhomogeneity of the discharge plasma in the generation parameters copper laser , Effective discharge metal vapor lasers. Tomsk, Institute of Atmospheric Optics SB RAS. 1978. P.189-196.

78. Batenin V.M., Vokhmin P.A., Painters V.S., et al., Teplofizika vysokikh temperatur1979. V.17. No.1. P.208-209.

79. Soldatov A.N., Shaparev N.Y., Kirilov A.E.,et al., Izv. VUZ. Fizika. 1980. V.23. No.10. P.38-43.

80. Klimkin V.M., Maltsev A.N., Fadin L.V.,Study the stability limits of pulsed gas discharge with a high repetition rate current pulse., Effective discharge in metal vapor lasers. Tomsk, Institute of Atmospheric Optics SB RAS. 1978. P.116-132.

81. Batenin V.M., Klimovskii I.I., Selezneva L.A., Teplofizika vysokikh temperatur 1980. V.18. No.4. P.707-712.

82. Eletskii A.V., Palkina L.A., Smirnov V.M., Transport phenomena in low ionized plasma. Moscow: Atomizdat. 1975.

83. Diachkov L.G., Kobzev G.A., Zhurnal tekhnicheskoi fiziki. 1978. V.48. No..11. P.2343-2346.

84. Gulani V.F. Zhilinskij A.P.,Sakharov I.L., Basics of plasma physics, Moscow: Atomizdat. 1977.

85. Raiser Y.P., Basics of modern physics discharge processes. Moscow: Nauka. 1980.

86. Baille P., Chang Jen-Shin, Claude A. et al., J.Phys. B: AV. Mol.Phys. 1981. Vol.14. No.9. P.1485-1495.

87. Batenin V.M., Burmakin V.A., Vokhmin P.A., et al., Teplofizika vysokikh temperatur. 1978. V.16. No.5. P.1145-1151.

88. Selezneva L.A., Influence of discharge parameters on the characteristics of the generation of self-heating copper vapor laser. PhD Thesis. Moscow:1980.

89. Isaev A.A., Kneipp, H., Rentsch M., Kvantovaya elektronika. 1983. Vol.10. No.5. P.967-973.

90. Kelman V.A., Klimovskii I.I,. Konoplev A.N., et al. Teplofizika vysokikh temperatur 1984. V.22. No.1, P.168-170.

91. Zharikov V.M, Zubov V.V., Forest M.A., et al., Kvantovaya elektronika. 1984.

V.11. No.5. P.918-923.

92. Fuchko V.Y., Examination of physical processes in the active medium repetitively pulsed copper vapor laser, and factors influencing its cardinality, and resource characteristics. PhD Thesis. Uzhgorod. 1987.

93. Zapesochnyi I.P., Kelman V.A., Klimovskii I.I., et al., Teplofizika vysokikh temperatur. 1988. V.26. No.4. P.671-680.

94. Hogan G.P., Webb C.E., Whyte C.G., Little CE Experimental studies of CVL kinetics., Proceedings of the NATO Advanced Research Workshop on Pulsed Metal Vapor Lasers – Physics and Emerging Applications in Industry, Medicine and Science, S.V. Andrews, U.K., Aug. 6-10, 1995: C.E. Little and N.V. Sabotinov, EdP. Dordrecht: Nato ASI Series, Kluwer Academic Publishers. 1996. P.67-72.

95. Isaev A.A., Kazaryan M.A., Petrash G.G., Kvantovaya elektronika. 1973 No.6 (18). P.112-115.

96. Nagibina I.M., Prokofiev V.P., Spectral instruments and spectroscopy. - A .: Engineering. 1967.

97. Ginzburg V.L., The propagation of electromagnetic waves in a plasma. Moscow: .Nauka, 1967.

98. Klimovskii I.I., Selezneva L.A., Teplofizika vysokikh temperatur. 1985. V.23. No.4. P.667-672.

99. Ilyushko V., Karabut E.K., Kravchenko V.F., Mikhalevsky V.S., Kvantovaya elektronika. 1985. vol.12. No.10. P.2185-2187.

100. Chang J.J., Warner B.E., Boley C.D., Dragon E.P., High-pover copper vapour lasers and applications , Proceedings of the NATO Advansed Research Workshop on Pulsed Metal Vapor Lasers - Physic and Emerging Applications in Industry, Medicine and Science, SV. Andrews, U.K., Aug. 6-10, 1995: eds. C.E. Little and N.V. Sabotinov, Dordrecht: NATO ASI Series, Kluwer Academic Publishers. 1996. P.101-112.

101. Miller J.L., Kan V., J. Appl. Physics. 1979. Vol.50. No.6. P.3849-3851.

102. Smilanski I., Investigation of the copper vapor kinetics. Proc. Intern. Conf. Lasers'79. Orlando, Florida, Ed. V.J. Corcoran. 1979. P.327-334.

103. Smilanski I., Levin L.A., Erez G., Optics Letters. 1980. Vol.5. No.1.P.93-95.

104. Tenenbaum J., Smilanski I., Lavi P., et al., Optics Communications. 1981. Vol.36. No.5. P.391-394.

105. Litvinenko A.Y., Kravchenko V.I., Egorov A.N., Ukr. Fiz.Zh. 1982. V.27. No.6. P.947-948.

106. Litvinenko A.Y., Kravchenko V., Egorov A.N., Kvantovaya elektronika. 1983. No.10. No.6. P.1717-1717.

107. Kalman V.A., Klimovskii I.I., Kononlev A.N., et al., Kvantovaya elektronika. 1984. V.11. No.11. P.2191-2196.

108. Isaev A.A., Petrash G.G., Ponomarev I.V., Relaxation of metastable atoms in the afterglow of a copper vapor laser Preprint FIAN No.271. Moscow: 1985.

109. Kelman V.A., Klimovskii I.I., Selezneva L.A., Fuchko V.Yu., The mechanisms that determine the population of the metastable levels SuI in repetitively pulsed copper vapor lasers. Inverse population and the laser transitions in atoms and molecules. Theses reports. Tomsk: TSU SFTI. 1986. Part I. The active medium and lasers on transitions of atoms and small molecules. P.133-134.

110. Zapesochnyi I.P., Kelman V.A., Klimovskii I.I., et al. Effect of irregularities settings repetitively pulsed high-voltage discharge in a mixture of copper vapors with inert gases on the kinetics of relaxation metastable atoms of copper. III All-Union Conference on Physics gas discharge. Abstracts. (Kiev, 21-23 October 1986). P.98-100.

111. Berwick E.B., Isaev A.A., Mihkelsoo V.T., et al. Spectroscopy active medium copper vapor laser. Preprint FIAN No.251. Moscow:1986.
112. Isaev A.A., Petrash G.G., Ponomarev I.V., Kvantovaya elektronika. 1986. V.13. No.11. P.2295-2301.
113. Isaev A.A., Kazakov V.V., Forest M.A., et al., Kvantovaya elektronika. 1986. V.13. No.11. P.2302-2309.
114. Isaev A.A., Mihkelsoo V.T., Petrash G.G., et al. Experimental study of the kinetics of the populations of atomic levels in a vapor laser Copper Works Institute of Physics, Academy of Sciences of the Estonian SSR. 1987. V.60. P.90-107.
115. Isaev A.A., Mihkelsoo V.T., Petrash G.G., et al. Spatial and time kinetics of excitation and relaxation of the atomic levels in plasma pulsed copper vapor laser. Preprint FIAN No.171. Moscow: 1987.
116. Batenin V.M., Zapesochnyi I.P., Kelman V.,A and others. The radial inhomogeneity of the plasma parameters in the interpulse period of self-heating copper vapor laser. Preprint IVTAN No.5-210. Moscow: 1987.
117. Isaev A.A., Mihkelson V.T., Petrash G.G., et al., Kvantovaya elektronika. 1988. V.15. No.12. P.2510-2513.
118. Isava Y., Shimotsu V., Yamanaka Ch. et al. Density measurements of the lower level of a copper vapor laser , Proc. SPIE, Metal Vapor, Deep Blue, and Ultraviolet Lasers, J.J. Kim, et al., eds.,1989. Vol.1041. P.19-24.
119. Brown D.J.W., Kunnemeyer R., McIntoch A.I. Radial excited state density effect in a small-bore copper vapor laser, ibid. P.25-33.
120. Isaev A.A., Mihkelson V.V., Petrash G.G. et al. Atomic levels population and depopulation kinetics in Cu-vapor lasers , ibid. P.40-46.
121. Brown D.J.W., Kunnemeyer R., McIntoch A.I., IEEE J. Kvantovaya elektronika. 1990. Vol.26. No.9. P.1609-1619.
122. Petrash G.G., Isaev A.A.,Pulsed gas discharge lasers. Optics and Lasers (Proceedings of the Lebedev Physics Institute. V.212). 1991. P.93-108.
123. Isaev A.A., Petrash G.G., Kinetics of excitation and physical processes in active media of copper vapor and copper bromide vapor lasers , Metal Vapor Lasers and Applicanions: CIS Selected Papers, Petrash G.G., Editor, Proc. SPIE 2110. 1993. P.2-45.
124. Carman R.L., Browm D.J.W., Piper J.A., IEEE J. Kvantovaya elektronika. 1994. Vol.30. No.8. P.1876-1895.
125. Webb C.E., Hogan G.P. Copper laser kinetics - a comparative study , Pulsed Metal Vapour Laser. Proceedings of the NATO Advansed Research Workshop on Pulsed Metal Vapor Lasers — Physic and Emerging Applications in Industry, Medicine and Science, SV. Andrews, U.K., Aug. 6-10, 1995:Ch. 4. repetitively pulsed lasers C.E. Little and N.V. Sabotinov, Eds. P. Dordrecht: Nato ASI Series, Kluwer Academic Publishers. 1996. P.29-42.
126. Petrash G.G. Kinetics of Metal Vapour and Metal Halide Lasers. Ibid. P.43-54.
127. Smilanski I. Plasma Parameters of Metal Vapour Lasers. Ibid. P.87-99.
128. Kazakov V., Markova S.V., Petrash G.G., Kvantovaya elektronika. 1982. v.9. No.4. P.688-694.
129. Kelman V.A., Shpenik J.O., Zapesochnyi I.P., The study of relaxation of metastable 6p32D03/2 Bi atoms in bismuth vapor laser. Preprint KIYAI Ukrainian Academy of Sciences. No.91-3. Kiev. 1991.
130. Shpenik Yu.O., Kel'man V.A. Relaxation of the 6p32D03/2metastable Bi atoms in the bismut vapor laser. XX International Conference on Physics of Ionized Gases. Contributed Papers 6. Pisa. 1991. P.1221-1222.
131. Kazakov V.V., Markova S.V., Petrash G.G., Kvantovaya elektronika. 1983.

Vol.10. No.4. P.787-792.

132. Kazakov V.V., Markova S.V., Petrash G.G., Kvantovaya elektronika. 1984. V.11. No.5. P.949-956.

133. Prokopiev V.I., Solomonov V.I., Kvantovaya elektronika. 1985 vol.12. No.6. P.1261-1269.

134. Rozdestvenskij D.S., Work on the anomalous dispersion of metal vapor. – Moscow–Leningrad Publishing House of the USSR Academy of Sciences. 1951.

135. Frisch S.E., Determination of the concentration of normal and excited oscillator strengths of atoms and methods of emission and absorption of light. Spectroscopy of gas-discharge plasma. Leningrad. Nauka. 1970. P.7-62.

136. Borowicz B.L., Yurchenko N.I.,, Kvantovaya elektronika. 1984. V.11. No.10. P.2081-2095.

137. Diachkov L.G., Kobzev G.A., Zhurnal tekhnicheskoi fiziki. 1977. V.47. Vol.3. P.527-528.

138. Moore C.E. Atomic Energy Levels. Washington: NBS-467. 1952. Vol.2.

139. Scheibner K.F., Hazi A.V., Henry R.J., Physical Review A. 1987. Vol.35. No.11. P.4869-4872.

140. Weinstein L.A., Sobel'man I.I., Yukov E.A.,The excitation of atoms and broadening of the spectral lines. Moscow: .Nauka. 1979.

141. Kasabov G.A., Eliseev V.V.,Spectral table for the low-temperature plasma. Moscow: Atomizdat. 1973.

142. Frank-Kamenetskiy D.A., Lectures on plasma physics. Moscow: Atomizdat.1964.

143. Klimovskii I.I., Selezneva L.A., Teplofizika vysokikh temperatur. 1987. V.25. No.4. P.773-777.

144. Carman R.J. Computer modelling of longitudinally excited elemental copper vapour lasers, Pulsed Metal Vapour Lasers. Proceedings of the NATO Advanced Research Workshop on Pulsed Metal Vapor Lasers – Physics and Emerging Applications in Industry, Medicine and Science, SV. Andrews, U.K., Aug. 6-10, 1995: C.E. Little and N.V. Sabotinov, EdP. Dordrecht: Nato ASI Series, Kluwer Academic Publishers. 1996. P.203-214.

145. Batenin V.M., Burmakin V.A., Vokhmin P.A., etc., Kvantovaya elektronika. 1977. V.4. No.7. P.1572-1575.

146. Winkler W., Annalen der Physik. 1973. Folge 7. Band 29. Heft 1. P.37-46.

147. Mnatsakanyan A.H., Naydis G.V., N. Stern, Kvantovaya elektronika. 1978 V.5. No.3. P.597-602.

148. Elaev V.F., Soldatov A.N., Sukhanov G.B., Teplofizika vysokikh temperatur. 1980. V.18. No.5. P.1090-1092.

149. Gridnev A..G., Evtushenko G.S., Elaev V.F., et al. The temporary nature of the emission spectrum of a pulsed discharge in a copper vapor laser. Effective discharge metal vapor lasers. Tomsk, Institute of Atmospheric Optics SB RAS. 1978. P.160-171.

150. Gridnev A..G, Gorbunova T..M, Elaev V.F., et al., Kvantovaya elektronika. 1978 V.5. No.5. P.1147-1151.

151. Naydis G.V , Zhurnal tekhnicheskoi fiziki. 1977. V.47. Issue 5.P.941-945.

152. Elaev V.F., Melchenko V.S., Pozdeev V.V., Soldatov A.N., Time course of the electron density in the afterglow of the discharge in the vapor lasers copper. Effective discharge metal vapor lasers. Tomsk, Institute of Atmospheric Optics SB AS USSR. 1978. P.179-188.

153. Batenin V.M., Klimovskii II, Forest A.M., Seleznev L.A., Kvantovaya elektronika. 1980. V.7. No.5. P.988-992.

154. Gorbunova T.M., Elaev V.F., Reutov T.A., et al. Radiation decaying plasma in copper vapor lasers, and the possibility of its diagnosis. Population inversion

and lasing at transitions in atoms and molecules. Abstracts. Part I: The active medium and lasers on transitions of atoms and small molecules. Tomsk: SFTI TSU. 1986. P.194-196.

155. Pyatnitsky L.N., Laser plasma diagnostics. Moscow: Atomizdat. 1978.

156. Elaev V.F., Soldatov A.N., Sukhanov G. Teplofizika vysokikh temperatur. 1981. V.19. Issue 2. P.426-428.

157. Diachkov L.G., Kobzev G.A., The electron temperature in mehimpulsnom glow of metal vapor lasers. II All-Union Seminar on physical skim processes in lasers (Uzhgorod, 15-17 May 1978). Abstract. Uzhgorod. 1978. P.148-150.

158. Mc Daniel J. Collision processes in ionized gases: Translation from English. ed. L.A. Artsimovich. Moscow: Mir. 1967.

159. Biberman L.M., Vorob'ev VS, Yakubov I.T., The physics of non-equilibrium low-temperature plasma. Moscow: Nauka. 1982.

160. Methods of plasma. Ed. Lochte-Holtgreven V. Moscow: Mir. 1971.

161. Biberman L.M., Norman G..E , Usp. Fiz. Nauk. 1967. V.91. Issue 2. P.193-246.

162. Maltsev A.N., Kinetics of repetitively pulsed lasing on copper vapor. Preprint No.1 IAO Tomsk branch of the Academy of Sciences of the USSR. 1982.

163. Bohan P.A., Metal vapor lasers with collisional de-excitation of the lower working states. PhD Thesis. Tomsk. 1988.

164. Batenin V.M., Galkin A.F., Klimovsky I.I., Radial distribution of plasma parameters in continuosly pulsed high-voltage discharge afterglow in the mixtures of metal vapor and rare gases. 15th International conference on phenomena in ionised gases. 1981. Minsk. P.1108.

165. Batenin V.M., Galkin A.F.,Klimovskii I.I., Teplofizika vysokikh temperatur. 1982. V.20. No.5. P.806-811.

166. Batenin V.M., Galkin A.F.,Klimovskii I.I., Kvantovaya elektronika. 1981 V.8. No.5. P.1098-1100.

167. Tenenbaum K., Smilanski I., Levin L.A., Optics Communications. 1981. Vol.36. No.5. P.395-398.

168. Burlakov V.D., Gorbunova T.M., Mickle J.P.,et al. The use of the hooks method for studies of copper vapor lasers. Izv VUZ. Fizika. 1984 Dep. VINITI 04.09.84, N 2856-84 Dep. (Izv VUZ. Fizika. 1994. No.12. P.3-27).

169. Burlakov V.D., Gorbunova T.M., Loboda S.A.,et al., Teplofizika vysokikh temperatur. 1987. V.25. Issue 2. P.394-396.

170. Tikhonov A.I., Samarskii A.A.,The equations of mathematical physics. Moscow: Gostehteorizdat. 1953.

171. Koshinar I., Kryukov N.A .,Redko ETC., Optika i Spektroskopiya. 1981. V.50. Issue 1. P.62-66.

172. Isaev A.A., Mihkelson V.T., Petrash G.G., et al. Kinetics of primary state of copper atoms in the plasma discharge repetitively pulsed laser copper vapor. Preprint No.98. Moscow: FIAN. 1988.

173. Isaev A.A., Mihkelson V.T., Petrash G.G., et al., Kvantovaya elektronika. 1989. V.16. No.6. P.1173-1183.

5

The results of analytical studies of self-contained lasers

The theoretical research of self-contained lasers was started in 1967 by Leonard [1]. The first analytical solution for the output characteristics of lasers of this type was obtained in [2]. The decisive factor of the possibility of analytical solutions in [2] of the problem of the characteristics of the lasing of the self-contained laser was the use in [2] of the saturated power approximation [3], based on the assumption of the equality of the populations of laser levels. In [3,4,5], this approach has been successfully used to calculate the characteristics of ultraviolet lasing of a nitrogen laser (337.1 nm band). The principal drawback of [2] is that in it, as in [1], the authors used a closed three-level scheme of the metal atom, which ignores the possibility of settlement of the laser levels by electrons up (to higher levels in the continuum). However, as was shown in [6], the population settlement process by electron impact of the upper laser levels up have a decisive influence on the lasing characteristics of the laser at self-terminating transitions of metal atoms, which makes the closed three-level scheme unacceptable to describe the kinetics of excitation and generation of self-contained lasers.

The analytical solution of the problem of the lasing characteristics of the self-contained lasers in the approximation of the saturated power, taking into account the kinetics of population and resettlement of the laser levels was obtained in [7–16]. We now consider the results of some of these works.

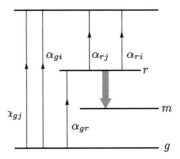

Figure 5.1. The open three-level scheme of the working atom of the laser at self-contained transitions. The thick arrow denotes the induced transitions. Thin arrows show the processes of population and resettlement of the resonant level by electron impact, as well as the process of resettlement of the ground state to the levels lying above the resonance and in continuum. Next to the arrow there are corresponding rate constants of these processes: α_{gr}, α_{gj}, α_{gi}, respectively, are the rate constants for the resettlement of the ground state to the resonance level, to the levels lying above the resonance level and in the continuum; α_{rj}, α_{ri} are respectively the rate constants for the resettlement of the resonant level to the higher-lying levels and in the continuum.

5.1. The characteristics of the lasing pulse of the self-contained lasers

In order to obtain a clear idea of the nature of the relationship of the characteristics of the lasing pulse with the processes that determine the kinetics of the population of levels in different self-contained lasers, the problem of the characteristics of the laser pulse was solved in [11] analytically. In this case, firstly, we used an open three-level scheme of the working atom, as shown in Fig. 5.1. Second, it was taken into account that during the excitation pulse, the main processes determining the kinetics of the population of laser levels are the processes of excitation and ionization of metal atoms, and reverse processes (de-excitation and recombination of electrons) practically do not play any significant role. Thirdly, it was assumed that the ratio of the rate constants of the processes determining the kinetics of population and resettlement of the laser levels is constant during the excitation pulse. In essence, this assumption is equivalent to the assumption of the constancy of the electron temperature during the excitation pulse, or, in other words, the excitation pulse means a rectangular pulse of the electron temperature. And finally, the fourth, the electron concentration was assumed to be an arbitrary function of time.

Given the assumptions made, the system of rate equations describing the kinetics of the population and the resettlement of the laser levels in the time interval from the beginning ($t = 0$) of the excitation pulse to the beginning ($t = t_1$) of the lasing pulse can be written as [11]:

$$dn_r / dt = \alpha_{gr} n_g n_e - \alpha_r n_r n_e, \tag{5.1}$$

$$dn_g / dt = -\alpha_g n_g n_e, \tag{5.2}$$

$$\alpha_r = \alpha_{ri} + \sum_j \alpha_{rj}, \quad \alpha_g = \alpha_{gr} + \sum_j \alpha_{gj} + \alpha_{gi}.$$

The initial conditions for the system (5.1), (5.2) at $t = 0$: $n_g = n_{g0}$, $n_r = 0$, $n_m = n_{m0}$. Here and below the second lower index (e.g. n_{m0}, n_{m1}) means that the given value of the concentration of certain particles refers to the appropriate point in time 0, t_1, etc. The solution of the system (5.1), (5.2) has the form:

$$n_g = n_{g0} \exp\left(-\int_0^t \alpha_g n_e dt \right), \tag{5.3}$$

$$n_r = \frac{\alpha_{gr} n_{g0}}{\alpha_r - \alpha_g} \left[\exp\left(-\int_0^t \alpha_g n_e dt \right) - \exp\left(-\int_0^t \alpha_r n_e dt \right) \right] \tag{5.4}$$

and holds up to time $t = t_1$, at which the populations of the resonant and metastable levels are compared, i.e. $n_{r1} = (g_r/g_m) n_{m0}$. Equation (5.4) for $t = t_1$ defines the time delay of the start of the lasing pulse to the excitation pulse $\tau_{del} = t_1$ and provided $n_{m0} \ll n_{g0}$ converts to

$$\tau_{del} \approx \frac{n_{m0} \xi}{\alpha_{gr} \bar{n}_e n_{g0}}, \quad \xi = g_r / g_m, \tag{5.5}$$

where the bar over the n_e denotes an average over time.

From the time $t = t_1$ until time $t = t_2$, corresponding to the end of the laser pulse, the system of rate equations for the concentration of atoms in the main, metastable and resonance levels is recorded in [11] in the approximation of the saturated power, as

$$dn_r / dt = \alpha_{gr} n_g n_e - \alpha_r n_r n_e - N, \tag{5.6}$$

$$dn_m / dt = N, \tag{5.7}$$

$$dn_g / dt = -\alpha_g n_g n_e, \tag{5.8}$$

$$n_r = \xi n_m, \tag{5.9}$$

where N — the specific lasing generation in the number of transitions. The initial conditions at $t = t_1$: $n_m = n_{m1} = n_{m0}$, $n_r = n_{r1} = \xi n_{m0}$, $n_g = n_{g1}$.

Solving the system (5.6)–(5.9), we have [11]:

$$n_g = n_{g1} \exp\left(-\int_{t_1}^{t} \alpha_g n_e dt \right), \tag{5.10}$$

$$n_r = \xi n_m = C_1 \exp\left(-\int_{t_1}^{t} \alpha_g n_e dt \right) - C_2 \exp\left(-\int_{t_1}^{t} \frac{\alpha_r}{\mu} n_e dt \right), \tag{5.11}$$

$$C_1 = \frac{n_{g1} \alpha_{gr}}{\alpha_r - \mu \alpha_g}, \ \mu = 1 + \frac{1}{\xi}, \ C_2 = C_1 - n_{r1} = C_1 - \xi n_{m1},$$

$$N = \frac{n_e}{\xi} \left[C_2 \frac{\alpha_r}{\mu} \exp\left(-\int_{t_1}^{t} \frac{\alpha_r}{\mu} n_e dt \right) - C_1 \alpha_g \exp\left(-\int_{t_1}^{t} \alpha_g n_e dt \right) \right]. \tag{5.12}$$

Given that at $t > t_1$ and using the assumptions made in [11], the lasing energy in the number of induced transitions is equal to $\theta(t) = n_m(t) - n_{m0}$, and from (5.11)

$$\theta(t) = \frac{C_1}{\xi} \exp\left(-\int_{t_1}^{t} \alpha_g n_e dt \right) - \frac{C_2}{\xi} \exp\left(-\int_{t_1}^{t} \frac{\alpha_r}{\mu} n_e dt \right) - n_{m0}. \tag{5.13}$$

From (5.12) provided that $t = t_2$, $N = 0$, we obtain a relation for the limiting (maximum) lasing pulse duration $\tau_{g\,lim}$

$$\int_{t_1}^{t_2} \alpha_g n_e dt = \alpha_g \bar{n}_e \tau_{g\,lim} = \frac{1}{1 - \alpha_r / (\mu \alpha_g)} \left(\ln \frac{C_1 \mu \alpha_g}{C_2 \alpha_r} \right). \tag{5.14}$$

In view of (5.14), from (5.11) and (5.12) we obtain that for the limiting specific lasing energy corresponding to the maximum lasing pulse duration,

$$W_{lim} = W_0 \frac{n_{g1}}{n_{g0}} \left(\frac{C_1}{C_2} \right)^{\mu \alpha_g / (\alpha_r - \mu \alpha_g)} - h\nu_1 n_{m0}, \tag{5.15}$$

$$W_0 = hv_1\theta_0 = hv_1 n_{g0} \frac{\alpha_{gr}}{\xi\alpha_r}\left(\frac{\mu\alpha_g}{\alpha_r}\right)^{\mu\alpha_g/(\alpha_r - \mu\alpha_g)}, \tag{5.16}$$

where W_0 and θ_0 are the limiting specific lasing energies at the zero prepulse concentration of metastable atoms in units of energy and in the number of transitions. The value θ_0 also determines the concentration of metastable atoms formed as a result of lasing at the maximum lasing pulse duration.

In the derivation of the expression that defines the limiting prepulse concentration of metastable atoms $n_{m0\ lim}$, at which lasing is impossible, in [11] the authors used the obvious reason that this concentration should be related by (5.9) with the highest concentration of atoms on the resonant level, which is realized in the absence of induced transitions. By differentiation with respect to time of (5.4) and equating the derivative dn_r/dt to zero we can determine time t'_1, at which in the absence of stimulated emission the concentration n_r reaches its maximum value

$$\int_0^{t'_1} \alpha_0 n_e dt = \frac{\alpha_g}{\alpha_r - \alpha_g} \ln \frac{\alpha_r}{\alpha_g} \tag{5.17}$$

and the corresponding point in time the relationship between exponential appearing in (5.4),

$$\exp\left(-\int_0^{t'_1}\alpha_g n_e dt\right) = \frac{\alpha_r}{\alpha_g}\exp\left(-\int_0^{t'_1}\alpha_r n_e dt\right). \tag{5.18}$$

In view of (5.17), (5.18), the desired expression for $n_{m0\ lim}$ follows from (5.4) subject to $t_1 = t'_1$

$$n_{m0\ lim} = \frac{\alpha_{gr}}{\xi\alpha_g}n_{g0}\left(\frac{\alpha_\mu}{\alpha_g}\right)^{\alpha_r/(\alpha_r - \alpha_g)}. \tag{5.19}$$

Assuming that the average electron energy is constant during the excitation pulse and the atoms of the working metal in the states lying above the resonance are immediately ionized by electron impact and, in addition, by neglecting the stepped ionization buffer gas atoms and energy losses by electrons in elastic collisions with heavy particles, we can write fairly simple expressions for the power density Q, input to the discharge at time t ($t_1 \le t \le t_2$), and for the specific energy G supplied to the discharge at this point of time t:

$$Q(t) = \alpha_g n_g(t) n_e I_i - N(t)(\xi + 1)(I_i - E_r) + An_a n_e, \tag{5.20}$$

$$G(t) = \left[n_{g0} - n_g(t)\right] I_i - (I_i - E_r) \times$$
$$\times n_r(t) - (I_i - E_r)\ \theta(t) + An_a F, \tag{5.21}$$

$$A = \sum_k \alpha_{ak} E_{ak}, \quad F = \int_0^t n_e dt, \quad n_r = \frac{g_r}{g_m}(\theta + n_{m0}), \tag{5.22}$$

where E_{ak} and a_{ak} are respectively the energy and the rate constant of the k-th inelastic interaction of electrons with the buffer gas atoms, n_a is the concentration of buffer gas atoms.

With reference to equations (5.21) and (5.22) it should be noted that the first terms on the right sides of these equations determine the amount of energy transferred by the electrons of the metal atoms, on the assumption that every act of excitation of these atoms is accompanied by ionization of the atoms. The second term on the right side of equation (5.20) and the second and third terms on the right side of equation (5.21) determine, so to speak, the energy savings determined by the first terms, due to the fact that some of the atoms of the working metal remain at the resonance and metastable levels. The last term on the right sides of the equations (5.20) and (5.21) determines the amount of energy used for the excitation and ionization of the buffer gas atoms.

It should also be noted that when calculating the value of $Q(t)$ in the first term of the right hand side of (5.20), it seems to be more accurate to use, instead of the ionization potential I, the sum of $I_i + (3/2)\ kT_e$, thereby taking into account that the energy introduced into the discharge, is used not only for the ionization of atoms, but also for heating the electrons. At the same time, in the first term of the right hand side of (5.21) it is not necessary to produce such a substitution because the assumption of a rectangular excitation pulse (a rectangular pulse T_e) corresponds to the condition that the electron temperature at the end of the lasing pulse is equal to its prepulse value $T_{e0} \ll T_e$.

In view of (5.13) and (5.21), (5.12) and (5.20), the relations for the physical (I.10) and dynamic (I.11) efficiencies of the laser at self-terminating transitions of metal atoms take the form

$$\eta_p = h\nu_1 \theta(t) / G(t), \tag{5.23}$$

$$\eta_t = P(t) / Q(t), \tag{5.24}$$

$$P(t) = h\nu_1 N(t). \tag{5.25}$$

In view of (5.15), (5.21), (5.22), the physical efficiency, corresponding to the limiting specific lasing energy (5.15) equals

$$\eta_p = W_{\lim} / \left[\left(n_{g0} - n_{g2} \right) I_i - \left(I_i - E_r \right) \times \right.$$
$$\left. \times n_{r2} - \left(I_i - E_r \right) \frac{W_{\lim}}{h\nu_l} + A n_a F \right], \tag{5.26}$$

$$n_{r2} = \frac{g_r}{g_m} \left(\frac{W_{\lim}}{h\nu_1} + n_{m0} \right), \tag{5.27}$$

$$F = \int_0^{t_2} n_e dt.$$

Having expressions for calculating the basic characteristics of lasers based on self-terminating transitions, we consider another condition [11] which is associated with the finite length of the active medium and limits the scope of the above relations. This condition ($\tau_g \gg \tau_s$) defines the relationship between the duration of the lasing pulse τ_g and the time τ_s needed to create an inversion of the population inversion of the laser levels at which the saturation of the laser transition induced by radiation takes place.

According to [17], stimulated emission saturates the laser transition at

$$\kappa_0 l_a \geq 30. \tag{5.28}$$

Under the assumption of zero prepulse concentration of metastable atoms and taking into account (5.28), τ_s can be defined as the time required to achieve a concentration of resonantly excited atoms equal to

$$n_r \approx 30 / \sigma_0 l_a, \tag{5.29}$$

where σ_0 is the stimulated transition cross section at the centre of the line.

Under the assumption that by the time the condition (5.28) is satisfied, first, the ionization of working atoms takes place by direct electron impact, and their concentration decreases slightly and, secondly, the processes of destruction of the resonant level up do not play any significant role, the rate equations, describing the kinetics of ionization and excitation of the upper laser level in an open three-level scheme, shown in Fig. 5.1, will have the form

$$dn_e / dt = \alpha_{gi} n_e n_g,$$
$$dn_r / dt = \alpha_{gr} n_e n_g,$$

(5.30)

with the initial conditions at $t = 0$: $n_e = n_{e0}$; $n_r = 0$.

Using (5.29) and a solution of (5.30), condition $\tau_g \gg \tau_s$ transforms to

$$\tau_g \gg \left(\alpha_{gi} n_{g0}\right)^{-1} \ln\left(\frac{30\alpha_{gi}}{\sigma_0 l_a \alpha_{gr} n_{e0}} + 1\right).$$

(5.31)

5.2. Lasing characteristics of copper vapour lasers

In calculating the lasing characteristics of a copper vapour laser, the authors of [11,12,16] used the open five-level scheme, shown in Fig. 5.2, consisting of two three-level schemes (Fig. 5.1) with a common lower level.

The function of electron energy distribution was assumed in [11,12,16] to be Maxwellian. The excitation cross sections of resonance levels $^2P^0_{1/2}$, $^2P^0_{3/2}$, from the ground $^2S_{1/2}$ and the corresponding metastable $^2D_{3/2}$ and $^2D_{5/2}$ levels, the cross sections of resettlement of all these levels in the states lying above the resonance ones, and also the ionization cross sections of the resonance-excited and metastable atoms were calculated by the Grizinskii approximation formula [19] proposed in [18]. In calculating the constants α_g, α_m, α_r transitions with the oscillator forces $f \geq 0.01$ were taken into account. It should be noted that the comparison the excitation of the cross section of resonance term $^2P^0_{1/2, 3/2}$ calculated in this way with the corresponding cross sections, measured in [20–22], shows that the calculation results are close to the most reliable results of measurements [22].

The ionization cross section σ_{gi} of the copper atom from the ground state is calculated by the semi-empirical formula proposed in [23], and is in good agreement with the experimental data [24]. The rate constants α_{gm} for the excitation by electrons of the metastable levels $^2D_{5/2}$ and $^2D_{3/2}$ from the ground state were calculated as follows. Initially, the measured [20] excitation cross sections of the resonant term $^2P^0_{1/2, 3/2}$ and the levels $^2D_{5/2}$ and $^2D_{3/2}$ for each pair of levels were calculated from the dependence of the ratios of rate constants of excitation $\alpha_{gr}(T_e)/\alpha_{gm}(T_e)$ on the electron temperature T_e. Then the desired value $\alpha_{gm}(T_e)$ for a given value of T_e was determined by dividing the value of the corresponding constant α_{gr}, calculated by the previously described method, by the corresponding ratio of the constants $\alpha_{gr}(T_e)/\alpha_{gm}(T_e)$.

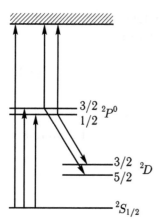

Figure 5.2. Open five-level scheme of the copper atom.

A comparison shown in Fig. 5.3 of the results of calculation of the rate constants of various elementary processes in the open five-level scheme of the copper atom (Fig. 5.2) shows that for values of $T_e > 1.5$ eV, the mixing of the working levels and excitation of the metastable levels from the ground state can be neglected. In addition, to simplify the calculations, in [9,11,12,16] no account is made of the resettlement of the metastable levels to the states lying above the resonance levels and in the continuum, since the neglect of this process can lead only to a certain underestimation of the value of specific lasing energy W_0 in comparison with the actual values.

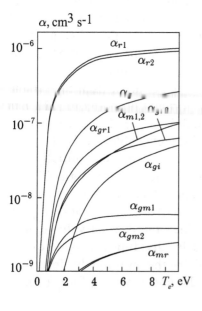

Figure 5.3. The rate constants of the population band resettlement of the lower levels of the copper atom. Index 1 corresponds to the levels of $^2D_{5/2}$, $^2P^0_{3/2}$, index 2 to levels $^2D_{3/2}$, $^2P^0_{1/2}$.

The probability of spontaneous decay of the levels $^2P^0_{3/2}$ and $^2P^0_{1/2}$ in laser transitions corresponding to the oscillator strengths given in [25] is $1.96 \cdot 10^6$ s^{-1} and $1.96 \cdot 10^6$ s^{-1}. The corresponding values of the lifetimes are about 510 and 595 ns. As in all experiments with a copper vapour laser with conventional resonators consisting of a dead end mirror and a glass or quartz plate, the duration of the laser pulse on the base is at the level of 50–60 ns, i.e., significantly less than the given values of lifetimes, in the calculations [9,11,12,16], the spontaneous decay of resonantly excited atoms by laser transitions was not considered, and the lasing characteristics of the copper vapour lasers were calculated using the formulas in the previous section 5.1.

The calculation results [16] of the limiting energy density generated by θ_0 (5.16) for green and yellow lines are shown in Fig. 5.4. The excess observed in almost the entire temperature range of the electrons was approximately 1.7 times for the specific lasing energy generation θ_0 on the 510.5 nm line above the specific lasing energy on the line 578.2 nm and is in good agreement with measurements of average lasing power and energy of copper vapour lasers in these lines at low pulse repetition frequencies.

Comparison of the calculation results shown in Fig 5.5 of the specific lasing energy W_0 on the line 510.5 nm and the results of measurements of the specific lasing energy of the copper vapour laser with a GDT 7 mm in diameter [26] shows that the theoretical value is more than three times higher than the value measured in the experiment. According to [11, 16], the main reason for the decrease in the specific lasing energy of copper vapour lasers, operating in the

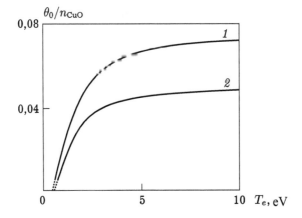

Figure 5.4. The dependence of the maximum lasing energy density of the copper vapour laser for green (1) and yellow (2) lines on the electron temperature.

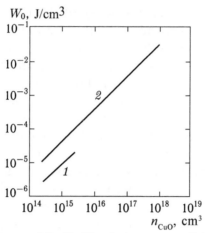

Figure 5.5. Dependence of the limiting lasing energy of the copper vapour laser on the concentration of copper atoms. 1 – experiment [26], 2 – calculation by (5.16) for the green lasing line, T_e = 5 eV.

single-pulse mode, compared with its limiting value lies primarily in the presence of the inductance of the discharge circuit. The voltage drop across this inductance reduces with time of the strength of the electric field in the discharge and results in a corresponding decrease in the electron temperature and interruption of lasing as a result of this.

The nature of the influence of the prepulse population of the metastable levels on limiting specific lasing energy, on the lasing pulse duration and its delay time relative to the start of the excitation pulse is illustrated by the dependences presented in Fig. 5.6 [11]. It follows from these dependences that at T_e = 5 eV, the maximum allowable prepulse concentration of copper atoms on the level $^2D_{5/2}$ is 0.088 of the prepulse concentration of ground-state atoms, and the corresponding maximum permissible temperature of the population of the metastable level is $T_{m\ lim} \approx 4500$ K. In accordance with (5.19), with a decrease in the electron temperature in the discharge, the maximum allowable concentration of metastable atoms and the corresponding value $T_{m\ lim}$ will also decrease.

In view of (5.12) (5.13) (5.20)–(5.22), as well as the presence of two lines of generation, and the assumption of zero prepulse concentration of metastable atoms expressions for the dynamic (5.24) and physical (5.23) efficiency of a copper vapour laser be rewritten as

$$\eta_{ti} = P_j / \left[\alpha_g n_g(t) n_e I_{Cu} - \sum N_j (\xi_j + 1)(I_{Cu} - E_{rj}) + \right.$$

$$\left. + \sum_k \alpha_{ak} \varepsilon_{ak} n_a n_e \right], \tag{5.32}$$

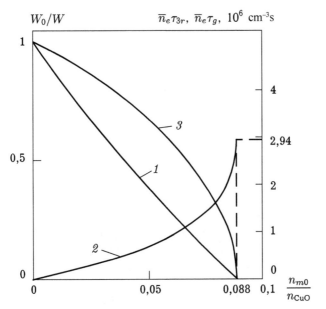

Figure 5.6. Effect of prepulse concentration of metastable atoms on the characteristics of the lasing pulse on the green line of a copper vapour laser for T_e = 5 eV. 1 – the specific lasing energy generation, $2 - \bar{n}_e \tau_{\mathrm{del}}$, $3 - \bar{n}_e \tau_g$.

$$\eta_{t\Sigma} = \sum_j \eta_{tj}, \tag{5.33}$$

$$P_j = h\nu_{1j} \frac{\alpha_{gr} n_{g0} n_e}{\xi_j (\alpha_{rj} - \mu_j \alpha_g)} \times$$
$$\times \left[\frac{\alpha_{rj}}{\mu_j} \exp\!-\!\left(\frac{\alpha_{rj}}{\mu_j} F \right) - \alpha_g \exp\!\left(-\alpha_g F \right) \right], \tag{5.34}$$

$$\Gamma_{\Sigma} - \sum_j \Gamma_j, \tag{5.35}$$

$$\eta_{pj} = W_j(t) / \left[\left[n_{g0} - n_g(t) \right] I_{Cu} - \sum_j (\xi_j + 1) W_j(t) \left(I_{Cu} - E_{rj} \right) + \right.$$
$$\left. + \sum_k \alpha_{ak} \varepsilon_{ak} n_a F \right], \tag{5.36}$$

$$\eta_{p\Sigma} = \sum_j \eta_{pj}, \tag{5.37}$$

$$W_j = h\nu_{1j} \frac{n_{g0} g_{grj}}{\xi_j \left(q_{grj} - \mu_j q_g\right)} \left[\exp\left(-q_g F\right) + \exp\left(-\frac{q_{rj}}{\mu_j} F\right)\right],$$ (5.38)

$$W_\Sigma = \sum_j W_j,$$ (5.39)

$$F = \int_0^t n_e \, dt = \bar{n}_e \tau_{\text{ex}},$$ (5.40)

where j takes the values 1 and 2, corresponding to the green and yellow lines of lasing.

The calculation of (5.32)–(5.40) was performed in [11, 12, 16], assuming that the buffer gas is neon. The value of $\sum \alpha_{ak} \varepsilon_{ak}$ in the ratios was approximately calculated as the sum (5.41)

$$\sum_k \alpha_{ak} \varepsilon_{ak} = \alpha_{ab} \varepsilon_{ab} + \alpha_{ai} I_{\text{Ne}},$$ (5.41)

where α_{ab} and α_{ai} are respectively the rate constants of excitation and ionization of neon atoms, calculated for a Maxwellian electron energy distribution and excitation and ionization cross sections of the neon atoms by electron impact from the ground state, carried out in [27]; ε_{ab} is the excitation energy of the neon atom assumed to be 19.2 eV; I_{Ne} is the ionization energy of the neon atom.

The results of calculations [12, 16] of the values θ/n_{g0}, η_{r1}, η_{p1} and θ_Σ/n_{g0}, $\eta_{r\Sigma}$, $\eta_{p\Sigma}$, presented in Figs. 5.7–5.9, indicate that copper vapour lasers with specific lasing energies which are less than the limiting values by an order of magnitude can be achieve values of the total physical and dynamic efficiency of 15–20%. Such an increase in the physical and dynamic efficiency, accompanied by a decrease in the proportion of atoms involved in the lasing, is due to lower electron energy losses in excitation and ionization of resonance-excited copper atoms. From a comparison of the dependences shown in Fig. 5.9, we see that for $T_e = 5$ eV the increase in the neon buffer gas pressure is accompanied by a marked decline in physical efficiency.

Figure 5.10 shows results of calculations [11, 16] by (5.36)–(5.38) for the physical efficiency of the green lasing line and the overall efficiency of the copper vapour laser, corresponding to the limiting values of specific energy W_0 (5.16) and maximum lasing pulse duration τ_g (5.14). As can be seen, the maximum value of the total physical efficiency $\eta_{p\Sigma}$ is slightly less than 8%.

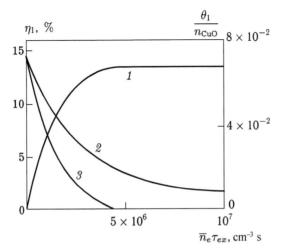

Figure 5.7. Depending on the specific lasing energy (1), physical (2) and dynamic (3) efficiency of a copper vapour laser on the parameter $\bar{n}_e \tau_{ex}$, $\lambda_1 = 510.5$ nm, $n_{m0} = 0$, $T_e = 5$ eV.

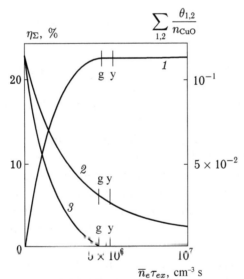

Figure 5.8. Dependence of the total specific lasing energy (1), physical (2) and dynamic (3) efficiency of a copper vapour laser on the parameter $\bar{n}_e \tau_{ex}$. Prepulse concentration of metastable atoms $n_{m10} = n_{m20} = 0$, $T_e = 5$ eV. Vertical lines denote the moments of completion of lasing on the green (g) and yellow (y) lines.

Neglecting the ionization of the buffer gas atoms, and the non-simultaneity of lasing pulses on the green and yellow lines, and assuming that the prepulse concentrations of resonance-excited and metastable atoms are equal to zero, the electron concentration $n_{e\,lim}$, formed by ionization of copper atoms at the end of the lasing pulse with duration limit $\tau_{g\,lim}$, can be estimated by the formula

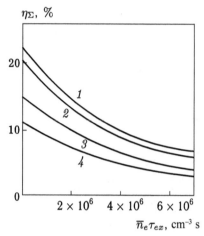

Figure 5.9. The dependence of the overall physical efficiency of the copper vapour laser on the values of the parameter for $\bar{n}_e \tau_{ex}$, $T_e = 5$ eV, the buffer gas – neon, the ratio n_{Cu}/n_{Ne}: $1 - \infty$; $2 - 10^{-1}$; $3 - 2 \cdot 10^{-2}$; $4 - 10^{-2}$.

$$n_{e\,\lim} = n_{Cu0} - n_{Cu\,\lim} - \sum_j n_{r_j\,\lim} - \sum_j n_{m_j\,\lim}, \qquad (5.42)$$

in which the values $n_{Cu\,\lim}$, $\Sigma n_{r_j\lim}$, $\Sigma n_{m_j\lim}$, are calculated by relations (5.9) and (5.10) (5.14) and (5.16).

The results of calculation of $n_{e\,\lim}$ from (5.42) and $n_{Cu\,\lim}$, $\Sigma n_{r_j\lim}$, $\Sigma n_{m_j\lim}$, from these relations are shown in Fig. 5.11 and suggest that in the case of limiting lasing time the concentration of copper atoms in the ground state is 65–40% of its prepulse value. At the same time, 10–20% of the atoms are in resonance and metastable levels, and 25–40% are ionized. This degree of ionization is sufficiently large, capable of having a significant impact on the recovery time of the prepulse properties of the active medium and, consequently, on the maximum repetition frequency of excitation pulses in repetitively pulsed lasers. In addition, this degree of ionization can lead to a significant mismatch of the resistance of the discharge and the discharge circuit impedance and, consequently, a decrease in lasing energy as compared to its limit.

5.3. Analytical solution of the characteristics of lasing pulse of copper vapour lasers with the development of ionization taken into account

The analytical solution of the lasing characteristics of the copper-vapour lasers is given in [13]. The problem was solved using the open five-level scheme of the copper atom. In the absence of lasing, the

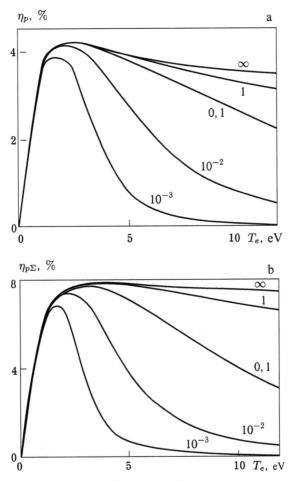

Figure 5.10. Dependence of the physical efficiency of a copper vapour laser at the limiting lasing energy W_0 on electron temperature. a: for the green lasing line, and b: total physical efficiency. The numbers at the curves characterize the relationship n_{Cu}/n_{Ne}.

kinetics of the populations of the different levels is described in [13] using the following equations:

$$dn_g / dt = -\alpha_g n_g n_e,$$ (5.43)

$$\frac{dn_{mk}}{dt} = \alpha_{gmk} n_g n_e - \alpha_{mk} n_{mk} n_e - \sum_{j=1}^{2} \alpha_{mkrj} n_{mk} n_e,$$ (5.44)

$$\frac{dn_{rk}}{dt} = \alpha_{grk} n_g n_e + \sum_{j=1}^{2} \alpha_{mkrj} n_{mk} n_e - \alpha_{rk} n_{rk} n_e, \quad k = 1, 2,$$ (5.45)

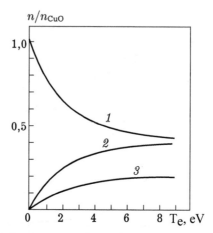

Figure 5.11. The relative concentrations of atoms of copper (1), electrons (2) and atoms in the metastable and resonance levels (3), realized in a copper vapour laser at the end of the lasing pulse at limiting lasing energy.

$$\frac{dn_e}{dt} = \alpha_{gi} n_g n_e + \sum_{j=1}^{2} \left(\alpha_{rj} n_{rj} + \alpha_{mj} n_{mj} \right) n_e, \tag{5.46}$$

$$\alpha_g = \alpha_{gi} + \sum_{j=1}^{2} \left(\alpha_{gmj} + \alpha_{grj} \right), \tag{}$$

where the subscripts 1 and 2 refer to the pairs of levels $^2D_{5/2}$, $2P^0_{3/2}$ and $^2D_{3/2}$, $2P^0_{1/2}$.

Substituting the value of n_e (5.43) into (5.44) and integrating, we obtain the relation for $n_{mk} = f_k(n_g)$. Then, using the relations for n_{mk} and n_g from (5.43), we can integrate equation (5.45) and find an expression for $n_{rk} = \varphi_k(n_g)$. By introducing the dimensionless variables $x = n_g / n_{Cu}$, $y_k = n_{mk}/n_{Cu}$, $z_k = n_{rk}/n_{Cu}$, where n_{Cu} is the concentration of heavy particles, the solution of (5.44) and (5.45) is reduced to the form [13]:

$$y_k = \left(y_k^0 - \mu_k x^0 \right) \left(x / x^0 \right)^{\beta_k} + \mu_k^x, \tag{5.47}$$

$$z_k = \frac{\alpha_{grk} + \sum_{j=1}^{2} \mu_j \alpha_{mj\,rk}}{\alpha_g (\gamma_k - 1)} x + \\ + \left(\frac{x}{x^0} \right)^{\beta_k} \sum_{j=1}^{2} \frac{\alpha_{mj\,rk} \left(y_j^0 - \mu_j x^0 \right)}{\alpha_g (\gamma_k - \beta_j)} + C_k x^{\gamma_k}, \tag{5.48}$$

$$\mu_k = \alpha_{gmk} \left/ \left(\alpha_{mk} - \alpha_g + \sum_{j=1}^{2} \alpha_{mkrj} \right) \right. ,$$

$$\beta_k = \frac{1}{\alpha_g} \left(\alpha_{mk} + \sum_{j=1}^{2} \alpha_{mkrj} \right), \quad \gamma_k = \alpha_{rk} / \alpha_g .$$

The value of C_k can be found from (5.48) if we put $x = x^0$, $z_k = z_k^0$ (x^0, y^0, z^0 are the initial values of variables x, y, z).

Taking into account the the law of conservation of the number of particles following from (5.43)–(5.46)

$$n_e + n_g + \sum_{j=1}^{2} \left(n_{mk} + n_{rk} \right) = n_{Cu}, \tag{5.49}$$

expressing n_e through n_g and the known functions n_{mk} and n_{rk}, substituting n_e in (5.43) and integrating, we obtain the following relation [13]

$$\alpha_g n_{Cu} t = - \int_{x^0}^{x} \frac{dx}{\left[1 - x - \sum_{k=1}^{2} \left(y_k(x) + z_k(x) \right) \right]} . \tag{5.50}$$

The explicit analytic form for the function $x(t)$ can be obtained from (5.50) only for some parameters β_k and γ_k. In general, to determine the dependence $x(t)$, the integral (5.50) should be tabulated numerically, and then, using expressions (5.47)–(5.49) to determined the level distribution of electrons for each time point.

When lasing takes place the equations (5.43)–(5.46) change. In [13], the authors considered the case when there lasing takes place at transition $r1 \rightarrow m1$ ($^2P^0_{3/2} \rightarrow {}^2D_{5/2}$) In the right-hand side of equation (5.44) for $k = 1$ we add the term N_1, and the right side of equation (5.45) N_1 is subtracted; N_1 describes in the number of transitions of the lasing power in the transition $r1 \rightarrow m1$. Under the assumption of saturated power ($n_{r1}/g_{r1} = n_{m1}/g_{m1}$) a new system of equations is solved like the system (5.43)–(5.46). In this case x, n_e, y_2, z_2 are still defined by (5.47)–(5.50), and

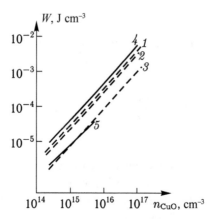

Figure 5.12. Dependence of the specific lasing in the line 510.5 nm on the concentration of copper atoms with T_e = 5 eV: 1 – T_0 = 0.2 eV; 2 – 0.3; 3 – 0.4; 4 – calculation [11], 5 – experiment [26].

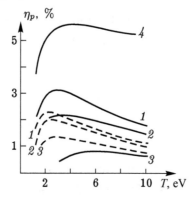

Figure 5.13. The dependence of the physical efficiency of a copper vapour laser on the electron temperature in the discharge line for the 510.5 nm line (solid lines) and for the line 578.2 nm (dashed lines): 1 – T_0 = 0.2 eV; 2 – 0.3; 3 – 0.4; 4 – results of calculations [11] for the 510.5 nm line.

$$z_1 = \frac{y_1}{\xi_1} = \frac{u_{gr1} + u_{gm1} + \mu_2 u_{m2r1}}{\alpha_{r1} + \xi_1 (\alpha_{m1} - \alpha_g) - \alpha_g} x + c'_1 \left(\frac{x}{x^0} \right)^{\gamma'_1} -$$

$$- \frac{\alpha_{m2r1} (y_0^2 - \mu_2 x^0)}{(1 + \xi_1)(\beta_2 - \gamma'_1)\alpha_g} \left(\frac{x}{x^0} \right)^{\beta_2}, \tag{5.51}$$

where $\gamma'_1 = (\alpha_{r1} + \xi_1 \alpha_{m1}) / [(1 + \xi_1)\alpha_g]$ and takes into account that $\alpha_{m1r2} = 0$. The value of C'_1 is obtained from (5.51) and the values of x and z_1 at the moment of time corresponding to the beginning of the lasing.

From the kinetic equations in the lasing mode we obtain the expression for the specific power of stimulated radiation in the number of transitions at the transition $r1 \rightarrow m1$

$$N_1 = \frac{n_e n_{Cu}}{(1+\xi_1)} \left\{ \left(\xi_1 \alpha_{gr1} - \alpha_{gm1} \right) x + \right.$$

$$\left. + \xi_1 \left[\alpha_{m1} - \alpha_{r1} + (1+\xi_1) \, \alpha_{m1 \, r1} \right] z_1 + \xi_1 \alpha_{m2 \, r1} y_2 \right\}. \tag{5.52}$$

When the lasing takes place at the transition $r2 \rightarrow m2$ ($^2P^0_{1/2} \rightarrow {}^2D_{3/2}$) in the corresponding kinetics equations we most consider term N_2, which describes the specific lasing power at this transition. In [13] it is indicated that the solution in this case is analogous to the solution described above. However, the results of the solution in view of their complexity are not given.

Figures 5.12 and 5.13 present the results of the calculation [13] by the above formulas of the values of specific lasing energy and the efficiency of a copper vapour laser. The initial populations of the levels of the copper atom were selected equilibrium with the temperature T_0. During the excitation pulse, the electron temperature is assumed to be constant and equal to some value of T_e. The rate constants of elementary processes were calculated by the formulas given in [19, 20] (the energy distribution function of the electrons was assumed to be Maxwellian). We used the approach of the instantaneous ionization levels lying above the resonance.

The physical efficiency of the laser, corresponding to the maximum lasing energy, was calculated from the ratio

$$\eta_p = W_k \Big/ \left(G - G_0 + \sum_{j=1}^{2} A_j \right), \tag{5.53}$$

$$W_k = \int_0^{\tau_{exk}} h\nu_{1k} N_k dt,$$

$$G = \sum_{j=1}^{5} \varepsilon_j n_j + \left(I_{Cu} + \frac{3}{2} kT_e \right) n_e,$$

where W_k is the specific energy in the lasing pulse, τ_{exk} is the duration of the excitation pulse, which coincides with the maximum lasing pulse duration on the k-th line, G is the energy left over after the lasing pulse (excitation), G_0 is the initial energy in the plasma, calculated

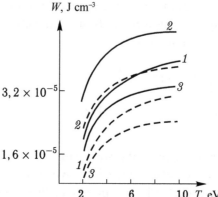

W, J cm^{-3}

$3,2 \times 10^{-5}$

$1,6 \times 10^{-5}$

2   6   10 T, eV

Figure 5.14. The influence of population and settlement processes of the metastable level $^2D_{5/2}$ on the specific lasing energy of the copper vapour laser on the line 510.5 nm ($T_0 = 0.2$ eV – solid lines, 0.3 – dashed): 1 – take into account the processes of population and resettlement, 2 – do not take into account the processes of population, 3 – processes of settlement not taken into account.

using the formula for G, in which the values of n_j and n_e are taken at the initial (prepulse) time, and $T_e = T_0$.

As can be seen from Figs. 5.12 and 5.13, the physical efficiency and specific lasing energy of the copper vapour laser depend substantially on the prepulse population of the metastable levels, which agrees well with the calculated results presented in Fig. 5.6. The lower values of physical efficiency, calculated in [13], compared with those calculated in [11, 12, 16] are explained by the authors of [13] by the fact that this paper takes into account the both prepulse population of the metastable levels of the copper atom, and the processes of population and settlement of the metastable levels. However, the most significant cause of the marked discrepancy is probably mainly in the fact that in [11, 12, 16], the calculation of physical efficiency was carried out under the assumption that the electron temperature in the discharge at the end of the excitation pulse instantaneously decreases to its prepulse value $T_e(\tau_{ex}) = T_{e0} << T_e$, while in the calculations of physical efficiency in [13], the electron temperature $T_e(\tau_{ex}) = T_e$.

Figure 5.14, taken from [13], illustrates the relative importance of the mechanisms of population and resettlement of the metastable levels on the example of the calculation results of the specific lasing energy of the copper vapour laser at the 510.5 nm line. It is seen that at low electron temperatures the dominant role is played by the process of population of the metastable level, but already at $T_e \geq 5$ eV the effect of both processes becomes comparable and they mutually compensate each other. Thus, the results of various analytical solutions of the problem of lasing characteristics of a single-pulse copper vapour laser are in good agreement with each other and indicate that even with the limiting specific laser energy output the total physical efficiency of the copper vapour laser can reach 5–8%. A reduction of the specific laser energy output should lead to an increase in physical efficiency.

5.4. The influence of self-absorption of stimulated emission on the lasing characteristics

In [28] in the numerical analysis of the effect of various factors on the lasing parameters of the copper vapour lasers it is concluded for the first time that a significant portion of the laser radiation can be lost in the active medium as a result of self-absorption. The condition in which the negative effect of self-absorption on the energy characteristics of copper vapour lasers can be neglected is written in [29] as

$$\frac{1+\rho_1}{1-\rho_1}n_{Cu}l_r \leq 3\cdot10^{17} \text{ cm}^{-2}, \tag{5.54}$$

where ρ_1 is the reflection coefficient of the output mirror (the reflection coefficient of the second mirror ρ_2 set equal to 1), n_{Cu} is the concentration of copper atoms, l_r is the length of the resonator.

The conclusion, drawn in [28], it is quite important, as in the existing copper vapour lasers the typical length of GDT and the concentration of n_{Cu} are respectively 100 cm and 10^{15} cm^{-3} and their product reaches a value close to the limit, as determined by the condition (5.54). Consequently, attempts to increase the lasing output power of the copper vapour lasers by significantly increasing the concentration of n_{Cu} are futile.

In [28] the mechanisms leading to annihilation of the induced quanta are not taken into account. At the same time, to explain their annihilation can not be explained only by self-absorption, since the latter can slow down the output of stimulated radiation from the active medium of the self-contained lasers and extended pulse generation, but can not lead to a reduction of its energy. In addition, in [28] nothing is said about how to change the condition (5.54) for the atoms of other metals, for which the cross section of absorption of radiation by the active transition varies quite widely[1].

In [15, 16] it is attempted to analyze the mechanisms responsible for the weakening of the stimulated radiation at its self-absorption, and to

---

[1]V.V. Buchanov holds a different view of the process of self-absorption of laser radiation in pulse metal vapour lasers (PMVL) in comparison with the view described here.

obtain conditions similar to (5.54) and applicable for various lasers at self-contained transitions. The analysis is based on the assumption that when the length of the active medium l_a is much higher than the mean free path of the induced photon, the laser radiation propagating along the active medium and absorbed by the metal atoms, converts them from the metastable state to the resonance state, thus compensating for the loss of the resonance-exited atoms as a result of their ionization, excitation and spontaneous decay in the underlying states. The characteristic time of the destruction of the resonance level is

$$\tau_r = \left(q_r n_e + A_r\right)^{-1}, \tag{5.55}$$

where A_r is the probability of spontaneous decay of the resonance level to the underlying states.

Obviously, in the case where the characteristic time τ_r is shorter than time τ_{mean} during which the stimulated radiation passes through the active medium, the self-absorption will greatly reduce the laser power. If the inequality

$$\tau_{mean} < \tau_r, \tag{5.56}$$

is fulfilled, then even in the presence of self-absorption, it should not lead to a significant decrease in lasing power.

Thus, the condition for the existence of self-absorption of stimulated radiation in lasers at self-contained transitions

$$\left(\sigma_n n_m\right)^{-1} < l_a, \tag{5.57}$$

where σ_n is the absorption cross section of the working transition, is necessary but not sufficient for determining the concentration of metal atoms and the length of the active medium in which there should be a deterioration of the lasing characteristics. When equation (5.57) is satisfied, the inequality depicts the small effect of self-absorption of laser radiation on the lasing power of lasers at self-contained transitions.

Time τ_{mean} can be calculated from the ratio

$$\tau_{mean} = \frac{l_{a\,eff}}{v_g}, \tag{5.58}$$

where $l_{a\,eff}$ is the effective length of the active medium through which passes the laser light before coming out of the resonator; v_g is the group velocity of this radiation.

As is known, the number of double passes n_{pass} through the cavity, leading to the release of part of this radiation through the resonator mirrors to reduce its intensity by a factor e, is

$$n_{pass} = \frac{1}{\ln(\rho_1\rho_2)}.$$ (5.59)

In view of (5.59), the effective length $l_{a\,eff}$ can be calculated as

$$l_{a\,eff} = l_a(1 + 2n_{pass}).$$ (5.60)

According to [29], the group velocity of light v_g is equal

$$v_g = \frac{c}{1+\beta}.$$ (5.61)

For Doppler and Lorentz contours of the spectral line parameter β has a value of

$$\beta_d = c\kappa_0(\ln 2)^{1/2}\pi^{3/2}\Delta v_d,$$ (5.62)

and

$$\beta_l = c\kappa_0 / 2\pi\Delta v_l,$$ (5.63)

where Δv_d and Δv_l are the widths of the Doppler and Lorentz contours.

Given (5.54) (5.58) and (5.61), we can transform (5.56) to the condition

$$(\alpha_r n_e + A_2)^{-1} > \frac{1}{c}l_a\left[1 - \frac{2}{\ln(\rho_1\rho_2)}\right](1+\beta),$$ (5.64)

which defined for the known values of κ_0 and n_e the maximum permissible length of the active medium.

To estimate the values of κ_0, realized in the lasers based on self-contained transitions, we use the results of [17], according to which in the resonatorless lasers saturation of the working transition induced by radiation takes place when the condition $l_a\kappa_0 \geq 30$, which for the examined case in view of (5.60), becomes

$$l_a\left[1 + \frac{2}{\ln(\rho_1\rho_2)}\right]\kappa_0 \geq 30.$$ (5.65)

is fulfilled. The correctness of the condition (5.65) can be tested in comparison with the results of [30, 31]. In [30], in addition to the gain factor on the green line of the copper vapour laser (κ_0 = 58 dB/m) there are also reflection coefficients of mirrors (ρ_1 = 0.3, ρ_2 = 1) and the length of the active medium (l_a = 80 cm). In [31], similar data are shown for manganese vapour lasers with lasing at a wavelength of

534.1 nm. Calculation of the values $\kappa_0 l_{a\,\text{eff}}$, conducted for the conditions [30, 31], taking into account (5.59) and (5.60), gives for a copper vapour laser $\kappa_0 l_{a\,\text{eff}} \approx 18$, and for a manganese vapour laser $\kappa_0 l_{a\,\text{eff}} \approx 21$. Both values are in quite satisfactory agreement with the condition (5.65). Therefore, to estimate the gain in the laser at self-contained transitions, we can use the relation

$$\kappa_0 \approx 30 / l_{a\,\text{eff}}. \tag{5.66}$$

Assuming that the contour of the gain line is of the Doppler type, and in view of (5.66), the relation (5.64) takes the form

$$l_{a\,\text{eff}} < \frac{\tau_r c}{2}\left[1+\left(1+\frac{120(\ln 2)^{1/2}}{\tau_r \pi^{3/2}\Delta v_d}\right)^{1/2}\right]. \tag{5.67}$$

Transforming (5.67) to the condition for the small influence of self-absorption stimulated radiation on the maximum lasing of the copper vapour laser on the green line and using the values $n_{\text{Cu0}} \approx 10^{15}$ cm^{-3}, $n_e \approx 0.3\, n_{\text{Cu0}}$, $\alpha_r = 3 \cdot 10^{-7}$ cm^3 s^{-1}, $A_r = 1.96 \cdot 10^6$ s^{-1} and $\Delta v_d \approx 6 \cdot 10^9$ Hz, we obtain the condition

$$n_{\text{Cu0}} l_a\left[1-\frac{2}{\ln(\rho_1\rho_2)}\right] < 3.5 \cdot 10^{17}\ \text{cm}^{-2}, \tag{5.68}$$

which satisfactorily agrees with (5.54), which once again confirms the correctness of the above-formulated assumptions about the mechanisms that lead to a decrease in laser radiation in the presence of self-absorption of this radiation.

Using (5.68) and the data, shown in Figure 5.4, and assuming that $\rho_1 = 0.08$, $\rho_2 = 1$, one can easily estimate the limiting lasing energy for a copper vapour laser at the output of the resonator $S_{\text{lim}} \approx 8 \cdot 10^{-3}$ J/ cm². Consequently, for a GDT with a diameter of 10 cm, the lasing energy of the copper vapour laser can reach values of about 0.6 J.

Reducing the influence of self-absorption on the energy characteristics of pulsed lasers at self-contained transitions can, in principle, be achieved through the excitation of the active medium of such lasers by moving discharged or ionization waves. In this case, the length of discharges (ionization waves) in the direction of their motion must satisfy the condition (5.67), and their velocity v_i must be less than the group velocity v_g of stimulated emission.

Currently, moving discharges or ionization waves are not used for excitation of lasers at self-contained transitions, using metal vapours as the active medium. However, there are studies (e.g. [32]), in which the ionization wave is used to excite the active medium of a nitrogen laser, which is a type of lasers at self-contained transitions. In [32] ionization waves were formed in a discharge tube (length 45 cm, internal diameter 9.5 mm), which formed a central conductor of the coaxial line, using a pulse of negative polarity with an amplitude of 300 kV and a duration of 40 ns. Filling the space between the discharge tube and a metal screen with water reduced the velocity of the ionization wave to about $3 \cdot 10^9$ cm/s.

In [32] it was found that the propagation velocity of the front of the laser radiation formed behind the ionization wave front, is about 10^{10} cm/s. In addition, in [32] it is reported that the transition from excitation of the nitrogen lasers with the waves with velocities of about $2 \cdot 10^{10}$ cm/s to ionization waves with velocities of about $3 \cdot 10^9$ cm/s leads to an increase of peak lasing power. Both these facts can be regarded as a manifestation of the effect of self-absorption on the formation of the laser radiation parameters in [32], whose results confirm not only the possibility but also the feasibility of the research and development of pulsed metal vapour lasers with excitation of the active medium by an ionization wave.

In concluding the results of the analytical solution of the characteristics of lasing of lasers at the self-contained transitions, it is necessary to note once again that this solution is based on the assumption of the constancy of the electron temperature during the excitation pulse, or, equivalently, on the assumption of a constant electric field in the discharge during the excitation pulse. Strictly speaking, for most real lasers this assumption is to some extent not satisfied. Nevertheless, the advantages of the analytical solution of the lasing characteristics of the self-contained lasers are obvious, as they not only demonstrate the correlation of the lasing characteristics with various elementary processes, but also determine the limiting values of lasing energy and the efficiency of these lasers, thereby directing the efforts of theoreticians and experimentalists to search for such operating conditions of the lasers that provide specific lasing energy and the efficiency of real lasers close to the limit.

5.5. Similarity relations for pulsed metal vapour lasers

A characteristic feature of real pulsed metal vapour lasers (PMVL) is the numerous and significant non-stationarity of physical processes that affect their work. Lasing times, population of atomic levels, the passage of light in the resonator and discharge of the storage capacitor through GDT are values of the same order. This greatly complicates the analysis of experimental data and makes it almost impossible to study theoretically the real PMVLs operating in the repetitively pulsed regime with the help of analytical formulas, because in this mode we must additionally consider the effect on the output characteristics of these lasers of the kinetics of recombination of the plasma in the interpulse interval.

The method which combines to some extent the clarity of the analytical solutions and at the same time is based on the results of physical or numerical experiments is the method of criterial similarity parameters and output characteristics of various lasers. The development of similarity criteria for PMVLs and their application to determine the characteristics of these lasers is the subject of an appreciable number of papers [28, 33–40].

In this section, following the results of [28], we discuss the similarity relation of the PMVL, which allow, in the first place, systematize the laser parameters, secondly, to simplify the analysis of full-scale and numerical experiments and, thirdly, to some extent indicate the ways to increase specific laser energy output and total energy per pulse. With the help of the similarity relations we can formulate quite general and at the same time useful conclusions about the features of PMVLs.

The similarity relations, relating to the period of the excitation pulse, were obtained in [28] based on the analysis of the system of equations describing the dynamics of PMVLs and including the equation of for the EEDF, the kinetics, one-dimensional transport of laser radiation and the external electric circuit.

Given the fact that during the excitation pulse the decisive role is played by binary collisions between particles when the velocity of the elementary processes is determined by the product of their concentrations, the above described original equations can be written in the new variables by dividing the right and left sides of the equations for the EEDF, the kinetics and radiation transport on n_M^2. In this case, the new variables are the relative concentrations of particles n/n_M (n is the electron concentration of excited atoms at different levels and photons) which depend on the reduced time $t' = tn_M$ and coordinates $x' = xn_M$. The solution of the systems of equations listed earlier in the reduced variables is determined by the following set of coefficients, referred to as similarity relations or parameters:

$$a_1 = U_0 / l_d n_M; \quad a_2 = L_c n_M S_d / l_d; \quad a_3 = C_s n_M l_d / S_d;$$

$$a_4 = n_M / n_B; \quad a_5 = l_a n_M; \quad a_6 = l_r / l_a; \quad a_7 = p_1; \quad a_8 = p_2; \tag{5.69}$$

$$a_9 = \{A_{ij} / n_M\}; \quad a_{10} = n_{e0} / n_M; \quad a_{11} = n_{m0} / n_M; \quad a_{12} = n_{p0} / n_M,$$

where in addition to the previously used notation n_B is the concentration of atoms of the buffer gas, and the prepulse concentration of the electrons, the metastable levels of the metal atoms and photons are marked by the index '0'. Parameter a_9 means the combination of relations A_{ij}/n_M. In the case of using a long cable line for the power supply to the electrodes of the discharge gap $a_3 = ZS_d / l_d$ (Z is the impedance of the cable line).

Lasers with the same set of parameters (5.69) are defined in [28] as the class of such lasers for given components of the mixture The reduced output radiation intensity I_l/n_M is a function of these parameters and the reduced time:

$$I_1 / n_M = F(a, tn_M),   (5.70)$$

where $\alpha = (\alpha_1, \alpha_2, ..., \alpha_{12})$.

The expression for the reduced specific laser energy output has the form

$$W / n_M = G(a),   (5.71)$$

where

$$G_a = \frac{1}{a_5} \int_0^\infty F(a,t') \, dt'.$$

According to (5.71) in such PMVLs for which the value of $G(a)$ is the same, specific laser energy output is proportional to the concentration of the metal atoms. In this case, the practical efficiency $\eta_r = 2Wl_d S_d /(C_s U_0^2) = 2G(a)/(a_1^2 a_3)$ in all such systems is the same. In addition, they have the same EEDF, the relative concentration of electrons, photons, and the level populations at similar time points $t = t'/n_M$. A corresponding analysis of the parameters (5.69), carried out in [28], shows that they have the following physical meaning. The parameters u_1 and u_1 determine the rate of energy input to the discharge, the EEDF and related characteristics. Specific energy input is proportional to the square of a_1. The parameter a_2 reflects the impact of the discharge circuit inductance on the characteristics of the discharge. Specific laser energy output is an increasing function of the parameter a_3, which together with a_1 defines the stored energy in the storage capacitor per atom of metal.

When designing PMVL we must take into account the limitations on the discharge circuit inductance and charging voltage on the capacitor. Accordingly, a_1 and a_2 determine the allowable geometric dimensions of the active volume and pressure of the gas. When the dimensions of the GDT are changed, parameter a_3 in accordance with the essential requirements can easily be 'tuned' because of the ability to change the capacitance.

The parameters a_5-a_8 determine the threshold conditions for lasing and laser energy output per one atom of the working metal. With a significant increase in the values of these parameters the laser energy output is reduced. The most significant of the parameters under consideration is $a_5 = l_a n_M$, which in practice changes in a wide range.

Usually, when modelling lasers with the lasing time much larger than the passage time of radiation in the resonator, instead of the equations of radiation transfer we use the balance equation for the photon density averaged over the resonator length. Numerical experiments, carried out at the Astrofizika company, have shown that this equation also applies when the photon lifetime in the resonator is comparable to the duration of the lasing pulse. In this case, the parameters a_5, a_7, a_8 are replaced by a single parameter

$$a_5' = n_M l_r / \ln\left(\frac{1}{\rho_1 \rho_2}\right). \qquad (5.72)$$

Parameter a_9 characterizes the effect of spontaneous decay of atomic levels, taking into account the reabsorption of radiation. Estimates show that usually we can restrict ourselves to the spontaneous decay of the working levels only. Specific energy and efficiency are decreasing functions of the parameter a_9. At high concentrations n_M when the lasing pulse duration is less than the lifetime of the upper laser level with respect to the decay of the laser transition, the influence of a_9 can be neglected. The above condition is satisfied in most cases in the vapour lasers on Cu, Au, Ba, etc. vapours.

The parameters $a_{10}-a_{12}$ reflect the similarity of the initial conditions for the prepulse concentrations of the electrons n_{e0}, metastable atoms n_{m0} and seed photons np_0. The values of W/n_M and η_r are decreasing functions of a_{10} and a_{11}. Changing the level of 'seed' radiation n_{p0} with all other parameters varying with reasonable ranges has no effect on the characteristics of lasers because the radiation power in the lasing process increases by many orders for a negligible time. Therefore, the influence of the parameter a_{12} on the output characteristics of lasers can in most cases be neglected.

When considering the specific mixtures the parameter $a_4 = n_M/n_B$ can be disregarded thereby extending the class of such systems. In particular, in a mixture of Cu–Ne the energy loss by ionization and excitation of neon are small in almost all existing lasers. The influence of neon affects mainly the drift velocity and average electron energy. In this case, the similarity parameters can be rewritten as follows:

$$a_1 = U_0 / \left\{ l_d \left[n_{Cu} n_{Ne} (1 + \gamma x) \right]^{1/2} \right\},$$

$$a_2 = S_d L_c n_{Cu}^2 / \left[l_d n_{Ne} (1 + \gamma x) \right],$$ (5.73)

$$a_3 = C_s (1 + \gamma x) \, l_d n_{Ne} / S_d,$$

where $x = n_{Cu}/n_{Ne}$; $\gamma \approx 20$ is the estimation ratio of transport cross sections of the copper atoms and neon. The other parameters remain unchanged. When using the cable line in (5.73) $a_3 = S_d Z_c n_{Cu} / [l_d n_{Ne} (1 + \gamma x)]$.

Estimates of the parameters a_1 and a_2 in (5.73), for lasers with transverse and longitudinal discharges, show that despite significant differences in the partial contents of the components, for both types of discharges we have the approximate similarity of the processes taking place.

Similarity relations define a class of lasers that have the same efficiency and specific laser energy output per one metal atom. At the same time within a class there can vary significantly the geometry, the concentration of atoms, the output energy, the parameters of the circuit. Therefore, having in possession the characteristics of the particular laser, we can calculate them for any other kind of laser. Examples of specific use of the similarity criteria to analyze the characteristics of copper vapour laser are contained in section 8.4.5.

References

1. Leonard D.A. , IEEE J. Quantum Electronics. 1967. Vol. QE-3. No.9. P.380-381.
2. Isaev A.A, Petrash G.G., Research pulsed gas lasers atomic transitions. Pulsed gas discharge lasers on transitions of atoms and molecules. (Proceedings of the Lebedev Physics Institute, V.81). Moscow: Nauka, 1975. P.3-87.
3. Gerry E.V., Appl.Physics Letters. 1965, Vol.7, No.1, P.6-9.
4. Ali A.W., Kolb A.C., Anderson A.D., Appl. Optics. 1967, Vol.6, No.12, P.2115-2119.
5. Ali A.W., Appl. Optics. 1969. Vol.2. No.5. P.993-996.
6. Eletskii A.V., Zemtsov J.K., Rodin A.V., Starostin A.N., Dokl. AN SSSR. 1975. V.220. No.2. P.318-321.
7. Batenin V.M., Vokhmin P.A., Klimovskii I.I., Kobzev G.A., Teplofizika vysokikh temperatur. 1976. V.15. No.6. P.1316-1319.
8. Klimovski I.I., Vokhmin P.A., Connection of the copper vapour laser emission pulse characteristics with plasma parameters. Proceedings of the XIII International Conference on Phenomena in Ionised Gases 1977. Berlin. 1977. Contributed Papers. Part II. p.635-636.
9. Vokhmin P.A., Klimovskii I.I., Selezneva L.A.,On the question of the efficiency of copper vapour lasers. II All-Union Seminar on physical processes in gas lasers (Uzhgorod, 15-17 May 1978). Abstracts. – Uzhgorod, Uzhgorod State. University. 1978. P.140-142.

10. Batenin V.M., Vokhmin P.A., Klimovskii II Influence of different processes on the parameters elmentarnyh pulse lasers to generate self-terminating. II All-Union Seminar on physical processes in gas lasers (Uzhgorod, 15-17 May 1978): Abstracts. Uzhgorod: Uzhgorod State. University. 1978. P.140-142.

11. Vokhmin P.A., Klimovskii I.I. , Teplofizika vysokikh temperatur. V.16. No.5. P.1080-1085.

12. Batenin V.M., Vokhmin P.A., Klimovskii I.I., Selezneva L.A., Teplofizika vysokikh temperatur. 1982. V.20. No.1. P.177-180.

13. Voronyuk L.V., Grechko L.G., Komarov O.V., et al., Teplofizika vysokikh temperatur. 1984. V.22. No.2. P.243-247.

14. Voronyuk L.V., Grechko L.G., Pinkevich I.P., et al., Kvantovaya elektronika. National Interdepartmental collection of scientific papers. Kiev. Naukova Dumka, 1985. Vyp.28. P.99-101.

15. Klimovskii I.I., Selezneva L.A. Effect of induced absorption radiation on the lasing characteristics of self-terminating transitions. Population inversion and lasing on transitions of atoms and molecules: abstracts. Part I: The active medium and lasers on transitions of atoms and small molecules. Tomsk: SFTI, TSU. 1986. P.141-142.

16. Klimovskii I.I., Self-contained lasers PhD Thesis. Moscow. 1991.

17. Casperson L.W. , J. Appl. Physics. 1977. Vol.48. No.1. P.256-262.

18. Green A.E., Rocketry and Astronautics (Russian translation). 1966. No.5. P.3-11.

19. Grysinski M., Phys. Rev. 1965. Vol.138. No.2A. P.305-321.

20. Trajmar P., Williams W., Srivastava P.K., J. Phys. B: Atom. Molec.Phys. 1977. Vol.10. No.16. P.3323-3333.

21. Borozdin V.S., Smirnov Y.M, Sharonov Y.D., Optika i Spektroskopiya. 1977. V.43. Issue 2. P.384-386.

22. Alexakhin I.S., Borovik A.A., Starodub V.P., Shafranyosh I.I., Journal of Applied Spectroscopy. 1979. V.30. Issue 2. P.236-239.

23. Lotz W.Z., Physik. 1970. Bd. 232. Heft 2. Z.101-107.

24. Pavlov S.I., Rakhiv V.I., Fedorov G.M., Journal of Experimental and Theoretical Physics. 1967. V.52. Issue 1. P.21-28.

25. Kasabov G.A., Eliseev V.V., Spectroscopic tables for low-temperature plasma. Moscow: Atomizdat. 1973.

26. Bohan P.A., Nikolaev V.N., Solomonov V.I., Kvantovaya elektronika. 1975. V.2. No.1. P.159-162.

27. Brown P., Elementary processes in plasma of gas discharge. Translationfrom English ed. Frank Kamenetskii. Moscow: Gosatomizdat. 1961.

28. Arlantsev S.V., Borovichi B.L., Buchanov V.V., et al , Kvantovaya elektronika 1983. Vol.10. No.8. P.1546-1552.

29. Casperson L., Yariv A., PhyS. Rev. Letters. 1971. Vol.26. No.6. P.293-294.

30. Walter W.T., Solimene N., Piltch M., Gould G., IEEE J. Quantum Electronics. 1966. Vol.QE-2. No.9. P.474-479.

31. Piltch M., Walter W.T., Solimene N., Gould G., Appl. Physics Letters. 1965. Vol.7. No.11. P.309-310.

32. Abramov A.G., Asinovsky E.I., Vasilyak L.M., Kvantovaya elektronika. 1983. Vol.10. No.9. P.1824-1828.

33. Arlantsev S.V., Borovichi B.L., Buchanov V.V., et al. Comparison of experimental studies rezultatov copper vapour laser with numerical calculations. Report of the NGO "Astrophysics". 1980.

34. Kravchenko V.F., Discharges for pulsed gas lasers. Manuscript submitted to VINITI, No. 5534-81. Dep. 4.12.1981.

35. Kravchenko V.F., The method of physical modeling of pulsed gas discharge lasers. Preprint FIAN No.271. Moscow. 1983.
36. Kravchenko V.F. , Izv. VUZ. Fizika. 1983. No.11. P.111-112.
37. Kravchenko V.F., Similarity criteria and ways of optimization of pulsed gas discharge lasers. Preprint IO AN No.216. 1984.
38. Kravchenko V.F., Prokhorov A.M., Electronic equipmens. Ser. II. Laser. equipment and optoelectronics. 1986. No.3 (327). Pulsed lasers. Moscow: Central Research Institute of Electronics P.10-11.
39. Yurchenko N. Theoretical study of copper vapour lasers, Metal Vapour Lasers and Their Applications. CIS Selected Papers, G.G. Petrash, Editor. Proc. SPIE. 1993. Vol.2110. P.78-89.
40. Kravchenko V.F., J. Russian Laser Research. 1994. Vol.15. No.1. P.83-89.

Numerical studies of pulsed metal vapour lasers

Currently, most interest in terms of practical applications is attracted by pulsed metal vapour lasers (PMVL) constructed on the circuit with stagnant excitation of longitudinal or transverse electrical discharges. In both cases, the problem of constructing PMVL with the required radiation characteristics must be solved on the basis of minimizing the economic costs associated mainly with the need for careful preliminary study of the working processes in these lasers. Discharge plasma in metal vapours is characterized by significant non-stationarity of many physical parameters, a detailed analysis of which in a broad range of experimental conditions is often extremely difficult. Taking into account that the metal vapour lasers for some metals have not as yet been sufficiently investigated, and the construction of their existing models involve various costs, it can be considered most appropriate to study these lasers by numerical simulation of processes in their active medium or, in other words, by numerical experiments aimed at optimizing and predicting the characteristics of lasers.

It should also be noted that the numerical experiments have a number of important advantages compared with full-scale experiments. Primarily, this relates to the mobile study of the influence of a particular factor or group on the dynamics of processes and output characteristics of the studied object, the relatively low cost of numerical experiments, the possibility of analyzing a large number of choices of initial data for a small period of time in comparison with preparation and similar full-scale experiment, etc.

Apparently, the first studies [1–3] for the numerical simulation of the lasing characteristics of pulse-periodic metal vapour lasers were carried out in 1975–1981. Subsequently, these studies were continued and, in addition, studies appeared in which the calculations were performed using a simplified scheme, dividing the period of operation of the laser into excitation and relaxation periods. Moreover, during the excitation period calculations were carried out to determined the kinetics of only the ground state and laser levels, high-lying levels were considered in the approximation of their instantaneous ionization by electron impact[1]

Chapter 6 presents mostly the results of numerical studies of the pulsed metal vapour lasers carried out in 1975–1990 at the Astrofizika Company. As required, these results are supplemented by the results of similar studies conducted in other organizations.

6.1. Physical and mathematical formulation of the problem

As will be seen later, in this chapter are two basic methods of numerical modelling of electric discharge lasers based on the continuous counting of excitation and lasing pulses without dividing it into two time periods.

For both methods in the physical formulation of the problem [4] the following processes are taken into account: excitation, deactivation and ionization by the electron impact and in interaction of heavy particles with each other, spontaneous and induced transitions, photoionization and photorecombination, electronic recombination in triple collisions, particle diffusion. To these processes, considered for the metal atoms and buffer gas, we added the processes of radiation transfer taking into account the line profiles and the possible capture and also dissociative recombination of ions of the buffer gas their recharging on metal atoms.

As the results of calculations performed by the Astrofizika Company, and comparison with the experimental data, the model EEDF can be the Maxwell distribution of electron energy. Therefore, the rate constants for electron–atom collisions were averaged over the Maxwellian distribution with a variable average electron energy, which was found by solving the balance equation for the electron energy.

The mathematical formulation of the problem was to write the system of equations describing the processes occurring in the discharge circuit, and in the active medium, and determining its lasing characteristics:

---

[1]Studies ot this type [70–76] were carried out at the Institute of High Temperatures of the Academy of Sciences of the USSR in 1980–1982. A special feature of these studies is that they took into account approximately for the first time the effect of inhomogeneities of the radial distributions of the pre-pulse plasma parameters on the characteristics of the metal vapour pulsed–periodic lasers (see chapter 7), editor's comment

$$U(t) = L\frac{dI_d}{dt} + \left(R_d(t) + R\right) I_d;$$

$$\frac{dn_{ik}}{dt} = \sum_{j=1}^{N_k} \left\{ \delta_{1i} v_{jk}^{da} n_{jk} + \left[n_e \left(v_{jik}^{ea} - v_{ijk}^{ea} \right) + B_{jik} Q_{jik} \Delta_{jk} \right] \mu_{ij} \right\} -$$

$$- \sum_{j=1}^{j-1} A_{ijk} n_{ik} + \sum_{j=j+1}^{N_k} A_{jik} n_{jk} - \left(v_{ik}^{da} + \sum_v v_{ik}^{iv} \right) n_{ik} -$$

$$- \omega_{ik} n_e + \theta_{ik} - \psi_{ik\Sigma} + \delta_{1i} v_k^{d+} n_{ik}^+ +$$

$$+ \sum_{s=1}^{2} \sum_{j=1}^{N_k} \sum_{l=1}^{N_k} n_{lk} \left(\alpha_{jikls}^{aa} n_{jk} - \alpha_{ijkls}^{aa} n_{ik} \right) \mu_{ij};$$

$$\frac{dn_k^+}{dt} = \sum_{i=1}^{N_k} \left(\omega_{ik} n_e + \sum_v v_{ik}^{iv} n_{ik} - \theta_{ik} + \psi_{ik\Sigma} \right) - v_k^{d+} n_k^+;$$

(6.1)

$$\frac{dQ_{jik}}{dt} = h v_{jik}^0 \left(\Delta\Omega A_{jik} F_{jik}^0 n_{jk} + B_{jik} Q_{jik} F_{jik}^0 \Delta n_{jik} \right) q - v_{jik}^v Q_{jik};$$

$$\frac{d(n_e \varepsilon)}{dt} = \frac{I_d^2 R_d}{V} - \sum_{k=1}^{2} \left\{ D_k + \sum_{j=1}^{N_k} \left[\theta_{ik} \varepsilon + n_e \left(v_{ik}^{ie} - v_{ik}^{re} \right) G_{ik} - \right. \right.$$

$$- \sum_v v_{ik}^{iv} \left(h v - G_{ik} \right) n_{ik} - \psi_{ik\Sigma} H(\gamma) \gamma +$$

$$+ \sum_{j=1}^{N_k} n_e v_{ijk}^{ea} \mu_{ij} \Delta E_{jik} \right\} - n_e v_w^{ea} \varepsilon;$$

$$\frac{d(n_\Sigma \varepsilon_a)}{dt} = \sum_{k=1}^{2} \left\{ D_k + \sum_{i=1}^{N_k} \left[n_e v_{ik}^{ra} G_{ik} + \theta_{ik} \left(G_{ik} + \varepsilon \right) \right] \right\} -$$

$$- \frac{2\lambda_g I}{3kr^2} \left(\varepsilon_a - \varepsilon_{aw} \right),$$

where δ_{ij} is the Kronecker symbol, H the Heaviside function, Γ is the geometric factor for the transitions for which the resonator is tuned;

$$\mu_{ij} = 1 - \delta_{ij}, \quad n_e = \sum_{k=1}^{2} n_k^+, \quad \Delta n_{jik} = n_{jk} - n_{ik} \frac{g_{jk}}{g_{ik}},$$

$$\omega_{ik} = \left(v_{ik}^{ie} - v_{ik}^{re} - v_{ik}^{ra} - \sum_{v} v_{ik}^{rv} \right), \quad \theta_{ik} = \delta_{2k} v_{ik}^{rd} n_k^+,$$

$$\psi_{ik\Sigma} = \sum_{s=1}^{2} \sum_{j=1}^{N_k} \alpha_{ikjs}^{ia} n_{ik} n_{js}, \quad D_k = k_k^{ea} v_k^{ea} \left(\varepsilon - \varepsilon_a \right) n_e,$$

$$R_d = \frac{l_r m_e}{s_y e^2 n_e} \sum_{k=1}^{2} \left[v_k^{ea} + \sum_{i=1}^{N_k} \sum_{j=1}^{N_k} v_{ijk}^{ea} + \sum_{i=1}^{N_k} v_{ik}^{ie} + v_{ik}^{re} + \right.$$

$$\left. + v_{ik}^{ra} + \frac{\theta_{ik}}{n_e} + \sum_{v} v_{ik}^{rv} \right], \quad r = \min\left(r_x, r_y \right),$$

$$\Delta E_{ijk} = E_{ik} - E_{jk}, \quad G_{ik} = E_k - E_{ik}, \quad \gamma = E_{js} - G_{ik},$$

$$q = l_{gz} / l_r, \quad \Delta\Omega = v_g / \left(l_{gz} \cdot \pi l_r^2 \right),$$

$$v_{jik}^{v} = \frac{c}{l_r} \left(\frac{1 - \rho_+}{1 + \rho_+} + \frac{1 - \rho_-}{1 + \rho_-} \right),$$

or

$$v_{jik}^{v} = \frac{c}{2l_r} \ln\left(\rho_+ \rho_- \right)^{-1},$$

where l_r is the cavity length (the distance between the mirrors); for the rest of the transitions:

$$q = 1, \quad \Delta\Omega = \frac{1}{4\pi}, \quad v_{jik}^{v} = \frac{c}{2r}.$$

$t, n, \varepsilon, g, k, \lambda, v, \alpha$ is the time, concentration, mean energy, statistical weight, the proportion of energy transfer, thermal conductivity, the frequency, the rate constant of the collision process; N_k, E, Q, F is the number of levels taken into account, excitation energy, the spectral density of the contour lines of emission (absorption); U, I, R, L is the voltage, current, electrical resistance, inductance; r, $\Delta\Omega$, ρ_-, ρ_+ – the distance measured from the centre of the active zone to the wall, the solid angle of propagation of radiation, the reflection coefficients of semitransparent mirrors and the blind mirror; l, S, V are the geometric mean dimensions: length, area, volume.

Subscripts: e, a, Σ, i, j, l refer to the electrons, atoms, the total values, energy levels, characters k, s correspond to the components under consideration, 1, 2 refer to the metal atoms and buffer gas atoms in the ground state, combination of symbols ij determines the appropriate transition; indices d, w, r, g, x, y, z denote the discharge, the wall of GDT or GDC, the resonator, the active zone, the size on the coordinate axes (z – axis of the resonator).

Superscripts: +, v, 0 refer to the ions, photons, the centre of the emission line; ea, aa, ie, ia, iv, rv, re, ra, rd, da, d^+, de relate to the processes of electron–atomic and interatomic collisions, electron impact ionization and impact of heavy particles (atoms), photoionization, photorecombination, and recombination with electrons and atoms, dissociative recombination, the diffusion of atoms, ions and electrons. The remaining notations are standard.

The numerical analysis of the model of the pulsed periodic metal vapour lasers was varried out by the steady-state approximation to calculate the temperature of the atoms and ions of $T_a = T_a(\varepsilon_a)$, $T^+ \approx T_a$:

$$T_a^{i+1,j+1} = \overline{T}_w + \frac{E_{qv}^{1(i)}}{12\overline{\lambda}_g} \cdot l_g^2 \cdot f, \tag{6.2}$$

where $\overline{\lambda}_g = \lambda_g(T_a^j)$. Here i is the number of the calculated excitation pulse interval $\Delta t = 1 / f$, where f is the frequency of excitation pulses; j is the number of iteration; $E_{qv}^{1(i)}$ is the the energy for heating the unit volume of the medium during the i-th interval Δt

$$E_{qv}^{1(i)} = \int_0^{\Delta t} \left\{ \sum_{k=1}^{2} n_e \left[\kappa_k^{ea} \left(\alpha_k^{ea} n_k + v_k^{e+} \right) \left(\varepsilon - \varepsilon^{(i)} \right) + \right. \right.$$

$$\left. \left. + \beta_{n_k}^{ra} n_k^+ \left(F_N^{ie} - F_{lo_k} \right) + \delta_{n_k} \beta_{N_k}^{raa} n_k \frac{n_k^+}{n_e} \left(E_k^{ie} - E_{N_k} \mid o \right) \right] \right\} dt.$$

$$l_g^2 \cong \frac{l_x^\perp l_y^\perp}{l_x^2 + l_y^2}.$$

The relation (6.2) determines the average cross-sectional temperature difference between the medium and the wall for the case of the arbitrary cross-section of the active zone.

The initial population of the energy states of atoms required for the solution of the system of equations and defined as the initial conditions, was assumed to be thermodynamically equilibrium with respect to the ambient temperature prior to supplying the first excitation pulse. The initial concentration of ions n_k^+ was given in the original set of data (such as equilibrium with respect to the temperature of the environment, or the corresponding steady state).

6.2. The calculated dependences of the coefficients, cross sections, rate constants and frequencies of elementary physical processes

1. *The cross section for elastic collisions of electrons with atoms*

In the theoretical calculation of σ_k^{ea} based on the data [5], the equation for calculating the ratio of the cross section of an atom at $\varepsilon = 0$ can be written as:

$$\sigma_k^{ea}(0) \approx 4\pi \hat{L}_k^2,$$

where $\hat{L}_k$ is the scattering length defined in [6]; $k = 1, 2$ is the index of the corresponding component;

$$\hat{L}_k^2 = 1 / \left[\hat{\gamma}_{ak} 1 - \frac{1}{3} \left(\alpha_{ak}^p \gamma_{ak} / a_0 \right)^{1/2} \right],$$

Here a_0 is the Bohr radius; α_{ak}^p is the polarizability of the atom of the k-th component.

Parameter $\hat{\gamma}_{ak}$ is calculated from the ratio of the energy of the affinity of the atom for the electron:

$$E^s = \frac{h\hat{\gamma}_{ak}}{8\pi^2 m_e}.$$

Here $\mathbf{E}^s$ was defined in the basic data set on the component in question. For the case of the absence of the affinity of the atom for the electron, assessment was carried out using the formula:

$$\sigma_k^{'''}(0) \approx \pi \frac{d_{ak}^2}{1}.$$

where d_{ak} is the effective diameter of the k-th atom.

Polarizability of the atom was determined from the relation [7], which gives good agreement with experimental values for α_{ak}^p:

$$\alpha_{ak}^p \cong \frac{4}{3} \xi_k a_0 \left\langle r^2 \right\rangle \left(R_y / E_k^{ie} \right)^2,$$

where ξ_k is the number of equivalent electrons;

$$\left\langle r^2 \right\rangle \cong \frac{5}{2} n_k^{*4} a_0^2,$$

where $n_k^* = (Ry/E_k^{ie})^{1/2}$ is the effective principal quantum number.

To calculate the cross section at high energies ($\varepsilon \geq 30$ eV), we used the Born approximation:

$$\sigma_k^{ea} = \frac{6 \cdot 10^{-15} z_{an}^{4/3}}{\varepsilon} \text{ cm}^2,$$

where z_{an} is charge of the atom nucleus.

Directly in the energy range of the incident electron $\varepsilon = 0 \div 30$ eV the cross section σ_k^{ea} was approximated by a linear relation with the corresponding nodal values

$$\sigma_k^{ea} = \sigma_k^{ea}(0) + \frac{\sigma_k^{ea}(30) - \sigma_k^{ea}(0)}{30} \varepsilon,$$

2. The frequency of the elastic collisions of electrons with ions of the components

Frequency v_k^{e+} was determined by the ratio obtained in [8]:

$$v_k^{e+} = 3.64 \cdot 10^{-16} n_k^+ L^* / T_e^{3/2},$$

Here $L^* = \ln\left(1.27 \cdot 10^{-7} T_e^{3/2} / n_e^{3/2}\right)$ is the Coulomb logarithm, T_e [K], n [m^{-3}].

According to the theory of elastic collisions of particles, for the above processes 1) and 2) the fraction of transferred energy is equal to

$$\kappa_k^{e+} \approx \kappa_k^{ea} = 2m_e / m_{ak}.$$

3. Cross sections and excitation constants of atoms by electron impact
   a) Optically forbidden transitions

Theoretical calculation of the excitation constants α_{ijk}^{ea} for optically forbidden transitions was carried out, as for the process (1), through the calculation of cross sections σ_{ijk}^{ea}. Excitation of electronic states, in which the electric dipole transitions are forbidden, can be attributed to the orbital electron being knocked out by the incident one, after which the latter is captured into an excited level of the atom. Based on the analysis of this electronic exchange, the complete cross section for excitation of the optically forbidden transition was derived in [9]

$$\sigma_{ijk}^{ea}(\varepsilon) = \begin{cases} \dfrac{\pi e^4 \left(\varepsilon - E_{ijk}\right)}{G_{ik}\left(\varepsilon + G_{ik}\right)\left(\varepsilon + G_{ik} - E_{jik}\right)}, & E_{jik} < \varepsilon \leq E_{lik}; \\[4mm] \dfrac{\pi e^4 \left(E_{lik} - E_{ijk}\right)}{\left(\varepsilon + G_{ik}\right)\left(\varepsilon + G_{ik} - E_{jik}\right)\left(\varepsilon + G_{ik} - E_{lik}\right)}, & \varepsilon > E_{lik}, \end{cases}$$

$$G_{ik} = E_k^{ie} - E_{ik}; \; l = j+1.$$

Table 6.1.

χ	0.01	0.02	0.04	0.1	0.2	0.4
$p(\chi)$	1.16	0.956	0.758	0.493	0.331	0.209

1	2	4	10	>10	
0.1	0.063	0.04	0.023	$0.066 \cdot \chi^{-1/2}$	

It should be noted that the results for these relations are in satisfactory agreement with experimental data.

b) *The optically allowed transitions*

Theoretical calculation of the constants of the excitation of the optically allowed transitions was carried out directly by the formula recommended in [7]:

$$\alpha_{ijk}^{ea} = 32 \cdot 10^{-8} f_{ijk} \left(R_y / \Delta E_{jik} \right)^{3/2} \chi^{1/2} e^{-\chi} p(\chi), \qquad (6.3)$$

where f_{ijk} is the oscillator strength for absorption of the k-th element,

$$\chi = \Delta E_{jik} / kT_e.$$

When $\chi \ll 1$ the value of the multiplier $p(\chi) = -\dfrac{3^{1/2}}{2\pi} E_i(\chi)$.

In this case the value of the integral exponential function is accurately calculated as:

$$E_i(\chi) = 0.5772157 + \ln(\chi) - \chi + \frac{1}{4}\chi^2.$$

For other values of χ the factor $p(\chi)$ is presented in Table 6.1.

The relation (6.3) is obtained using the van Regermoter equation for the cross section of the optically allowed transition, built on the basis of the Bethe model [7] and the Maxwellian electron energy distribution function.

Constants of deactivation (quenching process of excitation) by electron impact for both the optically forbidden (*a*) and for the allowed (*b*) transitions were determined according to the principle of detailed equilibrium:

$$\downarrow \alpha_{jik}^{ea}(\varepsilon) = \uparrow \alpha_{ijk}^{ea}(\varepsilon) \exp\left(\Delta E_{jik} / kT_e\right) \frac{g_{ik}}{g_{jk}}.$$

4. *The ionization constants of the various states of atoms by electron impact*

The theoretical calculation of the ionization constants of the atoms in the shell $n_0 l_0^\xi$ (n, l are the principal and orbital quantum numbers, ξ is the number of equivalent electrons) is carried out using the Seaton equaton [10], due to its simplicity and satisfactory accuracy:

$$\alpha_{ik}^{ie} = 4.3 \cdot 10^{-8} \xi_{ik} \left(R_y / G_{ik} \right)^{3/2} \chi^{-1/2} e^{-\chi}, \quad \chi = G_{ik} / kT_e,$$

where ξ_{ik} is the number of equivalent electrons in the state 'ik'.

5. Constant of the dissociative recombination of ions of inert gas

Reducing the concentration of electrons in the active medium of the pulsed periodic metal vapour lasers may be due to a two-step process

$$\begin{cases} N^+ + N + A \rightarrow N_2^+ + A, \\ N_2^+ + e \rightarrow N^* + N, \end{cases}$$

where N is the atom (N^* is the excited atom) of the buffer gas, which plays the role of a third body in reactions, N_2^+ is the molecular ion. In typical PMVLs, i.e. at concentrations of buffer gas atoms of $<10^{18}$ cm^{-3}, the resulting rate of deionization of the plasma is determined by the first of these reactions, the conversion of the ions. It can be assumed that the dissociation of the molecular ion occurs instantaneously. The resulting rate constant of deionization of the plasma through dissociative recombination of molecules of the inert gas was estimated by the formula

$$\beta^{raa} \approx 10^{-30} n_k, \text{ cm}^3\text{s}^{-1}.$$

If necessary, this constant can be adjusted according to [11].

At high pressures it is necessary to consider both reactions and kinetics and molecular ions, respectively. The corresponding rate constants are presented in [12].

6. The rate constants of electron–ion recombination in triple collisions

The electron–ion recombination in triple collisions, taken into account in the PMVL model and associated with the transfer of energy to the atom or electron, is most intense at low temperatures T_e and not too low densities of particles. The recombination rate constant at a fixed level for three-body collisions in the presence of an electron is determined by the rate constant of ionization, based on the principle of detailed balance [13]

$$\hat{\beta}_{ik}^{re} = \alpha_{ik}^{ie} \left(\frac{n_{ik}}{n_k^+ n_e} \right) \text{ cm}^6/\text{s}.$$

It is known that the recombination flux in the interpulse period is mainly determined by the wandering of the atomic electron on high-lying levels due to collisional processes. The model assumes that the entire flux passes through the upper of the separately considered levels, located slightly below the so-called bottleneck [14] for the flux of the recombining electrons. In the preferred mode of recombination, which occurs in plasma in the PMVL in the interpulse period, the effective recombination rate constant on the upper considered level is is determined by the formulas from [14]

$$\hat{\beta}^{re}_{N_k} = \frac{4(2\pi)^{1/2} \pi e^{10} \Lambda}{9 m_e^{1/2} (kT)^{9/2}},$$

for recombination with the electron regarded as the third particle and

$$\beta^{ra}_{N_k} = \frac{16 e^6 (2\pi m_e)^{1/2} kT_a}{3 m_k^+ (kT_e)^{7/2}},$$

with the atom taken into accout. The Coulomb logarithm in the first formula can be taken as $\Lambda = 0.2$ [14].

7. *Frequency of the process of ambipolar diffusion of ions (electrons)*
The algorithm for calculating the process of ambipolar diffusion of ions, used in numerical studies, takes into account in a general case the motion of particles in a mixture consisting of two components.

Ambipolar diffusion coefficient $D^+_{am_k}$ is determined by the diffusion coefficient of ions of the k-th component D^+_k by the well known relation:

$$D^+_{am_k} \approx D^+_k (1 + T_e / T_a), \qquad (6.4)$$

where
$$\frac{1}{D^+_k} = \frac{1}{D^+_{k-k}} + \frac{1}{D^+_{k-(3-k)}}.$$

Here D^+_{k-k} is the coefficient of diffusion of ions of the k-th component in a gas (k-th component); $D^+_{k-(3-k)}$ is the diffusion coefficient describing the motion of the k-th component ($k = 1, 2$) in another $(3-k)$-th component.

The diffusion coefficient is associated with the mobility of the ion b^+ by Einstein's relation:

$$D^+ = \frac{b^+ kT_a}{e^+},$$

where e^+ is the ion charge.

The calculation of the mobility of ions in the gas is based on the work [6.15]:

$$b_{k-k}^+ = \frac{0.331 e_k^+}{n_k \left(m_k^+ k T_a\right)^{1/2} \sigma_k^{res} \left(2.24 v_k^+\right)} \quad \frac{cm^2}{V \cdot s}$$

here σ_k^{res} is the cross section of the resonance recharge of the positive ion on its own atom.

σ_k^{res} is calculated using the theoretical dependence [15]:

$$\sigma_k^{res} = \frac{I}{2}\left(\ln \frac{100}{v_{ak}}\right)^2 \frac{E_k^{ie}}{m_e} \cdot \frac{h}{2m_e E_k^{ie}} \ .$$

The maximum error of (6.4) for many components does not exceed 50%. The mobility of ions in a foreign gas, related to the normal concentration of atoms $\tilde{n}_a = 2.69 \cdot 10^{19}$ cm^{-3}, is presented in [15]. Since the mobility is inversely proportional to the particle density, for the specific concentration of atoms of the $(3-k)$-th component, we have:

$$b_{k-(3-k)}^+ \approx \frac{35.9}{\left(\alpha_{a(3-k)}^p m_{k-(3-k)}^{+a}\right)^{1/2}} \cdot \frac{\tilde{n}_a}{n_{(3-k)}} \ .$$

Here $m_{k-(3-k)}^{+a}$ is the reduced mass of the ion of the k-th component of the atom and $(3-k)$-th component in units of proton mass.

Defined by (6.4) $D_{am_k}^+ (T_e, T_a)$, where D_k^+ is computed by $D_{k-k}^+ (b_{k-k}^+, T_a)$ and $D_{k-(3-k)}^+ (b_{k-(3-k)}^+, T_a)$, it is easy to find the frequency of the process of ambipolar diffusion of ions of the k-th component:

$$v_{am_k}^{d+} = \frac{6 D_{am_k}^+}{r^2} \ ,$$

where r is the characteristic size of the diffusion.

In this case, the frequency of ambipolar diffusion of electrons was calculated taking into account the share of the ion concentration of each component:

$$v_{am}^{d+} = \sum_{k=1}^{2} v_{am_k}^{d+} \frac{n_k^+}{n_e} \ .$$

The equations of the mathematical model took into account the effective frequencies of the process of ambipolar diffusion of ions of the components to the wall of GDT (GDC) $\hat{v}_{am_k}^{d+}$

$$\hat{v}^{d+}_{am_k} = v^{d+}_{am_k} \left[1 + \left(2v^{r+}_{\Sigma_k} / v^{d+}_{am_k} \right)^{1/2} \right],$$

where

$$v^{r+}_{\Sigma_k} = v^{re}_{\Sigma_k} + v^{ra}_{N_k} + v^{raa}_{N_k},$$

and the effective frequency of the ambipolar diffusion of electrons:

$$\hat{v}^{de}_{am} = \sum_{k=1}^{2} \hat{v}^{d+}_{am_k} \frac{n^+_k}{n_e}. \tag{6.5}$$

8. *Cooling of electrons due to their diffusion in the wall of GDT*
Numerical experiments carried out in the Astrofizika Scientific and Production Organization showed that in contrast to the metal vapour pulsed lasers (MVPL) with a transverse discharge, in a longitudinal excitation scheme a noticeable effect on the kinetics is exerted by electron cooling because of their ambipolar diffusion on the wall of the discharge tube. The effective frequency of electron cooling in relation to the equations (6.1) of the mathematical model of MVPL is given by:

$$v^{de}_w = \left(\hat{Z} - 1 \right) \hat{v}^{de}_{am},$$

where the coefficient $\hat{Z}$ depends on the size of the active zone and is determined on the basis of [16,17] by the relation:

$$\left(\hat{Z} - 1 \right) \approx \frac{2}{3} \left[3 + \ln \left(\left(l_{x,y} / D_i \right) \left(kT^+ / m_e \right)^{1/2} \right) + \ln \frac{m_i}{m_e} \right],$$

Here $l_{x,y} = r/2.4$ is the characteristic size; $D_i = v^{di} r^2/6$. The frequency of the electron diffusion v^{de} is calculated similarly to (6.5):

$$v^{de} = \sum_{k=1}^{2} v^{d+}_k \frac{n^+_k}{n_e}.$$

1) $A_0 + B^*_m \rightarrow A^*_m + B_0,$ $\sigma \approx 10^{-22}$ cm^2,

2) $A_0 + B^*_m \rightarrow A^*_r + B_0,$ see below

3) $A^*_m + B_0 \rightarrow A_0 + B_0,$ $\sigma \approx 10^{-20}$ cm^2,

4) $A^*_r + B_0 \rightarrow A_0 + B_0,$ $\sigma \approx 10^{-19}$ cm^2,

5) $A_0 + A^*_m \rightarrow A_0 + A_0,$ $\sigma \approx 10^{-20}$ s^2,

6) $A^*_r + A_0 \rightarrow A_0 + A_0,$ see below

7) $B_0 + B_m^* \rightarrow B_0 + B_0$, $\sigma \approx 10^{-20}$ cm^2,

8) $A_0 + B_m^* \rightarrow B_0 + A^+ + e$,   see below,

9) $A_0 + B_r^* \rightarrow A_m^* + B_0$, $\sigma \approx 10^{-22}$ cm^2,

10) $A_0 + B_r^* \rightarrow A_r^* + B_0$,   see below,

11) $A_0 + B^{**} \rightarrow A^+ + B_0 + e$,   see below,

12) $A_r^* + B^{**} \rightarrow A^+ + B_0 + e$,   see below,

13) $A_r^* + B_m^* \rightarrow A^+ + B_0 + e$, $\sigma \approx 10^{-20}$ cm^2,

14) $A_r^* + B^{**} \rightarrow A^{**} + B_m^*$,   see below,

15) $A^{**} + B_r^* \rightarrow A^+ + B_0 + e$,   see below,

16) $A^{**} + B^{**} \rightarrow A^+ + B_0 + e$,   see below,

17) $A^{**} + B_0 \rightarrow A^+ + B_0 + e$,   see below

9. The frequency of the diffusion of atoms and the thermal conductivity coefficient of the medium

An exact account of the diffusion of neutral atoms of different components in the plasma of the discharge gap as a function of its physical parameters is a complicated independent problem that goes beyond the model of calculation of the MVPLs. Therefore, the frequency of the diffusion of neutral particles was determined in the present case using the elementary theory of collisions. This approximation proved to be suitable, as shown by numerical calculations, for predicting the kinetics of the laser processes. Thus, we have:

$$v_k^{da} \approx \frac{2v_{ak}}{\left(\sigma_{k,k}^{aa} n_k + \sigma_{3-k,k}^{aa} n^{3-k} \right) r^2} \, ,$$

where $v_{ak} = v\,(T_a)$ is the average velocity of atoms of k-th component $\sigma_{k,l}^{aa} = \pi \cdot \tilde{d}_{k,l} / 4$ is the effective collision cross section of atoms of k-th and l-th component

$$\tilde{d}_{k,l} = \left(d_{ak}^2 + d_{al}^2 \right)^{1/2} .$$

In this case, the thermal conductivity of the medium can be written as:

$$\bar{\lambda}_G = 0,5 \sum_{k=1}^{2} \frac{n_k v_{ak}}{\left(\sigma_{k,k}^{aa} n_k + \sigma_{3-k,k}^{aa} n_{3-k}\right)} .$$

10. *Photoionization and photorecombination*
The effective capture cross section at the level n of the free electron with initial energy $\varepsilon = mv_e^2 / 2$ was calculated by the formula [18]:

$$\sigma^{rv} = \frac{128\pi^4}{3\sqrt{3}} \cdot \frac{z^4 e^{10}}{mc^3 h^4 v_e^2 v} .$$

The Kramers equation was used for the photoionization cross-section

$$\sigma_n^{iv} = \frac{64\pi^2}{3\sqrt{3}} \cdot \frac{m_e e^{10}}{c \cdot h^6} \cdot \frac{1}{n^5 v^3} .$$

11. *Inelastic interactions of heavy particles*
The processes of inelastic interactions in collisions of heavy particles with each other occur at a much lower efficiency than in collisions with electrons. The results of the present experimental and theoretical studies of inelastic interactions of heavy particles are very far from being able to provide a choice of sufficiently reliable calculation formulas. Apparently, the cross section of heavy particles is of the order of 10^{-20}–10^{-22} cm^2. However, at a small change in the energy of the atoms at collisions there can be resonant transfer of the interaction energy, and the cross section of these processes can reach values close to the values of the gas–kinetic cross sections.

Here are some basic processes of interaction of the atoms of the metal and buffer gas, taken into account in this model. The data on the collision cross sections are taken from [7,19-23].

Here A – the metal atom, B – the buffer gas atom, * – an excited state, ** – excitation at levels above the resonance.

The interaction cross sections of the processes 2), 10), 14) were calculated by the formula:

$$\sigma = 2\pi \left(\lambda_n / v\right)^{2/(n-1)} \exp\left[-2\left(2\beta_n\right)^{1/2} \sin\left(\pi / n\right)\right] I_n(\beta),$$

where n is the parameter which determines the interaction potential; $\lambda_n = \left[\dfrac{s_{\kappa_1} s_{\kappa_2}}{\left(2\kappa_1 + 1\right)\left(2\kappa_2 + 1\right) g_1 g_2}\right]^{1/2}$ is the interaction constant; s_κ is the force of the line of transition of the $I_1 \rightarrow I_1'$ multiplicity κ_1 for the first particle; g_1 is the statistical weight of the initial level I_1,

$$\beta_n = v^{-2} \lambda_n^{2/n} w^{2(n-1)/n},$$

here v is the relative velocity of colliding particles,

$$\omega = \frac{E_1 - E_2}{h} 2\pi.$$

When $w = 0$ the exact resonance occurs. The values of the integral I_n (β_n) are tabulated in [7].

The interaction cross sections of processes 11), 12), 15), 16), 17) were calculated using the following formula:

$$\sigma_n^{ia} = 9\left(\frac{2\pi e^2}{hv_a}\right)^{2/5} \cdot \left(\frac{e^2}{hvn}\right)^2 \cdot (g \cdot f_{ro})^{2/5},$$

where f_{ro} is the oscillator strength for the transition of buffer gas atoms from the ground to the resonantly excited state, $hv = I_n$; g is the Gaunt factor.

To calculate the interaction cross sections of processes 6), 8), we use the following formula.

For the process 6):

$$\sigma = 1.56 \frac{hv_a g^{2/3}}{2\pi \Delta E^{4/3}},$$

where ΔE is the energy difference between the levels

$$v_a = \left(\frac{8kT}{\pi\mu}\right)^{1/2},$$

where μ is the reduced mass, g is the matrix element of the dipole moment operator,

$$g^2 = \frac{e^2 h^2 f_{or}}{8\pi^2 m_e \Delta E_{or}},$$

where ΔE_{or} is the energy difference between the ground and resonance levels.

For the process 8):

$$\sigma = \sigma_0 f(x), \quad x = \frac{2mv_a^2}{J};$$

$$f(x) = \begin{cases} \dfrac{128}{15\pi} x^{3/2}, & x \ll 1, \\[2mm] 1 - \dfrac{1}{x}, & x \gg 1, \end{cases}$$

where J is the electron binding energy, σ_0 is the total cross section for elastic scattering of slow electrons on the atom.

12. *The contour of lines of radiative transitions*

The probability of transitions taken into account in calculations and induced by radiation and hence the population inversion, inecessary for the maintenance of lasing, depends on the width and shape of the spectral lines. Very important is also the account of possible radiation trapping. Accurate numerical prediction and detailed study of the dynamics of lasing of the stimulated emission requires the simultaneous consideration of the inhomogeneous (Doppler) broadening and homogeneous (collisional) broadening of the spectral line. In addition, it is important to keep in mind the hyperfine and isotopic splitting of lines. In most cases, the models use the Doppler contours of the spectral line.

The probability of photon emission θ_λ beyond the plasma volume is connected with the geometry of the active zone and is determined for every radiative transition by the coefficient [24]:

$$1/\theta_\lambda^D = 4\tilde{k}_\lambda^D r \pi^{1/2} \left[\ln\left(\tilde{k}_\lambda^D r\right) \right]^{1/2},$$

where r is the characteristic linear dimension of the plasma, $\tilde{k}_\lambda^D r$ is the optical density of the mid-line.

One of the variants of taking into account the reabsorption of radiation is given in [25]:

$$\theta_\lambda^D = 1.7 \left[1 + 2.25 \tilde{k}_\lambda^D r \sqrt{\ln\left(\tilde{k}_\lambda^D r + 1\right)} \right]^{-1}.$$

For the case $\theta_\lambda^D < 1$ the calculation of the initial system of equations of the model of the MVPL was conducted taking into account radiation trapping with an effective emissivity coefficient:

$$A_\lambda^{pl} = A_\lambda \theta_\lambda^{\prime\prime}.$$

6.3. Methods of solution, software and results of numerical experiments

In the first stage of calculations we used a simple to implement, but very efficient numerical method. Equations (6.1) are presented in the form

$$\frac{dy_i}{dt} = c1_i\left(\xi_i\right) + c2_i\left(\xi_i\right) y_i, \quad i = f_1\left(y_1, \ldots, y_N\right).$$

Having determined the coefficients of the system at a point m in time $[c1_i\,(\xi_i)]\,m$ and $[c2_i\,(\xi_i)]\,m$, and organizing the iterative loop on j, the solution at the point $m + 1$ is sought in the form

$$\left(y_i\right)^{j+1}_{m+1} = y_{im}\,\exp\left(R2^j_i\,h\right) + \frac{R1^j_i}{R2^j_i}\left(\exp\left(R2^j_i\,h\right) - 1\right),\tag{6.6}$$

where h is the the time step,

$$\begin{aligned}R1^j_i &= 1/2\left\{\left[c1_i\left(\xi_i\right)\right]_m + \left[c2_i\left(\xi_i\right)\right]^j_{m+1}\right\},\\ R2^j_i &= \left(1-\varphi\right)\left[c2_i\left(\xi_i\right)\right]_m + \varphi\left[c2_i\left(\xi_i\right)\right]^j_{m+1},\end{aligned}\tag{6.7}$$

$\varphi = 0 - 1$ is the parameter of an implicit scheme.

The initial system of equations and method for its solution were used by V.V. Tykotsky to develop a basic universal program for calculating the lasing characteristics and working processes of MVPLs, called MOLOT-F4. Figures 6.1–6.3 show the results of calculations using the program of the key parameters of repetitively pulsed lasers based on mixtures of Au + Ne, Cu + Ne and Ba + Ne, respectively, described in [26], [27], [28].

The parameters of various MVPLs, using the MOLOT-F4 programme, are consistent with the available experimental data over a wide range of initial conditions of the lasers (the divergence of the integral characteristics does not exceed 50%).

Further development of basic universal program MOLOT-F4 was aimed at making it simpler. Program MOLOT-GF for the calculation of the lasing characteristics of MVPLs was developed in which the photoprocesses and collisions of heavy particles with each other were no taken into account (except for the impact ionization of heavy particles in the Penning effect), but the program took into account the elastic collisions of electrons with ions of the components (see section 6.2, paragraph 2).

In addition, in the program MOLOT-GF much attention was paid to the choice of the integration step. It turns out that the most sensitive to the value of the step are the equations for the electron energy and radiation density. The parameter of implicitness of the scheme $\varphi = 0.75$.

Program MOLOT-GF was used for the numerical study of lasing characteristics on the atoms of 14 elements: Al, Ba, Au, In, Ca, Li, Mn, Hg, Pb, Sr, Tl, Zn, Cu, Eu. Numerical calculations of lasing were mainly performed by setting the operating modes and parameters of laser structures with transverse pumping, implemented in various modifications at the Astrofizika Company, and with longitudinal

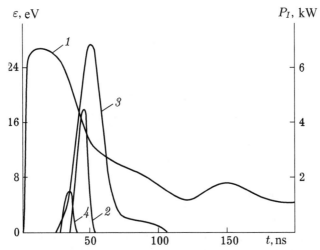

Figure 6.1. Changes of some physical parameters of a gold vapour laser (transverse discharge) with neon; p_{Au} = 1.3 kPa, p_{Ne} = 13 kPa, f = 9.1 kHz, 1 – ε, 2 – 4 – instantaneous radiation power P_I, 2 – 312.2, 3 – 627.8, 4 – 5065 nm.

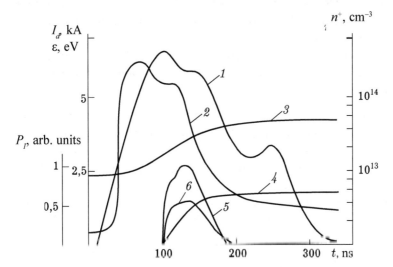

Figure 6.2. Changes in some physical parameters of a copper vapour laser (transverse discharge) with neon; p_{Cu} = 12.7 Pa, p_{Ne} = 60.8 kPa, f = 3 kHz; 1 – I_d, 2 – ε, 3 – n^+_{Cu}, 4 – n^+_{Ne}, 5, 6 – instantaneous output power in relative units (5 – 510.5, 6 – 578.2 nm).

pumping, implemented at the Physical Institute of Russian Academy of Sciences (FIRAN) and the Institute of Thermophysics, Siberian Branch, Russian AcademySciences (ITP SB RAS).

The results of calculations of lasers on gold, strontium, lead and manganese vapours, for typical operating conditions of the chambers with transverse pumping are presented in Figs. 6.4–6.7. The

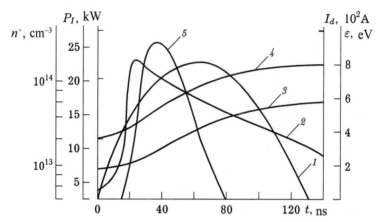

Figure 6.3. Changes in some physical parameters of a barium vapour laser (transverse discharge) with neon; p_{Ba} = 13.3 Pa, p_{Ne} = 2.93 kPa, f = 13.3 kHz, 1 – I_d, 2 – ε, 3 – n^+_{Ba}, 4 – n^+_{Ne}, 5 – the instantaneous radiation power P_I at a wavelength of 1500 nm.

calculations, carried out for the other elements (except for copper, europium and barium), showed no lasing in the typical transverse discharge conditions.

The following results were obtained directly for the elements for which lasing was observed and it is useful to compare them with the results of experiments.

1. Gold vapour laser (λ_g = 627.7 nm, 312.2 nm, 506.5 nm). Experiment [26.29]: laser tube diameter d_p = 6 mm, length l_p = 65 cm, f = 9.1 kHz, p_{Au} = 13.3 Pa, p_{Ne} = 1.33, 2.67 kPa, C_S = 2200 pF , U_0 = 20 kV. The total generation of UV and the red line of 1W was observed.
   In numerical studies for conditions p_{Ne} = 1.33 kPa, p_{Au} = 13.3 Pa, f = 8 kHz, d_p = 2 cm, l_p = 70 cm UV lasing with the power of ≈ 1.8 W, W ≈ 1μJ/cm³, was produced.
   The numerical prediction for p_{Au} ≈ 1.33 133.1 Pa, p_{Ne} ≈ 1.33 4.0 kPa, f ≈ 10–40 kHz typical laser tubes achieved the total average power of 1020 W for the UV and red lines.
2. Barium vapour laser (λ_g = 1/5 mm).
   Experiment [28]: laser tube d_p = 2.5 cm, lp = 85 cm, C_S = 2200 pF, p_{Ne} = 2.9 kPa, p_{Ba} ≈ 13.3 Pa. The maximum average output power of 12.5 W at f = 13.3 kHz was reached.
   Numerical studies confirm this experimentally observed efficient generation.
   Numerical prediction: in typical laser tubes at p_{Ne} ≈ 2.7–4.0 kPa, p_{Ba} ≈ 13.3–133.4 Pa, f ≈ 30-40 kHz the average output power of 50–60 W is achievable.

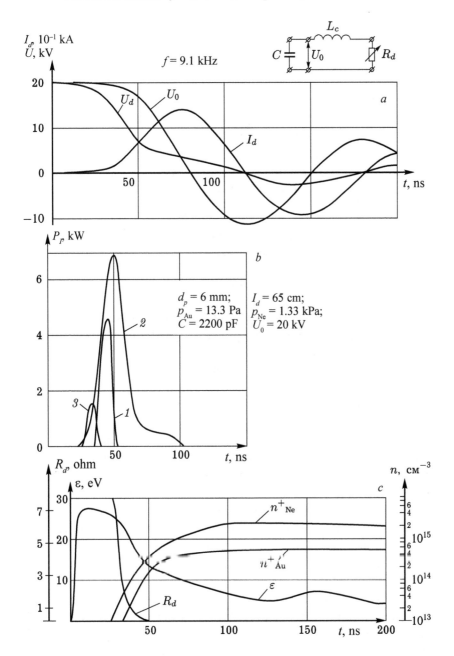

Figure 6.4. The characteristics of the working process in a gold vapour laser with neon. Time dependences of: a: U, I_d; b: instantaneous power P_l (1 – 312.2 nm; 2 – 627.8 nm, 3 – 506.5 nm), c: n^+_{Ne}, n^+_{Au}, ε and R_d.

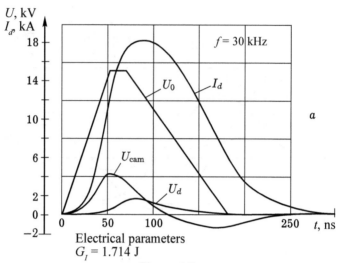

Figure 6.5.

3. Europium vapour laser (λ_g = 1.57–6.06 μm, 24 emission lines).
   Experiment [30] proved the possibility of generating infrared radiation in short GDTs.
   Numerical prediction: see section 6.5.
4. Manganese vapour laser (λ_g = 534.1–553.8 nm).
   Experiment [31]: lasing in the self-heating mode of the tube d_p = 7 mm, l_p = 260 mm, f = 200 Hz.
   Numerical studies for GDT for d_p = 2 cm, l_p = 70 cm under typical conditionns p_{Ne} = 2.67 kPa, p_{Mn} = 13.3 Pa, f = 16 kHz, C_S = 1500 pF, U_0 = 10 kV give the total average output power ≈ 2.8 W, W ≈ 0.8 μJ/cm³. The increase in the average power of radiation is achieved by increasing p_{Mn}, p_{Ne} and f.
   Experiment [32]: short GDT of small diameter working in the self-heating mode were used. The average radiation power 3.2 W.
   Numerical studies of GDT for d_p = 2 cm, l_p = 100 cm for typical initial conditions p_{Pb} = 13.3 Pa, p_{Ne} = 2.67, f = 8 kHz, C = 1600 pF, U_0 = 20 kV yielded the result P_l = 6 W, W = 2.42 μJ/cm³.
   Numerical prediction: in typical GDTs in the implementation of the laser operation mode p_{Pb} = 133.4 Pa, p_{Ne} = 6.67 kPa, f = 30 kHz the average output power of P_l ≈ 20-30 is achievable.
5. Copper vapour laser (see section 6.4).

As a result of the further improvement of programs MOLOT-F4 and MOLOT-GF the MVPL program was modified and presented in a separate standard module MOLOT-SM, independent of the particular method of numerical integration of the ordinary differential equations

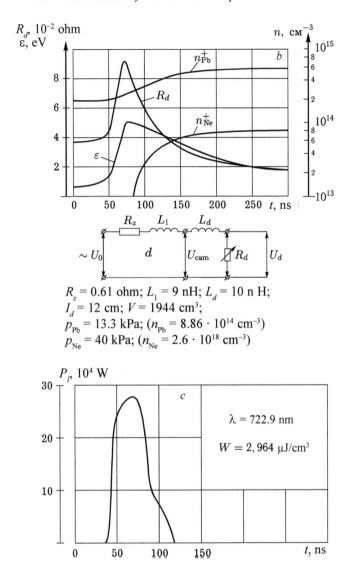

$R_z = 0.61$ ohm; $L_1 = 9$ nH; $L_d = 10$ n H;
$I_d = 12$ cm; $V = 1944$ cm³;
$P_{Pb} = 13.3$ kPa; $(n_{Pb} = 8.86 \cdot 10^{14}$ cm⁻³$)$
$P_{Ne} = 40$ kPa; $(n_{Ne} = 2.6 \cdot 10^{18}$ cm⁻³$)$

$\lambda = 722.9$ nm

$W = 2,964$ μJ/cm³

Figure 6.5 (continued). Characteristics of the working process in a lead vapour laser (transverse discharge) with Ne. Time dependences of: a: U, I_a; b: , n^+_{Pb}, n^+_{Ne} and R_d; c) P_l; d) equivalent electrical circuit.

of the MVPL, i.e. when the algorithm for solving systems of equations is determined and is given by an outside independent procedure. At the same time three efficient procedures for integrating the equations were developed and presented: a procedure that implements the method of the fifth-order Gir method (HOPE Program) for solving rigid systems of equations and RK2 and RK4 procedures, representing the Runge–Kutta second and fourth order methods with automatic step selection.

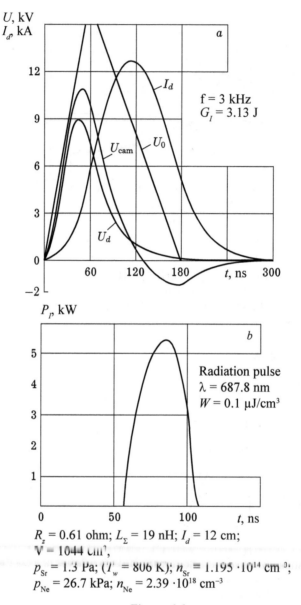

$R_z = 0.61$ ohm; $L_\Sigma = 19$ nH; $I_d = 12$ cm;
$V = 1044$ cm³,
$p_{Sr} = 1.3$ Pa; $(T_w = 806$ K); $n_{Sr} = 1.195 \cdot 10^{14}$ cm⁻³;
$p_{Ne} = 26.7$ kPa; $n_{Ne} = 2.39 \cdot 10^{18}$ cm⁻³

Figure 6.6.

Unlike the MOLOT-GF programs, in the MOLOT-SM program the cross sections for allowed transitions are given by the Drawin equation [33]:

$$\sigma_{ij}^{ka} = 4\pi a_0^2 \left(\frac{I_H}{\Delta E_{ij}}\right)^2 f_{ij} G(U),$$

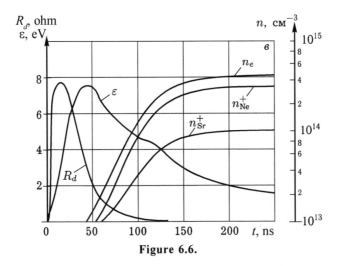

Figure 6.6.

where I_H is the ionization potential of the hydrogen atom,

$$U = \frac{\varepsilon}{\Delta E_{ij}}, \quad G(U) = \frac{U-1}{U^2}\ln(1.25\ U).$$

Numerous calculations of various MVPL showed that the best efficiency of the numerical studies in terms of speed and stability is provided by the use of the equations of the MVPL model during the generation of the RK2 or RK4 procedures, and during relaxation or dominant deactivation of the energy states of atoms – HOPE procedure. The version of the engineering program for calculating the lasing characteristics of MVPLs is called MTRH2. Under this program, the maximum difference between the calculated and experimental data does not exceed 60% in the integral characteristics (worst case). The MTRH2 mathematical model is suitable for the working modes with continuous lasing.

The program MIRH2 was used to calculate [34] the lasing characteristics of the MVPLs on Au, Ba, Cu, Mn and Pb vapours in order to determine the effect of prepulse electron density and population of the lower laser levels on the excitation pulse repetition rate. The calculations were performed for lasers with the neon buffer gas and a longitudinal discharge of the storage capacitor. In this case, the two most typical cases were studied:

1. FIRAN laser tube, for which calculations were carried out for Au, Ba, and Mn vapour lasers. The inner tube diameter d_p = 2 cm, the length of the active zone (electrode spacing) l_d = 70 cm, the distance between the mirrors l_r = 2 m, the reflection coefficient

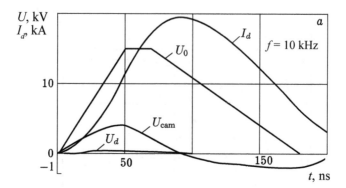

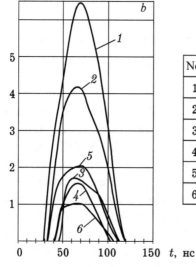

Radiation pulses λ:

No.	λ, nm	W, μJ/cm³	P, W
1	534,1	1,68	32,6
2	542,0	1,09	21,1
3	548,1	0,41	8,0
4	547,1	0,31	5,9
5	551,7	0,48	9,4
6	553,8	0,24	4,8

Figure 6.7.

of the output mirror $\rho = 0{,}08$, the storage capacitor capacitance $C_s = 1.5$ nF and the inductance of the system (discharge circuit) $L_C = 1.2$ µH.

2. A laser tube of the ITP SB RAS, for which Cu and Pb vapour lasers were calculated ($d_p = 2$ cm, $l_d = 1$ m, $l_r = 2$ m, $\rho = 0.08$, $C_S = 1.6$ nF, $L_C = 1$ µH).

The prepulse voltage on the capacitor U_0, the excitation pulse repetition frequency f, the neon pressure p_{Ne} and the metal vapour pressure p_M (or the concentration of its atoms) for these lasers are shown below in the description of the results.

Six energy levels were considered for the Ne atom. The sixth level combines the upper closely spaced (in energy) states. The calculations

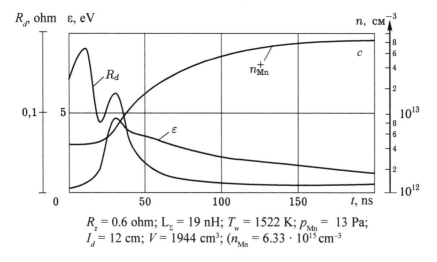

$R_z = 0.6$ ohm; $L_\Sigma = 19$ nH; $T_w = 1522$ K; $p_{Mn} = 13$ Pa;
$I_d = 12$ cm; $V = 1944$ cm^3; $(n_{Mn} = 6.33 \cdot 10^{15}$ cm^{-3}

Figure 6.7. (continued). The characteristics of the working process in Mn vapour laser without the buffer gas (transverse dicharge). Time dependences of: a: U, I_d; b: P_i; c: ε, n^+_{Mn}, R_d

for the atoms of Au, Ba, Mn and Pb were carried out for more than 11 energy levels, and for the Cu atom for nine energy states, with the last two levels including a large number of high-lying levels. In the calculation of the processes for the Cu and Ne atoms we used the known experimental ionization cross sections of the ground state σ^{ie} and elastic collision σ^{ea}.

The corresponding analysis has shown that in the closed MVPL model taking into account the large number of energy states of atoms of the active medium it is advisable to use the above relations for the calculation of the cross sections of various elementary processes, as well as the rate constants of excitation and ionization of the atoms by the electrons. Pronounced deviations from the classical relationships (forms) of sections $\sigma^{ie}_{ij}(\varepsilon)$ or excitation constants $u^{aa}_{ij}(\varepsilon)$ lead to a large distortion in the final calculation results. This is illustrated in Table 6.2 and Figure 6.8 (see below). The figure shows the calculated dependences of the average output power of the copper vapour laser (for both lasing lines) on the pulse repetition rate of excitation. The dependences, as well as the data in Table 6.2 marked by superscript I, correspond to the use in the MVPL models of theoretical relations proposed by Ochkur and Drawin respectively for all optically forbidden and allowed transitions in the Cu atom. Index II is used in the calculation of the excitation rate constant of metastable levels of the Cu atom with the unconventional cross section [35].

Table 6.2. Cu vapour laser ($\lambda_1 = 510.5$ nm, $\lambda_2 = 578.2$ nm, $U_0 = 20$ kV, $p_{Ne} = 2.67$ kPa, $n_{Cu} = 10^{15}$ cm^{-3})

f, kHz	Counter mode	$\overline{P_1}(\lambda_1)$, W	$W(\lambda_1)$, μJ/cm³	$\overline{P_1}(\lambda_2)$, W	$W(\lambda_2)$, μJ/cm³	$\overline{P_\Sigma}$, W	W_Σ, μJ/cm³
14	A^I	10	2.28	5.68	1.29	15.68	3.57
	B^i	12.75	2.9	8.18	1.86	20.93	4.76
	B^I	23.46	5.34	14.81	3.37	38.27	8.71
14	A^{II}	~0	~0	~0	~0	~0	~0
	B^{II}	7.63	1.74	4.8	1.09	12.43	2.83
	B^{II}	16.1	3.66	9.53	2.17	25.63	5.83
10	A^I	12.31	3.92	7.82	2.49	20.13	6.41
	B^i	13.6	4.33	8.5	2.71	22.1	7.04
	B^I	18.9	6.02	11.84	3.77	30.74	9.79

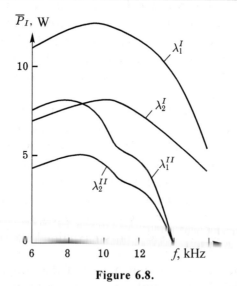

Figure 6.8.

The numerical experiments, undertaken to determine the effect of prepulse electron density and the population of the metastable levels on the output characteristics of lasing, were as follows. For certain parameters of the laser usual calculations were conducted using the program MTRH2 – counting mode 'A'. In establishing the working processes (after passage of a series of successive intervals of pumping and relaxation of the medium) the output energy and average power of laser radiation were calculated. Then, at the same laser parameters the

Table 6.3. Characteristics of the investigated atoms

Atom	The energy of the lower laser levels
Au	1.36, 2.66
Ba	1.41
Cu	1.39, 1.64
Mn	2.12, 2.145, 2.166, 2.181, 2.189
Pb	2.66

Table 6.4. Pb vapour laser (λ_g = 722.9 nm, U_0 = 20 kV, p_{Ne} = 2.67 kPa, p_{Pb} = 13.3 Pa)

f, kHz	Counter mode	$\bar{P_1}(\lambda)$, W	W, $\mu J/cm^3$
8	A	6.08	2.42
	B	6.39	2.54
	C	6.17	2.45
30	A	1.25	~ 0.13
	B	23.64	2.5
	C	4.2	0.45

counting mode 'B' was applied, in which at the time of the excitation pulse we artificially simulated the removal of 90% of the electrons and ions with converting them into the neutral metal atoms and inert gas in the ground state (while maintaining the balance of the appropriate particles); this is equivalent to the decrease in the concentration n_e by 10 times. Comparison of the results of calculations in the modes A and B allowed to determine the effect of prepulse concentration n_e on the output lasing power.

In analyzing the impact of the populations of metastable levels on the lasing characteristics calculations were carried out in the mode A and B. In the latter mode artificial instant transfer of 90% of the population of lower laser levels to the ground state of the metal atom was performed just before supplying a voltage pulse to the electrodes. Thus, at the beginning of excitation the population of each of these levels was reduced 10 times. Table 6.3 lists the elements and gives the energies of the lower laser levels for which the output laser characteristics were calculated. The comparison of the calculation results for the output characteristics of lasers in the A and B modes allows us to estimate the influence of the prepulse concentration of metastable atoms on the lasing characteristics.

Table 6.5. Ba vapour laser ($\lambda = 1.5$ mm, $U_0 = 8$ kV, $p_{Ne} = 2.9$ kPa, $p_{Ba} \approx 13.3$ Pa)

f, kHz	Counter mode	$\bar{P}_1(\lambda)$, W	W, $\mu J/cm^3$
8	A	6.64	3.78
	B	7.88	4.48
	C	7.51	4.27
16	A	6,23	1.77
	B	13.32	3.79
	C	10,1	2.87

Table 6.6. Mn vapour laser ($\lambda = 534.1$–553.8 nm, $U_0 = 10$ kV, $p_{Ne} = 2.67$ kPa, $p_{Mn} \approx 13.3$ Pa, $f = 16$ kHz)

Mode	$\bar{P}_\Sigma$, W	W_Σ, $\mu J/cm^3$
A	2.79	0.79
B	3.74	1.06
C	5.66	1.61

Table 6.7. Au vapour laser ($\lambda_1 = 312.2$ nm, $\lambda_2 = 627.8$ nm, $U_0 = 20$ kV, $p_{Ne} = 1.3$ kPa, $p_{Au} \approx 13.3$ Pa)

f, kHz	Counter mode	$\bar{P}_1(\lambda_1)$, W	$W(\lambda_1)$, $\mu J/cm^3$	$\bar{P}_1(\lambda_2)$, W	$W(\lambda_2)$, $\mu J/cm^3$	$\bar{P}_\Sigma$, W	W_Σ, $\mu J/cm^3$
8	A	~ 0	~ 0	1.82	~ 1.04	~ 1.82	~ 1.04
30	A	0.11	0.017	1.053	0.16	1.163	0.177
	B	0.225	~ 0.034	6.854	1.039	7.079	1.073
	C	$0.27 \cdot 10^{-4}$	~ $0.41 \cdot 10^{-5}$	2.7	0.406	~ 2.70	~ 0.406

The calculation results of the output characteristics of lasers based on vapours of Cu, Pb, Ba, Mn, and Au in the A, B and C modes are presented in Tables 6.3–6.7.

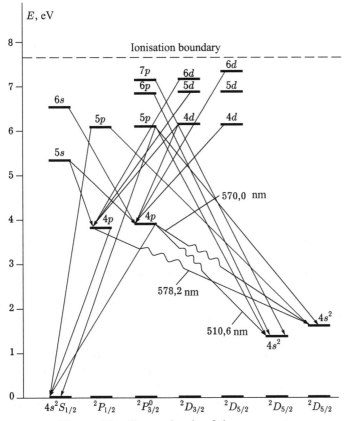

Figure 6.9.   Energy levels of the copper atom.

The analysis of the results of numerical studies presented in Tables 6.3–6.7 for the Pb vapour laser the restriction on the average lasing power occurs due to the prepulse electron concentration, i.e. their controlling role in limiting the pulse repetition rate of the pump is evident here. These calculations confirm the results of studies [36] and conclusions [37] for Pb-vapour laser. For others, ILPM in the calculation of the output characteristics of the mechanism for limiting the radiation power is complex, as it reflects on prepulse electron density and the concentration of atoms at the lower laser level.

6.4. Copper vapour laser

6.4.1. Monopulse operating mode of a copper vapour laser with a Maxwellian electron energy distribution function

The levels of the the copper and neon atoms, considered in the calculations, are given respectively by Figs. 6.9 and 6.10.

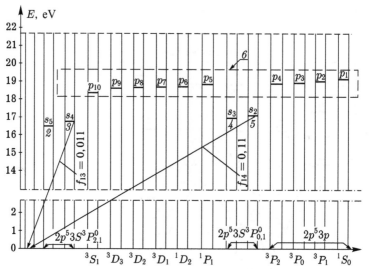

Figure 6.10. Energy levels of the neon atom.

As a result of a series of numerical experiments for the conditions of transverse discharge using the MOLOT-F4 program to determine the conditions for obtaining the maximum output energy in the lasing of a copper vapour laser with neon it was found that the lasing characteristics are influenced by all parameters of the active zone and its structural features. The greatest influence is exerted by pressure, the temperature of the active zone, the shape and amplitude of the excitation pulses.

In addition, it was found that depending on how many electrons involved in the pumping of the laser levels formed by the ionization of the buffer gas atoms it is possible to use different ways to obtain the maximum output energy per unit volume of the active zone. For example, when working with the neon buffer gas at pressures of 6.67–93.4 kPa and the temperature range $T_w = 1650$–1800 K the effective energy output is ensured at the initial parameters of the system, in particular, the form of the excitation pulse, at which in the section of the decrease of the strength of the electric field the buffer gas becomes, due to stepped ionization, the main supplier of free electrons, which (when their number is sufficient) provide effective pumping of the laser levels of copper. This effect is achieved with the other parameters constant by varying the length and shape of the excitation pulse.

In the section of the initial growth of the strength of the electric field (electron energy increases), the main source of electrons, providing a preliminary pumping of the laser levels and excited neon levels, are metal vapours. Upon reaching the peak amplitude value of the strength of the electric field E_{max} and in the decreasing part of the voltage pulse the main source of electrons is the buffer gas. It is established that the mean electron energy $\bar{\varepsilon}$ does not exceed $\approx 90\%$ of the ionization energy of the metal atom. Otherwise, the ionization instability is fixed (avalanche-like increase in the number of electrons) in the section of the growth of the excitation pulse by a simultaneous strong ionization of both components (atoms of the working metal and a buffer gas).

Maximum pumping of the laser levels takes place in the area of the decrease of the field strength, when the electron energy is reduced to a value of 5.3 eV. The efficiency of the electrons in exciting the neon levels and the length of the secition of decrease of the electron energy determine obtaining the maximum energy output.

The calculations showed that in the above neon pressure range the specific energy output is about 10 $\mu J/cm^3$.

The results of calculations also showed that a marked increase in energy output can be obtained with the combinations of input parameters at which the main source of free electrons throughout the ioniation process and the metal (copper) vapours. In this case, to obtain the energy output of 10 or more $\mu J/cm^3$ the neon pressure should be in the range 0–4.0 kPa or 80.0–146.7 kPa at a temperature of the active medium of 2000–2300 K and the electric field strength in the interelectrode gap of more than 10^3 V/cm.

To obtain the specific energy of 102 $\mu J/cm^3$ and more it is necessary to create a system that effectively operates at high concentrations of copper atoms (Cu vapour pressure greater than 4.8 kPa), i.e., at temperatures of the active zone above 2300 K and small additions of neon or even without neon. The duration of the excitation pulse must be less than with 50–100 ns. The results of calculating the lasing characteristics of a copper vapour laser in a transverse discharge without the buffer gas are shown in Fig. 6.11.

6.4.2. Calculation of single-pulse operating mode a of copper vapour laser with the non-Maxwellian electron energy distribution function

Calculations of the output characteristics PMVL relate to the definition of the dynamic parameters of electrons: their energy and concentration. Deviation of the mean electron energy by at least a factor of 1.5 leads to a drastic change in the pattern of lasing, since the weakly ionized plasma, formed in the active medium of PMVL, is characterized by the exponential dependences of the rates of the processes on electron temperature.

In [38] the authors calculated the stationary EEDF, which is realized in mixtures of inert gases and metal vapours. This EEDF can be used

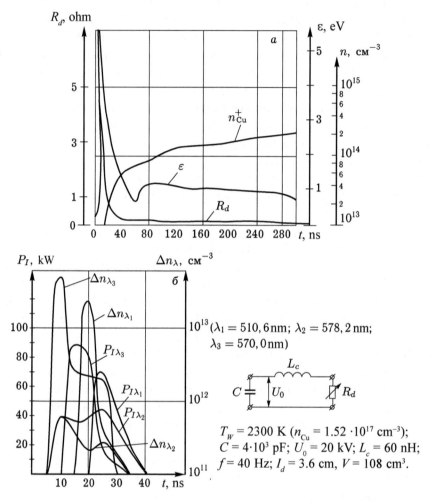

Figure 6.11. Characteristics of the working process of the single-module copper vapour lasers in pumping the active medium without the buffer gas (transverse discharge). Time dependences of: a: R_d, ε, n_{Cu}, б: P_l and the inverse population.

at slightly varying fields, when the characteristic time in the dynamics of the discharge is much greater than the relaxation time of the EEDF, which, generally speaking, is not always satisfied. It should also be noted that in [38] attention was given to the EEDF subjected to low impact of the electron–electron collisions. However, in the copper vapour lasers of these processes can not be neglected, as electron concentrations are high here (up to $5 \cdot 10^{14}$ cm^{-3}) electrons. According to [39], the characteristic time of electron–electron collisions τ_{ee} can be estimated by the following formula and is comparable to the duration of lasing or is greater:

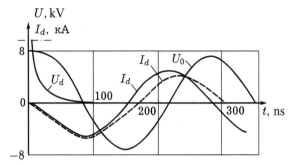

Figure 6.12. Calculated and experimental dependences of current (I_d) and voltage on time ($f = 40$ Hz, $p_{Ne} = 40$ kPa) (the solid curves – calculations, the broken curve – experiments).

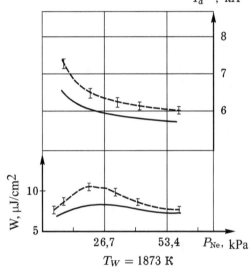

Figure 6.13. Dependence of the maximum current intensity (I_d^{max}) and energy output W on the pressure of neon (the solid curves – calculations, broken curve – experiments).

$$\tau_{ee} \approx 3.5 \cdot 10^4 \cdot T_e^{3/2} / n_e \approx 10^{-8} \text{ s},$$

Strongly varying electric fields and electron concentration require calculating the kinetics of the discharge and lasing based on the unstable electron energy distribution function. The non-stationary EEDF[2] was calculated in conjunction with the solution of kinetic equations for the basic, resonant, metastable levels and the external electric circuit equations for the experimental conditions, performed under the supervision of O.I.l Buzhinskii. The comparison of the

---

[2]Calculations were carried our usin the equation for EEDF identical with that derived for the first time in [77]. Editor's note

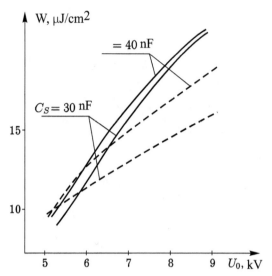

Figure 6.14. Dependence of energy output on the voltage in the capacitor (solid curves – calculations, dotted curves – experiments).

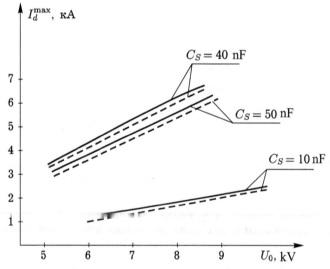

Figure 6.15. Dependence of maximum current and voltage in the capacitor (solid curves – calculations, dashed curves – experiments).

experimental and calculated data is shown in Figs. 6.12–6.16. Figure 6.12 compares the calculated and experimental time dependences of current at prepulse voltage on the storage capacitor of $U_0 = 8$ kV, the wall temperature of the GDC $T_w = 1873$ K, neon pressure $p_{Ne} = 40.0$ kPa and the capacity and the storage capacitor $C_S = 30$ µF.

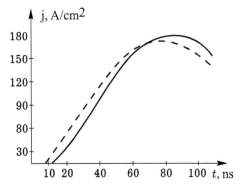

Figure 6.16. Dependence of the current density on time (solid curve – calculations, the dotted curve – experiments).

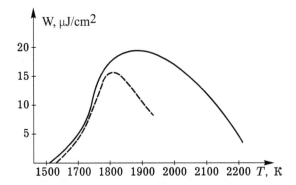

Figure 6.17. Dependence of the specific energy output on the wall temperature (solid curves - calculations, the dotted curve – experiments).

Figure 6.13 shows the dependence of specific energy output on of neon pressure. The difference between the calculated dependence and the experimental results is that the experimental curve has a strongly pronounced maximum, while the calculated curve has a flat maximum. Figures 6.14 and 6.15 show the dependences of the energy output and maximum current on the voltage on the capacitor, and Fig. 6.16 the time dependence of current density. Figure 6.17 also shows the point obtained from the calculation using the model in section 6.4.1 and Maxwellian EEDF. It is seen that at temperatures up to 2179 K the difference between the two calculation models and the experiment is small.

In general, the comparison of the calculation results for the dynamics of a copper vapour laser based on a model that takes into account the actual EEDF with the experimental results showed that there is satisfactory agreement between the two. Differences in the time

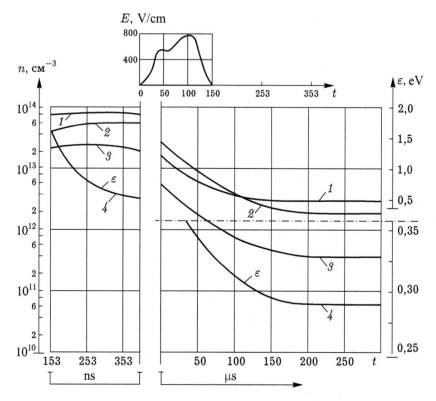

Figure 6.18. The variation of the main parameters of the operation of the laser (Cu + Ne) with time (p = 81 kPa, T = 1773 K): $1 - n_e$; $2 - n_m(3d^9 4s^2\ ^2D_{5/2})$; $3 - n_m (3d^2 4s^2\ ^2D_{3/2})$; $4 - \varepsilon$. The upper part of the figure shows the dependence of E on t.

dependences of current and voltage do not exceed 20%. Energy output does not differ by more than two times.

Comparison of the models of the PMVLs with the Maxwellian and actual EEDFs shows that in the experimental range of the parameters the difference between the models is small. It is possible that for the experimental conditions that are significantly different with respect to temperature T_w and the densities of discharge currents, it is important to take into account the actual EEDF. However, this issue requires further investigation.

6.4.3. Frequency operating mode of a copper vapour laser with neon in the mixture

The dynamic mathematical model, designed to calculate the frequency modes of the laser, is based on the mathematical model using the MOLOT programs. As an example of calculation results using the

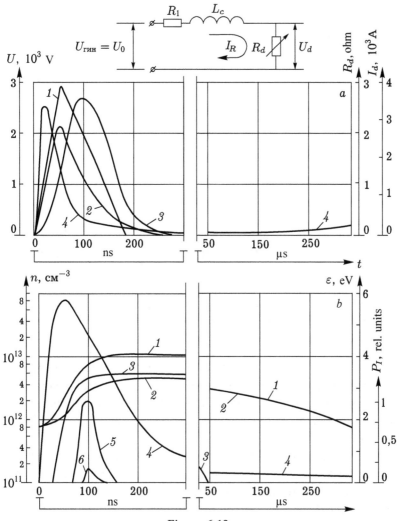

Figure 6.19.

dynamic model, Fig. 6.18 shows the changes of the concentration and energy of the electrons, the concentration of metastable atoms during the pulse of the copper vapour laser and its aftermath. The figure shows that after 200 ms, all these parameters reach a stationary level. Based on this fact, we can conclude that the maximum pulse repetition frequency in the considered conditions is approximately 5 kHz.

As part of the dynamic model calculations were carried out of the frequency modes for conditions when the electric field strength in the discharge is determined by the parameters of the discharge gap and the equivalent electrical circuit shown in Fig. 6.19 a. The following reference results were obtained for the input parameters: R_1 = 0.22

Table 6.8. $ki = 6$ – steady state. $W = 0.269$ μJ/cm³, $\overline{P}_l \approx 1.67$ W, $\lambda = 510.5$ nm.

t, ns	0	20	50	90	180	$0.5 \cdot 10^6$ beginning $ki = 7$
U_0, V	0	1200	3000	2077	0	0
U_d, V	0	1160	2392	1135	91.76	0
j, A/cm²	0	0.165	1.736	7.13	1.9	0
I_d, A	0	82.3	867.8	3567	951	0
R_d, ohm	1.188	14.1	2.757	0.318	0.0965	1.18
ε, eV	0.237	5.67	7.419	5.617	2.476	0.237
T_A, K	–	–	–	–	–	1773.44
$n_e \cdot 10^{12}$ cm⁻³	0.1132	0.1241	0.7787	5.488	9.368	0.1135
$n^+_{Cu} \cdot 10^{12}$ cm⁻³	0.1121	0.1185	0.2272	1.344	3.595	0.1135
$n^+_{Ne} \cdot 10^{10}$ cm⁻³	$1 \cdot 10^{-9}$	0.5614	55.2	414.4	577.3	$1 \cdot 10^{-9}$

ohms; $L_1 = 5$ nH, the width of the electrode $b = 5$ cm, the length of the electrode gap $l_d = 5$–6.2 cm, the length of the electrode $l_e = 100$ cm, the cavity length $l_r = 200$ cm, the reflectivity of the output mirror $\rho = 0.08$; parameters of the voltage pulse U_0 (t) at the output: $\tau_{ex} = 50$ ns, $\tau_{rel} = 130$ ns, $U_{0\,max} = 3000$ V.

Figure 6.19 a, b shows the change in the basic electrical and physical parameters of a pulsed copper vapour laser with neon when operating in the frequency mode ($p_{Ne} = 53.4$ kPa, $T_w = 1773$ K, $f = 2$ kHz).

Below we analyze some results of numerical calculations at various neon pressures, the excitation pulse repetition frequency and the voltage at the input U_0. Here 'ki' specifies the number of the counted pulse (number of pulses or pulse intervals).

1. $p = 53.4$ kPa, $T_w = 1773$ K; $l_d = 6.2$ cm; $f = 3$ kHz, $\lambda = 510,5$ nm
   The results of the count:
   $ki = 1$–3 lasing pulse was absent,
   $ki = 4$, $W = 0.047$ μJ/cm³,
   $ki = 5$, $W = 0.2275$ μJ/cm³,
   $ki = 6$, $W = 0.258$ μJ/cm³.
   After the sixth pulse the stationary regime of laser operation was established and was characterized by the reproducibility from pulse to pulse of the prepulse and postpulse plasma parameters (active mode).

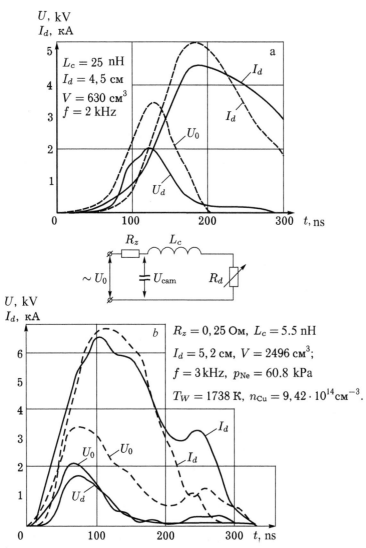

Figure 6.20. Typical characteristics of the laser (Cu+Ne): a) stand No. 1; b) stand No. 2 (solid curves – calculations; dotted curves – experiments).

2. $p = 40.0$ kPa, $T_w = 1773$ K; $l_d = 6.2$ cm; $f = 5$ kHz, $\lambda = 510.5$ nm

   The results of the count:

   $ki = 1-3$ lasing pulse was absent,

   $ki = 4$, $W = 0.184$ μJ/cm^3,

   $ki = 5$, $W = 0.077$ μJ/cm^3,

   $ki = 6$ no distinctive lasing recorded,

   $ki = 7$ – stationary operation mode of the laser (marked generation was not recorded).

3. $p = 26.7$ kPa, $T_w = 1773$ K; $l_d = 6.2$ cm; $f = 2$ kHz, $\lambda = 510.5$ nm

Table 6.9. Here $t(I_d^{max})$, $t(P_{l\,st})$, $t(P_{l\,max})$, $t(P_{l\,end})$ - measured from the beginning of the excitation pulse times to achieve the maximum discharge current, the start of generation, reaching the maximum of generation and the end of generation.

CVL object	I_d^{max}, kA $t(I_d^{max})$, ns		λ	$\lambda_1 = 510.5$ nm; $\lambda_2 = 578,2$ nm W, μJ/cm³		$t(P_{l\,st})$, ns		$t(P_{l\,max})$, ns		$t(P_{l\,end})$, ns		$\bar{P}$, W	
	calc.	exp.		calc.	exp.	calc.	exp.	calc.	exp.	calc.	exp.	calc.	exp.
Stand number 2	6.5	6.96	1	1.1	1.8	100	108	135	150	190	190	8.2	13
	110	115	2	0,63	1	102	138	135	158	198	205	4,7	7,6
Stand number 1	4.68	5.76	1	3	2.62	120	110	158	130	205	170	3.82	3.3
	189	190	2	2	–	117	120	161	150	210	200	2.54	–
Single-module CVL	5.23	5.35	1	9.7	9.67	15	20	43	47	78	80	–	–
	83.6	85	2	5.38	5.33								
	6.1	6.2	1	Σ 7.73	Σ 10.0	22	–	47	–	75	–	–	–
	94.3	–	2			24	–	49	–	81	–		

The results of the count:
ki = 1–2 lasing pulse was absent,
ki = 3, W = 0.226 μJ/cm³,
ki = 4, W = 0.278 μJ/cm³,
ki = 5 W = 0.271 μJ/cm³,
ki = 6 – stationary mode (the main results of the calculation, see Table 6.8).

The studies in which there is no continuous counting and the counting is divided into two regions (excitation and relaxation periods) will not be considered. The materials from such counts are considered in dissertations by N.I. Yurchenko, A.N. Mal'tsev and A.N. Timoshenko and AN, and published in [40–42].

Figure 6.20 shows the calculated and experimental data for two multi-module stands constructed at the Astrofizika company, and Table 6.9 gives the same data, but with the addition of the results for the single-module laser. The stands 1 and 2 differ mainly in the volume of the active zone and cavity design and details the voltage pulse oscillator. The chamber of stand 2 had the volume of the active zone ot $3.5 \cdot 10^3$ cm⁻³ (size of the zone 60 × 60 × 1000 mm), ten independent high-voltage inputs with an inductance of 50 nH, the length of the resonator 3 m. The chamber of stand 1 had the active

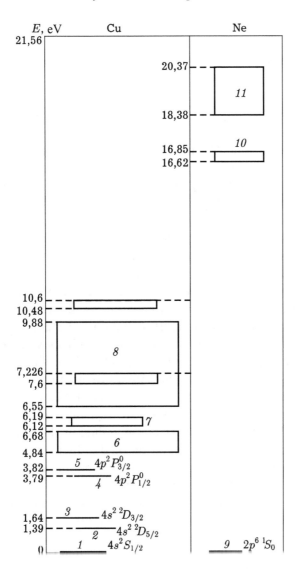

Figure 6.21. Scheme of levels of copper and neon atoms.

zone volume of 750 cm³ (size 40 × 60 × 300 mm), three independent high-voltage inputs, the length of the resonator 2.5 m. From these graphs and tables it can be seen that the agreement between the calculations and experimental data is quite satisfactory.

Having considered the results of numerical studies of copper vapour lasers, performed at the Astrofizika company, we consider the studies of the Institute for High Temperatures, Russian Academy of Sciences [43–45], because these studies are closely adjacent to the above works. The purpose of the work carried out in the Institute for High Temperatures, Russian Academy of Sciences, was to calculate the output characteristics of the laser by simultaneous solution of equations of the kinetics of populations of different levels, the equations of stimulated emission in an optical cavity and the equations of the electrical circuit. At the same time, a large number of excited levels and spectral lines with known transition probabilities were taken into account (124 level and 211 lines for the copper atom, 23 levels and 55 lines for the neon atom). The reabsorption of spontaneous radiation was taken into account using in the calculations new approximation formulas for the probability of photon emission [25].

The model [43–45] takes into account the following processes: Joule heating of the electron, ionization and triple recombination for all the relevant levels of copper and neon, photorecombination, conversion and dissociative recombination of molecular ions of neon, the spontaneous and stimulated emission, excitation and quenching by electron impact of various states of the copper and neon atoms, the elastic energy loss of electrons in collisions with the neon and copper atoms and their ions and the process of Penning-type ionization in collisions of excited neon atoms with the copper atoms in the ground state, recharging between neon ions and copper atoms, ambipolar diffusion of electrons and two ion species (approximately), the Doppler and Lorentz broadening of spectral lines. In calculating the rate constants of excitation and ionization by the electrons the effect of reducing the values of these constants associated with the deviation of the distribution function of the electron energy from the Maxwellian distribution in an environment with easily ionized additive is taken into account.

The diagram of the levels of the copper atoms and neon, incorporated in the blocks and used in [43–45], is shown in Fig. 6.21. The concentration of atoms belonging to the same block, is denoted by n_k, where k is the number of blocks. The copper atom corresponds to $k = 1$–8, the neon atom $k = 9$–10. The electrons at the threshold levels of blocks are in equilibrium with the free electrons.

The voltage pulse generator consists of a storage capacitor with an electrical capacitance C_s, charged to a voltage U_0, the discharge chamber of the laser which is connected in parallel with the peaking capacitor, which increases the steepness of the leading edge of the excitation pulse, and a thyratron switching the circuit. The inductance of the storage circuit sections and the discharge tube (chamber) is L_C and L_d, respectively.

The concentration of copper atoms n_{Cu} is defined in [43–45] on the basis of the equilibrium vapour pressure near the wall with a temperature T_w and converted into the tube volume at a temperature T_a:

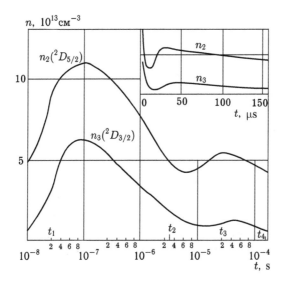

Figure 6.22. Variation of the concentration of metastable copper atoms n_2 and n_3 with time (p_{Ne} = 27 kPa, T_w = 1700 K).

$$n_{Cu} = 1.95 \cdot 10^{23} \, (T_w / T_a) \cdot \exp(-33160 / T_w), \, cm^{-3}.$$

The wall temperature T_w and the neon pressure p_{Ne} are considered given and constant.

Calculations were carried out for the discharge chamber with a ring-shaped working volume formed by two coaxial cylinders of radius r_1 and r_2 ($[(r_2-r_1)/r] < 1$) and the length l_d, as well as for the camera in the form of a cylinder (r_1 = 0). Electric discharge current and laser radiation were directed along the axis of the cylinders. The following constant parameters were used: r_2 = 30.5 cm, r_1 = 29.5 cm, l_d = 100 cm, C_S = 0.25·10⁻⁷ F, L_C = 0.5·10⁻⁸ H, L_d = 0.2·10⁻⁷ H, the length of the optical resonator was 150 cm, the reflection coefficients 0.95 and 0.05. Under these conditions, the volume of the chamber is 1.9·10⁴ cm³.

The calculations were performed for two values of the neon pressure and temperature of the wall: variant 1 – p_{Ne} = 2.67 kPa, T_w = 1700 K, variant 2 – p_{Ne} = 53.4 kPa, T_w = 1850 K. The excitation pulse repetition rate f and the prepulse voltage U_0 varied in both cases like the parameters of the calculation. The numerical calculations yielded a number of results for judging the role of various physical processes in the copper vapour laser (CVL) and the impact on the results of calculation of the assumptions made in the model.

Taking into account the large number of levels and the emission lines of copper and neon atoms and the application of a new model of radiation reabsorption allowed the authors [43–45] to obtain more reliable and detailed, than in other computational studies, information on the spontaneous emission, the electron energy balance and the dynamics of the concentration of excited atoms in the CVL. Figure 6.22 shows the results of calculation of concentrations of copper atoms in metastable states.

The main results of [45] can be summarized as follows.

In the conditions corresponding to the calculation conditions the energy of the spontaneous emission can reach large values and have a significant effect on the electron energy balance and gas temperature.

The bulk of the energy of the spontaneous emission is taken out of the plasma during its relaxation, which is caused by the recombination population of the upper levels of the copper atom followed by spontaneous settling to lower levels. Partial settling of the levels by electron impact also takes place.

The recombination flux supports large concentrations of the high lying levels, causes the appearance of second maximum values of the concentration of atoms at these levels in the interpulse period.

The computational model used in [43–45], leads to two characteristic times of the decrease of the concentrations of metastable levels of copper n_m in a relaxing plasma of the CVL in the interpulse period. The first decline of the concentration levels n_m due to a rapid decrease of T_e is related to their ionization and electron de-excitation (characteristic times 10^{-6}–10^{-5} s). The subsequent decline of n_m is much slower (characteristic time of the order of 10^{-4} s), which is associated with intense spontaneous levelling of the metastable levels from the upper block level, which in turn are populated by the recombination flux.

In order to reduce the prepulse concentrations of the metastable levels and improve the output characteristics of the CVL the authors of [43–45] recommend to add to the working mixture additives which would remove during the plasma recombination period the excitation from the upper levels of the copper atom, whose energies lie in the range of $E \geq 5$–6 eV.

6.4.4. Operation of a copper vapour laser at high buffer gas pressure

The calculations carries out using main program MOLOT-F4 show that the average output power of a copper vapour laser increases with increasing neon buffer gas pressure. This has not been confirmed by experiments, so there was a need for additional studies that were published in [46] and are described below, following mainly [46].

The dependences of the lasing energy of the copper vapour laser on most of the parameters (the concentration of copper atoms, the frequency of the pulse-periodic mode, voltage, charge and capacity of the excitation circuit drive, etc.) have been studied sufficiently and coincide, at least qualitatively, for different experimental setups. The only parameter whose influence on the energy output of the laser is ambiguous is the neon buffer gas pressure; with increasing pressure the energy output can decrease monotonically [47], increase [48], pass through one [49] or two [50] maxima. This ambiguity of the dependences is associated with a variety of mechanisms of the effect of pressure on the energy output. Analysis of the results of numerical calculations of the kinetics of the CVL, as well as some experimental data in [51], provide a basis to identify and interpret the most important of these mechanisms.

First, increasing pressure decreases the effective parameter of inductance of the device $X = (L \cdot s_d) / (l_d \cdot p)$ [51, 52] (L is the total inductance of the excitation circuit; s_d, l_d are the cross section and length of the discharge), which leads to an increase in the specific energy G and the corresponding specific energy output W.

Secondly, the optimal (with respect to the specific energy output W) prepulse field strength $E_0 = U_0/l_d$ (U_0 is the charging voltage of the storage) is strongly dependent on pressure, increasing with it. This means in particular that if E_0 for some equipment is bounded from above, the opportunities to increase pressure and thereby reduce the parameter X are also reduced.

Third, increasing pressure increases the fraction of energy input into the discharge, resulting in heating the gas, while at low pressures, the gas heating is usually irrelevant.

Fourthly, the pressure function which is crucial for the pulse-periodic regimes enables to regulate the rates of elastic and diffusion cooling of the electron gas in the interpulse period and, consequently, the relaxation rates of the plasma parameters.

In addition, the effect of pressure on the lasing characteristics can be exterded through the processes not recorded in the model, such as the skin effect which restricts the energy input to the discharge at low pressures, the cathode potential drop, the uniformity and stability of the discharge, retention of the copper vapours.

Specific results of the calculations of the output lasing power dependence on buffer gas pressure in various conditions and their interpretation are given in [51]. It should be noted that numerical calculations by the method proposed in [4, 40] describe well the experimental dependence of the energy output on the buffer gas pressure only to the pressure range at which the experiment has a maximum output energy. In the calculations, the monotonic increase of the energy output usually goes continues to a considerably higher pressure than in the experiment. Moreover, the largest discrepancies between theory and experiment were observed in direct numerical optimization of the laser parameters with respect to the lasing power, which led to a pressure at least several times larger than the experimental value, and to significantly higher lasing power.

The most marked difference is clearly seen in experiments with a laser with the section of the active zone (AZ) in the form of a highly elongated rectangle measuring 13 × 1.5 cm [53]. The laser with the slit discharge gap of this type was constructed in order to reduce heating of the gas, as well as to reduce the impact of the discharge circuit inductance on the lasing performance by reducing the the parameter X' ($X' = L \cdot s_d / l_d$). Measurements showed that at the extreme operating conditions the gas temperature exceeds the temperature of the walls of the GDC by no more than 100–200 K. Under optimal conditions, the total lasing power in both lasing lines was significantly higher than in lasers with GDC with a square cross-section, and reached 65 W at $f = 3$–4 kHz and the length $l_d = 110$ cm.

When using the CVL with the AZ with a transverse dimension 13 × 1.5 cm as the amplifier the average lasing power was 120 W at $f = 4$–5 kHz and the input signal with a power of 7 W. This fact indicates that, apparently, at large transverse dimensions of the AZ a significant size effect on the lasing characteristics is exerted by the loses in the non-axial beams of the induced radiation as a result of superluminescence and lasing due to parasitic scattering. These losses are significantly reduced in the transition to the 'master oscillator–amplifier' circuit due to saturation of the amplifying medium by a sufficient large input signal.

To be able to compare the results of the experiment and calculation over a wide range of conditions for the CVL having the active zone with a transverse dimension of 13 × 1.5 cm, a series of measurements was taken with varying buffer gas pressure and pulse repetition rates and with the optimization of the temperature of the chamber walls in each registered mode. The voltage pulse was applied to the electrodes through a cable transformer. As in [27,40,53,54], the shape of this pulse was close to a triangle, with the duration of the base of approximately 200 ns. The pulse amplitude $E_{max} = 500$ V/cm was close to the optimum for operating pressures $(0.3$–1$) \times 10^5$ Pa, and the specific energy of the storage $G_0 = 3.5$ µJ/cm^3 was in the saturation region, that is, its increase does not cause a noticeable increase in the lasing power.

It should be noted that this circumstance was confirmed in both the experiments and in the calculations.

A significant result of these experiments was a significant decrease in the optimal neon pressure to $(0.2–0.5) \cdot 10^5$ Pa compared to [27, 54] which used the discharge gap with the size of about 6×6 cm and the optimum pressure was in the range $> 10^5$ Pa. Thus, in the discussed experiment the discrepancy between the experimental and calculated values of the optimum buffer gas pressure p_{opt} was increased and reached the same order. The corresponding difference in the calculated and experimental values of the lasing power also became unacceptably high.

Thus, in this experiment with lasing power the optimum was achieved for all controlled parameters, pressure, and the results show the possibility of significant increase of power by increasing the buffer gas pressure. Obviously, the reason for this discrepancy, that is the mechanism for limiting the pressure of neon in the CVL from above, is of fundamental importance to determine the limiting capabilities of CVL with the transverse and longitudinal discharge.

Analysis of the results of the calculations undertaken to ascertain the causes leading to an increase in the calculation of output energy with increasing buffer gas pressure showed that the monotonic increase in power with pressure is due to to two factors at the same time: weak acceleration of the relaxation of metastable levels caused by increasing the frequency of quenching collisions, and increase of the efficiency of excitation of the upper laser levels associated with a decrease in the parameter X'.

It is also necessary to consider that the discussed discrepancy between the calculated and experimental results may be caused by errors in calculation that can occur due to the neglect of a particular event or factor: for example, increasing the width of the lasing line with increasing neon pressure, the inaccuracies in determining the energy losses due to excitation and ionization of the neon buffer gas atoms, neglecting the cathode potential drop, etc. To identify possible causes of discrepancies in the results of calculation and the experiment a special series of calculations was carried out, in which the parameters associated with these and some other processes and factors were varied, the rate constants, etc. The dependence $W = f(p_{Ne})$ of course changed, but values close to the experimental reduction of specific energy out at high pressures were obtained only after the introduction of additional recombination processes to the model of the CVL.

With increasing pressure, the role of triple collisions in the kinetics of plasma increases which, in particular, leads to acceleration of its decay. With regard to the CVL the following processes should be considered [23]:

$$Cu^+ + e + Ne \xrightarrow{q_1} Cu^* + Ne; \quad (1)$$

$$Cu^+ + Cu + Ne \xrightarrow{q_2} Cu_2^+ + Ne; \quad (2)$$

$$Cu_2^+ + e \xrightarrow{q_3} Cu^* + Cu; \quad (3)$$

$$Cu + Cu + Ne \xrightarrow{q_4} Cu_2 + Ne; \quad (4)$$

$$Cu_2 + e \xrightarrow{q_5} Cu^- + Cu; \quad (5)$$

$$Cu^- + Cu^+ \xrightarrow{q_6} Cu^* + Cu. \quad (6)$$

The processes (1)–(6) give rise to three recombination chains: a) triple recombination of the copper ion Cu^+ and the electron e; b) dissociative recombination via the formation of the molecular ion of copper Cu_2^+ and c) recombination via the formation of Cu_2 molecules, and then the negative copper ion Cu^-. Each of these chains ends in the relaxation of the highly excited Cu^* ion to the ground state by electron impact or spontaneous emission, and recombination currents, produced by these three chains, can be summed up (that is assumed to be independent), since the binding energy of Cu^* in all cases is consideably higher than the binding energy characteristic of the interpulse period of the electron temperature.

Triple recombination (1) is usually taken into account when calculating the characteristics of the CVL [4], but the classical value of the speed of this process q_1 [23] is very small, so that it is negligible even under these conditions, characterized by elevated neon pressure. At the same time there are a number of data [54], which may suggest that the magnitude of q_1 can be much larger, in particular, provides a practical formula for it:

$$q_1 = 10^{-26} \, (T/300 \text{ K})^{-2.5} \text{ cm}^6/\text{s}$$

glvlng one to two orders of magnitude higher values than it follows from [23].

The rate constants for the rest of these processes are more or less undefined. If we evaluate them based on the data [23, 55, 56], we can obtain the following values: $q_2 \approx 10^{-30}$ cm^6/s, $q_3 \approx 2 \cdot 10^{-7}$ cm^3/s, $q_4 \approx 10^{-31}$ cm^6/s, $q_5 \approx 10^{-9}$ cm^3/s, $q_6 \approx 10^{-7}$ cm^3/s at $T_a = 0.16$ eV, and the characteristics of their temperature dependence.

The input of the processes (1)–(6) to the model [51] leads to a qualitative change in the results of calculations at high pressures due to a sharp acceleration of the plasma decay: the optimal neon pressure is several times smaller. Under the assumption that the physical cause

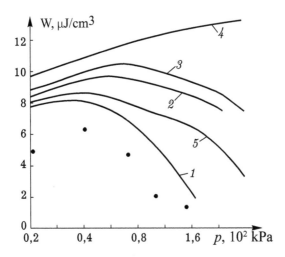

Figure 6.23. Calculated dependence of the energy output of the CVL on the neon pressure at $f = 3$ kHz: 1 – taking into account the processes (1)–(6); 2 – at $q_1 = 0$; 3 – when q_1, $q_4-q_6 = 0$, 4 – at $q_1 - q_6 = 0$; 5 – with q_2 three times higher. Points – experiment.

of the experimental dependence of $W(p)$ is found, the values q_1-q_6 were varied to obtain a good quantitative agreement with experiment. It turned out that the acceleration of recombination provided by the given values of q_1-q_6 still does not provide such agreement, and because these three chains act in parallel, when selecting the correct values for the rates of the processes q_1-q_6 and the corresponding values of the constants it is important to carry out a detailed comparison of the results of the calculations and measurements in a wide range of conditions.

Satisfactory agreement with experimental data is obtained in the case if the values q_1, q_2, q_5 and q_6 are tripled relative to the values given above, while maintaining the values of q_3 and q_4. The dependence $W - f(p_{Ne})$ obtained with a new set of values of q_1, q_2, q_5 and q_6 is shown in Fig. 6.23 (curve 1). Also shown are similar relationships obtained by gradually including in the CVL modes processes (1)– (6). In the calculation of all the dependences the concentration of the copper atoms was assumed to be $n_{Cu} = 1.2 \cdot 10^{15}$ cm^{-3}. It is seen that for the given choice of the values of q_1-q_6 the most significant impact on the results is exerted by the processes (1)–(3) and the processes (4)–(6) make a minor amendment.

The calculated pressure dependences of the most important integral parameters characterizing comprehensively the operation of the CVL in the selected mode are shown in Fig. 6.24. The concentration of copper atoms in this case is also constant and close to optimal for the

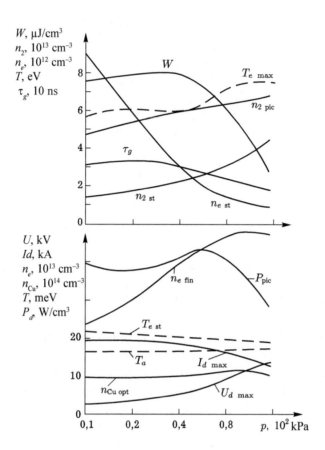

Figure 6.24. Dependence of the main results of calculations of CVL on neon pressure at $f = 3$ kHz, $n_{Cu} = 10^{15}$ cm^{-3}.

energy output for the entire area under considered pressure range. The following notations are used in Fig. 6.24: $U_{d,max}$ – the maximum voltage across the discharge, n_{plc} the peak lasing power; τ_g – the lasing pulse duration at the 0.1 level.

It is easy to see that the initial electron concentration n_{est} (indices st and fin correspond to the beginning and end of the excitation pulse) when pressure drops rapidly. As a result, at high pressures the development of ionization is hindered, resulting in the maximum discharge current $I_{d\ max}$ significantly decreases, which leads to some decrease in pump efficiency, which in this case, as in [51], is estimated on the basis of the population of metastable levels n_{2pic} at the end of lasing.

The considered mechanism of the influence of prepulse electron concentration n_{est} on the specific energy output is more important at frequencies f below 3 kHz. The main reason for the fall of the specific energy output with increasing pressure of neon in the calculation is considered a rapid increase with pressure of the prepulse concentration of metastable atoms, rather than its reduction, obtained in calculations for values of $q_1-q_6 = 0$. This is because the decrease in n_e during the interpulse period decreases the frequency of quenching collisions of electrons with metastable atoms, i.e., the mechanism of their destruction disappears. In other words, at high pressures, the rate of recombination of the CVL plasma is much higher than the optimal value, and thus taking into account the processes (1)–(6) leads to the same effect as a sharp decrease in the frequency of quenching collisions of metastable states.

A feature of the investigated variant of the CVL is the small transverse dimension of the active zone (1.5 cm) (with its relatively large volume of $2 \cdot 10^3$ cm^3) which allows almost completely eliminate the significant heating of the gas. For this reason, the optimum pressure of neon in this laser has decreased compared to the pressure in the lasers studied in [27, 54] and with the GSC with a square cross-section. The point is that the efficiency of recombination of chains (1)–(6) is reduced by heating the gas, leading to a decrease in the rate constants of some of these processes and equilibrium concentrations, Cu_2^+ and Cu_2 and Cu^-. We can assume that at high buffer gas pressures it is not always necessary to strive to reduce the heat, because, apparently, there is a certain optimum temperature of heavy particles at which, on the one hand, the harmful effect of the processes (1)–(6) is weakened and, on the other hand, the equilibrium population of the metastable levels is still quite small.

It should be noted that since the processes (1)–(6) have an impact on energy output, mainly by reducing the value of n_{est}, then the above choice of values of q_1-q_6 is partly due to the rate constants of some other process adopted in the CVL model under consideration. This is primarily related to the rate constants of deactivation of metastable atoms in collisions with the electrons and buffer gas atoms, which directly affect the relaxation of the metastable atoms. The effect associated with the tripling of the value of q_2 with respect to the value 10^{-16} cm^3/s accepteds in the model, is relatively small (see curve 5 in Fig. 6.23).

It should be added that the excitation and ionization of neon in the experimental conditions are small, but at higher pressures the influence of these processes increases. For $p \geq 105$ Pa, the total concentration of ions and excited neon atoms at the end of the pulse reaches 10% of the total number of the ions.

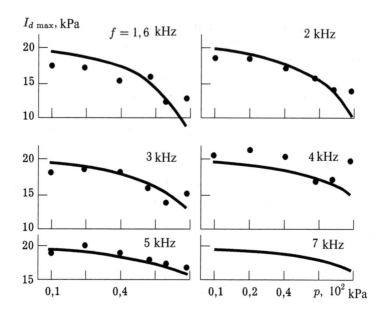

Figure 6.25. Calculated dependence of the maximum discharge current in the CVL on neon pressure (curves), points – experiment.

Figures 6.25 and 6.26 shown the transformation of the dependence $I_{d\,max} = f\,(p_{Ne})$ and $W = f(p_{Ne})$ when the excitation pulse repetition frequency changes. The calculated curves $I_{d\,max} = f\,(p_{Ne})$ agree well with the experimental data. The decrease of the current amplitude at high neon pressures is particularly noticeable at low frequencies f, therefore, it is precisely due to a decrease in n_{est}, and not by increasing pressure. It follows from the calculations that at $p_{Ne} \geq 10^5$ Pa the n_{est} value is very sensitive to the concentration of copper atoms and the gas temperature, since these parameters affect the intensity of the reactions (1)–(6). We can therefore assume that the random fluctuations in the concentration n_{Cu} from one mode of laser operation to another, without going beyond the optimal concentration n_{Cu}, must cause marked fluctuations in the amplitude of the discharge current. The experimental results are consistent with this hypothesis: the scatter of $I_{d\,max}$, observed during the transition from one mode to another at all neon pressures of neon, is markedly increased at $p_{Ne} \geq 10^5$ Pa.

With increasing frequency, specific energy output W decreases in both the calculation and the experiment (see Fig. 6.26), but in the latter case this decrease is much faster.

Thus, on the basis of the numerical studies we can conclude that the increase in the CVL volume at the expense of the cross-section, while optimizing the remaining parameters, should lead to an increase in the optimal buffer gas pressure. The result is a change in the

W, mJ/cm^2

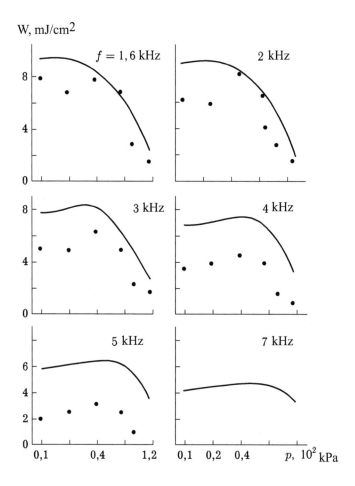

Figure 6.26. Estimated energy output depending on the pressure of neon CVL (lines) points – experiment. Optimal concentration of copper in each mode.

mechanism of electron cooling in the interpulse period from the diffusional to collisional and the additional recombination processes due to triple collisions in the plasma and limiting at the top the operating pressure of the buffer gas and the attainable output energy become important. Increasing buffer gas pressure increases its optimal temperature, and the optimal excitation pulse repetition frequency is reduced. We can assume that the findings made are of equal importance for lasers with transverse and longitudinal discharges.

6.4.5. A parametric study of the effect of initial data on the energy characteristics of a CVL

In [52] the authors obtained similarity relations for a single pulse, namely, it was shown that lasers with the same generic parameters mode

$$\frac{n_{Ne}}{n_{Cu}}; \quad \frac{U_0}{l_a \cdot n_{Cu}}; \quad \frac{a^2}{l_a \cdot C_S \cdot n_{Cu}}; \quad \frac{a^2 L \cdot n_{Cu}}{l_a}; \quad l_a \cdot n_{Cu}; \quad \tau_R n_{Cu}; \quad \frac{a^2 R_{st}}{l_a},$$

where U_0 is the initial voltage of the accumulator a, l_a is the transverse size and the length of the active zone for a rectangular section, τ_R is the characteristic time of activation of the voltage pulse oscillator, R_{st} is the prepulse discharge resistance, have the same efficiency and the W/n_{Cu}.

It should be noted that the further discussion of the results of parametric study of the effect of the initial parameters on the output characteristics of a CVL will be conducted in accordance with the thesis of N.I. Yurchenko [57].

The last four of the above parameters describe the effect on the lasing characteristics of the inductance of the discharge circuit, the finite time of the transfer of radiation along the axis of the laser and the non-ideality discharge switch. If we neglect this effect, the similarity relations indicate the possibility of the unlimited increase of the energy output of the laser at a fixed volume. Otherwise, there are a number of similar CVLs in which the growth of energy in the lasing pulse is associated with changes in other parameters as follows [57]:

$$\begin{cases} V_a W \sim a^2 l_a W \sim a^2 \sim l_a \sim V_a^{1/3} \sim 1/n_{Cu} \sim 1/n_{Ne} \sim C_S \sim \tau_R \\ \text{at } U_0, R_{st}, L/l_a = \text{const}, \end{cases}$$

or

$$\begin{cases} V_a W \sim a^2 l_a W \sim a^2 \sim l_a^2 \sim V_a^{2/3} \sim 1/n_{Cu}^2 \sim 1/n_{Ne}^2 \text{ and so on.} \\ \text{at } U_0, L = \text{const}. \end{cases}$$

These similarity series shows that increasing the energy output requires the growth of the laser volume, and this growth becomes greater as the dependence of inductance on the size becomes greater But this is not the only way to increase the lasing energy. If we neglect the excitation of the buffer gas atoms, which is almost always true for the CVL, it becomes possible to increase the energy output by increasing WV_a by increasing the neon pressure and the cross-sectional area of the active zone at the same values of n_{Cu}, l_a, τ_R, L and C_S

$$V_a W \sim V_a \sim a^2 \sim n_{Ne} \sim U_0^2$$

or if in addition we neglect the influence of the parameter $l_a \cdot n_{Cu}$ (neglecting it is acceptable either at $l_a \ll c_{rg}$, or in the amplification mode) by increasing the neon pressure and energy input at a fixed geometry of the active zone (a, l_a, L = const):

$$V_a W \sim W \sim n_{Cu} \sim n_{Ne}^{1/2} \sim U_0^{2/3} \sim 1/\tau_R \sim C_S^{-1/2}.$$

In the latter case, growth of the energy output an increase in U_0 energy output and reducing τ_R, which can be a constraint in the real operating conditions of lasers.

Thus, the similarity relations allow to qualitatively identify ways to increase the pulse energy (growth in the active zone volume and buffer gas pressure), but do not provide quantitative information on the characteristics achievable in certain specific conditions. This information can be obtained using the approximate formulas [51] which also allow us to compare the modes of the CVL, which are not similar and have different efficiency, and to determine the requirements for such important technical parameters as E_0, τ_R, n_{Cu}.

The results of calculations of the maximum achievable (for the given efficiency) given energy in the lasing pulse of the CVL with longitudinal and transverse discharges depending on the size of the cross section of the active zone, show that the longitudinal discharge has an advantage over the transverse discharge at small active zone cross-sections. This advantage of the longitudinal discharge is due to the dependence of the parameter $X' = LS_d/l_d$ on the geometry of the active zone. However, its implementation requires sufficiently high values of U_0 and small τ_R. Particularly high requirements are imposed on the values of U_0 and τ_R in the longitudinal discharge in the transition to the region at atmospheric pressure of neon or to the region of higher energy output. Therefore, the pressure in the longitudinal discharge is much lower than in the transverse discharge.

When taking into account the actual technical constraints to achieve the same output characteristics of the lasing energy input $W \cdot a_2$ and practical efficiency η_r in the longitudinal excitation of the active zone the required volume of this zone is several times greater than in the transverse excitation, if the required lasing energy output is $W \cdot a^2 \geq$ 10–20 mJ/m or approximately the same volume if $W \cdot a^2 \geq$ 5–10 mJ/m. In the latter case the technically simpler design with longitudinal excitation is preferred.

The validity of the relations obtained in [52] and presented above is limited to the single-pulse mode, but most of the existing CVLs operate in the pulse-periodic mode. It can be shown that if in the derivation of the similarity relations we take into account in the kinetics equations of the population additional terms responsible for the slow recombination and diffusion processes acting in the interpulse period, and also take into account the heating of the gas by the discharge, it is necessary to add to the the system of the previously presented similarity parameters the parameters f/n_{Cu}, $a \cdot n_{Cu}$, Aij/n_{Cu} and $q_{ij} \cdot n_{Cu}$, where A_{ij} are the effective probabilities of spontaneous transitions, and q_{ij} is the rate of recombination processes in the triple collisions. Since A_{ij} and q_{ij} are not external parameters of the problem, then many of these modes of operation of the CVLs with different n_{Cu} are reduced to one, and further analysis is impossible (the monopoly of binary collisions breaks down).

However, if we do not consider the range of the high pressures of the buffer gas ($p_{Ne} > 10^5$ Pa), where, besides impact recombination flowing at a rate of $q_{imp.rec}$, there are other triple processes, then in a certain range to obtain qualitative conclusions we use the approximate set of similarity parameters, eliminating the parameters Aij/n_{Cu} and $q_{ij} \cdot n_{Cu}$.

The reason for eliminating the parameter $q_{ipm.rec}$ n_{Cu} is that in the quasi-stationary conditions typical of the CVL plasma decay, due to much stronger dependence of $q_{ipm.rec}$ on T_e than the dependence on the cooling rate of electrons, the rate of recombination in the first approximation is determined by the cooling rate, rather than by the value of $q_{ipm.rec}$, which is simply adjusted with respect to the specified recombination flux.

Neglecting a set of parameters A_{ij}/n_{Cu} leads to a redistribution of flows of recombination in the copper and neon atoms and a change in the effective recombination potential in scaling. As shown by numerical calculations, the variation of the latter within reasonable limits, as well as variation in the numerical factors in front of $q_{ipm.rec}$, A_{ij}, has a relatively little effect on the final results and this is a basis for ignore the parameter A_{ij}/n_{Cu}.

After the introduction to the system of the similarity parameters of the relations f/n_{Cu} and $a \cdot n_{Cu}$ it is possible to construct a number of such regimes restricted only by the assumption of constancy of the discharge circuit inductance:

$$V\overline{P_l} \sim a^2 l_a Wf \sim l_a \sim a \sim V_a^{1/2} \sim 1/n_{Cu} \sim 1/n_{Ne} \sim$$
$$\sim 1/f \sim C_S^{1/2} \sim \tau_R \sim 1/R_{st}, \quad U_0, L = \text{const.} \tag{6.8}$$

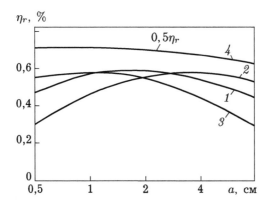

Figure 6.27. Change of η_r of the CVL in a series of similar modes of operation: $1 - f \sim a^{-1}$, $2 - f \sim a^{-1.1}$, $3 - f \sim a^{-0.9}$, $4 - f = 0$.

Figure 6.27 shows the results of numerical calculations characterizing the degree of precision of the conservation of practical efficiency which determines the efficiency of conversion of the energy stored in the   at the beginning of the discharge, to the lasing energy while moving along the series (6.8) (curve 1). The leading parameter is the diameter of the tube, and as the initial mode – the standard mode of [51]. It is easy to see that the decrease η_r with increasing transverse size a in the repetitively pulsed and single-pulse modes is similar. It is due to the influence of spontaneous transitions. The decrease of η_r with decreasing parameter a occurs only in the repetitively pulsed mode, it is due to the approximate nature of a number of similarity parameters, namely, the weak growth of n_{2st}/n_{Cu} and n_{est}/n_{Cu} with frequency. If we change the frequency f in scaling parameters not exactly in accordance with (6.8) and slightly faster or slower (curves 2, 3), the slope of the η_r (a) dependence changes dramatically. This indicates that the series (6.8) is close to an accurate one and may be used for qualitative analysis.

The gain of stimulated emission depends on the width of the lasing line Δv equal to about three Doppler lines (ΔvD), in accordance with the measurements [58,59]. Due to the rather deep saturation of the laser transition the results of calculations at $\Delta v \leq 10 \, \Delta v_D$ are weakly dependent on the choice of Δv (Fig. 6.28).

Below are the results of numerical studies of the effect on the excitation and lasing of the CVL of the main parameters of the pump source: U_0, C_S and L [51]. The results of calculations presented here and later are a set of integral parameters, fully characterizing the operation of the laser in the selected mode such as W, η_p, η_r, n_{2st}, n_{2fin}, T_a, $T_{e \, max}$, τ_R. Subscripts st and fin denote quantities taken at

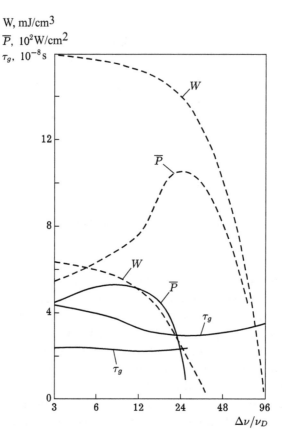

Figure 6.28. Sensitivity of simulation results to the value of the considered width of the lasing line of the CVL in the pulse-periodic (solid line) and single-pulse (dotted line) modes; Δv_D is is the Doppler width of an isolated absorption line.

the beginning and end of the excitation pulse, the index max – the maximum values.

The dependence of the basic characteristics of a CVL on the specific energy in terms of [60] are shown in Fig. 6.29 (diameter 2.4 cm, length 30 cm, $L = 1$ μH, $C_S = 1.25$ nF, $p_{Ne} = 26.7$ kPa, $f = 4$ kHz). The growth of the field strength, the electron temperature, the pumping efficiency and output energy saturates quickly, and this happens both at a constant concentration n_{Cu}, and in optimization of this contribution for each value of energy input.

The reason for this saturation is a drop in the efficiency of energy input η_d to the discharge caused, in turn, by a decrease the plasma resistance and inductance of the discharge circuit. Indeed, since the current pulse duration in this case is almost unchanged, at the maximum rate of increase of its amplitude the constant discharge efficiency can be achieved only if the resistance of the discharge is

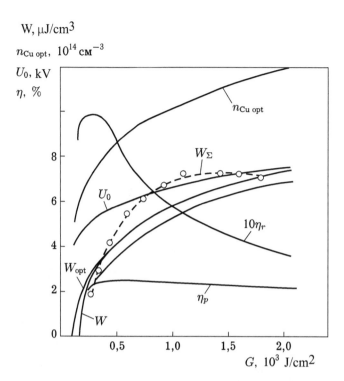

Figure 6.29. The dependence of W, $n_{Cu\ opt}$, U and η on the specific energy in the CVL in the woring condition in [60] (W_{opt} is the energy output at the optimum n_{Cu}, W_{Σ} is the specific energy at all radiation wavelengths. The dotted line – experiment at a constant T_w).

independent of the energy input. But the real increase in energy input to the discharge means a corresponding increase in the ionization fluxes and hence the conductivity of the plasma, so that to preserve η_d requires a more rapid increase in current than $I_d \sim G^{1/2}$, an impossible value for a constant L.

The dependences of the basic energy characteristics of a CVL on the storage capacity are shown in Fig. 6.30, and on the inductance of the discharge circuit in Fig. 6.31. As can be seen from the figure, the value of L, for which the energy output reaches its maximum value, is very small and, apparently, is practically unattainable in the actual conditions. The energy output in the standard mode of operation of the laser ($L = 500$ nH) and modes close to the standard mode in actual CVLs is about 0,5 W, where W is the energy output at zero inductance.

Figure 6.32 shows the dependence of the average lasing power on the most important parameter of the CVL mode – the concentration of copper atoms. Figure 6.33 compares the dependence on the energy output of the CVL on the pressure of He, Ne and Ar, calculated for the experimental conditions [60].

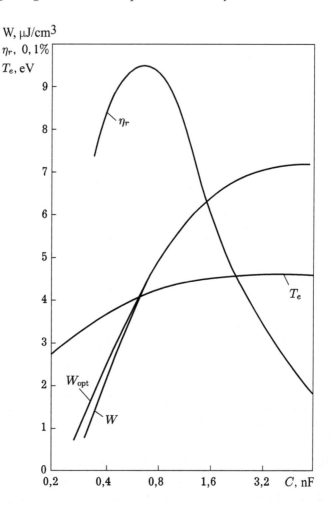

Figure 6.30. The dependence of the main energy characteristics of the CVL on the storage capacity in the standard mode for $n_{Cu} = 7 \cdot 10^{14}$ cm^{-3} (W_{opt} is the optimal energy output for a given capacitance).

Thus, the results presented in this section allow us to establish a complete qualitative and satisfactory quantitative agreement between the calculated (using the similarity criteria of the lasing characteristics of the CVL) with the experimentally measured values.

6.5. Europium vapour laser

In [30] the authors obtained experimentally for the first time lasing in an europium vapour laser in the infrared lines $\lambda_l = 1.66$; 1.76; 2.58; 2.72 µm. In addition, in [30] an assumption was made about the possibility of efficient lasing at the lines lying in the range of 4.3–5 µm. However, further investigation of the europium vapour laser

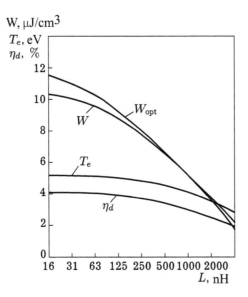

Figure 6.31. The dependence of the main energy characteristics of the CVL on the excitation circuit inductance in standard mode when $n_{Cu} = 7 \cdot 10^{14}$ cm^{-3} (W_{opt} the energy output optimal for a given inductance).

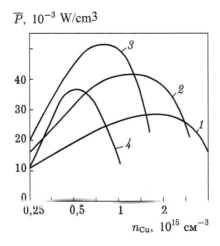

Figure 6.32. The dependence of the average specific output lasing power of the CVL on the concentration of copper atoms at different frequencies: 1 – 2 kHz, 2 – 4 kHz, 3 – 8 kHz, 4 – 16 kHz.

were focused mainly on obtaining powerful lasing on Eu ions [61], and the assumptions about the availability of efficient lasing of at self-contained europium atom transitions were not developed further.

To accelerate the research of various CVLs it is appropriate to conduct a preliminary numerical study of these lasers. The Astrofizika company developed a MTRH2 universal engineering program intended

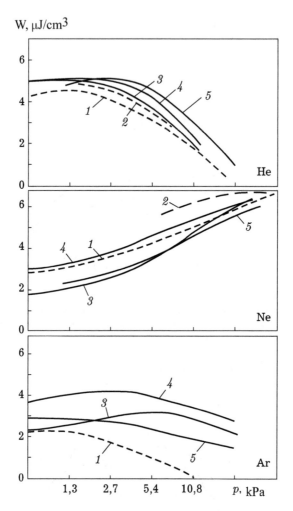

W, μJ/cm³

Figure 6.33. Dependence on the CVL energy output on the pressure of the buffer gas (helium, neon and argon). The experiment [60] at a constant T_w (1) and at optimum T_w (2). Calculation at constant $n_{Cu} = 7 \cdot 10^{15}$ cm^{-3} (3) and at optimum n_{Cu} (4, 5), $R_{ш} = 100$ ohms (3,4), 0.01 ohms (5).

for numerical studies and projections of the parameters of various CVLs to investigate the lasing characteristics of the europium vapour lasers.

The physical and mathematical model of the MTRH2 program is described in Section 6.3. This program uses two methods of formation of voltage pulses to the electrodes of the GDT (GDC): supply of a voltage pulse from the voltage pulse oscillator (VPO)) through a cable line, and the discharge through the discharge gap of the storage capacitor.

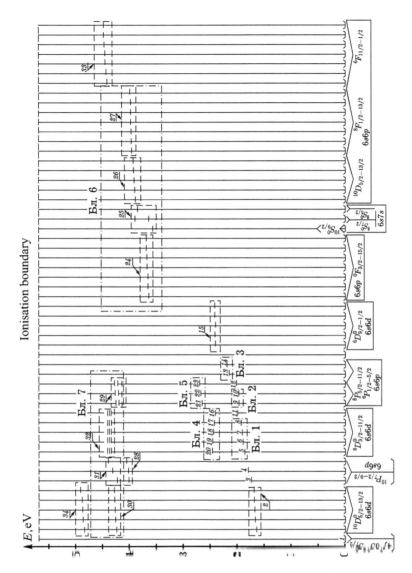

Figure 6.34. The energy spectrum of the europium atom.

The frequency of electron–atom collisions with He atoms was calculated using the well-known experimental ionization cross sections of the He atom from the ground state and elastic collision. The cross sections and rate constants of elementary processes for the He atom as well as all the processes for the Eu atom were calculated by semi-empirical dependences. In calculating the rate constants taken into account in calculating the transition from the lower lying i-th levels to higher lying j-th levels we used the operation of merging levels with similar energy states in the levels–blocks. In this case the energies E

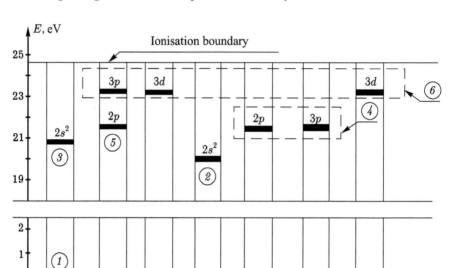

Figure 6.35. Energy spectrum of the helium atom.

Table 6.10.

Laser transition No. (i, j)	$i \to j$	λ_p, μm	ΔE_{ij}, eV	ΔE_{ij}, cm^{-1}
1	5 – 21	1.587	0.775	6300
2	6 – 21	1.616	0.76	6190
3	6 – 22	1.575	0.785	6350
4	7 – 21	1.661	0.74	6020
5	7 – 22	1.618	0.765	6180
6	7 – 23	1.577	0.785	6340
7	8 – 22	1.689	0.733	5920
8	8 – 23	1.644	0.753	6080
9	9 – 18	2.618	0.471	3820
10	9 – 19	2.584	0.476	3870
11	10 – 17	2.717	0.451	3680
12	10 – 18	2.66	0.465	3760

Laser transition No. (i, j)	$i \rightarrow j$	λ_p, µm	ΔE_{ij}, eV	ΔE_{ij}, cm^{-1}
13	10 – 23	1.761	0.704	5680
14	12 – 16	3.521	0.354	2840
15	12 – 17	3.311	0.371	3020
16	16 – 23	4.329	0.286	2310
17	17 – 22	5.076	0.25	1970
18	17 – 23	4.695	0.27	2130
19	18 – 21	5.78	0.21	1730
20	18 – 22	5.291	0.235	1890
21	18 – 23	4.878	0.255	2050
22	19 – 21	5.952	0.205	1680
23	19 – 22	5.435	0.23	1840
24	20 – 21	6.061	0.201	1650

Table 6.11.

GDT No.	d_p, cm	l_a, cm	l_r, cm	L, nH
1	2	100	200	1200
2	1.5	30	100	210

of all levels, grouped in a block, were set equal and corresponding to the average energy levels included in the block of levels l.

The transition rate constant a_{ij}^{ea} for each individual unit level was determined by the distribution of the calculated value of the excitation constant for a single level-block (with the total statistical weight of the combined levels) in proportion to the statistical weight of a separate state. With this approach, the transitions between the levels, conventionally combined into a block level, were excluded from consideration.

Figure 6.34 shows a scheme of 34 levels of Eu considered in the calculations (with conventional numbering of the ground state and increasing energy levels). The energy states, designated by the numbers 2, 15, 24–34, are the blocks of levels and combine the group of levels

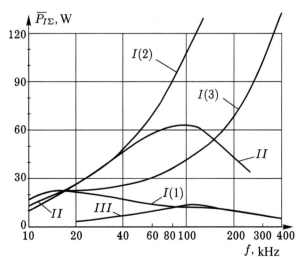

Figure 6.36. The dependence of the total average lasing power of the Eu vapour laser on frequency. Curves I correspond to the vapour pressure of europium p_{Eu} = 133 Pa; II – p_{Eu} = 13.5 Pa at p_{He} = 2.67 kPa; III – p_{Eu} = 1, 3, Pa at p_{He} = 133 and 667 Pa; I (1) – frequency mode,I (2) – dual pulse mode; I (3) – triple puls0e mode.

closely spaced in energy. For the He atom we take into account six energy levels indicated in the diagram in Fig. 6.35. The fourth and second levels are the blocks of levels and combine the upper closely spaced (in energy) states.

In the numerical experiments attention was given to the possible (unknown) lasing in the infrared lines of the Eu atom, that is, we analyzed the characteristics of lasing at 24 lines lying in the wavelength range λ_1 = 1.57–6.06 µm. Conditional numbers of laser transitions and some of their characteristics are listed in Table 6.10.

Calculations of the energy characteristics of lasers based on Eu vapours were performed for different laser tubes, helium buffer gas and dual pumping systems of the active medium; the discharge of the storage capacitor C_S (pump system 'A'), supply of the voltage pulse from the voltage pulse oscillator (VPO) through the cable line with wave impedance R (pump system 'B'). For the second pump system the pulse shape at the output of the VPO was given in the form of a triangle. The characteristics of GDT, used in the calculations, are given in Table 6.11.

In the calculations the following parameters were varied: T_w, p_{Eu} (n_{Eu}), p_{He} (n_{He}), R, f, U_0 (for pumping system B $U_0 = 2U_{VPO\ max}$).

The numerical studies of the Eu-vapour laser with the GDT No. 1 were performed for pumping system A, C_S = 1500 pF, and the following sets of values of U_0 and p_{He}: U_0 = 8 kV, p_{He} = 2.67 kPa;

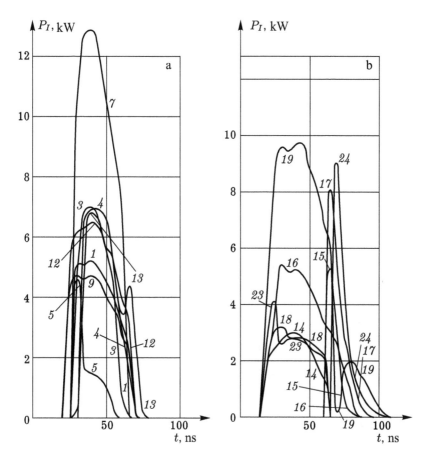

Figure 6.37. The dependence of the instantaneous radiation power on time for different lasing lines for conventional numbers of laser transitions (Table 6.10) for the tube No. 1; a) the beginning, b) continuation.

$U_0 = 4$ kV, $p_{He} = 667$ Pa; $U_0 = 4$ kV, $p_{He} = 133.4$ Pa. The studied range of the excitation pulse repetition frequency $f = 10 \ 500$ kHz.

Figure 6.36 shows results of calculation of the dependence (all lines) of the total average lasing power of the excitation pulse repetition frequency $\bar{P}_{I\Sigma} = \bar{P}_{I\Sigma}(f)$. Curves I correspond to the Eu vapour pressure $p_{Eu} = 133$ Pa, II $- p_{Eu} = 13.3$ Pa at $p_{He} = 2.67$ kPa, III $- p_{Eu} = 1.33$ Pa at $p_{He} = 133.4$ and 667 Pa.

For $p_{Eu} = 133.4$ Pa the curve I (1) defines the parameters of the calculation of the steady-state (frequency) mode after the passage of many pulses, and the dependences I (2) and I (3) characterize the same parameters after passing the 2nd and 3rd interpulse intervals of time, i.e. the characterize the operation of the laser in the regime of double and triple excitation pulses. This mode is sometimes used to predict the parameters of lasers with self-contained transitions operating in

Table 6.12.

Laser transition No.	λ_p, µm	$\bar{P}_l$, W	W, µJ/cm³	η_r, %
1	1.587	8.307	0.529	0.098
2	1.616	1.328	0.085	0.016
3	1.575	9.209	0.586	0.109
4	1.661	9.603	0.612	0.114
5	1.618	2.252	0.143	0.027
7	1.689	18.23	1.161	0.216
9	2.618	8.556	0.545	0.101
12	2.66	12.023	0.766	0.142
13	1.761	9.943	0.633	0.118
14	3.521	4.08	0.26	0.048
15	3.311	2.414	0.154	0.029
16	4.329	11.88	0.757	0.141
17	5.076	5.318	0.339	0.063
18	4.695	5.977	0.381	0.071
19	5.78	22.556	1.437	0.267
22	5.952	1.275	0.081	0.015
23	5.435	6.049	0.385	0.072
24	6.061	3.504	0.223	0.042
Σ	–	142.504	9.077	1.609

the pulse-periodic regime. However, as shown by calculations, the regime of double and triple excitation pulses and the corresponding methodology for calculating the output characteristics of the PMVL are unacceptable for predicting the output characteristics of the Eu-vapour laser with high concentrations of Eu atoms and at $f > 20$ kHz. At low Eu vapour pressures ($p_{Eu} = 13.3$ Pa and $p_{Eu} = 1.3$ Pa) the techniques of double and triple excitation pulses are more suitable for evaluating the characteristics of the repetitively pulsed lasers, but also become erroneous at $f \geq 50$ kHz.

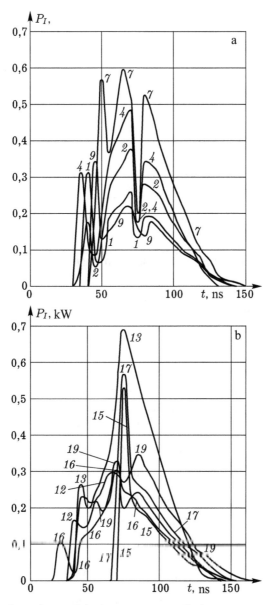

Figure 6.38. The dependence of the instantaneous radiation power on time for different lasing lines for conventional numbers of laser transitions (see Table 6.10) for the tube No. 2 at p_{Eu} = 133.4 Pa, a: start, b: continuation.

Calculations show that for the typical working conditions of the laser the maximum (total for all lines) lasing power is reached at $\bar{P}_{l\Sigma}$ ≈ 61.4 W at p_{Eu} = 13.3 Pa, and f ≈ 100 kHz (maximum of curve II). The effective lasing is fixed at 18 radiative transitions with total values W_{Σ} ≈ 1.96 µJ/cm and $\eta_{r\Sigma}$ ≈ 1.282%. The most intensive lasing

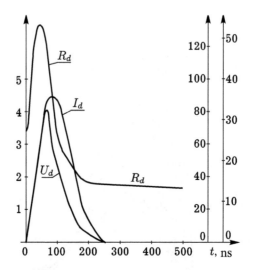

Figure 6.39. The dependence of the physical parameters of the discharge tube from time to time for the number 2.

Table 6.13.

Laser transition No.	λ_p, μm	$\overline{P}_l$, W	W, μJ/cm³	η_p, %
1	1.587	0.258	0.244	0.135
2	1.616	0.366	0.346	0.191
4	1.661	0.451	0.426	0.236
7	1.689	0.584	0.551	0.305
9	2.618	0.235	0.222	0.123
12	0.66	0.001	0.010	0.172
13	1.761	0.567	0.535	0.296
14	3.521	0.082	0.078	0.043
15	3.311	0.18	0.17	0.094
16	4.329	0.295	0.278	0.154
17	5.076	0.272	0.257	0.142
18	4.695	0.054	0.051	0.028
19	5.78	0.416	0.393	0.217

Laser transition No.	λ_l, μm	$\bar{P}_l$, W	W, μJ/cm³	η_l, %
20	5.291	0.079	0.074	0.041
22	5.952	0.017	0.016	0.009
23	5.435	0.119	0.112	0.062
24	6.061	0.085	0.08	0.044
Σ	–	4.391	4.145	2.293

is recorded here for 9 and lines, which account for $\bar{P}_{l\Sigma} \approx 47.206$ W, $W_\Sigma \approx 1.503$ μJ/cm, $\eta_{r\Sigma} \approx 0.983\%$.

For a typical GDT No. 1 with the pumping system A numerical calculations predict achievingt $\bar{P}_{l\Sigma} \approx 142.5$ W ($W_\Sigma \approx 9.077$ μJ/cm³) when $C_S = 1500$ pF, $U_0 = 15$ kV, $p_{Eu} = 133$ Pa, $p_{He} = 13.3$ kPa, $f = 50$ kHz. For this mode, the dependences of the instantaneous radiation power P_l for lasing lines are shown in Fig. 6.37 a, b (for conventional numbers of laser transitions), and the energy parameters of radiation are given in Table 6.12. The results of numerical experiments with short GDTs ($l_d \leq 6.48$ cm) showed that the highest output characteristics are obtained with the GDT No. 2 (see Table 6.11) when $p_{Eu} = 133$ Pa, $p_{He} = 13.3$ kPa and $U_{0\,max} = 8$ kV ($U_{VPO} = 4$ kV) at $f = 20$ kHz. Using these parameters, we have the best agreement of the pulses oscillator with the load. Figure 6.38 a, b shows the calculated dependences of the radiation power P_l for the lasing lines, Fig. 6.39 shows the results of calculation of the parameters of the plasma discharge gap as a function of time, and Table 6.13 gives the results of calculation of the lasing energy characteristics.

For the same set of the parameters the characteristics of the laser were calculated at an increased input voltage of $U_{0\,max} = 14$ kV. It was found that increasing the voltage on the VPO does not increase the radiation power.

In the Eu vapour laser with GDT No. 2 the bulk of the lasing power is provided by 10 lasing lines. The laser pulse duration is 100–110 ns. The most powerful laser pulses are peak shaped.

Thus, the results of numerical studies of the Eu vapour laser have shown that for different operating conditions lasing occurs mainly at the 18 emission lines in the infrared range. The highest values of the average output power are recorded in nine lines of the Eu atom: $\lambda_l = 1.587; 1.616; 1.661; 1.689; 1.761; 2.660; 4.329; 5.076$ and 5.780 μm. In this case the radiation power on the lines $\lambda_l = 2.660$ μm and

λ_l = 4.329 μm are close in magnitude for almost all investigated GDTs and modes of the laser and the laser pulse duration on these lines (base) is ≈ 100–110 ns.

Voltage change U_{VPO} at the same energy input to the active zone for long and short GDTs has different effects on the radiation characteristics of the lasers. In large tubes increase of U_{VPO} provides a significant increase in the total average output power for all lines. For the shorter GDTs the output power does not increase.

Numerical calculations show that in the europium vapour laser with the GDT No. 1 (see Table 6.11) at p_{Eu} = 133 Pa, p_{He} = 13.3 kPa, f = 50 kHz, C_S = 1500 pF and U_0 = 15 kV. the total average lasing power can reach approximately 140W.

6.6. Optimization of parameters of metal vapour lasers

When producing pulsed metal vapour lasers (PMVL) with the required radiation characteristics their parameters are most effectively optimized using mathematical modelling and computers. The calculations show that the lasing characteristics have a significant effect on at least ten parameters. These include the concentration of atoms of the working metal and the buffer gas, the geometrical dimensions of the active zone and the cavity, the reflection coefficients of mirrors, the electrical parameters of the excitation circuit, the duration of the leading edge and the pulse amplitude and the voltage across the electrodes of GDT and the pulse repetition rate. The list of the parameters adequately characterizes the work of PMVL. Neglect in the physical–mathematical model of at least one of them significantly narrows the range of application of the optimization results.

The mathematical formulation of the optimization problem is represented as the problem of finding the largest or smallest values of the function of many variables

$$\Psi - \Psi(x_1, x_2, ..., x_n)$$

where the function Φ is a quantitative estimate of the object of optimization of the target function. The target function are laser output parameters or their combination, such as efficiency, pulse energy, average radiation power, etc.

As the number of local optima of the target function is not known in advance, the difficulties that can arise in finding the global optimum in the multifactorial (n-dimensional space) become apparent. In optimization it is expedient to use the most efficient methods based on the theory of experiment planning. The following tasks are very important: identifying the influence on the target function of each of the factors acting in the system when they vary, minimizing the total number of numerical experiments, the use of mathematical tools that formalize the numerical experiments, the choice of a clear strategy to make informed decisions after each series of numerical experiments.

In the numerical experiments planned for the optimization of various output parameters of the PMVL it is required at the beginning to identify the degree of influence on the optimized function of the initial values of the vector of input parameters to identify the most important input parameters (see section 6.4.5). Here, however, it should be emphasized that the process of ranking of input parameters using the Spearman rank correlation coefficient or the Kendall coefficient of agreement [62] with respect to PMVL was not even discussed. Based on the experience of field experiments, we can assume that the main significant factors are: the evaporator temperature, which determines the concentration of metal atoms, buffer gas pressure, excitation pulse repetition frequency, the size of the discharge gap, the amplitude of the exciting voltage pulse, the parameters of the circuit of the excitation system and of the cell. The range of possible changes (the region of variation) of each initial parameter is defined on the basis the design and actual technological capabilities.

A variety of methods for solving tasks of non-linear programming (planning) of experiments indicate the complexity of the problem of finding the optimum and the difficulties associated with assessing the effectiveness of using any method. The investigation of the appropriate studies suggests that, for the multivariate numerical experiments with restrictions on the range of variation of the initial parameters, the effective method of finding the optimum of the target function is the Box complex method [63].

The essence of the Box method is to modify the simplex search which uses figures consisting of $q \geq n + 1$ vertices. They are called complexes. The first peak of the initial complex is the point belonging to the interior of the permissible region. As in the unconditional simplex search procedure, at each iteration we replace the 'worst vertex' of the complex. The count ends if the target function values vary little over five successive iterations, and the centre of gravity of a polyhedron is the solution of the problem. The method of constructing the initial complex is a simple method for determining the allowable polyhedron of the reasonable size and shape, relieving the need to follow this operation each time independently. If the solution of the problem obtained by this method is found in the optimum range, it can be refined further by calculations using the Gelfand–Tsetlin method [64, 65].

Table 6.14.

Number of optimization cycles	Number of optimization steps	Type of VPO	Parameters of the best modes of CVL												The main design characteristics of the CVL							
			Independent (input)								Arbitrary											
			n_{Cu}	n_{Ne}	f	U_0	R	τ_U	a	l_a	W	W_1	$\bar{\eta}_{II}$	P_1	η_d	η_p	W	T_a	T_e^{st}	τ_g	$I_{d\,max}$
			D^{14} cm^{-3}	10^{18} cm^{-3}	kHz	kV	ohm	ns	cm	cm	J/l	J	kW/m	W	%	%	μJ/cm^3	meV	meV	ns	kA
1	10	B	49	10	3.5	3.6	0.33	200	6	100	0.71	2.6	9	56	50	1.3	4.5	256	–	41	13
2	10	B	44	5.9	3.5	3.4	0.35	200	6	100	0.62	2.2	7.8	49	44	1.5	1.5	248	–	41	13
3	7	B	41	5	1.9	4.9	0.261	200	23	100	0.12	6.2	11	197	70	2.5	2	261	–	40	15
4	12	B	35	5	5.2	2.7	0.097	200	6	366	0.38	5	7	128	50	1	1.9	236	–	41	35
5	9	B	25	3.1	3.4	3.8	0.193	200	15	205	011	5	8.2	232	67	2.1	1.5	238	–	38	23
6	8	B	35	3.8	3.3	4.7	0.3	200	11	200	0.1	5	8.4	250	78	1.9	1.6	220	–	40	17
7	8	B	29	5.3	2.6	5.8	0.45	200	21	147	0.08	5	8.7	220	78	2.2	1.3	239	–	43	13
8	11	A	63	3	5.7	8.6	0.322	200	6	100	4.2	15	86	91	13	0.8	4.4	254	–	23	40
9	9	A	57	4.2	2.1	5.5	0.256	200	19	100	0.22	7.9	17	251	56	2.7	3.4	240	–	48	21
10	8	A	..	3.3	4	7.3	0.184	69	13	100	0.42	6.7	27	301	35	3.2	4.7	251	–	21	27
11	9	A	15	5	7.3	11	0.199	44	6	100	2.2	8.1	59	196	16	2.1	7.4	260	–	14	28
12	4	B	12	4.6	5.1	26	44	200	1.7	100	3.5	1	5.1	12	57	0.42	8.2	228	237	28	0.6
13	9	B	88	5.8	3.2	32	34	200	3.5	100	1.6	2	6.3	32	50	1	8.2	237	241	38	0.9
14	9	B	61	7.3	3.2	38	39	200	6	100	0.67	2.4	7.7	62	56	1.5	5.4	246	249	41	1
15	10	B	43	4.9	2.2	36	23	200	10	100	0.36	3.6	7.9	90	51	2.3	4.1	245	247	49	1.5
16	10	B	44	5.4	1.7	41	23	200	15	100	0.22	4.9	8.2	124	51	3	3.3	246	248	52	1.8
17	8	B	34	6.3	2	41	23	200	20	100	0.12	4.8	9.5	150	59	2.6	1.9	256	257	40	1.8

The developed method of numerical computer calculations which aim at predicting and optimizing the characteristics of the lasers is based on the use of the universal program of numerical experiments for the PMVL and simultaneous application of the algorithm of the multivariate optimization of the output parameters, which was constructed on the basis of the Box complex method the Gelfand–Tsetlin method. Calculations made by this method for a copper vapour laser in neon have shown that in the optimization of the geometry of the laser active zone and the parameters of the circuit (including the voltage pulse and excitation frequency) we can confine ourselves to the use of the Box method. The optimum of the target function with respect to the concentration of the metal atoms and buffer, found using this algorithm, requires further clarification by the Gelfand–Tsetlin method. As an example [66] we present the results of a three-parameter optimization of a copper vapour laser with neon and transverse excitation in terms of [50, 51]. The optimized (variable) parameters were the concentration of copper atoms $n\text{Cu} = 10^{14}$–$5 \cdot 10^{15}$ cm^{-3}, neon $n_{\text{Ne}} = 5 \cdot 10^{17}$–$10^{19}$ cm^{-3} and the excitation pulse repetition frequency $f = 2$–10 kHz. The remaining input parameters were considered unchanged. The voltage pulse at the output of the cable transformer was in the form of a triangle with a base of 150 ns, the duration of the leading edge 75 ns and an amplitude of 14 kV. The optimization of the laser parameters shows that the optimal parameters for maximum lasing power are the following: $n_{\text{Cu}} = 4.1 \cdot 10^{15}$ cm^{-3}, $n_{\text{Ne}} = 8.1 \cdot 10^{18}$ cm^{-3}, $f = 5.1$ kHz. The average lasing power is thus about 150 W.

As an example of the multifactor optimization of a copper vapour laser are presented below in the form the results presented by N.I. Yurchenko [57]. The physical and mathematical model of the laser [57] takes into account the main processes that affect the operation of the laser. In this model, the excitation pulse repetition period is divided into two parts: the period of active medium excitation and its relaxation time. The calculations determine the basic output characteristics of the laser. The input parameters are the direct laser parameters that are either easy to vary or have certain technical limitations. The CVL model under consideration allows the optimization at the same time of all or some or all of the laser parameters, and the high performance of the algorithm can do this automatically.

Multiparameter optimization methods are well known [65]. The method of the 'steepest descent', used in [57] can be described as follows. In the original random point in space of the optimized parameters Π we determine gradient of the target function $\Phi(\Pi)$, followed by motion (a step change in the parameters) in the opposite direction to the gradient until the value of Φ decreases. This process constitutes the first stage of optimization. The algorithm then gradually change the input parameters of the CVL, seeking the 'steepest descent' to minimize the target function.

The effectiveness of this method of optimization is known to depend on the shape of the surface $\Phi(\Pi)$. If it has a well defined minimum, the 'descent' to it is fast, usually within 2–3 stages of optimization. However, at a large number of parameters to be optimized the surface $\Phi(\Pi)$ can take the form of so-called 'gully', when in the space Π there is a curve passing through the minimum point ('bottom of the gully') and in motion along this curve Φ changes very slowly (a significant change in the parameters leads a slight decrease in Φ) and the lines of the $\Phi(\Pi)$ level are stretched along the curve.

If there are 'gullies' on the surface $\Phi(\Pi)$, this optimization method provides a rapid descent to the bottom of the 'gully', but the subsequent movement on it is slow, since the exact determination of the direction of the gradient at the bottom of the 'gully' is difficult. In this case, for accelerating the convergence of the count it is necessary to use so-called 'gully' method [67].

However, even without resorting to the 'gully method', based on a series of successive stages of optimization, the trends of the parameters can be used to characterize by the direction of elongation of the 'gully', the position of the optimum region and attainable values of Φ. The repetition of such optimization cycles ('gradient descent') from different starting points for various additional conditions to the fixing of a certain parameter provides sufficient information for the practical purposes of the shape of the surface $\Phi(\Pi)$, although its exact optimum may not be achieved in the presence of the 'gully' in each cycle.

Table 6.14 presents the main results of the optimization of n_{Cu}, n_{Ne}, f, U_0, impedance R, excitation pulse time τ_U, the transverse size of the GSC (GDT) a and l_a for the average lasing power in the entire GDS volume in excitation of the active medium by transverse (cycles 1 11) or longitudinal (cycles 12–17) discharges formed by VPU with voltage pulses of amplitude U_0 and duration τ_U through the cable line with wave resistance R.

The values of the parameters that are fixed (not optimized) in a given optimization cycle are underlined. It is assumed that the VPO either absorbs the pulse reflected from the load pulse (VPO-A type) or fully reflects it (VPO-B type). The VPO pulse is triangular, the AZ (active zone) section, determined in the plane perpendicular to the axis of the resonator for the longitudinal and transverse discharges is assumed to be square with side length a, and the length of the resonator and the circuit inductance are associated with the length of AZ as follows: $l_r = la + 1$ m, $Ll_a = 10$ nH·m in cross-excitation and

L/l_a = 103 nH·m in longitudinal, which is typical for real designs of lasers. In all optimization cycles but the 6th cycle, the AZ section is a square with the side size a. In the sixth cycle, the AZ cross section is a rectangle with the side ratio b/a = 2, where b is the distance between the electrodes (discharge length l_d).

In the cycles 1, 2 (Table 6.14), differing in the position of the initial point, five main laser parameters are optimized: n_{Cu}, n_{Ne}, f, U_0 and R (impedance of the cable line), the rest (τ_a, a, l_d) are very close close to the constant parameters of the experimental setup, investigated in detail in [27]. It is easy to see that regardless of the starting point the optimization algorithm leads to the same region of parameters ($n_{Cu} \approx$ 5 · 10^{14} cm^{-3}, $n_{Ne} \geq 6 \cdot 10^{18}$ cm^{-3}, $f \approx 3$ kHz, U_0 $\approx$ 3–4 kV, $R \approx 0.3$ ohms), which usually includes the experimental optimum of the given laser parameters. The maximum lasing power on the green line $\overline{P}_{l_{max}} \approx 50$ W is close to a record high value P_l reached in the experiment [27]. Consequently, in the laser system [27] it is not necessary to 'force' the regime, only technical refinement. It should be noted that, according to calculations, the range of the optimum parameters is extended to the parameter p_{Ne} ('gully'): in the range p_{Ne} $\approx$ (1–3)·10^5 Pa, lasing power varies slightly.

For the correct interpretation of the optimization results of the CVL parameters given in Table 6.14 it should be noted that according to the above calculation results, virtually anywhere at the optimum parameters $p_{Ne} \geq 10$ Pa and $T_a \approx 2800 \pm 200$ K. The high pressure of neon is necessary for the CVL to work effectively with the transverse dicharge discharge and satisfactorily with the longitudinal discharge, if there is no restriction on the parameters $E_0 = U_0/l_d$ and $G_0 = (C_s U_0^2 /2)/V_d$, which are the initial field strength in the discharge and the specific energy input in the discharge (specific stored energy) at the discharge of the storage capacitor to the matched load. At the same time, as shown previously, the magnitude of p_{Ne} is limited at the top by the harmful effect on the output characteristics of the CVL of the recombination processes in triple collisions at $p_{Ne} \geq 10^5$ Pa. To partially compensate for this effect, the gas temperature is maintained extremely high in terms of the equilibrium population of the metastable level, and consequently, the level of energy input to the discharge which determines the heating of the gas must be in the range of optimal values. Therefore, the optimization of the CVL parameters $\overline{P}_{l_{max}}$ virtually requires either optimizing the laser parameters in order to obtain the maximum practical efficiency $\eta_r = \eta_p \cdot \eta_d$, where η_r is the efficiency of the discharge (for VPO of type B), or increasing the length of GDT.

As shown above, to increase η_d it is necessary to reduce the specific energy margin G_0, i.e., or increase either the volume of the AZ or the excitation pulse repetition frequency f. At the same time to increase the physical efficiency η_p it is necessaty to reduce the role of the initial concentration of metastable atoms, that is, reduce the ratio n_{2st} /n_{2peak}, which show that, conversely, it is necessary to increase the pump (increase the parameter G_0), because the value of n_{2st} is weakly dependent on it, and the value of n_{2peak} increases with G_0, or reduce the frequency f. These requirements conflict with each other and intuitively it is difficult to give preference to any of them.

Numerical optimization shows that the simultaneous growth of η_d and η_p in the case of increase in the AZ volume while maintaining the pressure at the level of the atmosphere and with some decrease in the frequency f. For example, if the number of optimized parameters includes the transverse size of the AZ (Table 6.14, cycle 3), then during the successive stages of the calculations the size monotonically increasing up to very large values, providing growth of $\bar{P}_{l_{max}}$, while other parameters are quickly stabilized at the previous level.

Optimization of the length of the AZ was carried out at a fixed pulse energy of the VPO without considering the major factor limiting the length of AZ – the loss of lasing energy in the non-axial rays. If the cross-sectional dimensions of the AZ are fixed a = const (Table 6.14, cycle 4), the length of the AZ increases dramatically, but the optimum volume and with it the lasing power remain much smaller than in the previous cycle. It is easy to see that the increase in length in this case is limited by the decrease of T_a. This was evidenced by a clear trend of increasing frequency f higher than the normal optimal value and the decrease of physical efficiency η_p.

If we include in the number of parameters to be optimized simultaneously the length and diameter of the AZ (Table 6.14, cycle 5), we observed the same trend of growth, but the length of the AZ no longer rises much above the optimum value, which appears to be close to 2 m at the total stored energy $V_d G_0$ = 5 J. The overall increase in the volume of the AZ slows down when the specific energy reaches $G_0 \approx 0.1 \cdot 10^{-3}$ J/cm³, indicating the approach to the lasing threshold.

Changing the shape of the section of the AZ (Table 6.14, cycle 6, b/a = 2), on the one hand, reduces the negative impact on the lasing characteristics of the inductance of the discharge circuit (parameter $X' = LS_d / l_d$), but on the other hand reduces the temperature of the gas. Under these conditions, these factors cancel each other out, the power for the equal volumes of AZ is approximately the same. At smaller length of the AZ this change of the shape of its cross section would lead to a gain in power, as it would weaken the influence of superheating the gas.

Cycle 7 (Table 6.14) characterizes the sensitivity of the results directly to an increase in the inductance of the discharge circuit: lasing power at the optimum of the variable parameters is reduced by 15–20% when L increases by 1.5–2 times.

When replacing the type B VPO by type A it is possible to significantly increase the energy of the excitation pulse, since only a part of this energy falls in the AZ and heats the gas, and the growth of the parameter G_0, as is usually the case in the CVL, decreases η_d, so that the heating increases slowly non-linearly with G_0. For example, if $a = 6$ cm $=$ const the optimal gas temperature is reached only after an increase in the VPO power by an order of magnitude (Table 6.14, the cycle 8). The lasing power increases at the optimum by only 70% as compared to identical cycle 1, indicating a failure of the type A VPO while limiting the volume of the AZ: at the real value of the energy it is insufficient for the required heating of the gas.

As the AZ volume increases during the optimization (Table 6.14, cycle 9) the efficiency of the discharge η_d increases to 50–70%, so that the results of calculations of the output characteristics of lasers obtained for the VPOs of both types converge. The restriction of the lasing power associated with superheating of the gas comes to the fore and the VPO of type A is preferred, although it requires greater power of the source.

Increasing the energy input into the discharge can be fully effective only if it is accompanied by a reduction in the duration of the excitation pulse τ_U. Fixing the value τ_U negates the potential benefits of the VPO of type A. When the duration τ_U is included in the number of parameters to be optimized the duration rapidly decreases during the optimization to 40–70 ns which, with other conditions being equal, leads to the doubling of output lasing power (Table 6.14, the cycles 10, 11). In the VPO of type B, where the stored energy is limited mainly by gas superheating, the effect of the inclusion of τ_U is significantly lower (Table 6.14, cycle 6).

It should be noted that the increase in lasing power with a decrease of τ_U is accompanied by a significant shortening of the radiation pulse, which leads to a significant loss of the lasing energy on the walls of the GDC at the expense of non-axial rays. If we consider the related phenomenon of decrease of the energy output, the optimization leads to significantly higher values of τ_U (80–100 ns), if these values are variable, and smaller quantities of $\bar{P}_{l_{max}}$. To eliminate the above-mentioned loss of stimulated radiation the laser must be transferred to the oscillator-amplifier operating mode. In addition, for small τ_U stricter requirements are imposed on the timing of discharges along the length of the AZ, so that the implementation of the growth prospects of lasing power with reduced τ_U is associated with certain technical difficulties. If these difficulties are not overcome the effect may be counterproductive.

It is easy to see that in the case of the type B VPO the increase the lasing power is first approximately quadratic and then increases linearly with the transverse dimension of the AZ. In this case the required excitation power is almost constant, the frequency f decreases slowly and the energy of VPO increases. The type A VPO provides the lasing power 1.5–2 times greater, but at a much lower efficiency, especially when for small volumes of the AZ.

In the cycles 12–17 (Table 6.14) the parameters of CVL with a longitudinal discharge are optimized. The principal difference between this CVL in the absence of restrictions on the variable parameters is reduced to the fact that, if in the case of the transverse discharge under the above assumptions about the inductance the parameter $X' = LS_d/l_d$ does not depend on the cross section of the AZ, then in the longitudinal discharge this parameter increases is proportion to this cross section: $X' \approx a^2$. When $a = d = 10$ cm the values of the parameter X' for lasers with longitudinal and transverse discharges are the same, and the optimization results of course coincide.

For $a > 10$ cm a definite advantage in the parameter X' is found for the transverse discharge, but this advantage has little effect, since with increasing transverse dimensions of the AZ the optimal value of specific energy G_0^{opt} decreases and with it the influence of the discharge circuit inductance on the output energy.

For $a < 10$ cm, on the contrary, the longitudinal discharge is more advantageous, and because of high values of G_0 the difference in the energy output of lasers with longitudinal and transverse discharges becomes more noticeable, and for $a \approx 3.5$ cm it reaches 30–40%.

Thus, the qualitative difference between the longitudinal and transverse excitation circuits of the CVL remains the same as in the single-pulse mode, but quantitatively it is much less clearly expressed, as also the dependence of the achievable lasing power on the volume of the AZ. In general, the results of optimization of the parameters discussed in this section are similar for both excitation circuits, so that all the previous findings can be transferred to the CVL with a longitudinal discharge.

The calculated and experimental results are in good agreement, despite the significant difference in the structures of the experimental devices, their optimal modes of operation, etc. This leads to the conclusion that in the longitudinal discharge, as well as in the transverse discharge, to increase the average lasing power it is necessary to generate an approximately proportional increase in the diameter of the active zone with parallel optimization of other parameters.

It should be noted that a similar conclusion about the effectiveness of increasing the laser volume was reached by the authors of the experimental work [68], testing tubes of different diameters, and by the authors of [69] in the course of experiments and calculations in the specific conditions of the CLV design developed by them.

References

1. Harstad K.G., Proposed Computer Model for Electric Discharge Atomic Vapor Lasers. Jet. Prop. Labor. NASA. 1977. No.77-11.
2. Harstad K.Q., LEEE J. Kvantovaya elektronika. 1980. Vol.QE-16. No.5. P.550-558.
3. Kushner M.J., IEEE J. Quantum Electronics. 1981. Vol.QE-17. No.8. P.1555-1565.
4. Arlantsev S.V., Buchanov V.V., Vasiliev L.A., et al. Dokl AN SSSR. 1981. V.260. No.4. P.853-857.
5. Smirnov B.M., Atomic Collisions and Elementary processes in plasma. Moscow: Atomizdat. 1968.
6. Smirnov B.M., Physics of weakly ionized gas. Moscow: Nauka. 1972.
7. Weinstein L.A., Sobel'man I.I., Yukov E.A., The excitation of atoms and broadening of the spectral lines. Moscow. GIFML Science, 1979.
8. Mitchner M., Kruger Ch. Partially ionized gases. Moscow: Mir. 1976.
9. Ochkurov V.I., Petrunkin A.L., Optika i Spektroskopiya. 1963. V.14. Issue 4. P.457-464.
10. Seaton M.I., Planet. Space Sci. 1964. Vol.12. P.55-74.
11. Smirnov B.M., Usp.Fiz.Nauk. 1977 T.121. P.231-258.
12. Eletskii A.V., Elementary processes in gases and plasmas. Physical quantities. Directory. Moscow: Atomizdat. 1991. P.391-411.
13. Malkin O.A., Relaxation Processes in plasma. Moscow: Atomizdat. 1971.
14. Biberman L.M., Vorob'ev V.S., Yakubov I.T., Kinetics of nonequilibrium low-temperature plasma. Moscow: Nauka. 1982.
15. Lausanne E.D., Firsov O.B., The theory of the sparks, Moscow: Atomizdat 1975.
16. Golant V.E., Zilinskij A.P., Sakharov S.A., Basics of plasma physics. Moscow: Atomizdat. 1977.
17. Diachkov L.G., Kobzev G.A., Zhurnal tekhnicheskoi fiziki. 1978. V.48. Vyp.11. P.2343-2346.
18. Zel'dovich J.B., Raiser J.P., Physics of Shock Waves and High-Temperature Hydrodynamic Phenomena. Moscow: Nauka. 1966. Moscow: Fizmatlit. 2008.
19. Harsted J., The physics of atomic collisions. Moscow: Mir. 1965.
20. Smirnov B.M., Atomic Collisions and Elementary processes in plasma. Moscow: Atomizdat. 1968.
21. Mc Daniel J., Collision processes in ionized gases. Moscow: Mir. 1967.
22. Smirnov B.M., Asymptotic methods in the theory of atomic collisions. Moscow: Atomizdat. 1973.

23. Smirnov B.M., Ions and excited atoms in the plasma. Moscow: Atomizdat. 1974.
24. Biberman L.M., Vorob'ev V.S., Yakubov I.T., Usp. Fiz, Nauk 1978 V.128. Issue 2. P.233-272.
25. Direktor L.B., Malikov M.M., Fomin V.A., Zhurnal tekhnicheskoi fiziki. 1987.V.57. Issue 1. P.28-36.
26. Markova S.V., Petrash G.G., Tserezov V.M., Kvantovaya elektronika. 1978 V.5. No.7. P.1585-1587.
27. Artemyev A.Y., Babeyko Yu.A., Borovichi B.L., et al., Kvantovaya elektronika. 1980. V.7. No.9. P.1948-1954.
28. Isaev A.A., Lemerman G.Y., Petrash G.G., Markov S.V., Kvantovaya elektronika. 1979 V.6. No.8. P.1942-1947.
29. Markova S.V., Tserezov V.M., Kvantovaya elektronika. 1977. V.4. No.3. P.614-618.
30. Bohan P.A., KlimkinV.M., et al., Kvantovaya elektronika. 1977. V.4. No.1. P.152-154.
31. Solomon V.I., Kvantovaya elektronika. 1979 V.6. No.6. P.1252-1257.
32. Kirilov A.E. Kuharev V.N., Soldatov A.N., Kvantovaya elektronika. 1979 V.6. No.3. P.473-477.
33. Drawin H.W., Z.Phys. 1961. Bd.164. No.5. P.513-521.
34. Borowicz B.L., Young E.I., Ryazan L.A., Tykotsky V.V., Kvantovaya elektronika. 1990. V.17. No.10. P.1265-1271.
35. Msezane A.Z., Henry R.J.W., Phys. Rev. A. 1986. Vol.A 33. No.3. P.1631-1639.
36. Xia Tiejun, Yao Zhixin, Wang Yuitszyan, Song Wei et al., Computer modeling lead vapor lasers. (Acta Optica Sinica). Guansyue syuevan. 1985. Novosibirsk GPNTIB SB RAS. Transfer number 13356.
37. Bohan P.A., Kvantovaya elektronika. 1985. V.12. No.5. P.945-952.
38. Biberman L.M., Mnatsakanyan A.K., Naydis G.V., The distribution function electron energy in mixtures of inert gases and metal vapors. Report IVTAN No. 43/76, 1976.
39. Gudzenko L.I., Yakovlenko S.I., Plasma lasers. Moscow: Atomizdat. 1978.
40. Arlantsev S.V., Buchanov V.V., Vasiliev L.A., et al., Kvantovaya elektronika. 1980. V.7. No.11. P.2319-2326.
41. Maltsev A.N., Kinetics repetitively pulsed lasing on copper vapor. Tomsk branch of the Academy of Sciences. Preprint No.1. Tomsk. 1982.
42. Artemyev A.J., Borovichi B.L., Vasiliev L.A., et al., Kvantovaya elektronika. 1983. Vol.10. No.7. P.1441-1452.
43. Direktor L.B., Malikov M.M. Fomin V.A., Shpil'rain E.E., The physical model and the method of calculating the parameters of a copper vapor laser. Preprint No.5-188 IVTAN. Moscow: 1986.
44. Direktor L.B., Malikov M.M., The physical model and calculation method parameters of a copper vapor laser. Preprint No.5-249 IVTAN. Moscow. 1988.
45. Direktor L.B., Malikov M.M.,The energy balance of electrons and excited atoms in the plasma of a copper vapor laser. Dep. VINITI number 3571-B-89 from 05/30/89.
46. Borowicz B.L., Nalegach E.P., Rybin V.M., Yurchenko N., Kvantovaya elektronika. 1984. V.11. No.12. P.2371-2479.
47. Bohan P.A., Silant'ev V.I., Solomonov V.I., Kvantovaya elektronika. 1980. V.7. No.6. P.1264-1268.
48. Batenin V.M., Klimovskii I.I., Selezneva L.A., TVT. 1980. V.18. No.4. P.707-712.
49. Isaev A.A. MA Kazaryan M.A., Kvantovaya elektronika. 1977. V.4. No.2.

P.451-453.

50. Babeyko Yu., Vasiliev L.A., Sokolov A.V., et al., Kvantovaya elektronika. 1978. V.5. No.9. P.2041-2042.
51. Borowicz B.L., Yurchenko N.I., Kvantovaya elektronika. 1984. V.11. No.10. P.2081-2095.
52. Arlantsev S.V., Borowicz B.L., Buchanov V.V., et al., Kvantovaya elektronika. 1983. Vol.10. No.5. P.1546-1553.
53. Borowicz B.L., Vasiliev L.A., Ryazan V.M., et al. Lasers on copper vapours with a transverse discharge. All-Union. Conference "Laser Optics-82." Abstracts. Leningrad. 1982. P.125-126.
54. Artemyev A.J., Borovichi B.L., Vasiliev L.A., et al., Kvantovaya elektronika. 1982. v.9. No.4. P.738-742.
55. The plasma lasers / Ed. J. Bekefi. Moscow: Energoizdat. 1982.
56. The plasma chemistry, No.7, Ed. Smirnov B.M., Moscow: Atomizdat. 1980.
57. Yurchenko N.I., Theoretical study of the basic laws of the kinetics of plasma, the dynamics of the radiation pattern and multiparametric optimization of copper vapor laser. PhD Thesis. Moscow. 1984.
58. Batenin V.M., Klimovskii I.I., Morozov A.V, Selezneva L.A., Teplofizika vysokikh temperatur. 1979. V.17. No.3. P.483-489.
59. Isaev A.A., Kvantovaya elektronika. 1980. V.7. No.3. P.599-607.
60. Smilanski I., Erez G., Kerman A., Levin L.A., Opt. Communications. 1979. Vol.30. No.1. P.70-74.
61. Bohan P.A., Metal vapor lasers with collisional de-excitation of the lower working conditions. Doctor Thesis. Sci. Sciences. Novisibirsk. 1988.
62. Planning for the experiment in the study of technological processes. / Ed. Letsko E.K., Moscow: Mir. 1977.
63. Numerical methods for constrained optimization. Moscow: Mir. 1977.
64. Himmelblau D. The analysis of the statistical methods. Moscow: Mir. 1973.
65. Application of Computational Mathematics in the chemical and physical kinetics. Moscow: Nauka. 1969.
66. Buchanov V.V., Young E.I., Tykotsky V.V., Kvantovaya elektronika. 1983. Vol.10. No.3. P.629-631.
67. Moses N.I., Ivanilov Y.N., Stolyarova E.M., Optimization techniques. Moscow: Nauka. 1978.
68. Smilanski I., Kerman A., Levin L.A, Erez Q., Optics Communications. 1978. Vol.25. No.1. P.79-82.
69. Kushner M.I., Warner B.E., J. Appl. Physics. 1983. Vol.54. No.6.P.2970-2982.
70. Batenin V.M., Klimovskii I.I., Selezneva I A., On the question of the optimal parameters of self heating copper vapor lasers. IVTAN M. Dep. VINITI 18 06.80 No. 2832-80.
71. Batenin V.M., Vokhmin P.A., Klimovskii I.I., Selezneva L.A., Dokl. AN SSSR. 1981 V.256. No.4. P.831-834.
72. Galkin A.F., Klimovskii I.I., Selezneva L.A., Teplozizika vysokikh temperatur. 1983. V.21. No.5. P.976-981.
73. Galkin A.F., Klimovskii I.I., Effect of radial inhomogeneity on the characteristics of the plasma generating repetitively pulsed copper vapor lasers with longitudinal discharge. IVTAN Preprint No. 5-220. Moscow: 1987.
74. Galkin A.F., Klimovskii I.I., Optimal parameters of a repetitively pulsed copper vapor lasers with a non-uniform distribution of plasma parameters over the section of GDT. IVTAN Preprint No. 5-228. Moscow: 1987.
75. Klimovskii I.I., Teplozizika vysokikh temperatur. 1989. V.27. No.6. P.1190-

1198.
76. Galkin A., Klimovskii I., Computer model of copper-vapor laser with average specific output pover above 1 W/cm². Metal Vapor Lasers and Their Applications: CIS Selected Papers, G.G. Petrash, Editor, Proc. SPIE 2110. 1993. P.90-99.
77. Ginzburg V.L., Gurevich V.L., Usp. Fiz. Nauk. 1960. V.70. No.1. P.201-255.

7

Numerical modeling of repetitively pulsed copper vapour lasers taking into account the heterogeneity of distribution of plasma parameters (inhomogeneity of discharge) over the cross section of the gas discharge tube

7.1. The place and role of model studies of repetitively pulsed copper vapour lasers, taking into account the radial inhomogeneity of the plasma parameters [1–6] in a number of other numerical studies of such lasers

The wide use of numerical methods for the study of repetitively pulsed self-contained lasers is due to at least two reasons. First, the numerical analysis in combination with experimental studies helps to identify the main mechanisms determining the evolution of plasma parameters over time and the lasing characteristics. Secondly, the method of empirical optimization of self-contained lasers on using the basic parameters (specific energy, type and pressure of the buffer gas, the shape and amplitude of the voltage pulse across the discharge gap, the radius of GDT, etc.) is much more complicated than the search for optimal conditions of operation of such lasers by numerical analysis.

At the time of publication of the monograph [7], most numerical studies of lasers with self-contained transitions of metal atoms was carried out with allowance for the radial transport of particles and energy in the zero-dimensional approximation (see previous chapter), that is for certain plasma parameters, averaged over the cross section of the GDT. Although the results of these numerical studies are very informative and practical, strictly speaking, the zero-dimensional approximation to a certain extent is not consistent with the results of experimental studies (see chapter 4). These results demonstrate the inhomogeneous distribution of most parameters of the plasma cross-section of GDT in repetitively pulsed CVLs, including significant temporary fluctuations in the concentration of copper atoms in the ground state in the axial zone of GDT. It is clear that ignoring these factors can lead to some uncontrolled error in the calculation of the output power of lasers and their efficiency in the zero-dimensional approximation.

An exception from the majority of the numerical studies of the CVLs was a small number of studies [1–6], which attempted to model a (not self-consistent) solution of the problem of the characteristics of repetitively pulsed CVLs, taking into account the radial inhomogeneity of the plasma parameters in the GDT. The lack of results of studies [1–6] in [7] is due to the fact that shortly before the publication of this book studies were published (e.g. [8, 9], in which the problem of the effect of radial inhomogeneity of the plasma on the lasing characteristics of the CVLs was solved in a much more rigorous mathematical formulation than in [1–6]. The presence of works such as [8, 9], and rapid progress in the growth of computational capabilities of computers allowed us to hope that the problem of the effect of radial inhomogeneity of the plasma on the characteristics of copper vapour lasers will be resolved in the shortest time and at a high mathematical level. However, for some not quite clear reasons this has not happened, and the results of [1–6] are, apparently, the only of its kind and have not lost the relevance to the present because they, first, provide a qualitative idea of the effect of radial inhomogeneity of the plasma parameters on the performance of repetitively pulsed CVLs in a wide range of neon pressures and the radii of the GDTs. Second, they are in good agreement with the results of experiments carried out before the publication of [4, 5], which clearly demonstrates the applicability of the modeling approach used in [1–6] to calculate the characteristics of real repetitively pulsed CVLs. Third, as shown in [10], the results of [1–6] not only correctly predicted the characteristics of industrial CVLs [11], but also allowed us to determine ways to further increase the efficiency of the CVLs. Fourth, given the data on the equilibrium constants, the rate constants of thermal dissociation and three-particle association of diatomic molecules [12], the results of [1–6] allow us to predict [13] ways to improve the characteristics of the CVLs with improved kinetics (see Chapter 8) and copper halide vapour lasers.

The following are the main results obtained in [1–6] by solving the problem of modeling the influence of the radial inhomogeneity of the plasma on the characteristics of repetitively pulsed CVLs.

7.2. The results of numerical studies of repetitively pulsed CVLs, assuming uniform distribution of plasma parameters over the cross section of the GDT [1–3]

Since the characteristic times of the processes determining the plasma parameters during the excitation pulses and between them vary considerably, the plasma parameters in [1–3] were calculated using two different systems of equations. In addition, given the approximate nature of calculations, they were given taking into account lasing only on the green line.

Calculation of the concentration of particles of various sorts during the excitation pulse was carried out using a system of equations [1–3]

$$\frac{d\langle n_e \rangle}{dt} = \alpha_{gi} \langle n_g \rangle \langle n_e \rangle + \alpha_r \langle n_r \rangle \langle n_e \rangle, \tag{7.1}$$

$$\frac{\langle dn_g \rangle}{dt} = -\alpha_g \langle n_g \rangle \langle n_e \rangle, \tag{7.2}$$

$$\frac{d\langle n_r \rangle}{dt} = \alpha_{gr} \langle n_g \rangle \langle n_e \rangle - \alpha_r \langle n_r \rangle \langle n_e \rangle - \langle N \rangle, \tag{7.3}$$

$$\frac{d\langle n_m \rangle}{dt} = \langle N \rangle, \tag{7.4}$$

$$\langle n_r \rangle = \frac{g_r}{g_m} \langle n_m \rangle, \tag{7.5}$$

where, besides the well-known notations, brackets $\langle \rangle$ denote average concentrations of particles over the cross section of the GDT, N is the specific lasing power in a number of transitions from a resonant to a metastable level.

The initial conditions at $t = 0$ for system (7.1)–(7.5): $\langle n_{Cu} \rangle = \langle n_{Cu} \rangle_0$, $\langle n_e \rangle = \langle n_e \rangle 0$, $n_r = n_m = 0$.

In the interpulse time interval the concentration and the electron temperature were calculated in [1–3] in the zero-dimensional approximation using a system of equations [14], written taking into account the cooling of electrons in collisions with copper ions:

$$\frac{d\langle n_e \rangle}{dt} = \frac{6D_a}{r_p^2}\langle n_e \rangle - \langle n_e \rangle^3 \beta + \langle n_e \rangle \langle n_g \rangle \alpha_i, \tag{7.6}$$

$$\langle n_e \rangle \frac{kd\langle T_e \rangle}{dt} = -\langle n_e \rangle \frac{6D_a}{r_p^2}\frac{2}{3}k\langle T_e \rangle \left(\ln \frac{r_p\sqrt{k\langle T_g \rangle}}{2,4D_a M_i} + \ln \sqrt{\frac{M_i\langle T_e \rangle}{m\langle T_g \rangle}} \right) -$$

$$- \frac{2m}{M_{Ne}}\langle n_e \rangle v_{eNe}k\left(\langle T_e \rangle - \langle T_g \rangle\right) - \frac{2m}{M_i}\langle n_e \rangle v_{ei}k\left(\langle T_e \rangle - \langle T_g \rangle\right) +$$

$$+ \left(\frac{2}{3}I_{Cu} + k\langle T_e \rangle\right)\left(\langle n_e \rangle^3 \beta - \langle n_e \rangle \langle n_g \rangle \alpha_i\right) \tag{7.7}$$

where v_{eNt}, v_{ei} and D_a refer to the quantities corresponding to the electron temperature, defined by the system (7.6), (7.7).

In calculating the parameters of the plasma during the excitation pulse and the interpulse interval in time [1–3] used the following assumptions:

1) rectangular voltage pulse was assumed that ends at the end of the generation,

2) the electron temperature $\langle T_e \rangle$ averaged over the cross section of the GDT and the rate constants of various elementary processes during the excitation pulse are constant, and

3) prepulse concentration of copper atoms $\langle n_{Cu} \rangle_0 = 10^{15}$ cm^{-3},

4) the gas temperature $\langle T_g \rangle$ is constant and equal to 2500 K.

Figure 7.1 shows the dependences of prepulse $\langle n_e \rangle_0$ and postpulse $\langle n_e \rangle_\tau$ electron concentrations found using equations (7.1)–(7.5) used to calculate the concentrations $\langle n_e \rangle_\tau$ at arbitrarily defined values $\langle n_e \rangle_0$, and the electron temperature during the excitation pulse $\langle T_e \rangle_{ex}$. Note that under the above assumption in the approximation of saturated power the beginning of lasing corresponds to the beginning of the excitation pulse and the end of the laser pulse corresponds to a maximum population of the resonant level.

The equations (7.6), (7.7) in [1–3] were used to calculate the dependences of the temperature $\langle T_e \rangle$ and concentration $\langle n_e \rangle$ in the interpulse interval for the buffer gases helium and neon in the pressure range from 1.3 to 13.3 kPa and the range of the diameters of the GDT from 0.3 to 2.5 cm. As an example Fig. 7.2 shows the time dependences of $\langle n_e \rangle$ and $\langle T_e \rangle$, which occur at the same conditions. It is seen that in helium the relaxation of $\langle n_e \rangle$ and $\langle T_e \rangle$ is much faster than in neon. In addition, attention is drawn to the fact that the extrapolation of the electron temperature (1) to the long time t range gives values $\langle T_e \rangle$ significantly lower than the temperature T_w of the wall of the GDT. This effect is not physically justified and does not indicate the incorrectness of the (7.6), (7.7). It is due to the fact that in equation

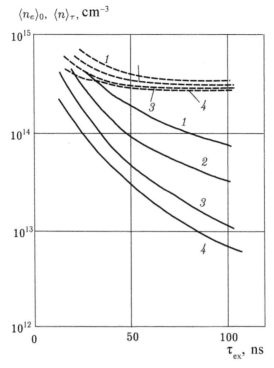

$\langle n_e \rangle_0, \langle n \rangle_\tau, \mathrm{cm}^{-3}$

Figure 7.1. Dependence of the prepulse $\langle n_e \rangle_0$ (solid lines) and postpulse $\langle n_e \rangle_\tau$ (dash-dotted line) electron concentration on the duration of the excitation pulse. $1 - \langle T_e \rangle_{ex} = 2$ eV. $2 - 3$ eV, $3 - 4$ eV, $4 - 5$ eV.

(7.7) the first term on the right hand of the equation of the temperature T_w is independent and is negative for any values $\langle T_e \rangle$, including at values $\langle T_e \rangle < T_w$, i.e. in cases in which the electrons will transfer energy from the wall of the GDT to its axial zone. The marked incorrectness can be eliminated, replacing in the first member of the right-hand side of equation (7.7) the temperature of the electrons $\langle T_e \rangle$, facing the parenthesis, by the temperature difference ($\langle T_e \rangle - T_w$). However, this substitution has no practical sense as it is shown by analysis that all the calculations of $\langle T_e \rangle$ end at $\langle T_e \rangle \cong 3000$ K substantially greater than the temperature of the wall of the GDT of the copper vapour lasers.

The dependences presented in Figs. 7.1 and 7.2, and similar dependences calculated for the other initial data were used in [1–3] to determine the excitation pulse repetition frequency f and the linear power input to the discharge of the CVL $\overline{Q}_{dl}$. The relationship between the excitation pulse repetition frequency f in [1–3] was determined from the condition that in steady-state operation of a laser, an increase of the electron concentration during the excitation pulse is compensated

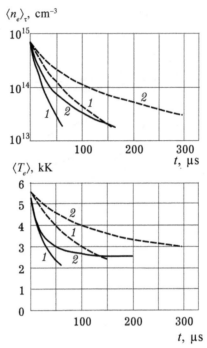

Figure 7.2. Depending on the concentration $\langle n_e \rangle$ and electron temperature $\langle T_e \rangle$ versus time for the buffer gases Ne (dash-dotted line) and He (solid lines) in the GDT 2 cm in diameter. *1* – buffer gas pressure $p = 1.33$ kPa, *2* – 6.7 kPa.

by its decrease in the interpulse interval. In the calculation of $\overline{Q}_{dl}$ it was assumed that the specific energy input in the discharge during the excitation pulse is constant which in the framework of the assumptions made in [1–3] indicates the constancy of the prepulse and postpulse electron concentrations. In [1–3] the prepulse electron concentration was assumed to be typical of a repetitively pulsed lasers value $\langle n_e \rangle_0 = 4 \cdot 10^{13}$ cm^{-3}, and the electron temperature during the excitation pulse $\langle T_e \rangle = 3$ eV.

As can be seen from Fig. 7.1 these values of $\langle n_e \rangle_0$ and $\langle T_e \rangle_{ex}$ correspond to the excitation pulse duration $\tau_{ex} \cong 75$ ns and the electron density $n_{ek} \cong 3 \cdot 10^{14}$ cm^{-3}. The given values of $\langle n_e \rangle_0$ and $\langle n_e \rangle \tau$ and using the dependences similar to those shown in Fig. 7.2 in [1–3] were used to determine the duration of interpulse interval T_{rel} and prepulse electron temperature $\langle T_e \rangle_0$.

It should be noted that the calculation of the concentration $\langle n_e \rangle$ and temperature $\langle T_e \rangle$ of the electrons through the system (7.6), (7.7) is valid, starting with the establishment of equilibrium between the electron concentration and temperature (establishing ionization equilibrium). This time does not match the end of the excitation pulse. However, the time interval between the end of the excitation pulse and the establishment of ionization equilibrium is a fraction of a microsecond [14], that is negligible compared to the interpulse interval.

In addition, at the time of the establishment of ionization equilibrium the electron concentration exceeds $\langle n_e \rangle_\tau$ due to the fact that after the end of the excitation pulse, the kinetic energy of the electrons is expended on ionization of Cu atoms. The corresponding estimates show that this excess may reach 50% of the values $\langle n_e \rangle_\tau$, whereby the interpulse interval will be longer than the interpulse interval calculated for the value of the electron concentration $\langle n_e \rangle_\tau$. However, according to the form of the time dependences of $\langle n_e \rangle_\tau$ shown in Fig. 7.2, this excess will not exceed a few microseconds, while the interpulse interval for the typical CVLs is tens microseconds or more.

Thus, the moments of the end of the excitation pulses identified in [1–3], and of the establishment of ionization equilibrium, as well as the neglect of increasing electron concentration after the end of the pulse and the corresponding increase in the duration of interpulse interval are quite correct and have no impact on the calculation of the duration of interpulse time interval in practice.

For the previously specified range of GDT diameters r_p and pressures p of Ne and He the prepulse electron temperature increases with height and is in the range from 3100 to 3300 K. If we assume that the temperature of population of the metastable level is $^2D_{5/2}$ $\langle T_m \rangle_0 = \langle T_e \rangle_0$ then the specified temperature range $\langle T_e \rangle_0$ corresponds to the range of concentrations of metastable atoms in the state $^2D_{5/2}$ from 1 to 3% of the concentration of copper atoms in the ground state.

Figure 7.3 shows the dependence [1–3] of the duration of interpulse interval T_{rel} and the corresponding frequency $f = 1/T_{rel}$ on the pressure of the buffer gases He and Ne.

Figure 7.4 shows the calculated linear input to the discharge $\bar{Q}_{dl}$ using the ratio

$$Q_{dl} = f \int_0^{\tau_m} \sigma E^2 \pi r_p^2 \, dt, \qquad (7.8)$$

$$\sigma = \frac{n_e e^2}{mv},$$

where, besides the well-known notations, E is the electric field intensity in the discharge required to maintain the electron temperature $T_e = 3$ eV in the discharge and calculated from the dependences of electron energy on the parameter $E / p\sqrt{10^3 n_{Cu} / n_a}$, given in [15]. This figure also shows the horizontal lines indicating the level of heat input power $\bar{Q}_{dl}$ at which the calculation of the ratio (4.18)–(4.20) at $r_d = r_p$ with the

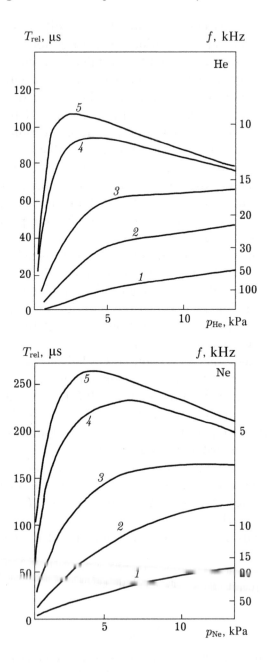

Figure 7.3. Dependences of the duration of interpulse interval T_{rel} and of the excitation pulse repetition frequency f, corresponding to this interval, in the repetitively pulsed CVL on the pressure of neon (Ne) and helium (He). $1 - r_p = 0.3$ cm, $2 - 0.6$; $3 - 1.0$; $4 - 2.0$; $5 - 2.5$.

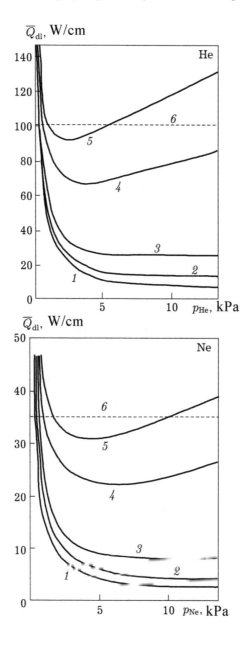

Figure 7.4. Dependences of the heat input to the discharge $\bar{Q}_{dl}$ in the repetively pulsed CVL on the pressure of neon (Ne) and helium (He). $1 - r_p = 0.3$ cm, $2 - 0.6$; $3 - 1.0$; $4 - 2.0$; $5 - 2.5$. Horizontal lines (*6*) indicate the level of heat input power $\bar{Q}_{dl}$, corresponding to the gas temperature in the axial zone of GDT calculated from (4.18), $T_g(0) = 4500$ K.

assumption that all the energy injected into the discharge goes into heating the gas, gives the gas temperature in the axial zone of GDT of $T_g(0) = 4500$ K.

Under the assumption that the physical efficiency of the laser η_p does not depend on the radius of the GDT, the optimum conditions of its work, providing the maximum average lasing power, are the conditions in which the unit power $\bar{Q}_{dl}$, introduced into the discharge, is close to its limiting value. According to the data presented in Fig. 7.4, at pressures of helium and neon from 1 to 15 kPa the unit power injected into the discharge increases with increasing radius of the GDT.

However, according to data presented in Fig. 7.4, at GDT radii of $r_p \geq 2.5$ cm temperatures $T_g(0)$ becomes so high that they virtually eliminate the possibility of lasing at the green line from the axial zone of the GDT (see Fig. 5.6). It should be noted, however, that this negative effect will be reduced under buffer gas pressures at which the alignment of the electron temperature T_e in the cross section of the GDT will be more efficient than the cooling of the electrons due to collisions with the buffer gas atoms and thus will satisfy the condition $\langle T_e \rangle < T_g(0)$. Nevertheless, it should be mentioned that the prepulse electron temperature in the GDT with a radius of around 2.5 cm and more at $\langle n_e \rangle_0 \cong 4 \cdot 10^{13}$ cm^{-3} can reach such values at which there will be a noticeable decrease in lasing generation.

The way out of this situation, allowing for increased lasing unit power at $\langle T_e \rangle < T_g(0)$, is to reduce the prepulse electron temperature by reducing the frequency f, as compared with the values of f, corresponding in Fig. 7.3 to the GDT radius of 2.5 cm or more. The

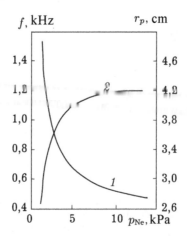

Figure 7.5. The dependence of the pulse repetition frequency f (*1*) and the radius r_p (*2*) on the neon pressure at $\bar{Q}_{dl} = 35$ W/cm, $\langle n_e \rangle_0 = 10^{13}$ cm^{-3}, $\langle n_e \rangle_\tau = 3 \cdot 10^{14}$ cm^{-3}.

change of the working conditions of a repetitively pulsed CVL with a decrease in prepulse electron temperature $\langle T_e \rangle$ to a value of 2500 K, is described by the dependence on f and r_p on the neon pressure, shown in Fig. 7.5 and calculated [1–3] under the following conditions: $Q_{dl} =$ 35 W/cm $n_{e0} = 10^{13}$ cm^{-3}, $n_{ek} = 3 \cdot 10^{14}$ cm^{-3}.

On the basis of calculations, presented in Figs. 7.4 and 7.5, we can conclude that for copper vapour laser at high lasing power with a single GDT with the buffer gas neon it is advisable to work with GDTs of large diameters (5–10 cm) in the frequency range f from 5 to 1 kHz. This conclusion is in good agreement with the experimental results given in [16, 17]. In [16] it is reported on the construction of a repetitively pulsed CVL with an average lasing power $\overline{P}_l = 85$ W with a GDT diameter of 3.5 cm at a sufficiently low pressure of neon of 4.6 kPa and the excitation pulse repetition frequency of 3.5–5 kHz. In [17] with a CVL with a GDT 8 cm in diameter at a frequency $f = 5$ kHz produced the average lasing power of 110 W.

The excess of the frequency f in the CVL with large diameter GDTs compared with the calculated results shown in Fig. 7.5 most likely due to the fact that in experiments [16,17] the prepulse electron density was greater than that used in the calculations. In addition, it is clear that for a CVL with lasing at two wavelengths, the frequency f, corresponding to the maximum output power will be somewhat higher than the corresponding frequency for generate only on the green line. This is due to the fact that the lower laser level $^2D_{3/2}$ is noticeably higher than the $^2D_{5/2}$ level and, consequently, at the same temperatures $\langle T_m \rangle_0$, its population will be lower than that of the level $^2D_{5/2}$. Finally, according to [4,5], in the repetitively pulsed CVL with the uniform distribution of the prepulse electron temperature T_{e0} in the GDT there may be conditions in which the lasing power increases when T_{e0} increases from 3000 K to 3500 K. This relationship permits rapid superheating of the gas at the GDT axis in comparison with 4500 K and, consequently, higher (in comparison with Fig. 7.5) excitation pulse repetition frequencies and lasing power $\overline{P}_l$.

Thus, according to data in [1–3], neglecting the inhomogeneity of the plasma parameters in the cross section of the GDT, although it can not reliably estimate the lasing power of the copper vapour lasers, it does allow us to reliably estimate the excitation pulse frequency for different radii of the GDT and the buffer gas pressure on the basis of the given values of $\langle n_e \rangle_0$, $\langle T_e \rangle_0$ and $\langle n_e \rangle_t$, $\langle T_e \rangle_t$.

This conclusion is quite important, since it cook allows in the calculation of the characteristics of repetitively pulsed CVLs to consider the effect of radial plasma inhomogeneities on such characteristics only during the excitation pulse.

7.3. Typical radial distributions of the prepulse plasma parameters in the repetitively pulsed copper vapour lasers

As shown by experimental studies of CVL carried out, for example, in [18–21], the lasing rate of such lasers and the radiation intensity gas-discharge plasma are distributed non-uniformly over the cross section of the GRT. The degree of inhomogeneity depends on many factors: buffer gas pressure, excitation pulse repetition frequency, charging current of the storage capacitor flowing through the discharge in the interpulse interval, etc. Various reasons leading to the nonuniform distribution of the intensities of lasing and plasma radiation have been more or less discussed in detail in several articles (e.g. [22, 23]). However, the number of papers devoted to the quantitative analysis of the effect of radial inhomogeneities of the plasma parameters on the lasing characteristics of the copper vapour lasers, slightly before the publication of [4, 5] was limited, in fact, to the article [24], in which a copper vapour lasers with GDT diameters more than 2 cm was used to analyze the influence of the skin effect on the lasing characteristics.

Strictly speaking, when determining the prepulse radial distributions of the plasma parameters for each specific repetitively pulsed CVL, characterized by a certain set of parameters such the type and pressure of buffer gas, excitation pulse repetition frequency, etc. it is necessary to solve of the balance equations for energy and particles, taking into account transport of both in the radial direction. This method is rather complicated and unsuitable for the analysis of the energy characteristics of copper vapour lasers, finding the optimal mode of operation throughout the range of possible changes in the values of n_p, p, n_{Cu}, f, $\bar{Q}_H$ and other parameters. Apparently, for this reason the majority of the calculations of the energy characteristics of pulse-periodic metal vapour lasers (see, for example, the previous chapter) were carries out for some plasma parameters averaged over the GDT cross section, and the transport of particles and energy to the wall of the GDT was taken into account in the zero-dimensional approximation.

In [4–6] in analysis of the influence of the radial inhomogeneity of the plasma parameters on the lasing characteristics of the CVLs the average (over the GDT cross section) concentration $\langle n_e \rangle$ and electron temperature $\langle T_e \rangle$ in the interpulse interval were calculated using equations (7.6), (7.7). In the transition from the mean values $\langle n_e \rangle_0$ and $\langle T_e \rangle_0$ to prepulse distributions $n_{e0}(r)$ and $T_{e0}(r)$, and also the determination of the prepulse distributions of $T_{g0}(r)$, $n_{g0}(r)$, $n_{m0}(r)$ in these studies the authors compared characteristic times of relaxation time and the characteristic lengths of the spatial alignment of the plasma parameters (see section 4.2) with the duration of the interpulse interval and discharge and with the GDT radius and the typical distributions of the prepulse plasma parameters were also determined.

1. The radial distribution of gas temperature in the GDT is weakly dependent on the time and and the prepulse distribution $T_{g0}(r) \cong T_g(t,r)$ is defined by (4.18)–(4.20)

2. The prepulse concentration distribution of copper atoms in the ground state $n_{g0}(r)$ was calculated in [4–6] on the basis of the distribution $T_{g0}(r)$ and the ratio

$$n_w T_w = n_{g0}(r)T_{g0}(r). \qquad (7.9)$$

Accounting for the possible existence of the nonuniform distribution $n_{g0}(r)$ is not appropriate, since a marked increase in heterogeneity of the distribution $n_{g0}(r)$ as compared with (7.9) is known to be detrimental to the lasing characteristics.

3. Prepulse distribution $T_{e0}(r)$ was assumed to vary depending on the ratio of the characteristic dimensions of the alignment of the electron temperature Λ_T (see section 4.2.2) and the GDT radius.

$$T_{e0}(r) = \text{const at } \Lambda_T \geq r_p, \qquad (7.10)$$

$$T_{e0}(r) = \begin{cases} T_g(r) + \Delta T_e, & r/r_p \leq 0.6 \\ T_g(r) + 2.5\Delta T_e(1 - r/r_p), & r/r_p \geq 0.6 \end{cases}$$

$$\text{at } \Lambda_T < r_p, \qquad (7.11)$$

The value ΔT_e was determined for a given value of the prepulse electron temperature $\langle T_{e0} \rangle$ averaged oved the GDT cross section.

4. The prepulse distribution of the population temperature of the metastable level $T_{m0}(r)$ was assumed identical with the distribution $T_{e0}(r)$, the prepulse distribution of the concentration of metastable atoms $n_{m0}(r)$ was calculated by the Boltzmann ratio

$$n_{m0}(r) = \frac{g_m}{g_r} n_{g0}(r)e^{-E_m/kT_{e0}(r)}. \qquad (7.12)$$

5. As the prepulse population of the resonance levels calculated from the Boltzmann relation is much smaller than $n_{m0}(r)$, in the calculations it is assumed to be zero.

6. As the prepulse distributions $n_{e0}(r)$ the authors of [4–6] used two model dependences: the diffusion distribution

$$\frac{n_{e0}(r)}{n_{e0}(0)} = J_0\left(\frac{\mu_1 r}{r_p}\right) \text{ at } \tau_D < \tau_r, \qquad (7.13)$$

where J_0 is the Bessel function of zero order, μ_1 is the first root of the zero-order Bessel function, τ_D and τ_r are the characteristic times of ambipolar diffusion of charged particles on the wall of the GDT and the bulk recombination of the electrons for the prepulse plasma parameters, and the trapezoidal distribution

$$\frac{n_{e0}(r)}{n_{e0}(0)} = \begin{cases} 1 & \text{at} \quad r/r_p \le 0.4 \\ \dfrac{10}{6}(1 - r/r_p) & \text{at} \quad r/r_p \ge 0.4 \end{cases} \text{ at } \tau_D \ge \tau_r. \qquad (7.14)$$

The conclusion made in [4–6] is similar to the conclusion in section 4.2.5, that for repetitively pulsed CVLs with typical GDT ($r_p \ge 0.5$ cm) and neon pressure ($p_{Ne} \ge 2$ kPa) there is no equalization of the temperature and electron concentration during the excitation pulse, makes it possible to solve the problem of the plasma parameters and lasing characteristics during the excitation pulse for each current value of the radius r. To simplify the problem, in [4–6] no account is made of the alignment of T_e in the cross section of the GDT in a situation where during the excitation pulse the condition $\Lambda_T > r_p$ is satisfied. However, as will be shown later, the changes in the lasing characteristics which may be caused by considering the alignment T_e during the excitation pulse, were evaluated.

7.4. The equations for calculating the lasing characteristics in the conditions of inhomogeneous distribution of plasma parameters over the cross section of the GDT

The system of equations for the plasma parameters and lasing characteristics at a wavelength of 510.6 nm was written in [4–6] in the approximation of saturated power and included the balance equation of the concentration of copper atoms in the ground n_g, metastable $^2D_{5/2}$ n_m, and resonance $^2P^0_{3/2}$ states, electron concentration n_e and the algebraic equations for T_e, representing the approximation of the results of calculations [15] of the average energy of electrons $\bar{\varepsilon}_e$ in a mixture of copper atoms with neon as a function of the parameters $E/p_0\sqrt{X}$ and E/p_{0Cu}, where, besides the known notations, p_0 and p_{0Cu} are the pressures of neon and copper vapours reduced to a temperature of 0°C, $X = 10^3 \cdot n_g/n_{Ne}$,

$$\frac{dn_r(x)}{dt} = \alpha_{gr}(x)n_e(x)n_g(x) - \alpha_r(x)n_e(x)n_r(x) - N(x), \quad (7.15)$$

$$\frac{dn_m(x)}{dt} = \alpha_{gm}(x)n_e(x)n_g(x) - \alpha_m(x)n_e(x)n_m(x) + N(x), \quad (7.16)$$

$$\frac{dn_g(x)}{dt} = -\alpha_g(x)n_e(x)n_g(x), \quad (7.17)$$

$$\frac{dn_e(x)}{dt} = \alpha_{gi}(x)n_g(x)n_e(x) + \alpha_r(x)n_r(x)n_e(x) + \alpha_m(x)n_m(x)n_e(x) +$$

$$+ \alpha_{P_{1/2}}(x)n_{P_{1/2}}(x)n_e(x) + \alpha_{D_{3/2}}(x)n_{D_{3/2}}(x)n_e(x), \quad (7.18)$$

$$x = r/r_p; \quad n_{P_{1/2}} = 0.5n_r; \quad n_{D_{3/2}} = 0.5n_m,$$

$$T_e = f(E/p_0\sqrt{X}), \quad T_e = f(E/p_{0Cu}). \quad (7.19)$$

Attention should be paid to equation (7.18), according to which in [4–6] the concentration of copper atoms in the levels $^2P^0_{1/2}$ and $^2D_{3/2}$ was estimated approximately and took into account only when calculating the electron concentration.

The initial conditions for the system were in the form of certain combinations (see below for them) of the prepulse distributions of various plasma parameters (4.18)–(4.20) and (7.9)–(7.14).

The system of equations (7.15)–(7.19) was solved successively for three time intervals: $0 - t_1; \ t_1 - t_2; \ t_2 - t_3.$

In the first time interval from the beginning of the excitation pulse to the start of lasing, corresponding in the approximation of the saturated power to the condition $n_m(t_1,x) = \frac{g_m}{g_r}n_r(t_1,x)$, the specific power generation $N(x) = 0.$

Time t_2 corresponds to the end of lasing. In the time interval $t_1 < t < t_2$ the equality $n_m(t,x) = \frac{g_m}{g_r}n_r(t,x)$ and the specific lasing power is defined by

$$N(x) = \frac{g_m}{g_m + g_r}[\alpha_{gr}(x)n_m(x)n_e(x) - \alpha_r(x)n_r(x)n_e(x)] -$$

$$- \frac{g_r}{g_m + g_r}[\alpha_{gm}(x)n_m(x)n_e(x) - \alpha_m(x)n_m(x)n_e(x). \quad (7.20)$$

Time $t_3 = \tau_{ex}$ corresponds to the end of the excitation pulse. When $t_2 \le t \le t_3$ the specific lasing power is $N = 0$.

The rate constants of various elementary processes were the constants shown in Fig. 5.3. The pulse shape of the electric field strength in the discharge $E(t)$, for brevity called 'the excitation pulse' was assumed to be given. The system of equations (7.15)–(7.19) was solved out for values $x = 0, 0.2, 0.4, 0.6, 0.8$, by the Runge–Kutta numerical integration method.

The radial distribution of specific lasing energy (specific energy output) was calculated in [4–6] by the formula

$$W(x) = h v_l \int_0^{t_3} N(t,x)dt = \int_0^{t_3} P(t,x)dt. \tag{7.21}$$

Lasing pulse energy W_l is related to the specific lasing energy $\langle W \rangle$ and $N(t, x)$, averaged over the GDT cross section, by the relations

$$W_l = \pi l_p r_p^2 \langle W \rangle = h v_l 2\pi l_p r_p^2 \int_0^{t_3} \int_0^1 N(t,x)xdxdt. \tag{7.22}$$

Time-averaged lasing power $\bar{P}_l$ and unit lasing power $\bar{P}_{ll}$ of the laser are

$$\bar{P}_l = l_p \bar{P}_{ll} = W_l f. \tag{7.23}$$

The radial distribution of laser intensity during the excitation pulse

$$I_l(t,x) = l_p h v_l N(t,x) = l_l P(t,x). \tag{7.24}$$

Instantaneous power of the laser radiation

$$P_l = 2\pi l_p r_p^2 h v_l \int_0^1 N(t,x)xdx. \tag{7.25}$$

The specific energy averaged over the GDT cross section per excitation pulse is calculated in [4–6] by the ratio

$$\langle G \rangle = 2 \int_0^{t_3} \int_0^1 \sigma(t,x)E^2(t,x)xdxdt, \tag{7.26}$$

$$\sigma(t,x) = \frac{n_e(t,x)e^L}{m[v_{eNe}(t,x)+v_{ei}(t,x)+v_{eCu}+v_{eCu^*}]}, \tag{7.27}$$

where, besides the well-known notations, v_{eCu} and v_{eCu^*} are the frequencies of the elastic and inelastic collisions of electrons with copper atoms.

The calculation results of v_{eCu} [15, 25] were approximated in [4–6] by the relation

$$v_{eCu} = n_g \cdot 2.28 \cdot 10^{-7} \cdot T_e^{0.19}, \tag{7.28}$$

where T_e is in eV. The frequency v_{eCu^*} was calculated taking into account the inelastic collisions with atoms in the ground state and the resonantly excited atoms

$$v_{eCu^*} = \alpha_g n_g + \sum_{j=1,2} \alpha_{r_j} n_{r_j}, \tag{7.29}$$

where the subscripts $j = 1, 2$ refer to the resonant levels $^2P^0_{3/2}$ and $^2P^0_{1/2}$. The physical laser efficiency was calculated in [4–6] as

$$\eta_p = \langle W \rangle / \langle G \rangle \tag{7.30}$$

7.5. Dynamics of lasing in a radially inhomogeneous active medium of a copper vapour laser

Despite the large number of papers published by the time of publication of the studies [4, 5] and concerned with experimental studies of repetitively pulsed CVLs, the results lacked a description of the experimental conditions, with simultaneous indication of all parameters necessary for correct calculations of the energy characteristics of these lasers, taking into account the inhomogeneous distribution of the plasma parameters. Therefore, in order to test the applicability of the above system of equations (7.1), (7.7), (7.15)–(7.19) for calculating the characteristics of the CVLs and to identify features of the lasing in an inhomogeneous active medium, in [4, 5] the authors calculated the energy characteristics of a self-heating copper vapour laser experimentally investigated in these papers.

These experiments were carried out using a self-heating CVL with a GDT radius of 1 cm and the active zone length of $l_a = 80$ cm. The rectifier output voltage $U_r = 6$ kV, current $I_r = 0.54$ A, the power consumed from the rectifier $Q_r \cong 3.2$ kW. The excitation pulse repetition frequency was $f = 14.5$ kHz. The duration of the voltage pulse at the base τ_{vA} was 180–200 ns, the duration of the leading edge $\tau_{ed} \cong$ 40 ns, neon pressure $p_{Ne} = 1.6$ kPa. The excitation unit included two 1150 pF storage capacitors which were discharged to the GDT at the same time through separate thyratrons.

The power injected into the discharge was about 50% of the power consumed from the rectifier. The linear power, introduced into the discharge was $\bar{Q}_{dl} \cong 20$ W/cm, and the specific energy per excitation pulse excitation $\langle G \rangle \cong 4.4 \cdot 10^{-4}$ J/cm^3. The increase in electron concentration $\Delta\langle n_e \rangle$ at the excitation pulse was calculated by the formula

$$\Delta\langle n_e \rangle \cong \frac{aQ_r}{f\pi r_p^2 l_p I_{Cu}}, \tag{7.31}$$

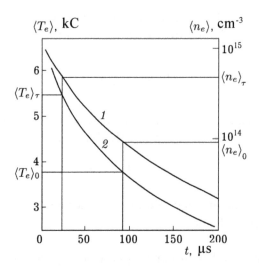

Figure 7.6. Time dependences of the averaged concentration of the cross section of GDT (*1*) and temperature (*2*) of the electrons in the interpulse interval for GDT radius of 1 cm at a neon pressure of 1.6 kPa. All other laser parameters are given in the text.

where $a = 0.5$ is a coefficient taking into account the transmission efficiency of the energy consumed by a rectifier, to the discharge and was $3.6 \cdot 10^{14}$ cm^{-3}.

Characteristics of lasing on both lines were as follows: $\bar{P}_{I\Sigma} \cong 19$ W the average specific energy $\langle W \rangle_\Sigma \cong 5.3$ µJ/cm^3, practical efficiency $\eta_{r\Sigma} \cong 0.6\%$, physical efficiency $\eta_{p\Sigma} \cong 1.2\%$. The relevant characteristics of the green laser line were equal $\bar{P}_I \cong 12.7$ W, $\langle W \rangle \cong 3.5$ µJ/cm^3, $\eta_r \cong 0.4\%$, $\eta_p \cong 0.8\%$. It should be noted that, in view of reflections on the optical windows of the emitter and the substrate, which served as the output mirror of the cavity, all the above characteristics should be increased by about 20%.

In the calculation in [4–6] the prepulse plasma parameters were determined using the time dependences $\langle n_e \rangle$ and $\langle T_e \rangle$ (Fig. 7.6) calculated by the system (7.6), (7.7) for a GDT with a radius of 1 cm and the neon pressure 1.6 kPa. The duration of the interpulse interval $T_{rel} = 1/f$ and the calculated electron concentrations values $\Delta \langle n_e \rangle$ were used to calculate the values $\langle n_e \rangle_\tau$ and $\langle n_e \rangle_0$ for the beginning t_1 and end t_2 of the interpulse interval $(T_{rel} = t_2 - t_1)$ and the corresponding values of $\langle T_e \rangle_\tau$ and $\langle T_e \rangle_0$. According to calculations, $\langle n_e \rangle_0 = 9 \cdot 10^{13}$ cm^{-3}, $\langle T_e \rangle_0 = \langle T_m \rangle_0 = 3750$ K. At this temperature of the metastable level population the specific output energy is reduced by 40% compared with the maximum possible energy.

The calculation in [4–6] of the characteristic times (see Section 4.2) of the three-body recombination τ_r of the electrons and their ambipolar diffusion on the wall of the GDT τ_a, the recombination electron heating τ_{er} and spatial relaxation (alignment) of the electron temperature τ_{eT} gave the following values: $\tau_r = 1.4\cdot10^{-4}$ s, $\tau_a = 9\cdot10^{-5}$ s, $\tau_{er} = 6.2\cdot10^{-6}$ s, $\tau_{eT} = 2\cdot10^{-7}$ s.

On the basis of the ratios of these characteristic times the distribution $n_{e0}(r)$ in the experiment conditions under consideration was assumed in [4–6] to be of the diffusion type, and the prepulse electron temperature T_{e0} and the temperature of the metastable level population were assumed to be constant over the cross section of the GDT. The shape of the voltage pulse, given in the calculations, correspond to the actual shape of the voltage pulse U_R at the active resistance of the discharge restored from the experimentally measured current and voltage pulses across the discharge gap. The duration of the leading edge of the voltage pulse $\tau_{ed} = 40$ ns, the pulse duration $\tau_{ex} = 180$ ns. The electric field in the discharge was calculated as $E = U_R(t)/l_l$. The concentration of copper atoms in the ground state near the wall of the GDT was assumed to be $n_{gw} = 2\cdot10^{15}$ cm^{-3}.

Figure 7.7 shows the calculated time dependences of the concentration of copper atoms in the ground state n_g, at the metastable ($^2D_{5/2}$) n_m, the resonance ($^2P^0_{3/2}$) levels, electron concentration n_e, specific lasing power on the green line P and the electron temperature T_e at the axis ($x = 0$) and near the wall of the GDT ($x = 0.8$). The calculated average energy deposited in the cross section $\langle G \rangle = 4.42\cdot10^{-4}$ J/cm^3. The time-averaged lasing on the green line $\bar{P}_l = 17.3$ W, the physical efficiency $\eta_p = 1.08\%$, specific energy $\langle W \rangle = 4.8\cdot10^{-6}$ J/cm^3.

Given the loss of the laser radiation power on the windows of the emitter and the output mirror the coincidence between the calculated and experimental results is more than satisfactory. This conclusion is not affected even by the fact that the concentration of copper atoms n_{gw} at the walls of GDT, included in the calculation [4–6], strictly speaking, may differ from the true concentration to higher or lower values. However, the results of calculations of the laser characteristics for the true concentration should be close to those obtained above, because their influence on the lasing characteristics the increase or decrease in the concentration of the copper atoms is compensated by a corresponding decrease or increase of the electron temperature, which ultimately contributes to the stabilization of the laser characteristics when the concentration of the copper atoms near the wall changes.

From a comparison of the dependences presented in Fig. 7.7 it follows that the inhomogeneity of the radial distributions of the prepulse plasma parameters leads to a nonuniform distribution of the lasing power in the cross section of GDT. For example, increasing

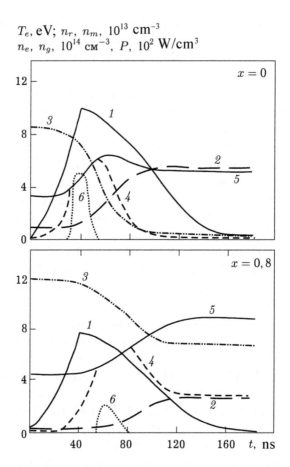

T_e, eV; n_r, n_m, 10^{13} cm^{-3}
n_e, n_g, 10^{14} см$^{-3}$, P, 10^2 W/cm^3

Figure 7.7. Time dependences of temperature (*1*) and concentration (*2*) of the electrons, the concentration of copper atoms in the ground (*3*), resonance (*4*) and metastable (*5*) levels, specific lasing power (*6*) in the centre of the GDT ($x = 0$) and near the wall ($x = 0.8$) during the excitation pulse $r_p = 1$ cm, $p_{Ne} = 1.6$ kPa, $n_w = 2 \cdot 10^{15}$ cm^{-3}.

the electron temperature in the axial zone of the GDT due to greater heating of the gas and the increase as a result of this heating of the parameter E/p_{Cu} leads to a higher rate of excitation of the resonance level and much earlier completion of lasing in comparison with the end wall regions of the GDT. This uneven progress in the completion of lasing leads to undesirable losses in the axial region of the GDT and, consequently, reduces the laser efficiency.

The possible equalization of the electron temperature in the cross section of the GDT during the excitation pulse not taken into account in calculations should lead to the equalization of the rates of excitation of resonance levels and the simultaneous beginning and end of the lasing in the entire cross section of the GDT and, ultimately, to an increase in the lasing energy and efficiency of the laser. Therefore, for those pressures and radii of GDT for which the equalization of T_e by electron heat conduction plays a significant role (see section 4.2.5), the calculated output power and efficiency should be considered as close to the minimum possible.

After receiving confirmation that the duration of the interpulse interval can be calculated by the system (7.6), (7.7), and the lasing energy and efficiency can calculated by the system (7.15)–(7.19) with the prepulse radial distributions of the plasma parameters as initial conditions, we can go to consider the findings of [4–6] in the analysis of the influence of inhomogeneities of the plasma parameters on the characteristics of the lasing of repetitively pulsed CVLs.

7.6. Effect of radial inhomogeneity of the plasma parameters on the efficiency and specific lasing energy and efficiency of the CVL

Analysis of the effect of the radial inhomogeneity of the plasma parameters on the specific lasing energy $\langle W \rangle$ and physical efficiency η_p of the CVL was carried out in [4–6] by comparing these characteristics realized in different radial distributions of the plasma parameters. THe radial distributions $n_e(t, x)$, $T_e(t, x)$ and of the lasing power $N_e(t, x)$ were calculated using the equations (7.15)–(7.19), and then averaging according to the formulas (7.21) and (7.26) was performed to calculate values $\langle W \rangle$ and $\langle G \rangle$ and the ratio (7.30) the value η_p.

The values of $\langle W \rangle$, $\langle G \rangle$ and η_p were calculated out for three combinations of the distributions of the prepulse plasma parameters, for convenience hereinafter referred to as homogeneous, partly homogeneous and inhomogeneous distributions. In the case of the homogeneous distribution all the prepulse plasma parameters T_{e0}, n_{e0}, T_g, n_{g0}, n_{Ne0}, were assumed in [4–6] to be constant in the cross-section of the GDT. In the case of partially homogeneous distribution the distribution $T_g(r)$ is determined by (4.18), the temperature T_{e0} is constant over the cross section of the GDT, distribution $N_{e0}(r)$ is of the diffusion type (7.13), the distributions $n_{Ne0}(r)$ and $n_{g0}(r)$ are defined by (7.9). The inhomogeneous distribution is different from the partially homogeneous distributions only by the distributions $T_{e0}(r)$ and $n_{e0}(r)$ described in this case by (7.11) and (7.14). The prepulse population temperature of the metastable level was assumed equal to the prepulse electron temperature $T_{m0}(r) = T_{e0}(r)$. The pulse shape of the electric field strength (hereinafter referred to as 'excitation pulse') in the discharge was assumed to be rectangular. The concentration of copper atoms near the wall of the GDT was assumed to be $n_w = 2 \cdot 10^{15}$ cm^{-3}.

In order to determine the effect of radial inhomogeneity of the plasma parameters on the lasing characteristics the values of $\langle W \rangle$, $\langle G \rangle$ and η_p, calculated for the above three combinations of the prepulse distributions of the plasma parameters, were compared out, firstly, with the condition of equality of the averaged (over the cross-section of the GDT) prepulse concentrations of copper atoms $\langle n_g \rangle_0$, the electron concentration $\langle n_e \rangle_0$, gas temperature $\langle T_g \rangle_0$, the temperature of the populations of the metastable level $\langle T_m \rangle_0$ and, secondly, under the condition of equality of the averaged (over the cross section of the GDT and time) electron temperatures of the during the excitation pulse $\langle \overline{T}_e \rangle_{ex}$.

$\langle n_e \rangle_0$, $\langle n_e \rangle_0$, $\langle T_g \rangle$, $\langle T_m \rangle_0$ were calculated from the relation

$$\langle \Pi \rangle_0 = 2 \int_0^1 \Pi_0(x) x \, dx, \qquad (7.32)$$

where $\Pi \equiv n_g$, n_e, T_g, T_e, and $\langle \overline{T}_e \rangle_{ex}$ was calculated from

$$\langle \overline{T}_e \rangle_{ex} = \frac{2 \int_0^1 \int_0^{\tau_{ex}} T_e(x,t) x \, dx \, dt}{\tau_{ex}}. \qquad (7.33)$$

In the calculation of $T_g(r)$ from (4.18)–(4.20) the unit power injected into the discharge was assumed to be $\overline{Q}_{dl} = 25$ W/cm. This value $\overline{Q}_{dl}$ corresponds to a temperature of $\langle T_g \rangle = 2530$ K and $\langle n_g \rangle = 1.4 \cdot 10^{15}$ cm^{-3}.

Figure 7.8 shows as an example the results of calculation of dependences $\langle W \rangle$, $\langle G \rangle$ and η_p on the duration of the rectangular excitation pulse, $\langle n_e \rangle_0 = 3 \cdot 10^{13}$ cm^{-3}, two values of the neon pressure (2 and 13.3 kPa), two values of $\langle T_m \rangle_0$ (0, 3200 K), $\langle T_e \rangle_{ex} = 4.2$ eV and various combinations of the radial distributions of the prepulse plasma parameters. A comparison of these dependences allows us to establish a number of qualitative relationships. Firstly, for the same duration of the excitation pulse, the prepulse distribution of the plasma parameters and the value $\langle T_g \rangle_{ex}$ the specific energy $\langle W \rangle$ does not depend on the neon pressure, and the efficiency decreases markedly with its growth. The heterogeneity in the prepulse distributions of the plasma parameters leads to a prolongation of the lasing pulse and a decrease of the specific energy output and efficiency η_p. A particularly strong impact on the reduction of $\langle W \rangle$ and η_p is shown by the heterogeneity in the prepulse distribution of the population temperature of the metastable level.

The dependences of the specific energy input $\langle G \rangle$, energy output $\langle W \rangle$ and physical efficiency η_p on the duration of the excitation pulse, calculated for $\langle T_e \rangle_{ex} = 4.2$ eV (including those in Fig. 7.8) were used in [4–6] to obtain the dependences of $\langle W \rangle$ and η_p on the neon pressure

$\langle G \rangle$, 10^{-4} J/cm^3

$\langle W \rangle$, 10^{-6} J/cm^3

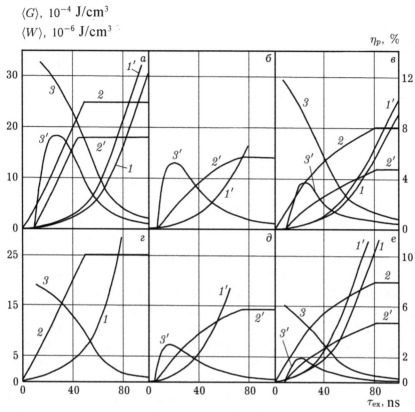

Figure 7.8. Dependence on the specific energy input $\langle G \rangle$ ($1, 1'$), the specific energy output $\langle W \rangle$ ($2, 2'$) and physical efficiency η_p ($3, 3'$) on the duration of the rectangular excitation pulse. $a, b, c - p_{Ne} = 2$ kPa, $c, d, e, - p_{Ne} = 13.3$ kPa, $a, d -$ homogeneous distribution of prepulse plasma parameters, $b, e -$ partly homogeneous distribution, $c, f -$ the inhomogeneous distribution; $\langle T_g \rangle = 2530$ K, $\langle n_g \rangle_0 = 1.4 \cdot 10^{15}$ cm^{-3}, $\langle n_e \rangle_0 = 3 \cdot 10^{13}$ cm^{-3}; $1, 2, 3 - \langle T_m \rangle_0 = 0$ K, $1', 2', 3' - \langle T_m \rangle_0 = 3200$ K.

for different combinations of the prepulse distributions of the plasma parameters

Based on the analysis of these curves in [4–6] it was found that the maximum possible specific energy $\langle W \rangle_{max}$ does not depend on the neon pressure, and in determined by the nature of the prepulse distribution of the plasma parameters and the level of the prepulse population of the metastable level. Table 7.1 gives examples of the results of calculations of $\langle W \rangle_{max}$ for $\langle T_e \rangle_{ex} = 4.2$ eV, $\langle n_g \rangle_0 = 1.4 \cdot 10^{15}$ cm^{-3}, $\langle T_g \rangle = 2530$ K at neon pressures from 0.67 to 13.3 kPa for the homogeneous and inhomogeneous distributions of the prepulse plasma parameters and three values of $\langle T_m \rangle_0$: 0, 3200 and 3500 K.

Table 7.1. Influence of the distribution of prepulse plasma parameters and $\langle T_m \rangle_0$ on the maximum possible energy output

Distribution of prepulse plasma parameters	Homoge-neous	Inhomoge-neous	Homoge-neous	Inhomoge-neous	Homoge-neous
$\langle T_m \rangle_0$, K	0	0	3200	3200	3500
$\langle T \rangle_{max}$, µJ/cm²	25	19.7	17.7	11.9	9.3

A visual representation of the strength of the influence of the prepulse population of the metastable level and the inhomogeneous distribution $T_{m0}(r)$ on the lasing characteristics at a constant specific energy input $\langle G \rangle$ is shown in Fig. 7.9 by the dependences of $\langle W \rangle$ and η_p on $\langle T_m \rangle_0$ calculated for the homogeneous and inhomogeneous distributions of the prepulse plasma parameters. From a comparison of these curves it follows that the combined effect of the prepulse population of the metastable level and the heterogeneity of its distribution can lead to a significant (up to five times) decrease in efficiency and specific energy output compared to the case with the case of homogeneous distribution of the prepulse plasma parameters and $\langle T_m \rangle_0 \leq 2000$ K.

It is interesting to compare the results of calculation of $\langle W \rangle$ and η_p shown in Fig. 7.9, not relating to any specific CVL, with the above described results of calculations for a specific CVL with a GDT with a radius of 1 cm and the length of the laser active medium $l_a = 80$ cm.

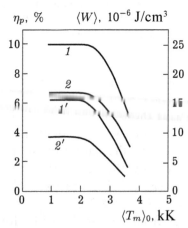

Figure 7.9. Dependence on the specific energy output $\langle W \rangle$ (*1,1'*) and physical efficiency η_p (*2, 2'*) on the prepulse population temperature of the metastable level $\langle T_m \rangle_0$ $p_{Ne} = 2$ kPa, $\langle G \rangle = 4 \cdot 10^{-4}$ J/cm³, $\langle T_e \rangle_{ex} = 4.2$ eV, $\langle T_g \rangle = 2530$ K, $\langle n_g \rangle_0 = 1.4 \cdot 10^{15}$ cm⁻³. *1, 2* – homogeneous, *1', 2'* – inhomogeneous distribution of prepulse parameters.

According to these calculations, $\langle T_m \rangle_0 = 3750$ K, $\langle W \rangle = 4.8$ µJ/cm³, $\eta_p = 1.08\%$. Comparing these data with the calculations shown in Fig. 7.9, we see that they are situated between the data for the homogeneous and inhomogeneous distributions of prepulse plasma parameters which should be the case since the calculation of the characteristics of the specific laser was carried out for the partially homogeneous distribution of the plasma parameters. This fact once again confirms the correctness of the physical and mathematical models used in [4–6] to account for the effect of the radial inhomogeneities of the plasma parameters on the performance of repetitively pulsed CVLs.

According to the data presented in Fig. 7.8, at $\langle T_e \rangle_{ex} = 4.2$ eV, $\langle T_m \rangle_0 = 3200$ K and the rectangular shape of the excitation pulse the maximum physical efficiency $\eta_{p\,max}$ is obtained at specific energy input $\langle G \rangle \cong (1 \div 2) \cdot 10^{-4}$ J/cm³ and excitation pulse duration $\tau_{ex} \cong 20 \div 30$ ns.

In contrast to the maximum specific energy output $\langle W \rangle_{max}$, the physical efficiency η_p is a strong function of the neon pressure. This is indicated, for example, by the dependence of the maximum physical efficiency $\eta_{p\,max}$ on p_{Ne} presented in Fig. 7.10. When evaluating on the basis of data presented in Fig. 7.10, the real increase of $\eta_{p\,max}$ with the neon pressure decreasing from 13.3 kPa to tenths of kPa it must be remembered that this pressure decrease will be accompanied be a change in the shape of the distribution of the prepulse distribution of the electron temperature and, correspondingly, the population temperature of the metastable level, since $T_{e0}(r) = T_{m0}(r)$. For example, according to [4–6], for a GDT radius of 0.8 cm the transition from the neon pressure of 13.3 kPa to 2.7 kPa and lower means a transition from the inhomogeneous distribution $T_{m0}(r)$ of a uniform distribution,

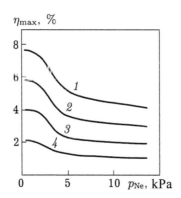

Figure 7.10. Dependences of the maximum physical efficiency $\eta_{p\,max}$ on the neon pressure $\langle T_e \rangle_{ex} = 4.2$ eV, *1* – homogeneous, *2* – partly homogeneous, *3, 4* – inhomogeneous distribution of the prepulse plasma parameters $\langle T_g \rangle = 2530$ K, $\langle n_g \rangle_0 = 1.4 \cdot 10^{15}$ cm⁻³. *1* – $T_{m0} = 3200$ K, *2, 3* – $\langle T_m \rangle_0 = 3200$ K, *4* – $\langle T_m \rangle_0 = 3500$ K.

which corresponds in Fig. 7.10 to the transition from curve *3* to curve *2*. That is, for a GDT with a radius $r_p \leq 0.8$ cm a reduction in the neon pressure of 13.3 to 2.7 kPa can increase $\eta_{p\ max}$ from about 2% to 6%. Thus, it can be concluded that in the conditions of the nonuniform radial distribution of the plasma parameters higher values of η_p can be obtained by working at low pressures $p_{Ne} \leq 1.3$ kPa and small GDT radii $r_p \leq 0.5$ cm.

The strength of the influence of the electron temperature during the excitation pulse and of the heterogeneity of its distribution on the lasing characteristics of the CVL can be seen from the results of the calculations [4–6] of $\langle W \rangle$ and η_p shown in Figs. 7.11 and 7.12. It should be noted that in the case of the partially homogeneous and inhomogeneous distributions of the prepulse plasma parameters the end of the lasing pulse in these calculations the end of lasing at $x = 0.8$.

As follows from the data presented in Fig. 7.11, irrespective of the nature of the distribution of the prepulse parameters the limiting (maximum possible) specific energy output $\langle W \rangle_{max}$ increases monotonically with increasing electron temperature $\langle T_e \rangle_{ex}$ and saturates at $\langle T_e \rangle_{ex} \cong 6$ eV. At the same time, with the increase of $\langle T_e \rangle_{ex}$ the initial growth of $\eta_{p\ max}$ is followed by its decrease, so that in the inhomogeneous distribution of the prepulse plasma parameters there is some optimal value $\langle T_e \rangle_{ex}$ at which $\eta_{p\ max}$ has the highest value.

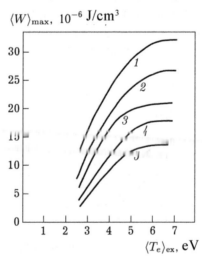

Figure 7.11. Dependence on the limiting energy output on the average (over the cross section of the GDT) the electron temperature $\langle T_e \rangle_{ex}$ during the excitation pulse ($\tau_{ex} \approx 50 \div 80$ ns), $p_{Ne} = 2$ kPa, $\langle n_g \rangle_0 = 1.4 \cdot 10^{15}$ cm^{-3}, $\langle T_g \rangle = 2530$ K, *1, 3* – homogeneous, *2, 4, 5* – inhomogeneous distributions of prepulse plasma parameters, Rectangular excitation pulse *1, 2* – $T_{m0} = 0$ K, *3* – $T_{m0} = 3200$ K, *4* – $\langle T_e \rangle_0 = 3200$ K, *5* – $\langle T_m \rangle_0 = 3500$ K.

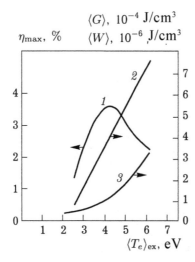

Figure 7.12. Dependence of the maximum physical efficiency $\eta_{p\ max}$ (*1*) and the corresponding specific energy output $\langle W \rangle$ (*2*) and energy input $\langle G \rangle$ (*3*) on the average (over the cross-sectional of the GDT) electron temperature $\langle T_e \rangle_{ex}$ during the excitation pulse. Inhomogeneous distribution of the prepulse plasma parameters, rectangular excitation pulse ($\tau_{ex} \approx 20 \div 30$ ns), $p_{Ne} = 2$ kPa, $\langle n_g \rangle_0 = 1.4 \cdot 10^{15}$ cm^{-3}, $\langle T_g \rangle = 2530$, $\langle T_m \rangle_0 = 3200$ K.

In the calculation conditions described in the legend to Fig. 7.12, the optimum value of $\langle T_e \rangle_{ex}$ is 4.2 eV.

7.7. Effect of the excitation pulse shape on the lasing characteristics of a CVL

Since the real excitation pulses are different in shape from rectangular ones, in [4–6] the authors compared the lasing characteristics for two pulse shapes: rectangular and triangular. Calculations were carried out for triangular excitation pulses as well as for rectangular pulses of the time dependences $\langle W \rangle, \langle G \rangle$ and η_p for different prepulse distributions of the plasma parameters and averaged (over the cross section of the GDT and the excitation pulse duration) electron temperature $\langle \overline{T_e} \rangle_{ex}$. Comparison of these characteristics for different forms of excitation pulses was carried out under the condition $\langle W \rangle \leq \langle W \rangle_{max}$ and specific energy supplied to the discharge $\langle G \rangle$. The condition $\langle W \rangle \leq \langle W \rangle_{max}$ limits the value of $\langle G \rangle$ at a level appropriate for the end of lasing for the rectangular excitation pulse (see Fig. 7.8).

As a result of this comparison it was revealed that for a triangular excitation pulse with given pulse duration τ_{ex} and the leading edge τ_{ed} and the maximum electric field strength E_{max} we can found such duration τ_{ex} and strength E for the rectangular pulse that the values

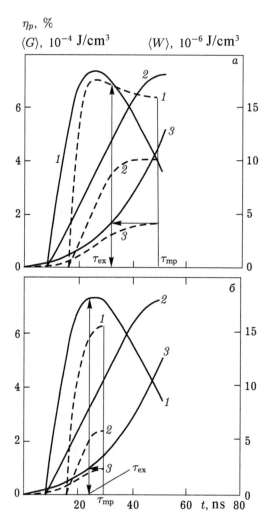

Figure 7.13. Time dependences of the physical efficiency η_p (*1*), the specific energy output W (*2*) and the specific energy input G (*3*) for rectangular (solid line) and triangular (dashed-dotted line) excitation pulses. Homogeneous distribution of the prepulse plasma parameters $p_{Ne} = 2$ kPa, $n_{g0} = 1.4 \cdot 10^{15}$ cm^{-3}, $T_g = 2530$ K, $n_{e0} = 3 \cdot 10^{13}$ cm^{-3}, $T_{m0} = 3200$ K, *a*: $\tau_{ed} = 20$ ns, $\tau_{ex_{tr}} = 50$ ns, the maximum electron temperature $T_{e\,max} = 5.2$ eV, *b*: $\tau_{ed} = 20$ ns, $\tau_{ex_{tr}} = 30$ ns, $T_{e\,max} = 5.6$ eV for a rectangular excitation pulse $T_{eex} = 4.2$ eV.

of $\langle W \rangle$ and η_p for the triangular and rectangular excitation pulses are close to each other.

As an example Figs. 7.13 and 7.14 show the time dependences of $\langle G \rangle$, $\langle W \rangle$ and η_p for a rectangular excitation pulse and two triangular pulses with the duration of 50 and 30 ns, at the leading edge duration $\tau_{ed} = 20$ ns for homogeneous (Fig. 7.13) and inhomogeneous (Fig.

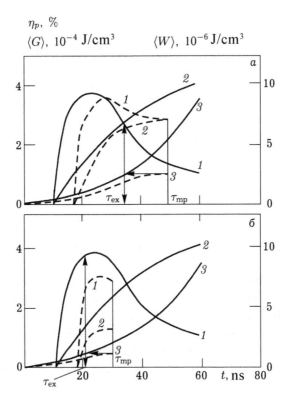

Figure 7.14. Time dependences of the physical efficiency η_p (*1*), the specific energy output $\langle W \rangle$ (*2*) and the specific energy input $\langle G \rangle$ (*3*) for a rectangular (solid line) and triangular (dashed-dotted line) excitation pulses. Inhomogeneous distribution of the prepulse plasma parameters $p_{Ne} = 2$ kPa, $\langle n_e \rangle_0 = 1.4 \cdot 10^{15}$ cm^{-3}, $\langle T_g \rangle = 2530$ K, $T_w = 1520$ °C, $\langle n_e \rangle_0 = 3 \cdot 10^{13}$ cm^{-3}, $\langle T_m \rangle_0 = 3200$ K, *a*: $\tau_{ed} = 20$ ns, $\tau_{ex_{tr}} = 50$ ns, $\langle T_{e\,max} \rangle_{ex} = 5.7$ eV, and *b*: $\tau_{ed} = 20$ ns, $\tau_{ex_{tr}} = 30$ ns, $\langle T_{e\,max} \rangle_{ex} = 6.4$ eV for a rectangular excitation pulse $\langle T_e \rangle_{ex} = 4.2$ eV.

7.14) distributions of the prepulse plasma parameters. The arrows in these figures show how, starting from the condition that the specific energy inputs during the triangular and rectangular shapes are equal, from the given value $\langle G \rangle_{tr}$ we define the lasing characteristics for a rectangular excitation pulse and its duration τ_{ex}. As follows from the data presented in Figs. 7.13 and 7.14, the lasing characteristics $\langle W \rangle$ and η_p differ from each other by about 20%. In practice, this means that by experimental optimization of the real excitation pulse we can obtain lasing characteristics practically corresponding to the lasing characteristics for the rectangular excitation pulse.

Thus, the main factor during the excitation pulse the lasing energy generation and the efficiency of copper vapour lasers are the specific energy $\langle G \rangle$ and the electron temperature $\langle \bar{T}_e \rangle_{ex}$. Moreover, with the same specific energy input $\langle G \rangle$ the electron concentrations $\langle n_e \rangle_\tau$ at the end of excitation pulses of triangular and rectangular shapes are practically identical. That is, the shape of the excitation pulse is not a parameter that significantly influences the achievable lasing performance. This conclusion is fundamental, since it allows in the calculation of the optimal modes of operation of pulse-periodic copper vapour lasers to use a model of the rectangular excitation pulse. In implementing the optimal mode of operation of a laser with a specific power supply, shaping the excitation of a certain form, we can choose empirically the duration and amplitude of the pulse such that they provide the same values of $\langle G \rangle$, $\langle W \rangle$ and η_p as in the calculations carried out for rectangular excitation pulses.

7.8. Lasing characteristics of repetitively pulsed CVLs

To calculate the lasing characteristics of the repetitively pulsed copper vapour lasers, the authors of [4–6] used dependences, similar to those shown in Figs. 7.2 and 7.8. The procedure for calculating the excitation pulse repetition frequency f, the radius of the GDT r_p, the specific energy input $\langle G \rangle$, the unit power supplied to the discharge $\bar{Q}_{dl}$, specific lasing power $\bar{P}_{ll}$ and physical efficiency η_p was as follows. In calculating the energy input $\langle G \rangle$, the prepulse population temperature of the metastable level $\langle T_m \rangle_0$ (3000, 3200, 3500 K), neon pressure (0.67, 2.0, 4.0, 13.3 kPa) and the type of distribution of the prepulse plasma parameters (homogeneous, partially homogeneous and inhomogeneous). Using the dependences similar to those shown in Fig. 7.2, the $\langle T_m \rangle_0$ was used to determined the prepulse value $\langle n_e \rangle_0$ and the time t_2 corresponding to Fig. 7.2. Assuming that almost all the energy supplied to the discharge is expended in the long run for the ionization of copper atoms, the value of $\langle G \rangle$ was used to calculated using the average (over the cross-section of the GDT) increment of the electron concentration in the excitation pulse

$$\Delta \langle n_e \rangle = \frac{\langle G \rangle}{I_{Cu}}. \tag{7.34}$$

From the known values $\langle n_e \rangle_0$ and $\Delta \langle n_e \rangle$ we calculated the average (over the cross section of the GDT) electron concentration after the excitation pulse

$$\langle n_e \rangle_\tau = \langle n_e \rangle_0 + \Delta \langle n_e \rangle \tag{7.35}$$

and the time dependence $\langle n_e \rangle_0$ and $\langle T_e \rangle$, similar to those shown in Fig. 7.2, were used to determine the time t_1 corresponding the determined value of $\langle n_e \rangle_\tau$. This time corresponds to the beginning of the interpulse interval; and the electron temperature $\langle T_e \rangle_\tau$ at this time was also determined.

The excitation pulse repetition frequency, corresponding to the interval of time it takes for the relaxation of the temperature of population of the metastable level $^2D_{5/2}$ to the value given by the calculation of the prepulse value, defined as

$$f_m = (t_2 - t_1)^{-1}. \qquad (7.36)$$

Using the dependences similar to those shown in Fig. 7.8, the value $\langle G \rangle$ was used to calculate the specific energy output $\langle W \rangle$ and physical efficiency η_p. Strictly speaking, the duration of the rectangular excitation pulse can be determined from the time dependences of $\langle G \rangle$, $\langle W \rangle$ and η_p (Figure 7.8) only when the prepulse electron concentration $\langle n_e \rangle_0$ is $3 \cdot 10^{13}$ cm^{-3}, that is, corresponds to the value that was used in calculations of the dependences similar to those shown in Fig. 7.8. In those cases, when the value $\langle n_e \rangle$ determined the specified value of $\langle T_m \rangle_0$ is not equal to $3 \cdot 10^{13}$ cm^{-3}, the duration of the excitation pulse τ_{ex}, defined by relationships similar to those shown in Fig. 7.8, using the initial value $\langle G \rangle$ should be regarded as approximate.

However, attention should be given to the fact that a change in the electron concentration $\langle n_e \rangle_0$ which takes place for the values of $\langle T_m \rangle_0$ used in the calculations, has effect on such parameters as $\langle G \rangle$, $\langle W \rangle$, η_p and $\langle \Delta n_e \rangle$.

Unit power $\overline{Q}_{dl}$ introduced into the discharge and the unit lasing power $\overline{P}_{ll}$ were determined in [4–6] respectively as

$$\overline{Q}_{dl} = \langle G \rangle f_m S_p \qquad (7.37)$$

and

$$\overline{P}_{ll} = \langle Q \rangle f_m S_p. \qquad (7.38)$$

As an example, Figs. 7.15 and 7.16 show part of the calculation results [4–6] of the frequency dependence f_m and physical efficiency η_p for the green lasing line from the average (over the cross section of the GDT) specific energy input $\langle G \rangle$ for the rectangular excitation pulse for different radii of the GDT, neon pressure, prepulse temperatures $\langle T_m \rangle_0$ and various combinations of distributions of the prepulse parameters. Also given are the results of calculations by (7.37), (7.38) of the unit lasing power $\overline{P}_{ll}$ and the unit power $\overline{Q}_{dl}$ to the discharge, for a frequency $f = f_m$.

From a comparison of the time dependences of $\overline{P}_{ll}$ and η_p, shown in Figs. 7.15, 7.16 and similar results in [4–6], it is concluded that

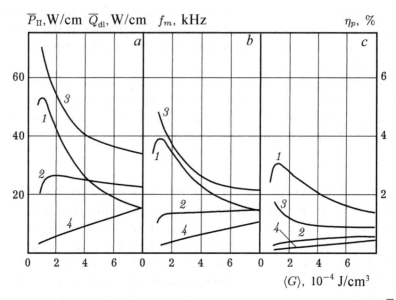

Figure 7.15. Dependence of the physical efficiency η_p (1), unit lasing power $\bar{P}_{II}$ (2), excitation pulse repetition frequency f_m (3), power supplied to the discharge $\bar{Q}_{dl}$ (4), on the specific energy input $\langle G \rangle$. Partially homogeneous distribution of prepulse plasma parameters, $r_p = 0.3$ cm, $T_{m0} = 3200$ K. The remaining prepulse plasma parameters are the same as for Fig. 7.8; a: $p_{Ne} = 2$ kPa, b: 4 kPa c: 13.3 kPa.

at $f = f_m$ there is a number of characteristics of the pulse-periodic copper vapour lasers. First, regardless of the neon pressure the maximum physical efficiency is realized in the specific energy input range $(1 \div 2) \cdot 10^{-4}$ J/cm³. Second, at neon pressures of about 4 kPa and less the maximum output power correspond to values $\langle T_m \rangle_0$, close to 3200 K, and at pressures of about 13 kPa to neon to values $\langle T_m \rangle_0$ close to 3500 K. Third, when the neon pressure is changing the character of the dependences of the unit output power $\bar{P}_{II}$ of the energy input $\langle G \rangle$ also changes, from a non monotonic dependence with a maximum at $\langle G \rangle \simeq 2 \cdot 10^{-4}$ J/cm³ at $p_{Ne} = 0.67$ kPa, up to a monotonically increasing with increasing $\langle G \rangle$ at $p_{Ne} = 13.3$ kPa. Fourth, at low neon pressures the position of the maxima of the dependences $\bar{P}_{II}$ and η_p is practically the same, ie. the maximum lasing power input $\bar{P}_{II\,max}$ is obtained at maximum physical efficiency $\eta_{p\,max}$. At pressures $p_{Ne} > 2$ kPa this correspondence between $\bar{P}_{II\,max}$ and $\eta_{p\,max}$ is not observed.

It should be mentioned here that in the case of the repetitively pulsed CVLs the restriction of the lasing power $\bar{P}_{II}$ with increasing excitation pulsed repetition frequency f is often attributed to the operation of the mechanisms leading to a decrease of the specific energy output

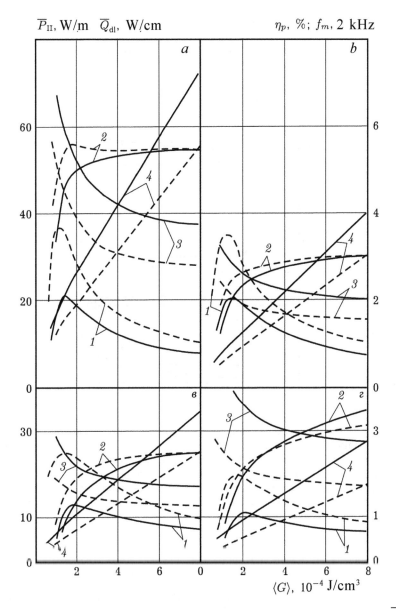

Figure 7.16. Dependence of the physical efficiency η_p (*1*), unit lasing power $\overline{P}_{11}$ (*2*), excitation pulse repetition frequency f_m (*3*), unit power to the discharge $\overline{Q}_{dl}$ (*4*), on the specific energy input $\langle G \rangle$. Inhomogeneous distribution of the prepulse plasma parameters, $r_p = 2$ cm, solid lines – $\langle T_m \rangle_0 = 3500$ K, dot-and-dash curves – 3200 K. The remaining prepulse plasma parameters are the same as for Fig. 7.8; *a*: $p_{Ne} = 0.67$ kPa, *b*: 2 kPa, in *c*: 4 kPa, *d*: – 13.3 kPa.

with increasing frequency *f*. One of these mechanisms is traditionally the increase of the population of the metastable level with increasing

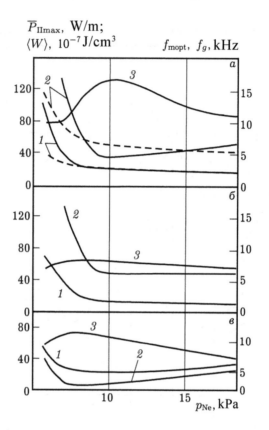

Figure 7.17. Dependence of the maximum heat input laser power generation $\overline{P}_{\text{||max}}$ (*1*)and the corresponding frequency of excitation pulses f_{mopt} (*2*) and the specific energy output $\langle W \rangle$ (*3*) on the neon pressure *a*: r_p = 1 cm, partially homogeneous distribution of prepulse plasma parameters; dot-and-dash – frequency f_g – excitation pulse repetition frequency determined by the time of restoration concentration of copper atoms in the axial zone of GDT (see below) and the corresponding unit lasing power $P_{g||}$; *b*: r_p = 1 cm; inhomogeneous distribution of the prepulse plasma parameters, *c*: r_p = 2 cm, the inhomogeneous distribution of prepulse plasma parameters.

frequency f. The data shown in Fig. 7.16 d indicate that the population of the metastable level is not a factor determining unambiguously the limiting excitation pulse repetition frequency and the corresponding maximum lasing power and that, strictly speaking, the maximum lasing power and the corresponding excitation pulse repetition frequency form as a result of the interaction of different processes and factors.

The dependences of $P_{||}$ and η_p on the specific energy input shown in Figs. 7.15, 7.16, and similar dependences in [4–6] were used to calculated the maximum unit lasing power $\overline{P}_{||max}$ (Fig. 7.17) and the corresponding optimal excitation pulse repetition frequency f_{mopt}

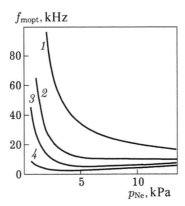

Figure 7.18. Dependence of the optimal excitation pulse repetition frequency f_{mopt} on neon pressure. $1 - r_p = 0.3$ cm, $2 - 0.6$; $3 - 1.0$; $4 - 2.0$.

(Fig. 7.18). It should be noted that in cases where the value $\bar{P}_{ll}$ depends only weakly on $\langle G \rangle$, in the calculations [4–6] f_{mopt} was represented by the value of f_m at which the efficiency was greater.

Comparison of the dependences of $\bar{P}_{ll\,max}$ on p_{Ne} (Fig. 7.17) shows that for a given length of the GDT the lasing power, with other things being equal, increases with increasing GDT radius, which agrees well with the results of experimental studies. It should be noted that according to [4–6], the slight increase of $\bar{P}_{ll\,max}$ with increasing neon pressure is accompanied by increase of the input power, $\bar{Q}_{dl}$ supplied to the discharge to a level above the maximum permissible level (about 35 W/cm). In practice, this would mean stabilizing the output lasing power $\bar{P}_{ll\,max}$ a copper vapour laser with a GDT radius of about 2 cm with increasing neon pressure at about 40 W/m.

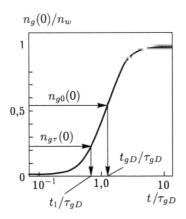

Figure 7.19. Recovery of the concentration of copper atoms in the axial zone of the GDT by their diffusion from the wall of the GDT.

In analyzing the possibility of constructing the repetitively pulsed copper vapour lasers with a repetition rate $f = f_m$ (see Fig. 7.18) and the corresponding lasing characteristics it must be remembered that in the interpulse interval there should be the recovery of not only the prepulse population of the metastable state, but the prepulse concentration of copper atoms in the ground state $\langle n_e \rangle_0$ which strongly decreases for during the excitation pulse.

In [4–6] the lower estimate of the recovery time t_{gD} of the prepulse concentration of copper atoms by their diffusion from the wall of the GDT was obtained from the dependence of the reconstruction of the relative concentration of copper atoms in the axial zone of the GDT on the parameter t/τ_{gDu} ($\tau_{gD} = r_p^2/6D_{Cu}$) calculated from (4.40) with the six first terms taken into account and shown in Fig. 7.19. The calculation of the dependence was carried out for the unit power input to the discharge, $\bar{Q}_{dl} = 25$ W/cm, the parabolic distribution of energy contribution G along the GDT radius, the GDT wall temperature $T_w = 1893$ K and the gas temperature in the axial zone of the GDT $T_g(0) = 3350$ K. It was assumed that the recovery of the concentration of the atoms in the axial zone of the GDT takes place from zero to the value

$$n_{g0}(0)/n_{gw} = T_w/T_{g0} \cong 0.54. \tag{7.39}$$

The sequence of calculation was as follows: first the T_w/T_{g0} ratio and the dependence, shown in Fig. 7.19, were used to determine t_{gD}/τ_{gD}, and then the given neon pressure and the GDT radius were used to calculate the recovery time of concentration $n_g(0)$ from zero to a given value (7.39). The results of calculation of the frequency of recovery of the concentration of copper atoms by their diffusion to the

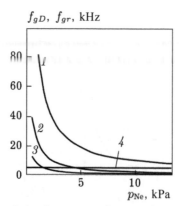

Figure 7.20. Dependence of the frequency of restoration of the copper atoms in the axial zone of GDT due to their diffusion from the wall of the GDT f_{gD} (*1 – 3*) and three-body electron recombination and f_{gr} (*4*) *1 – r_p = 0.3 cm, 2 – 0.6; 3 – 1.0.*

walls of the GDT $f_{gD}^{-1} = t_{gD}$ for different neon pressures and the GDT radii are shown in Fig. 7.20.

This figure also shows the results of estimating the frequency of recovery of the concentration of copper atoms as a result of three-body recombination of the electrons

$$f_{gr} = t_{gr}^{-1} = \beta_e n_e^2. \tag{7.40}$$

Assessment of the ratio (7.40), carried out in [4–6] for $n_e = 10^{14}$ cm^{-3}, $T_e = 4000$ K, yielded a value $f_{gr} \cong 5$ kHz. Strictly speaking, the estimation of f_{gr} by this ratio is a rough estimate of the order of magnitude, but the comparison of estimated values of f_{gr} with the optimal excitation pulse repetition frequencies at a time when the three-body recombination of electrons is the primary mechanism for the recovery of copper atoms in the axial zone of the GDT, shows that the values of these frequencies are in satisfactory agreement. For example, for the GDT diameter of 7 cm and 8 cm at a neon pressure of several kPa the optimal excitation pulse repetition frequency $f_{opt} \leq 5$ kHz [17].

Since the restoration of the concentration of copper atoms in the axial zone of the GDT included two processes, in [4–6] the authors introduced the total frequency of recovery of the concentration of copper atoms in the axial zone of the GDT

$$f_g = f_{gD} + f_{gr}. \tag{7.41}$$

When the excitation pulse repetition frequency is $f < f_g$ the prepulse inhomogeneity in the radial distribution of the concentration of copper atoms in the ground state is close to the minimum determined by nonuniform heating of the gas. When $f > f_g$ the inhomogeneity in the distribution of $n_{g0}(r)$ will increase, leading to a deterioration of the lasing characteristics, so that in terms of restoring the concentration of the copper atoms in the axial zone of the GDT the optimal excitation pulse repetition frequency f_{opt} should not exceed f_g. On the other hand, the interpulse interval should result not only in the recovery of the prepulse concentration of copper atoms, but also the prepulse population of the metastable level $^2D_{5/2}$. Therefore, the optimal frequency f_{opt} should be close to the minimum of the frequency f_{mopt} and f_g (in contrast to f_m the frequency f_g has no optimal value, its optimal value is the maximum possible value). The previously determined values $\bar{P}_l$ and $\bar{Q}_{dl}$ and the values of unit power $\bar{P}_{gl}$ and $\bar{Q}_{dlf}$, corresponding to the minimum of the frequencies f_{mopt} and f_g, determining the value f_{opt}, were recalculated carried in [4–6] using the formulas

$$\bar{P}_{gll} = \frac{f_g}{f_{mopt}} \bar{P}_{ll},$$

(7.42)

$$\bar{Q}_{gdl} = \frac{f_g}{f_{mopt}} \bar{Q}_{dl}.$$

(7.43)

From a comparison of the frequencies f_{mopt} and f_{gD}, f_{gr}, presented in Figs. 7.18 and 7.20, it follows that for a neon pressure of $p_{Ne} \le 2$ kPa and $r_p \le 1$ cm value f_{mopt} significantly exceeds the value f_g defined by (7.41).

In practice however, this difference can be much lower. The fact is that, as noted above, the calculation of f_{gD} in [4–6] was performed assuming the absence of copper atoms in the axial zone of the GDT immediately after the excitation pulse. However, in real terms, after the excitation pulse there is a distribution of $n_{gr}(r)$ the presence of which should lead (especially at low input energies $\langle G \rangle \le 3 \cdot 10^{-4}$ J/cm³) to an increase in the values f_{gD} compared with those given in Fig. 7.20.

For example, when $\langle G \rangle = 2 \cdot 10^{-4}$ J/cm³ the average (over the cross section of the GDT) decrease in the concentration of copper atoms $\Delta \langle n_g \rangle$ per excitation pulse is about $1.6 \cdot 10^{14}$ cm⁻³, which for a parabolic distribution of energy contribution $\langle G \rangle$ along the radius of the GDT corresponds to a decrease in the concentration of the copper atoms in the axial zone $\Delta n_g(0) \cong 3.2 \cdot 10^{14}$ cm⁻³. At $n_{gw} = 10^{15}$ cm⁻³ $n_{g0}(0) \cong 5.4 \cdot 10^{14}$ cm⁻³ the restoration of the concentration of copper atoms in the axial zone of the GDT starts from the value of $n_{gr}(0) = n_{g0}(0) - \Delta n_g(0) = 2.2 \cdot 10^{14}$ cm⁻³ at the time t_{1D}/τ_{gD}. In this case, the interval required for recovery of the copper atom concentration

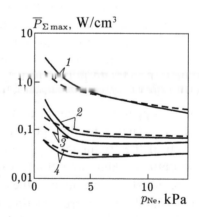

Figure 7.21. Dependence of the total (two wavelengths) maximum specific lasing power $\bar{P}_{max}$ on neon pressure. Solid lines – calculation for frequency f_m, the dot-dashed line – for frequency f_g. *1* – $r_p = 0.3$ cm; *2, 3* – 1.0; *4* – 2.0; *1, 2* – partially homogeneous distribution of the plasma parameters, *3, 4* – inhomogeneous distribution.

to $5.4 \cdot 10^{14}$ cm^{-3} is half of the time interval required for recovery $n_{g0}(0) = 0.54 n_{gw}$ under the condition of complete ionization of copper atoms per excitation pulse over the entire cross section of the GDT. This means that in case of partial ionization of copper atoms per excitation pulse the value f_{gD} is greater than the values f_{gD} shown in Fig. 7.20, and hence the discrepancy between the values f_{mopt} and f_g will be considerably less than is the case in Figs. 7.18 and 7.20 (especially at low pressures, the radii of the GDT and specific energy input).

The possible influence of the finite rate of recovery of the concentration of the copper atoms in the axial zone of the GDT was assessed in [4–6] with the help of (7.42), (7.43). As an example, Fig. 7.17 a also shows dependences of f_g on p_{Ne} and also shows how the linear lasing power $\bar{P}_{\parallel}$ changes if it is calculated for the frequency $f_{opt} = f_g$. It should be borne in mind that if $f_g < f_{mopt}$ the values η_p, shown in Figs. 7.15 and 7.16, and values $\bar{P}_{g\parallel}$ calculated by relation (7.42) represent the bottom estimates of the values of these quantities, since a $f_{opt} = f_g < f_{mopt}$ the prepulse population temperature $\langle T_m \rangle_0$ is below the value used in the calculation, and, consequently, physical efficiency η_p and linear power generation $\bar{P}_{\parallel \max}$ will be higher than the η_p values shown in Figs. 7.15, 7.16, and the values $\bar{P}_{\parallel f}$ calculated by the formula (7.42).

Figure 7.21 shows the calculated dependence of the total (both lines) specific lasing power, roughly defined in [4–6], $\bar{P}_{\parallel \Sigma} = \dfrac{3}{2} \bar{P}_{\parallel}$, on the neon pressure for GDTs of different radii and for different distributions of the prepulse plasma parameters. The calculation of $\bar{P}_{\parallel \Sigma}$ was performed in [4–6] for the conditions $f_{opt} = f_{mopt}$ and for the case $f_{opt} = f_g$.

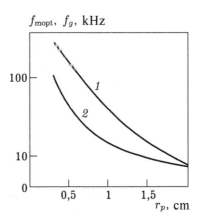

Figure 7.22. Dependence on the excitation pulse repetition frequency f_{mopt} (*1*) and f_g (*2*) in the copper vapour laser with a neon pressure of 0.67 kPa.

From a comparison of the dependences shown in Fig. 7.21 it follows that the difference between the frequencies f_{mopt} and f_g has a strong influence on the value $\bar{P}_{\Pi\Sigma}$ only at GDT radii $r_p < 1$ cm and low neon pressures. But regardless of the excitation pulse repetition frequency for which $\bar{P}_{\Pi\Sigma}$ was calculated, the results show that use of GDTs of small radii and low neon pressures can increase specific average lasing power $\bar{P}_{\Pi\Sigma}$ to 1 W/cm^3 and above.

The biggest difference between the values of f_{mopt} and f_g was observed at low neon pressures. Figure 7.22 shows as an example the dependence of the optimal frequency f_{mopt} and the corresponding frequency f_g on the of GDT radius at a neon pressure of 0.67 kPa. Because of the previously mentioned reasons, the true values of the excitation pulse repetition frequency will be situated at least higher than the values f_g. The calculated results shown in Fig. 7.22 indicate that the use of the GDT with radii less than 0.3 cm and at neon pressures of $p_{Ne} \sim 0.7$ kPa makes it is possible to create a pulse-periodic copper vapour lasers operating at excitation pulse repetition frequencies in the hundreds of kHz.

This conclusion, made more than 20 years ago, was confirmed in practice. Recently, copper bromide vapour lasers were built with an with a excitation pulse repetition frequency of 300–400 kHz of [26,27].

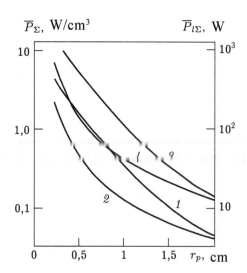

Figure 7.23. Dependence of the total specific lasing power $\bar{P}_\Sigma$ of a copper vapour laser (*1, 2*) and its corresponding total average lasing power $\bar{P}_{l\Sigma}$ (*3, 4*) taken from the volume of 314 cm^3, on the radius of the GDT Neon pressure 0.67 kPa, *1, 3* – calculation for frequency f_{mopt}, *2, 4* – for frequency f_g.

Figure 7.23 shows the dependence of $\bar{P}_{l\Sigma}$ on the radius of the GDT for $f_{opt} = f_{mopt}$ and $f_{opt} = f_g$ and the neon pressure of 0.67 kPa. There is also the dependence of the power of the laser with a total volume of 314 cm³ on the GDT radius. This volume is equal to the volume of the GDT with a radius 1 cm and a length of 100 cm. From this relation it follows that for the parallel operation of 11 discharge tubes with a radius 0.3 cm and a length of 100 cm the resultant total lasing power is 300 W. To ensure that the physical efficiency is 8%, it is necessary to operate at neon pressures of 1.3 kPa and lower, at a specific energy input of $\langle G \rangle \cong 1.5 \cdot 10^{-4}$ J/cm³ and provide heating of the GDT at unit powers supplied to the discharge on the level of several watts per centimeter.

7.9. Comparison of the results of calculations [4–6] and experiment

The results of calculation of the excitation pulse repetition frequency f_{mopt} and unit power $\bar{Q}_{dl}$ inputted to the discharge from the given values of the specific energy $\langle G \rangle$, the radius of the GDT r_p and the neon pressure p_{Ne} were compared in [4–6] with the results of experimental studies which achieved, at the time of publication of [4,5], the highest lasing power and efficiency.

In those cases when these works were missing the data on the shape and duration of the voltage and current pulses it was assumed in the calculations that the discharge received half the power required from the rectifier. If the experimental work cited specific energy output $\langle W \rangle$ and practical efficiency η_r, the energy input $\langle G \rangle$ was calculated by the ratio of $\langle G \rangle = 0.5 \langle W \rangle \eta_r$. In cases where the values of the average lasing power $\bar{P}_l$ and η_r were quoted, the power introduced into the discharge was initially determined

$$\bar{Q}_d = 0.5 \bar{P}_l / \eta_r,$$

and then specific energy input

$$\langle G \rangle = \frac{\bar{Q}_d}{fV_p}.$$

In the event that the neon pressure in the experiment coincided with one of the pressure values used in the calculations [4–6], the value sof f and $\bar{Q}_d$ were determined from the known values of specific energy input with dependences similar to those presented in Figs. 7.15, 7.16, for the neon pressures used in the experiment. Since the neon pressure in all experiments considered in [4–6] was in the range of $p_{Ne} \leq 4$ kPa, f_{mopt} and $\bar{Q}_{dl}$ were calculated for the values of the prepulse population temperature $\langle T_m \rangle_0 = 3200$ K at which, according to results of the calculations, this pressure range is characterized by the maximum lasing power $\bar{P}_{ll}$. It should be noted that under the assumptions adopted in the calculations [4–6], the value f_{mopt} does not depend on the nature of the distribution of prepulse plasma parameters and for given values of r_p and p_{Ne} is determined only by the magnitude of the specific energy $\langle G \rangle$ and temperature of the prepulse population $\langle T_m \rangle_0$.

In the event that the neon pressure in the experiment was very different from the values of the pressure used in the calculations of the last two values are chosen, one more and one less than the experimental pressure. For each of these pressures for a given energy input $\langle G \rangle$ described earlier were calculated frequency f_{mopt}. The calculation of the required quantity f_{mopt}, corresponding to the experimental pressure of neon was carried out by linear interpolation between two neon pressures and their corresponding frequencies f_{mopt}.

The results of calculations and measurements of f and $\bar{Q}_{dl}$ presented in Table 7.2, in good agreement with each other, which once again confirms the efficiency of the method used in [4–6] for calculating the optimal parameters of the pulse-periodic copper vapour lasers.

Because the optimal operating conditions of repetitively pulsed CVLs in different studies were chosen empirically, it is interesting to compare the value of the limiting average lasing power $\bar{P}_\Sigma$ determined by the calculation [4–6] and achieved in the most successful experiments.

Table 7.3 compares the values of specific output power $\bar{P}_\Sigma$, given by the calculation at a concentration of copper atoms near the wall of the

Table 7.2. Comparison of the results of calculations [4–6] and the experimental data on the excitation pulse repetition frequency and linear power introduced into the discharge in copper vapour lasers

Type of data	r_p, cm	p_{Ne}, kPa	$\langle G \rangle$, 10^{-4} J/cm^3	f, kHz	$\bar{Q}_{dl}$ W/cm	Literature
Experiment	1	2.7	2.8	8	7.0	[28]
Calculation		2.7		9	7.9	
Experiment		2 – 4		12.5	26	
Calculation	1	2	6.7	11	23	[29]
Calculation		3		9	19	
Calculation		4		6.7	14	
Experiment	1	3.46	8.3	6	15.5	[29]
Calculation		4		6	15.6	
Experiment	1.9	1.7 – 2	4.8	5	27	[30]
Calculation		2		4.4	24	

Table 7.3. Comparison of the results of calculations [4–6] and experimental data on the specific lasing power of CVLs

Type of data	r_p, cm	p_{Ne}, kPa	$\bar{P}_l$, W	V_p, cm³	$\bar{P}_\Sigma$, W/cm³	Literature
Calculation	1	2	-	-	0.16	
Experiment	1	2 – 4	23	230	0.10	[29]
	1	2.7	9.75	83	0.12	[28]
	1	–	38	314	0.12	[18]
	0.75	–	15	125	0.12	[31]
Calculation	1.4	4	-	-	0.13	
Experiment	1.4	4	43.5	500	0.087	[29]
Calculation	2	2			0.054	
Experiment	1.9	1.7 – 2	47	1500	0.031	[30]

GDT n_{gw} = 2 · 10¹⁵ cm⁻³ and achieved in various experiments for GDT radii of 1, 1.4 and 2 cm of the desired calculation of the values $\bar{P}_\Sigma$ held in [4–6] as follows. The known values of the neon pressure and the radius of GDT using the dependences presented in Figures 7.15, 7.16, and they are similar for partially homogeneous distribution of prepulse parameters determined by the specific value of the average laser power on the green line $\bar{P}$ and the corresponding value of $\bar{P}_\Sigma$ = 1.5 $\bar{P}$.

For example, according to calculations [4–6], with r_p = 1 cm and p_{Ne} = 2 kPa the linear output power is maximum at $\langle T_m \rangle_0$ = 3200 K and $\langle G \rangle$ = 1.5 · 10⁻⁴ J/cm³ is $\bar{P}_{ll}$ = 34 W/m, which corresponds to the specific mean lasing power of $\bar{P}$ = 0.11 W/cm³. In view of the lasing on the yellow line $\bar{P}_\Sigma$ = 0.16 W/cm³.

As seen in Table 7.3, the highest values of the specific average lasing power $\bar{P}_\Sigma$ realized in experiments with the GDT radius of 1 cm, are in the range from 0.1 to 0.12 W/cm³, that is in good agreement with the calculated value.

According to the data presented in Table 7.3, with increasing GDT radius the calculated values $\bar{P}_\Sigma$ are higher than the experimental increases and for r_p = 2 cm they are up to 1.7 times higher. The above pattern can be explained by the fact that the increasing GDT radius should increase the heterogeneity in the radial distribution of prepulse population temperature of metastable levels. In addition, a role in reducing the experimental average lasing energy in comparison with the calculated results can be played by the mismatch of the discharge and the discharge circuit impedance which is not taken into account in the calculations and which increases with increasing GDT radius and

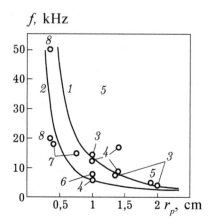

Figure 7.24. The dependence of the excitation pulse repetition rate on the radius of the GDT. Lines – calculation: *1* – 2 kPa, *2* – 4 kPa, points – experimental data (see Table 7.4).

Table 7.4. Experimental data used compared with the results of calculations [34] pulse repetition frequencies excitation in Fig. 7.24

Point number	r_p, cm	f, kHz	p_{Ne}, kPa	$\bar{P}_l$, W	Literature
3	1	13 – 16			
	1.35	7 – 8	1.9	-	[32]
	2	4 – 5			
4	1	6	3.5	14.5	
	1	12.5	–	2.3	[29]
	1.4	9	–	36.	
	1.4	16.7	4	43.5	
5	1.9	5	1.9	36	[30]
6	1	8	2.7	9.75	[28]
7	0.4	18	-	6	[31]
	0.75	15	-	15	
8	0.35	20	2	3.5	[33]
	0.35	50	1.3	1.5	

reduces the efficiency of energy transfer from the storage capacitor to the discharge, .

As can be seen from Tables 7.2 and 7.3 the neon pressures, most commonly used in obtaining the highest lasing power, are in range from 2 to 4 kPa. Figure 7.24 compares the results of the optimum frequency f_{mopt} for two pressures (2 and 4 kPa) and the experimental results (see Table 7.4). One can see good agreement between the calculated

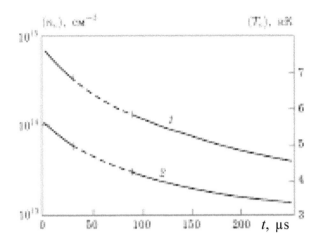

Figure 7.25. Time dependence of the concentration and temperature of the electrons calculated from the experimental conditions [29]. Dot-and-dash curve – the zone of the stationary operating mode of the laser.

results with the results of most experiments. The exception are the experimental results [29], in which average output power of 43.5 W was recorded for a short period of time.

The authors of [34] analyze in more detail the laser operating mode, investigated in [29]. Figure 7.25 shows the calculated relaxation of the electron concentration and temperature in the interpulse interval for the conditions of [29] (GDT diameter 2.8 cm, the pressure of neon 4 kPa, excitation pulse repetition rate 16.7 kHz). The same figure shows the changes of $\langle n_e \rangle$ and $\langle T_e \rangle$ corresponding to the steady-state (sustainable) mode of the laser in [29], at the level of power consumed by a rectifier of 4.5 kW. When calculating the range of the electron concentration from equation (7.34) it was assumed that because of the energy losses in the thyratron only 50% of the energy stored in the storage capacitor is supplied to the discharge.

As follows from the time dependence of the electron temperature (Fig. 7.25), in the steady-state operation of the laser [29] the electron temperature before the next excitation pulse is about 4200 K, so that in this mode the average lasing power should not be large. This conclusion is consistent with the results of [29] according to which during heating of the laser the average lasing power initially increases and then, as the laser changes to the stationary mode of operation, decreases.

More than 40 years have passed since the beginning of investigations of the copper vapour lasers. 4 monographs were published in this period, including [11], dealing with the results of development of the

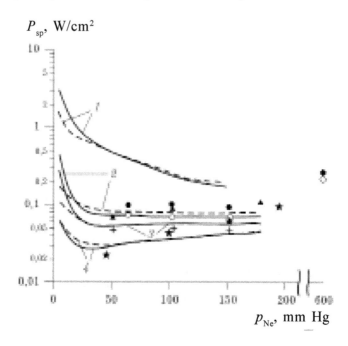

P_{sp}, W/cm²

P_{Ne}, mm Hg

Figure 7.26. Calculation of the total (at two wavelength) power $\bar{P}_{\Sigma}$ of different copper vapour lasers in relation to the neon pressure: solid lines – for the frequency equal to f_{mopt}, the dotted line – f_g; 1 – d_p = 6 mm, 2, 3 – d_p = 2 cm, 4 – d_p = 4 cm. 1, 2) partially homogeneous distribution of the prepulse plasma parameters, 3, 4 – inhomogeneous distribution of the prepulse plasma parameters. Experiments [11]: ◆ – LT-1Cu d_p = 0.7 cm, thyratron modulator (TM), storage capacitor C_S = 1500 Pf; ◇ – LT-1.5Cu d_p = 0.7 cm, TM, C_S = 1500 Pf; O– GL-201, d_p = 2 cm, TM, C_S = 1650 Pf; ● – at C_S = 3300 Pf, ★– the same, TM with the transformer doubling of voltage (see reference [11]; ▲ – LT-40Cu d_p = 2 cm, TM with doubled voltage, C_S = 1000 tF; + – LT50 Cu, d_p = 3.2 cm, TM with doubled voltage, C_S = 1000 pF.

industrial copper vapour lasers with the GDT of different diameters d_p and the radiation power of 1–100 W. Nevertheless, the question whether the limiting energy characteristics of these lasers (mean specific lasing power $\bar{P}_{\Sigma}$ and the efficiency η_p) have been achieved at this time has not been answered. Without this answer, it is not possible to answer unambiguously another important question of the methods of further improvement of the energy characteristics of the CVL.

In [10] the authors answered these questions on the basis of comparing the results of theoretical studies of the CVL [4– 6] and the energy characteristics of the industrial CVL available at the time [11]. Figure 7.26 compares the calculated power [4–6] (Fig. 7.21) and the power $\bar{P}_{\Sigma}$ recorded in practice [11]. According to the comparison of results, the resultant values $\bar{P}_{\Sigma}$, obtained in the industrial lasers indicated in the caption for the figure are close to the maximum power.

Table 7.5. Temperature dependence of the magnitude of dissociation of molecules of HBR, HCl and H_2

Molecule	T, кK n, cm^{-3}	1.5	1.7	1.9	2.1	2.3	2.5	2.7	2.9
HBr	10^{15}	0.01	0.08	0.26	0.57	0.84	0.96	0.99	1.00
	10^{16}	0.004	0.02	0.09	0.24	0.48	0.73	0.89	0.96
HCl	10^{15}	0.001	0.009	0.04	0.14	0.35	0.64	0.85	0.95
	10^{16}	0.0004	0.003	0.01	0.05	0.13	0.28	0.50	0.72
H_2	10^{15}	0.02	0.02	0.07	0.24	0.54	0.82	0.95	0.98
	10^{16}	0.006	0.005	0.02	0.08	0.22	0.45	0.71	0.88

This means that as a result of the primarily empirical optimisation of the operating conditions of the repetitively pulsed copper vapour lasers the developers of the industrial CVL [11], listed in the caption for Fig. 7.21, obtained for all these lasers the specific lasing power closed to maximum.

Thus, according to theoretical studies in [4–6] it was found that for high average lasing power of the repetitively pulsed copper vapour lasers with a single GDT it is rational to use GDTs with a diameter of 5 cm or more.

To increase the specific lasing power and efficiency of the repetitively pulsed copper vapour lasers 2–3 times higher than tyhe value achieved by the mid-1980s, it is necessary to work with low-pressure neon with GDT radius of 0.3 cm or less.

To construct repetitively pulsed copper vapour lasers with the physical efficiency of 8% GDTs with a diameter of 4 mm or less. The total unit lasing power will be approximately 20 W/m. Simultaneous use of GDTs with a diameter of 4 mm and 1 m long will create modular designs of the copper vapour lasers with a total average output power at 200 W.

The information, which is of considerable importance for the development of repetitively pulsed CVL with the modified kinetics, was published in [13]. In this study, the authors presented the results of calculations (Table 7.5) of the temperature dependence of the degree of dissociation of the diatomic molecules used as the addition to the active medium of the copper vapour lasers. According to the data in the table, the pyrolysis of the molecules HBr, HCl and H_2 in the copper vapour lasers with the gas temperature T_g in the axial zone of the GDT at 3000 K is the main mechanism of dissociation of these molecules into the atoms forming them. This conclusion is used to explain the data obtained in [11] according to which the increase of the GDT diameter increases the efficiency of the effect of hydrogen on the output characteristics of the copper vapour lasers. This increase is caused by the increase of the inhomogeneity of the distribution of the prepulse parameters with the increasing diameter of the GDT and, consequently, increasing T_g in the axial zone of the GDT resulting in the increase of the degree of dissociation of H_2.

The data presented in Table 7.5 indicate that to realise all the advantages of the CVL with the modified kinetics it is efficient to use the GDTs with diaphragms. The gas temperature in the discharge zone, restricted by the diaphragms, should be approximately 3000 K, and the working temperature of the GDT wall should be maintained by an external heater.

7.10. Self-consistent models, taking into account radial inhomogeneities

In the model, described above, the kinetic equations describe the period of the excitation pulse. The initial distribution of the radius of the particles was determined *apriori* on the basis of the theoretical and experimental studies. The reproduction of the radial profiles of the distribution of the particles from postal pulse was assumed. More accurate models require a large volume of computing studies and, in contrast to the model described above, they relate to the calculations of only individual variants of the copper vapour lasers.

The first numerical model of the copper vapour laser taking into account the radial inhomogeneites was developed in 1984 [24]. The inclusion of the skin effect in the model made it possible to investigate the radial evolution of the strength of the electric field in the plasma. The duration of penetration of the electric field from the wall region to the central region is estimated by $\tau \approx 2.7r^2$, where τ is the duration of penetration in ns, r is the tube radius in cm [24] and is comparable with the duration of the loading kinetic processes at the tube diameters

Table 7.6. Operating mode of the CVL [35] with optimum powder (6 W)

Active zone length	86 cm
Internal tube diameter	1.8 cm
Neon pressure	4 kPa
Wall temperature	1790 K
Charge voltage	9 kV
Pulse repetition frequency	7.85 kHz

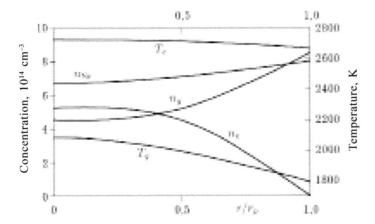

Fig. 7.27. The radial profiles of the plasma parameters prior to the application of the excitation pulse.

greater than 4 cm. The calculations by this model show that in the theoretical aspect the laser tubes should be divided into narrow tubes (with the diameter smaller than 2 cm), where the skin effect of the charge can be ignored, and wider tubes where the effect is considerable.

The self-consistent model was developed for the narrow tubes [8]. The model takes into account a relatively large number of the atomic levels and also a very large number of the processes of transfer of particles and their energies in the radial direction. This is carried out using the latest data for the sections and rate constants of the elementary processes. The electrons in the model are divided into two groups with the energy lower and higher than the first excitation potential of the neon of 16.7 kV. The calculations were carried out taking into account 15 atomic and ionic states, three temperatures, two lengths of laser radiation and 70 elementary processes. The model included the direct and stepped ionisation, electron–ion recombination, Penning collisions, diffusion of the particles and heat transfer. The main parameters of the calculations were selected for the operating mode of the copper vapour laser with the optimum power (6 W [35] and are presented in Table 7.6.

As a result, it was shown that the decrease of the copper atom concentration at the axis of the tube prior to the application of the excitation pulse is greater than the value calculated on the basis of temperature in accordance with the equation of state of the ideal gas (Fig. 7.27). In contrast to this, the neon atom concentration corresponds to the gas temperature, as determined assuming the pressure independent of radius. This effect may be attributed to the radial cataphoresis of the metal atoms in the afterglow when the large part of the copper atoms

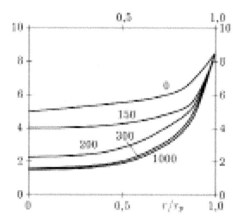

Fig. 7.28. Radial profile of the concentration of copper atoms in the ground state at different moments of time (indicated at the curve in ns).

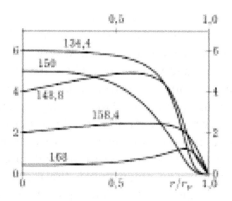

Fig. 7.29. The intensity of radiation at a wavelength of 5106 Å at a different moments of time (indicated at the curves in ns).

remain ionised and move in the direction to the walls as a result of ambipolar diffusion. The ions, neutralized on the wall, move towards the centre of the tube at lower velocities in accordance with the normal thermal diffusion. The radial profile of the electron temperature is relatively smoother in comparison with the gas temperature profile as a result of the high thermal conductivity of the electrons. At the end of the afterglow period, the electron temperature is several hundreds of degrees higher than the gas temperature. Evidently, this is associated with the superelastic collisions with the excited copper atoms.

The temporal radial behaviour of the populations of the ground states of the copper atoms in the first microsecond after activating the thyratron is shown in Fig. 7.28. The decrease of the copper atom concentration by more than 50% is noted in the large part of the tube radius with small changes with time in the early stage of the afterglow. The radial profile of the ions reflects the profile of the concentration of the copper atoms. At 1000 ns the electron concentration is higher than the ion concentration in the ground state. This is associated with the fact that a large part of the concentration of the ions is attributed to the metastable states of the ions (usually approximately 20% at the axis). The electron temperature is always lower at the wall in comparison with the axis. This may be explained by the increase of the frequency of inelastic collisions of the electrons with the copper atoms and the associated energy losses because the copper atoms in the ground state are distributed mostly at the wall.

Figure 7.29 shows the radial profiles of the intensity of radiation at different moments of time. The graph indicates that the generation of radiation starts in the vicinity of the tube axis and moves with time to the walls of the tube. This is in qualitative agreement with the previously obtained results.

References

1. Batenin V.M., et al., Doklady AN USSR. 1981. V. 256, No. 4. P. 831–834.
2. Batenin V.M., et al., The limiting parameters of self-heating lasers and their conditions of implementation. X Meeting on Spectroscopy (Population inversion and lasing on transitions of atoms and molecules): Abstracts. Tomsk: Tomsk University Publishing House, 1981.
3. Galkin A.F., et al., Teplofiz. Vysokikh Temperatur. 1983. V. 21, No. 5. P. 976–981.
4. Galkin A.F., Klimovsky I.I., Effect of radial inhomogeneity of the plasma on the characteristics of the generation of pulse-periodic copper vapour lasers with longitudinal discharge. Preprint IVTAN No. 5-220. Moscow, 1987, 40.
5. Galkin A.F., Klimovsky I.I., The optimal parameters of the pulse-periodic copper vapour lasers in inhomogeneous distribution of plasma parameters over the cross section of the GDT, Preprint IVTANNo. 5–228. Moscow, 1987.
6. Galkin A., Klimovskii I., Computed model of copper-vapour laser with the average specific output power above 1 W/cm³. Metal Vapour Lasers and Their Applications: CIS Selected Papers, G.G. Petrash, Editor, Proc. SPIE 2110, 1993. P. 90–99.
7. Batenin V.M., et al., Lasers on self-contained metal atom transitions. Moscow, Nauchnaya kniga, 1988. 544.
8. Carman R.L., Brown D.J.W., Piper J.A., IEEE J. Quant. Electronics. 1994. Vol. 30, No. 8. P. 1876–1895.
9. Carman R.J. Computer modeling of longitudinally excited elemental copper vapour lasers. Pulsed Metal Vapour Lasers. Proceedings of the NATO Advansed Research Workshop on Pulsed Metal Vapour Lasers – Physics and Emerging Applications in Industry, Medicine and Science, St. Andrews, U.K., Aug. 6-10, 1995. C.E. Little and N.V. Sabotinov, Eds. Dordrecht: NATO ASI Series, Kluwer Academic Publishers, 1996. P. 203–214.

10. Klimovsky I.I., Malikov M.M., in: Proc. All-Russian Symposium LPM-2008, P. 26

11. Grigor'yants A.G., Kazarian M.A., Lyabin N.A., Copper vapour laser: design characteristics and applications. Moscow, Fizmatlit, 2005.

12. Garelina S.A., Klimovskii I.I. Al'ternativnaya energetika i ekologiya. 2007. No. 5 (49).

13. Vasil'eva N.N., Garelin S.A., Klimovsky I.I., see Ref. 10, P. 45.

14. Dyachkov L.G. Kobzev G.A., Zh. Tekh. Fiziki.1987. V. 48, No. 11. P. 2343–2346.

15. Mnatsakanyan A.Kh., et al., Kvat. elektronika. 1978. T. 5, No. 3. P. 597–602.

16. Anderson R.S., et al., IEEE J. Quantum Electron. 1981. Vol. QE-17, No. 12. P. 50

17. Grove R.E., Laser focus. 1982. Vol. 18, No. 7. P. 45–50.

18. Kirilov A.E., et al., Metal-vapour lasers for atmospheric research. Measuring instruments for investigating the parameters of surface layers of the atmosphere. Tomsk Institute of Atmospheric Optics, USSR Academy of Sciences, 1977. P. 59–79.

19. Elaev V.F., et al., Effect of radial inhomogeneity of the discharge plasma parameters on the generation of a copper laser. Effective gas-discharge lasers on metal vapors. Tomsk Institute of Atmospheric Optics, USSR Academy of Sciences. 1978. P. 189–196.

20. Batenin V.M., et al., Teplofiz. Vysok. Temperatur. 1979. V. 17, No. 1. P. 208–209.

21. Soldatov A.N., et al., Izv. VUZ. Fizika. 1980. V. 23, No. 10. P. 38–43.

22. Burmakin V.A., et al., Kvant. Elektronika. 1978. V. 5, No. 5. P. 1000–1004.

23. Klimkin V.M., et al., Investigation of the stability limits of a gas discharge with a high pulse repetition DC frequency. Effective gas-discharge lasers on metal vapors. Tomsk Institute of Atmospheric Optics, USSR Academy of Sciences. 1978. P. 116–132.

24. Kushner M.J., Warner B.E., J. Appl. Phys. 1984. Vol.54, No. 6. P. 2970–2982.

25. Sergeev V.M., Teplofiz. Vysokikh Temperatur. 1977. V. 15, No. 1. P.199–201.

26. Yevtushenko G.S., et al., Kvant. Elektronika. 1999. V. 28. No. 3. P. 220–222.

27. Guharev F.A., et al., Izv. Tomsk. Polit. Univ. 2008. V. 312. No. 2. P. 106–107.

28. Bokhan P.A., Gerasimov V.A. Kvant. elektronika. 1979. V. 6. No. 7. P. 451–455.

29. Isaev A.A., et al. Kvant. Elektronika. 1977. V. 5. No. 7. P. 1413–1417.

30. Bokhan P.A., Zh. Tekh. Fiziki. 1981. V. 51. No. 1. P. 206–209.

31. Isaev A.A., et al., Pis'ma ZhETF. 1972. V. 16. No. 1. P. 40–42.

32. Bokhan P.A., et al., Kvant. Elektronika. 1980. V. 7. No. 6. P. 1264–1269.

33. Alaev M.A., et al., ibid, 1976. V. 3. No. 5. P. 1134–1136.

34. Klimovskii I.I. Lasers on self-contained metal transitions. Dissertation. Moscow, 1992.

35. Brown D.J.W., Experimental study of excited state densities in a copper vapour laser. PhD dissertation. Univ. New England, Armidale, Australia, 1988.

Simulation of copper vapour lasers. Kinetically enhanced lasers

8.1. Introduction

Copper vapour lasers (CVL) have been known for more than 40 years [1–4]. Modelling of the laser was conducted in a large number of papers. Nevertheless, many fundamental issues remain unresolved to this day and the solution of some of them there has been suggested only in recent times. In addition, the development and improvement of the lasers has been continuing. For example, in recent years many papers have been devoted to kinetically enhanced copper vapour lasers, whose lasing characteristics are improved by introducing small amounts (in percentage) of impurities in the active medium.

This chapter in some ways fills this gap. The rationale for the use of the Maxwellian distribution function of electrons (8.3) is presented, the restriction of the repetition rate of laser pulses (8.4) is considered, and the mechanisms to improve the lasing characteristics of the modified kinetics lasers with additions of hydrogen (8.5.1) and hydrogen chloride (8.5.2) to the active medium are described. We also consider the question of the level of power supplied to the input of the signal amplifier sufficient for complete removal of the inversion of the amplifier in high-quality radiation (8.6), which is important with regard to issues of using the CVLs in laser isotope separation (AVLIS).

8.2. The kinetic model

Studies [5–7] provide a detailed kinetic model of the active medium, which describes the time variation of the volume-averaged values of the populations of the levels of the copper and neon atoms, the density of the copper and neon ions, the electron temperature and laser radiation intensity on the green and yellow lines of copper. This model is constructed in the traditional way (see, e.g. [8–15]). Below we present some basic information about the model.

Kinetics of excited states

The basis of model building is the model considered in [16]. The kinetic model discussed below includes the kinetic equations of the balance of the populations of the nine states of the copper atom: $Cu(4^2S_{1/2})$, $Cu(4s^2\ ^2D_{5/2})$, $Cu(4s^2\ ^2D_{3/2})$, $Cu(4^2P_{1/2})$, $Cu(4^2P_{3/2})$, $Cu(5^2S_{1/2})$; for two levels, respectively linking three $Cu^* = (Cu(^4P^0,\ ^4D^0,\ ^4F^0))$ and four $Cu^{**} = (Cu(5^2P_{3/2},\ 5^2P_{1/2},\ 4^2D_{5/2},\ 4^2D_{3/2}))$ closely spaced excited levels, as well as for the ground state of the copper ion Cu^+. We take into account the ground and first excited states of the neon atom: Ne, Ne^*, as well as the ground state of the neon ion Ne^+.

To obtain the rate constants of the most important reactions (e.g., excitation and de-excitation of the levels of working transitions), the model used the standard Maxwellian energy distribution function. This approach is justified below in 8.3.

Reactions involving neon and their rates corresponded to those which we used earlier for modelling XeCl [17,106], ArF [18] and XeF [19] lasers (see also reviews [10,13–15]) and KrCl lamps [20–22] in mixtures containing a neon buffer gas (see also the review in [23]).

The influence of ambipolar diffusion on the electron density is taken into account by adding the appropriate term to the balance equation of

$$\frac{dN_e}{dt} = ... - \frac{0.929\mu T_e}{R^2\sqrt{T_g}} N_e + ...$$

Here $N_e = N_{iCu} + N_{iNe}$ is the electron density (in cm^{-3}), N_{iCu}, N_{iNe} is the density of the copper and neon ions; t is time (in seconds) $\mu = 7$ cm^2/(V·s) is the mobility of the copper ions in neon; T_e, T_g are the electron and gas temperatures (in eV); R is the radius of the tube (in cm).

Radiative transitions

The probabilities of spontaneous radiative transitions of excited states of copper were calculated according to [16,24]. For transitions to the ground state the re-absorption was taken into account by introducing radiation escape factor (see, e.g. [9]).

In accordance with the experimental widths of the transitions $Cu(P_{3/2})$ → $Cu(D_{5/2})$ ($\Delta v = 6$ GHz, 510.6 nm) and $Cu(P_{1/2})$ → $Cu(D_{3/2})$ ($\Delta v = 9$ GHz, 578.2 nm), given in [16], we used the following values of the gain cross sections for laser transitions on the lines with a wavelength $\lambda =$

510 and $\lambda = 578$ nm: $\sigma_{ph}(510.6\ nm) = 3.37 \cdot 10^{-14}$ cm^2, $\sigma_{ph}(578.2\ nm) = 2.81 \cdot 10^{-14}$ cm^2.

Description of laser radiation intensity

We considered two equations for the intensity $I(\lambda)$ of each lasing line in the zero-dimension approximation (for more detail see [10–15])

$$\frac{dI(\lambda)}{dt} = \sigma_{ph}(\lambda)c(N_b - \frac{g_a}{g_b}N_a)I(\lambda) - \gamma I(\lambda) - wI(\lambda) + AN_b hvc\frac{\Delta\Omega}{4\pi}, \quad (8.1)$$

where N_b, N_a are the populations of the upper and lower laser levels, $\gamma = \frac{c}{2l}\ln\frac{1}{r_1 r_2}$ is the inverse photon lifetime in the resonator, c is the speed of light, l is the length of the excited medium, r_i are the reflection coefficients of the resonator mirrors; w is the coefficient of radiation losses in the resonator due to absorption in optical elements, and the last term describes the lasing seeding from spontaneous emission with the cavity solid angle $\Delta\Omega$, h is the Planck constant, $v = c/\lambda$ is the photon frequency.

To study the inversion mode we also consider the weak signal amplification mode, in which the stimulated emission of radiation does not affect the level populations.

Heat balance

We used the heat balance equation for electrons in the form of:

$$\frac{d}{dt}\left(\frac{3}{2}N_e T_e\right) = -Q_{iCu} - Q_{iNe} - Q_{wall} - Q_{\Delta T} + \rho j^2(t). \quad (8.2)$$

Here Q_{iCu}, Q_{iNe} are the power densities spent on ionization and excitation of respectively copper and neon (these values are expressed in terms of the population of the excited atomic states [8–12], and they take into account the processes of excitation and de-excitation of the levels considered in the model);

$$Q_{wall} = \frac{5.41 \cdot 10^4 T_e^{1.5}}{R^2 \sigma_{eNe}}\frac{N_e}{N_{Ne}}\ W/cm^3,$$

is the power density of the heat sink on the wall (T_e is in eV, σ_{eNe} (T_e) is the transport cross section for elastic collisions of the electron with the neon atom – in 10^{-16} cm^2 (it depends only weakly on temperature in the range 2 eV and is approximately equal to $1.5 \cdot 10^{-16}$ cm^2 [25]), the radius of the tube R is in cm;

$$Q_{\Delta T} = 2 \cdot [(m_e/m_{Ne}) \cdot k_{Ne} \cdot N_{Ne} + (m_e/m_{Cu}) \cdot k_{ei} N_e] \cdot N_e \cdot (T_e - T_g)$$

is the power density spent on cooling of electrons due to elastic collisions with neon atoms and copper ions; k_{Ne}, k_{ei} are the rates of elastic collisions of electrons with the neon atoms and ions; N_{Ne} is the density of the neon atoms; m_e is the electron mass, m_{Ne}, m_{Cu} are the masses of the neon and copper atoms;

$$\rho = \frac{1}{\sigma} \approx \frac{7.456 \cdot 10^{-2}}{\sqrt{T_e}} \left(\frac{1}{T_e} + 5.181 \cdot 10^{-3} \frac{N_{Ne}}{N_e} T_e \right) \text{ohm} \cdot \text{cm}$$

is the resistivity of the plasma [26] (σ is the conductivity of the plasma), j is current density.

The balance equation for gas temperature was not considered because the integration time of the equations does not exceed the interpulse interval for which the gas temperature does not change.

Description of the energy input in the active medium

The electric current density j through the active medium in the model is described in two ways. In one case, the kinetic equations for the concentrations of various reagents and the balance equation for the electron temperature are solved with the Kirchhoff's equations for the circuit. In another case the experimental dependence of current on time is used directly. This was especially important when testing the kinetic model using the experimental data.

The self-consistent solutions of non-stationary equations for the concentrations of various reagents, the balance equation for the electron temperature, radiation intensity, and (if necessary) the Kirchhoff equations for the circuit (a total of 22 equations) was obtained by the PLAZER programs package [10,13–15]. A total of 107 kinetic reactions are taken into account in the model.

8.3. The electron distribution function

In the overwhelming number of papers the CVLs were modelled considering the Maxwellian distribution function of electron energy (MDFEE) [1–4, 27]. In most cases, a good agreement with experimental results was achieved (see also [5–7, 28–31]). In this section we consider the question of the legitimacy of using MDFEE [32].

Attempts to take into account DFEE in the simulation of CVLs were made in [16, 27,33–36]. In [33,35], the DFEE was calculated for the quasi-stationary conditions. In [16,34] calculations of DFEE were carried out in conjunction with the solution of kinetic equations for the concentrations of various reagents of plasma and Kirchhoff's equations for the excitation circuit, however, the shape of the DFEE was chosen as a bi-Maxwellian distribution form (see also [37–39]). The simulation results [33–35] have shown that the use of MDFEE usually leads to an overestimation of the calculated rates of excitation and ionization of the buffer gas by more than one order of magnitude. In [36] it was mentioned that the reaction rates were calculated on the basis of the calculated DFEE, but no data were published on the method of calculation or the results obtained. It was only mentioned that when calculating the rates of electron excitation for all transitions in a copper atom it is allowed to use MDFEE, while the reaction rate constants for electron excitation of the neon atoms in the calculations with are 2–3 times overestimated relative to the actual values. As noted in the same paper, in the calculation of the kinetics the rate constants of reactions calculated on the basis of MDFEE.

Thus, the form of the actual dependence of the electron distribution function on energy and time for the CLVs is not known. Note that there are common methods for calculating the DFEE [40], in addition, the DFEE was calculated with respect to the modelling of other (non-CVL) types of lasers [41].

Below are the results of calculation of the distribution function of electrons on the energy and time for the CVLs. The form of the DFEE was calculated for both the standard CVLs and the kinetically enhanced CVLs (see section 8.5 of this chapter) [32]. Additives were H_2, H_2+HCl, Cs, as the most common types of doping used in the experiment. The results of [16,33–35] to model the DFEE are briefly described, as well as the distribution function used there – a bi-Maxwellian and Druyvesteyn. The method for calculating the DFEE [32] and the results obtained are described.

Bi-Maxwellian DFEE

In [16,34] the kinetics of lasers on pure copper vapours and with the addition of hydrogen was studied using the bi-Maxwellian DFEE. Electrons were separated by the energy into two groups with a gap at $\varepsilon_1 = 16.6$ eV (the excitation potential of the first level of neon). This allowed the authors [16,34] to more accurately calculate the distribution of energy between the atoms of copper and neon without solving the

Table 8.1. The parameters of the GDT and prepulse concentrations of reagents used in the calculation of the DFEE (see text for details). Detailed description of the electrical excitation circuit is presented in [6]

Active medium	Cu–Ne [5,6]	Cu–Ne–H$_2$ [28,29]	Cu–Ne–H$_2$–HCl [30]	Cu–Ne–Cs [31]
$N_{H_2(v=0)}$ (cm^{-3})	–	$2.25 \cdot 10^{15}$	$1.38 \cdot 10^{15}$	–
N_H (cm^{-3})	–	$1.28 \cdot 10^{15}$	$2.88 \cdot 10^{15}$	–
$N_{HCl(v=0)}$ (cm^{-3})	–	–	$1.14 \cdot 10^{14}$	–
N_{Cs} (cm^{-3})	–	–	–	$9.73 \cdot 10^{13}$
N_{Ne} (cm^{-3})	$1.62 \cdot 10^{18}$			
N_{Cu} (cm^{-3})	$1.6 \cdot 10^{15}$			
N_{Cu+} (cm^{-3})	$1 \cdot 10^{13}$			
$Cu_{D_{5/2}}$ (cm^{-3})	$1.5 \cdot 10^{12}$			
$Cu_{D_{3/2}}$ (cm^{-3})	$2.55 \cdot 10^{11}$			
Length of active medium (cm)	40			
GDT diameter (cm)	2			
Repetition frequency of excitation pulses (kHz)	10			

Boltzmann equation. In this case the DFEE of electrons with energies below 16.6 eV had Maxwell's profile for the electron energy with temperature T_1:

$$f_{M1}(\varepsilon) = 2\beta N_e \left(\frac{\varepsilon}{\pi T_1^3} \right)^{1/2} \exp\left(-\frac{\varepsilon}{T_1} \right)$$

The form of the DFEE for the high-energy part of the electron was similar:

$$f_{M2}(\varepsilon) = 2\beta N_e S \left(\frac{\varepsilon}{\pi T_2^3} \right)^{1/2} \exp\left(-\frac{\varepsilon}{T_2} \right),$$

where $\beta^{-1} = \int\limits_0^{\varepsilon_1} f_{M1}(\varepsilon)d\varepsilon + \int\limits_{\varepsilon_1}^{\infty} f_{M2}(\varepsilon)d\varepsilon$ is the normalized factor for the electron density and S is the normalized factor in the energy of the gap ε_1, such that $f_{M1}(\varepsilon_1) = f_{M2}(\varepsilon_1)$:

$$S = \left(\frac{T_1}{T_2} \right)^{3/2} \exp\left(-\varepsilon_1 \left(\frac{1}{T_1} + \frac{1}{T_2} \right) \right),$$

temperature is in energy units (eV).

Using the bi-Maxwellian DFEE allows to account more accurately the contribution of high-energy electrons in the calculation of reaction rates, however, it says nothing about the actual form of the DFEE in the active medium.

Calculation of the quasi-stationary DFEE

[33] presents the results of calculation of quasi-stationary energy distributions of electrons in mixtures of copper vapours with neon and helium. Calculations in [33] were carried out without considering electron–electron collisions and collisions with excited Cu atoms. It is noted that they are insignificant when [Cu]/[Ne] = 10^{-3} and E/p = 10 V/(cm · Torr), where E is the electric field strength, p is pressure (there is a large class of CVLs with [Cu]/[Ne] = 10^{-2}, the ratio [Cu]/[Ne] = 10^{-3} is performed for lasers with a neon pressure of 300 Torr, and the ratio E/p = 10 V/(cm · Torr) corresponds to the voltage drop across the discharge tube not less than 120 kV, well above the typical values of voltage drop on the tube in the experiment). At the time of the study [33] there were no data on the cross sections of electron excitation of metastable levels of the copper atom. The reaction rate constants calculated using the DFEE and MDFEE were compared. At the electron temperature approximately equal to 2 eV, the discrepancy between the excitation rate constants of atomic copper increases up to 2–3 times for the excitation of high-lying levels (as much as tens of percent for the resonance levels and about 5 times for ionization) and decreases with increasing electron temperature. The calculated DFEE with MDFEE were not conducted.

The authors of [35] also calculated as the stationary DFEE for a copper vapour laser, with the first attempt to take into account the influence of additions of hydrogen halides (HBr impurity) on the form of the DFEE. In this work, as well as in [33] the reverse processes for electron excitation were not taken into account. Introduction to the laser active medium of hydrogen halide molecules leads to a substantial decrease in the electron density in the active medium [28–30], which can lead to distortion of the DFEE profile. In [35] it is argued that the introduction of the hydrogen bromide to the active medium of the laser transforms the DFEE from Maxwellian to the Druyvesteyn form (in the calculations the concentrations of HBr additions exceeded the optimum experimental value about two–three times).

A common shortcoming of [33,35] is that the DFEE was calculated in the stationary discharge conditions. However, it is known that metal-vapour lasers operate in essentially non-stationary conditions, and to calculate the rate constants of reactions it is important to know the behaviour of the DFEE in plasma with the time-varying electric field take into account.

The kinetic model [32]

The kinetic schemes used for calculating the kinetics of active media (Cu–Ne, Cu–Ne–Cs, Cu–Ne–H$_2$, Cu–Ne–H$_2$–HCl) of the CVLs have not been modified in comparison with [5–7,28–31] in which the reaction rate constants were calculated in the MDFEE approximation (see 8.2, 8.4–8.6 of this chapter). The developed detailed transient kinetic models [5–7,28–31] allow us to calculate and analyze the changes in average values (in the volume) of the population levels of the atoms of copper, neon and introduced impurities (Cs, H$_2$, HCl, respectively), the density of ions of these elements, the electron temperature and the laser intensity at two wavelengths.

Compared with [5–7, 28–31], the kinetic rate constants of reactions involving electrons were calculated now based on the cross sections of these reactions ($\sigma(\varepsilon)$) and the calculated DFEE (see below)

$$k = \sqrt{\frac{2 \cdot e}{m_e}} \int_0^\infty \varepsilon \cdot \sigma(\varepsilon) \cdot f(\varepsilon) \cdot d\varepsilon,$$

here $f(\varepsilon)$ is the DFEE, ε is the electron energy.

The system of stiff differential equations, including kinetic equations for the concentrations of various reagents of the active medium, the Kirchhoff equations for the electrical excitation circuit and the equations for calculating the DFEE was self-consistently solved using the PLAZER software package [10,13–15].

Calculation of the DFEE [32]

Generally speaking, when an electric field is present in the plasma the electron distribution function depends not only on the magnitude of electron velocity (electron energy) as in the absence of the field, but also on the vector of the velocity of the electrons. Moreover, since the direction of the electric field is selected, then the distribution function of electrons is usually sought in the form of an expansion in Legendre polynomials from the cosine of the angle, measured from the direction of the field. If the electron mean free path between collisions is small compared to the size of the active region (with a reserve that holds for CVLs because the range of operating pressures in the range from 10 to 300 Torr), then because of the large deviations of the electrons in collisions with heavy particles, their velocity distribution function is almost independent of the direction of the velocity. In this case, the expansion usually consider only polynomials of the zeroth and first degree and, accordingly, this approximation is called an almost an isotropic approximation

$$f(\mathbf{v}) = f_0(v) + (\mathbf{n}_v \mathbf{n}_E) f_1(v)$$

where $\mathbf{n}_v$ and $\mathbf{n}_E$ are the unit vectors in the direction of the electron velocity and the field. Including only two Legendre polynomials in the expansion of the distribution function is also associated with the fact that when the polynomials of the second and higher degrees are considered to find f_k, $k > 1$, we would have to know the angular dependence of the inelastic scattering cross sections, which are usually unknown. The electron distribution function is found by solving the Boltzmann equation, which splits into two equations (the terms of zeroth and first order in $\cos\theta = (\mathbf{n}_v, \mathbf{n}_E)$) [41,42]:

$$\frac{\partial f_1}{\partial t} - \frac{eE}{m_e}\frac{\partial f_0}{\partial t} = \left(\frac{\partial f_1}{\partial t}\right)_c,$$

$$\frac{\partial f_0}{\partial t} - \frac{eE}{3m_e v^2}\frac{\partial}{\partial v}(v^2 f_1) - \left(\frac{\partial f_0}{\partial t}\right)_c$$

These equations are collected in the equation [13,41–43]

$$\sqrt{\varepsilon}\frac{\partial}{\partial t}(n_e f_0) - \frac{\partial}{\partial \varepsilon}\left(n_e \varepsilon^{3/2}\left\langle \delta_m v_m(\varepsilon)\right\rangle\left[f_0 + T\frac{\partial}{\partial \varepsilon}f_0(\varepsilon)\right]\right) -$$

$$-\frac{1}{3}n_e\sqrt{\frac{2g}{m_e}}\sigma_{ee}(\varepsilon)\left[h_1(\varepsilon)f_0(\varepsilon) + h_2(\varepsilon)\frac{\partial}{\partial \varepsilon}f_0(\varepsilon)\right] -$$

$$-\frac{1}{3}n_e\frac{e^2}{g^2}E^2\frac{2g}{m_e}\frac{\varepsilon^{3/2}}{\left\langle v_m(\varepsilon)\right\rangle}\frac{\partial}{\partial \varepsilon}f_0(\varepsilon)\right) =$$

$$= n_e \sqrt{\frac{2g}{m_e}} \sum_{j,N} \left[(\varepsilon + \varepsilon_j) f_0(\varepsilon + \varepsilon_j) N \sigma_j(\varepsilon + \varepsilon_j) - \varepsilon f_0(\varepsilon) N \sigma_j(\varepsilon) + \right.$$

$$\left. + (\varepsilon - \varepsilon_j) f_0(\varepsilon - \varepsilon_j) N_j^* \sigma_{-j}(\varepsilon - \varepsilon_j) - \varepsilon f_0(\varepsilon) N_j^* \sigma_{-j}(\varepsilon) \right],$$

where

$$\langle \delta_m v_m(\varepsilon) \rangle = \sum_k \frac{2m_e}{M_k} \sqrt{\frac{2\varepsilon g}{m_e}} N_k \sigma_m^k(\varepsilon), \quad \langle v_m(\varepsilon) \rangle = \sum_k \sqrt{\frac{2\varepsilon g}{m_e}} N_k \sigma_m^k(\varepsilon),$$

$$\sigma_{ee} = \frac{2\pi e^4}{g^2} \ln \Lambda, \quad \Lambda = \frac{300 e}{2\bar{\varepsilon}} \sqrt{\frac{4\pi n_e e^2}{gT_e}},$$

$$h_1(\varepsilon) = 3 \int_0^\varepsilon u^{1/2} f_0(u) du, \quad h_2(\varepsilon) = 2 \int_0^\varepsilon u^{3/2} f_0(u) du + 2\varepsilon^{3/2} \int_0^\infty f_0(u) du.$$

In these relations ε, T_e – the energy of the electron, the gas temperature, measured in eV, and the other quantities are expressed in CGS units, $g = 1.6 \cdot 10^{-12}$ is the conversion factor from eV to ergs and $e/g = 300$, $\sigma_m^k(\varepsilon)$ is the elastic electron collision cross section on the k-th gas component, $\sigma_j(\varepsilon)$ is the cross section of the inelastic j-th process, N and N^* are the concentrations of components in the initial and final states of that process, $\sigma_{-j}(\varepsilon)$ is the cross section of the collisions of the second kind (superelastic processes), m_e is the electron mass, M_k is the to mass of the k-th gas component, n_e is the electron density [43]. The second and third terms in the above equation are the contributions of elastic electron collisions with gas components, the fourth and fifth are the contributions of electron–electron collisions [42], the right side of the equation is the contribution of inelastic collisions (excitation, deexcitation, ionization) (see also [13]). A numerical method for solving the above equation is described in detail in [44].

In general, the right hand side should also contain the term

$$S_0 = n_{\gamma} v_0 \delta(\varepsilon) + S(\varepsilon)$$

where the first term takes into account the arrival of electrons with zero energy: the ionization in atomic collisions, Penning ionization, electron detachment from halogens, etc. The second term $S(\varepsilon)$ takes into account external sources of electrons, for example, electron beam, photoionization, and others, S_0 is often insignificant and is not used in the calculations.

The strength of the electric field enters into the equation for the DFEE as a parameter, so that together with the Boltzmann equation it is necessary to solve Kirchhoff's equations describing the electrical circuit of the pump.

Description used in the calculation of cross sections

In calculating the DFEE used section shown in [5–7, 28–31], which was carried out modeling of the kinetics of the active media of Cu–Ne, Cu–Ne–H$_2$, Cu–Ne–Cs, Cu–Ne–H$_2$–HCl (in these studies to simplify the calculations we used the rate constants using MDFEE). For the excitation of the resonance and metastable levels of the copper atom the authors of [45,46] used the cross sections known with great accuracy. For the cross sections for transitions to highly excited states and between excited levels the modified van Regemorter formula [47] was used:

$$\sigma_i(x) = \frac{8\pi}{\sqrt{3}} \cdot \pi a_0^2 \cdot f_i \cdot \frac{Ry^2}{E_i^2} \cdot \frac{G_i(x)}{x},$$

where $x = E/E_i$, a_0 is the Bohr radius, Ry is the Rydberg constant, f_i is oscillator strength, E_i is energy and G_i is the Gaunt factor of the i-th level, which was calculated using the formula proposed in [48]:

$$G_i(x) \cong \frac{\sqrt{3}}{2\pi} \cdot \ln x.$$

Results and discussion

Neon was used as the buffer gas in all the investigated mixtures. Note that the DFEE in pure neon has the Druyvesteyn form [22]. This reflects the fact that the transport cross section for electron scattering by neon is almost constant (for the energy of the order of 1 eV to 100 eV, it is in the range $\sigma_{tr} = (2 \div 3)\cdot 10^{-16}$ cm^2), as the Druyvesteyn form of DFEE occurs under the condition of independence of the electron path on their energy [49].

As a result, comparison of the calculated DFEE was conducted both with MDFEE

$$f_M(\varepsilon) = \frac{4\pi \sqrt{\varepsilon}}{(2\pi T)^{3/2}} \sqrt{\varepsilon} \exp\left(-\frac{\varepsilon}{T_1}\right),$$

and the Druyvesteyn DFEE (DDFEE), which is given by the expression:

$$f_D(\varepsilon) \equiv \frac{2}{\Gamma(3/4)} \left(\frac{3m_e}{M}\right)^{3/4} \sqrt{\frac{\varepsilon}{\varepsilon_0^3}} \exp\left(-\frac{3m_e}{M}\frac{\varepsilon^2}{\varepsilon_0^2}\right).$$

Here M is the mass of gas particles (neon atoms); $\Gamma(3/4) = 1.225$; $\varepsilon_0 = eE/(\sigma_{tr}N)$ is the energy gained by an electron in an electric field with strength E for the distances $l = 1/(\sigma_{tr}N)$ from a single elastic collision to another.

When should we expect a substantial deviation of the DFEE from the Druyvesteyn form? The elastic scattering cross section of electrons on the copper atoms is approximately 100 times greater than the analogous cross section for the neon atoms and depends strongly on the electron energy [35]. Thus, approximately at a ratio of Ne:Cu > 100:1 the form of DFEE should be reconstructed. Such a relationship is typical of working mixtures used in the CVLs. Thus, the operating pressure of a large number of CVLs is several tens of Torr (the concentration of the neon atoms is of the order of several units from 10^{17} cm^{-3}), and the operating temperatures are around 1600 °C (the concentration of copper atoms is of the order of several units from 10^{15} cm^{-3}). So, for the active mixtures of the CVLs the DFEE may generally differ from DDFEE.

On the other hand, in high-intensity electron–electron collisions, the shape of DFEE to the first ionization potential is almost Maxwellian. This form of the DFEE is characteristic of, for example, plasma active media of exciplex lasers (pressure of these active media is of the order of several atmospheres), which arises when pumped with a hard ionizer (such as electron beams) [13].

Based on the foregoing, the DFEE in the active medium of a CVL may differ from DDFEE as well as from MDFEE. The degree of difference is difficult to determine analytically without numerical calculations.

To identify deviations of the calculated DFEE from MDFEE and DDFEE, all the three distribution functions are compared below. At the same time the electron temperature here means two-thirds of the average energy of the electron:

$$T_e = (2/3)\bar{\varepsilon} = (2/3)\int_0^\infty \varepsilon f(\varepsilon)d\varepsilon.$$

For DDFEE the temperature was also determined by this procedure, but the calculated DFEE was replaced by DDFEE:

$$T_D = (2/3)\bar{\varepsilon} = (2/3)\int_0^\infty \varepsilon f_D(\varepsilon)d\varepsilon.$$

Using this definition the temperature can be expressed for the DDFEE by the problem parameters. If the Druyvesteyn distribution function is written in the variables $y = \varepsilon^2$, and $y_0 = \dfrac{M\varepsilon_0^2}{3m_e}$, then

$$f_D(\sqrt{y}) = 2y^{1/4}C\exp(-y/y_0),$$

where

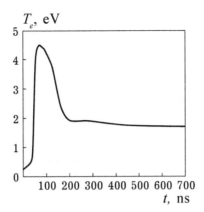

Figure 8.1. The time dependence of the electron temperature for the active medium of the Cu–Ne laser calculated taking the calculated DFEE into account. Parameters of the GDT and the prepulse concentrations of the reactants are given in Table 8.1

$$C = \frac{1}{\Gamma(3/4)y_0^{3/4}},$$

respectively

$$\bar{\varepsilon} = Cy_0^{5/4}\Gamma(5/4) = y_0^{1/2}\frac{\Gamma(5/4)}{\Gamma(3/4)} = \frac{3}{2}T_D$$

and

$$T_D = \beta y_0^{1/2},$$

where $\beta = \dfrac{2\Gamma(5/4)}{3\Gamma(3/4)} \approx 0.493$.

As an example, Fig. 8.1 shows the dependence of the electron temperature, taking into account the calculated DFEE.

The analysis of the DFEE presented in Figs. 8.2–8.5 shows that the calculated DFEE has noticeable differences from the Maxwellian and Druyvesteyn distributions both at the beginning of the excitation pulse (~20 ns) and at the time corresponding to the maximum lasing pulse (~100 ns). Let us discuss these differences in more detail.

For all active media (Cu–Ne, Cu–Ne–Cs, Cu–Ne–H$_2$, Cu–Ne–H$_2$–HCl) for which we calculated the DFEE in the present work during the beginning of the excitation pulse, the maximum value of the calculated DFEE is less than MDFEE, but more than DDFEE (Figs. 8.2–8.5). Note that the energies, which correspond to the maxima of the DFEE almost coincide with MDFEE, while for DDFEE the maximum is shifted to higher energies.

As well as for the time corresponding to the beginning of the excitation pulse, the DFEE calculated for the time corresponding to the

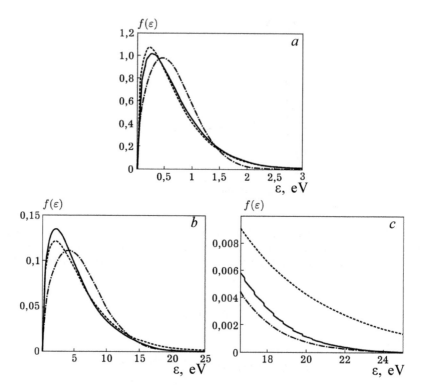

Figure 8.2. DFEE for the active medium of the Cu–Ne laser at a time *a*: 20 ns, *b*: 100 ns; *c*: 100 ns (high-energy part of DFEE) (solid curve corresponds to the calculated DFEE dashed to the MDFEE, dot-dash – DDFEE). GDT parameters and the prepulse concentrations of the reactants are given in Table 8.1. Electron temperatures for MDFEE and DDFEE at construction of these curves were chosen equal to the electron temperature for the calculated DFEE.

maximum of the lasing pulse is different from the profiles of MDFEE and DDFEE (Figs. 8.2–8.5). Thus, the maximum value of the calculated DFEE is higher than the corresponding values for MDFEE and DDFEE. Otherwise, the behaviour of the distribution function is similar to the behaviour described in the preceding paragraph. The energy values at which the DFEE reaches a maximum almost coincide with the energies at which the maximum MDFEE is reached. The DDFEE maximum is shifted somewhat to the high-energy region.

In general, during the pump pulse the DFEE profile is gradually broadened along the energy axis while the maximum is shifted to the high-energy side. Note that in the high-energy region (above 16.6 eV) there is a significant (several times) difference between the calculated DFEE and the Maxwellian DFEE profile. This leads to a significant decrease in the rate constants of excitation processes and ionization of

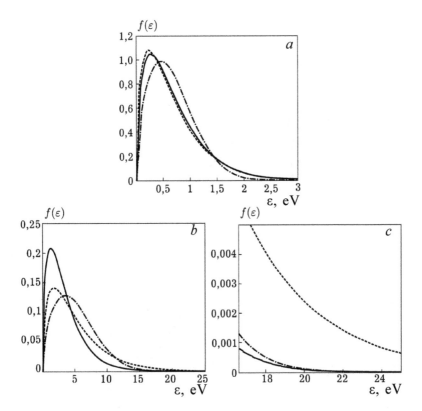

Figure 8.3. DFEE for the active medium of the Cu–Ne–Cs laser point at time *a*: 20 ns, *b*: 100 ns, *c*: 100 ns (high-energy part of DFEE) (solid curve corresponds to the calculated DFEE, the dashed curve – MDFEE, dot-dash curve – DDFEE). GDT parameters and prepulse concentration of the reagents are listed in Table 8.1. Electron temperatures for MDFEE and DDFEE in the construction of the curves were chosen equal to the temperature of the electrons for the calculated DFEE.

the buffer gas atoms, which in turn will reduce flows of energy lost in inelastic processes with the participation of the buffer gas, which should improve the overall efficiency of lasing. In addition, due to the energy loss reduction in excitation and ionization of the buffer gas atoms the part of the energy flux directed to the excitation of copper atoms should increase and may lead to an increase in the lasing energy of the laser.

As regards the rate constants with copper participation, since the energy levels in an atom of copper are significantly lower than the levels of the neon atoms, they differ to the extent similar to the difference between the calculated DFEE and MDFEE (Fig. 8.6). The dip in the rate constant of excitation of the $D_{5/2}$ level of the copper atom in Fig. 8.6, *b* is due to the fact that this constant has a maximum at the electron temperature approximately equal to 2.8 eV [6] (see

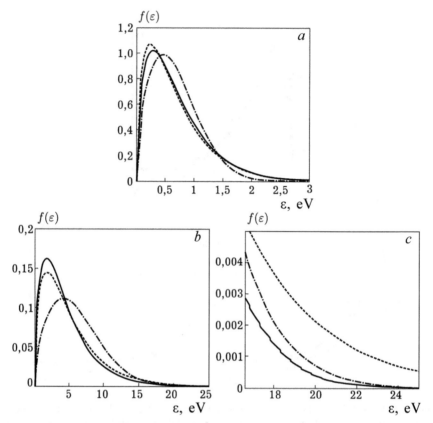

Figure 8.4. DFEE for the active medium of the Cu–Ne–H$_2$ laser at time *a*: 20 ns, *b*: 100 ns, *c*: 100 ns (high-energy part of DFEE) (solid curve corresponds to the calculated DFEE, dashed curve to MDFEE, dot-dash curve to DDFEE). GDT parameters and prepulse concentration of the reagents listed in Table 8.1. Electron temperatures for MDFEE and DDFEE when constructing of the curves were chosen equal to the temperature of the electrons for the calculated DFEE.

also Fig. 8.7), and the electron temperature in the calculations reaches 4.5 eV (Fig. 8.1).

Earlier, we noted that the results of calculations of DFEE in the CVL when the form of the DFEE is not fixed, were published only in [33,35]. Our calculations do not confirm the conclusions of [35] that the DFEE in energy in the active medium of a CVL has the   form. Although we have not directly modelled the active medium of a CVL with the addition of HBr, but only with the additions of H$_2$, H$_2$–HCl, Cs, we believe that the particular type of additive in this case will not affect our results.

Not confirmed are also the results of [33], in which the temperature of the electrons ~2 eV was accompanied by significant differences of reaction rate constants calculated for the real DFEE and MDFEE.

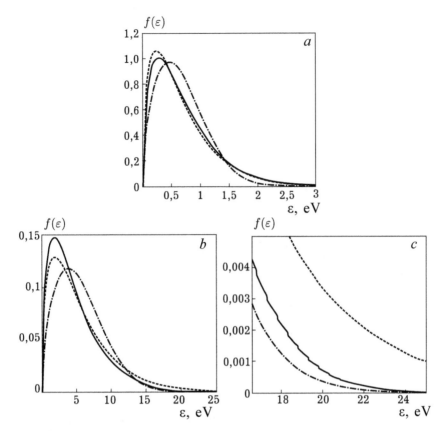

Figure 8.5. DFEE for the active medium of the Cu–Ne–H$_2$–HCl laser at time
a: 20 ns, *b*: 100 ns, *c*: 100 ns (high-energy part of DFEE) (solid curve corresponds
to the calculated DFEE, dashed curve to MDFEE, dot-dash curve to DDFEE). GDT
parameters and prepulse concentration of the reagents are listed in Table 8.1. Electron
temperatures for MDFEE and DDFEE in the construction of the curves were chosen
equal to the electron temperature for the calculated DFEE.

In [16,34] the form of the DFEE is not calculated and the bi
Maxwellian form is used for it. According to calculations of the DFEE
in this representation the tail of such MDFEE (with T_2) is significantly
depleted compared with MDFEE (with T_1). Our calculations confirm this
behaviour. The calculated DFEE tail is very similar to the behaviour
of the DDFEE tail and lies significantly below the high-energy part
of the MDFEE.

For example, Fig. 8.7 shows the rate constants for the excitation
of the most important levels – the resonance and metastable states of
copper atoms, which are respectively upper and lower levels of laser
transitions 510 and 578 nm in the CVL, calculated with MDFEE and
with bi-Maxwellian DFEE [6]. Total cross sections for excitation of the

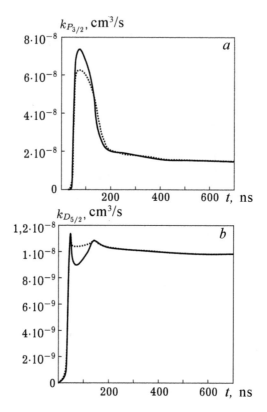

Figure 8.6. Time dependences of the rate constants of electron excitation. a – the upper laser level $P_{3/2}$ of the Cu atoms (510.6 nm); b – the lower laser level $D_{5/2}$ of the Cu atoms (510.6 nm), calculated using the MDFEE (dotted line) and calculated DFEE (solid line) in a Cu–Ne mixture. Parameters GRT prepulse and reactant concentrations are given in Table 8.1.

mentioned states by electron impact from the ground state of copper are given in [50]. As in [16], when calculating of excitation rates of the sublevels $P_{3/2}$ and $P_{1/2}$ of the resonant level and sublevels $D_{5/2}$ and $D_{3/2}$ of the metastable level it is assumed that the contribution to each cross section of these sublevels is proportional to the total cross section with the corresponding statistical weight of this sublevel. Work [16] presented the rate constants for excitation of the sublevels $P_{3/2}$ and $D_{5/2}$, so here we compare only these constants with the constants of [16]. In this range of (low) electron energies there is good agreement between the calculated constant rates.

Conclusion

The non-stationary DFEEs were calculated for the active media of the standard CVL and CVL with the addition of cesium, hydrogen,

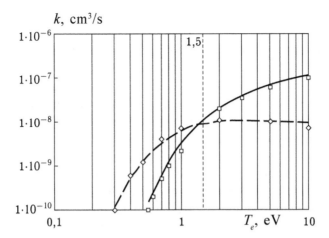

Figure 8.7. Comparison of the dependences of the excitation rates of resonance $P_{3/2}$ (solid line) and the metastable $D_{5/2}$ (dashed line) states of the copper atom on the electron temperature in the presented model with the dependences (circles) of [16].

and hydrogen chloride [32]. It is shown that DFEE differs from both MDFEE and from DDFEE.

The main difference between the real and Maxwellian DFEE is observed in the high-energy part of DFEE. This leads to a substantial (more than fivefold) decrease in the rates of excitation and ionization of the neon buffer gas atoms and confirms the findings of [16,34]; this fact is also noted in [36].

On the other hand, the DFEE differs also from DDFEE and this differs from the findings of [35]. The main difference between these distributions functions is found in the small- and medium-energy electrons. The difference between these functions decreases with increasing electron energy. Not confirmed are the results of [33], in which in the temperature range of the electrons ~2 eV, there are significant differences of the reaction rate constants calculated for the real DFEE and MDFEE. Up to energies of 7 eV the DFEE and MDFEE calculated in this work do not differ significantly (10%), so that the same small differences should also be expected in the rate constants calculated from these DFEE.

Thus, in the high-energy region (above 16.6 eV) there is a significant difference between the calculated DFEE and the Maxwellian DFEE profile. This leads to a significant decrease in the values of the constants of excitation and ionization of the buffer gas atoms, which in turn will reduce energy flux lost in inelastic processes with the participation of the buffer gas, and it should be generally necessary to increase slightly the lasing efficiency. In addition, due to the energy loss reduction in excitation and ionization of the buffer gas atoms the part of the energy flux directed to the excitation of copper atoms should increase which may also lead to some increase energy of the laser.

In the calculation of most of the key reaction rate constants that determine the kinetics of the CVL, the main contribution comes from low-energy electrons. The difference between the real DFEE and MDFEE in this case, according to the calculations, may not significantly affect their magnitude and is about 10%. Note that the large uncertainty in their values is caused apparently by uncertain cross sections of the interaction of electrons with the Cu atoms and molecular additives since the cross sections are known at best to within 50–100%. Thus, although the actual shape of the DFEE differs from MDFEE in the active medium of CVL, however the use of MDFEE in the CVL models is fully justified.

8.4. Restrictions on pulse repetition rate

The fundamental issue as to what limits the repetition rate of CVL is still a matter of controversy among experts. In some works the main limitation is associated with a large prepulse population of the lower metastable operating state of the atom, in other works – with a prepulse electron density. Background details are in review [51].

In [52] it is shown that there is a certain critical value of the electron density N_{ecr}. If the prepulse electron density is greater than N_{ecr}, then the electron temperature during the heating pulse of the plasma does not reach a certain critical value $T_{ecr} \approx 1.5 \div 2$ eV required for population inversion. A simple estimate of N_{ecr} through the experimental peak current density of the pump and the plasma parameters is given, as well as performed simple calculations confirm this estimate. Calculations within the framework of somewhat more detailed models [53,54] confirm the presence of prepulse electron density [52]. In [55] the existence of a critical electron density is questioned. In [5–7] (see section 8.2) a detailed model is constructed. The values of N_{ecr}, obtained in calculations using this model, agree well with the estimates of [52]. In a recent review of lasers with modified kinetics [56] it is stated that in the optimal conditions the output characteristics of these lasers can be increased by more than a half at a 5–10 fold decrease in the prepulse electron density, and they are practically not sensitive to a decrease in the prepulse density of metastable states of copper, leading only to a small increase power output.

Mechanisms that prevent lasing
The mechanism of constraints on the initial density of metastable

atoms is obvious. It stems from the fact that during the pump pulse the population of the resonant level has a maximum. Clearly, if the initial population of the metastable is already sufficiently high, then throughout the pulse its population may be greater than the population of the resonant state. However, it should be noted that the critical density of metastables was not previously estimated and not calculated.

Mechanism of constraints on the initial electron density is also fairly transparent [52]. It is due to the fact that for an inversion we require a relatively high electron temperature $T_{ecr} \approx 2$ eV, determined by the ratio of the rates of excitation of the resonance and metastable states (see Fig. 8.7). At $T_e < T_{ecr}$, the population of metastable atoms is dominant. Given that the power supplied to the medium is limited, for example, by the capabilities of the pumping system, for sufficiently large initial electron density $N_e(0) > N_{ecr}$ the critical temperature can not be attained.

This is clear even from the fact that increasing electron density increases the specific heat of the electron gas. Indeed, even if we assume that all the energy entered into the medium is spent on heating the electrons, at a limited power of the pump source, the electron temperature will not exceed a certain maximum value

$$T_{e\,max} \sim \rho j_{max}^2 \Delta t / N_e,$$

where Δt *is* the time scale of the pump pulse, ρ is resistivity of the plasma, j_{max} is the maximum current density in the discharge tube. Given that the energy introduced into the medium $\rho j_{max}^2 \Delta t$ is limited, the maximum temperature will be less than the critical temperature $T_{e\,max} < T_{ecr}$, if the electron density is high enough

$$N_e > \rho j_{max}^2 \Delta t / T_{ecr}. \qquad (8.3)$$

Of course, in reality the bulk of the introduced energy is not used for heating the electrons and is used for the excitation and ionization of the copper atoms. Therefore, the restriction on the electron density (8.3) is 'too soft'. It only demonstrates the existence of a critical density of electrons at a limited power of the pump source. For more accurate estimates there should be more detailed consideration to which we now turn.

The simplest model of the thermal regime of the CVL [52]

Consideration on the basis of a detailed kinetic model is preceded by discussing the simplest model. In the latter case it is easier to trace the nature of emerging constraints.

Due to the fact that the duration of the characteristic time of the pump is three orders of magnitude less than the time between pulses, it makes sense to consider two models of the temporal behaviour of the plasma characteristics of the CVLs. One model describes the development of plasma ionization under the influence of the heating field pulse, the second – the afterglow. The results of calculations within the framework of these models give initial conditions for each other.

The ionization of the plasma during pumping. The kinetic model of ionization of a mixture of copper vapours by the heating pulse includes the equation for the density of copper ions and the inert gas N_{iCu}, N_{iNe}, as well as the heat balance equation for the electron temperature T_e:

$$\frac{dN_{iCu}}{dt} = k_{iCu} \cdot N_e \cdot (N_{Cu} - N_{iCu}), \qquad \frac{dN_{iNe}}{dt} = k_{iNe} \cdot N_e \cdot (N_{Ne} - N_{iNe}),$$

$$\frac{d}{dt}\left(\frac{3}{2}N_e T_e\right) = -Q_{iCu} - Q_{iNe} - Q_{\Delta T} + \frac{1}{\sigma}j^2(t). \qquad (8.4)$$

Here $N_e = N_{iCu} + N_{iNe}$ is the electron density; k_{iCu}, k_{iNe} (in cm³/s) is the rate of ionization of respectively the copper and neon atoms (everywhere here the velocity k_X of the binary process X is the product of the cross section of this process σ_X by the relative velocity of the particles $k_X = \langle\sigma_X v\rangle$ averaged over the Maxwellian distribution); N_{Cu}, N_{Ne} are the densities of heavy particles (ions and neutrals) of copper and neon;

$$Q_{iCu} = J_{iCu}k_{iCu}N_e(N_{Cu} - N_{iCu}),$$
$$Q_{iNe} = J_{iNe}k_{iNe}N_e(N_{Ne} - N_{iNe})$$

is the power density spent on ionization of neon and copper, respectively; $J_{iCu} = 7.73$ eV, $J_{iNe} = 21.6$ eV are the ionization energies of copper and neon;

$$Q_{\Delta T} = 2\left[(m_e/m_{Ne})\cdot k_{Ne} \cdot N_{Ne} + (m_e/m_{Cu})\cdot k_{ei}N_e\right]\cdot N_e \cdot (T_e - T_g)$$

is the power density spent on cooling of electrons due to elastic collisions with neon atoms and copper ions; k_{Ne}, k_{ei} is the rate of elastic collisions of electrons with neon atoms and ions, m_e is the electron mass, m_{Ne}, m_{Cu} are the atomic masses of neon and copper, T_g is the gas temperature;

$$\sigma = \frac{e^2 N_e}{m_e}\left(k_{Ne}N_{Ne} + 1.96k_{ei}N_e\right)^{-1}$$

is the conductivity of the plasma.

The last term in the right-hand part of equation (8.4) of the thermal balance of electrons j^2/σ describes the Joule heating proportional to the square of current density. The dependence of current density on the time $j(t)$ is taken from experimental data.

The reaction rates. For the reaction rates of elastic collisions we use the expressions:

$$k_{Ne} = 8.9 \cdot 10^{-9} (T_e / eV)^{1/2} \text{ cm}^3/\text{s}, \quad k_{ei} = \frac{4\sqrt{\pi}}{3} \frac{e^4 \Lambda}{T_e^2} \sqrt{\frac{2T_e}{m_e}}$$

$$\Lambda = \frac{1}{2} \ln\left(1 + \frac{T_e^3}{2e^6 N_e}\right).$$

The cross section of the elastic collisions of the electron with the neon atom is set equal to $\sigma_{Ne} = 1.5 \cdot 10^{-16}$ cm², and for Coulomb collisions we use the well-known expression [9].

The ionization rates for copper and neon are expressed by:

$$k_{iCu} = 2 \cdot 10^{-7} \cdot F(E_{Cu}^*/T_e) \text{ cm}^3/\text{s},$$
$$k_{iNe} = 4 \cdot 10^{-10} \cdot F(E_{Ne}^*/T_e) \text{ cm}^3/\text{s},$$
$$F(x) = 0.5 \cdot e^{-x}/x^{1/2},$$

where $E_{Cu}^* = 3.8$ eV, $E_{Ne}^* = 16.6$ eV. In this case the ionization rate of copper and neon was considered equal to the rate of excitation of resonant states. This is true in the quasi-stationary mode of ionization, when every act of excitation is accompanied by an act of ionization of the excited state (for details see [8.9]). The fitting function $F(x)$ describes well the dependence of the rate of excitation of resonant states of copper $^2P_{3/2}$ on the electron temperature (see Fig. 8.7).

Generally speaking, under these conditions we can not be strictly introduce the rate of inelastic energy losses that does not depend explicitly on time. To adequately describe the process of ionization and thermal balance it is necessary to consider the balance equations of the populations of many excited states, as, for example, in section 8.2. However, it is clear that the minimum energy loss rate under these conditions is the rate of excitation of the resonance level, multiplied by the energy of its excitation $E_{Cu}^* = 3.8$ eV. In other words, the difference between the expression used here for the energy loss and the lowest possible energy loss is only a factor of $7.73/3.8 \approx 2$. Clearly, if the magnitude of energy losses is overstated, it is less than two-fold. In addition, the value J_{iCu} in the estimates (see below) is under the sign of the square root, so that the error due to possible overestimation of the energy loss is certainly not greater than 40%. This is quite possible for simplified models which aim only to reveal the essence of the mechanism of the restrictions.

Afterglow. The kinetic model of the afterglow is quite simple and well studied in relation to the analysis of the kinetics of plasma lasers [8]. It includes the equation for the density and electron temperature:

$$\frac{dN_e}{dt} = C_r T_e^{-9/2} N_e^3, \frac{d}{dt}\left(\frac{3}{2} N_e T_e\right) = E_r C_r T_e^{-9/2} N_e^3 - Q_{\Delta T}.$$

Here

$$C_r = \frac{4}{5} \cdot \frac{2^{5/2}\pi^{3/2}}{9} \cdot \frac{e^{10}}{\sqrt{m_e}} \cdot \Lambda = 5.8\cdot10^{-26}\cdot eV^{9/2}\cdot cm^6/s$$

is a constant characterizing the three-body recombination rate [8,57,58]; $E_r \approx J_{iCu} = 7.73$ eV is the energy released by an act of recombination.

Evaluation of the critical density of electrons. The power introduced into the medium is proportional to the resistance of the plasma, i.e., inversely proportional to the electron density. At the same time the 'cost' of ionization is directly proportional to the electron density (see 8.4)). Consequently, for a given current density and electron temperature there is a critical density of electrons from which the power introduced into the medium will be less the power spent on ionization.

The equation for the critical electron density N_{ecr} can be obtained by equating the power input to the medium at the peak current density j_{max}, and the power expended in the ionization of copper at the critical temperature of the electrons

$$[j_{max}^2 m_e/(e^2 N_{ecr})]\cdot(k_{Ne}(T_{ecr})\cdot N_{Ne} + 2k_{ei}(T_{ecr})\cdot N_{ecr}) =$$
$$= J_{iCu}\cdot k_{iCu}(T_{ecr})\cdot N_{ecr}\cdot N_{Cu}.$$

Hence, for the critical density we have the expression:

$$N_{ecr} = N_{ecr0}\cdot[a+(a^n+1)^{1/n}].\qquad(8.5)$$

Here

$$N_{ecr0} = \frac{j_{max}}{e}\left(\frac{m_e k_{Ne}(T_{cr})N_{Cu}}{J_{Cu}k_{iCu}N_{Ne}}\right)^{1/2}$$

is the critical density for the case where the conductivity is determined by collisions with neutrals $k_{Ne}(T_{ecr})\cdot N_{Ne} \gg k_{ei}(T_{ecr})\cdot N_{ecr}$;

$$a = k_{ei}(T_{ecr})\cdot N_{ecr0}/k_{Ne}(T_{ecr})\cdot N_{Ne}$$

is a dimensionless quantity, essential in cases where a significant contribution to the conductivity is provided by Coulomb collisions.

The boundary value of the degree of ionization at which the contribution of Coulomb collisions and collisions with neutrals is equal, $a = 1$ is given by:

$$N_{ecr0}/N_{Ne} = k_{Ne} = (T_{ecr})/k_{ei}(T_{ecr}) \approx 5 \cdot 10^{-4}.$$

If the initial electron density is greater than the critical $N_e > N_{ecr}$ lasing is impossible, even if the prepulse population of metastable states is for any reason negligible. Of course, this does not mean that when $N_e < N_{ecr}$ lasing unavoidably takes place.

The choice of initial data. When considering the distribution of the gas temperature at radius r of a long tube the expression [51] is used:

$$T_g(r) = \left[T_w^{b+1} + \left(1 + \frac{4r^2}{d^2}\right) \frac{W_d(b+1)}{4\pi A} \right]^{1/(b+1)}.$$

This assumes that the thermal conductivity of the gas is approximated by the expression of $\kappa = A \cdot T_g^b$ (for neon $A = 8.96 \cdot 10^{-6}$ W·cm^{-1}·K$^{-1.683}$, $b = 0.683$ [2]), T_w is the temperature of the walls; d is the tube diameter; W_d is the average power injected per unit length of the tube. It is believed that all of the input power goes into heating the gas.

When the input power is $W_d = 50$ W/cm (2 kW tube per 40 cm length) and wall temperature $T_w = 1590$ °C $= 0.161$ eV then for the ratio of the temperature on the axis of the tube to the wall temperature we have $T_g(0)/T_w = 2$. The calculations use the average gas temperature $T_g = 1.5 \cdot T_w$.

In [2,51,59,60] there are data for the populations of metastable N_m, in particular, for the initial moment of time. Assuming that the population of the metastable state is associated with the population of the ground-state by the Boltzmann distribution, for the copper vapour density we have:

$$N_{Cu} - N_m \cdot (2/6) \cdot \exp(1.389 \cdot eV/T_{e0}) \approx 4 \cdot 10^{15} \text{ cm}^{-3},$$

where $T_{e0} \approx 1.22T$ is the initial electron temperature (see below), $N_m \sim 1 \cdot 10^{13}$ cm^{-3} is the prepulse population of the metastable atoms. In the calculations, focusing on experimental data, we used a slightly lower value $N_{Cu} = 2 \cdot 10^{15}$ cm^{-3}.

For the problem of ionization the initial values of density N_{e0} and temperature T_{e0} of the electrons were taken from calculations of the afterglow. For calculations of the afterglow the initial density N_{e0} and temperature T_{e0} followed from solving the problem of ionization. Iterative calculations are performed to ensure that the initial and final values of the density and electron temperature in these problems are consistent.

As a result, the following values were obtained for the initial density and electron temperature in front of the pulses, following after 100 μs: $N_{e0} \approx 2.3 \cdot 10^{13}$ cm^{-3}, $T_{e0} = 1.22 \cdot T$ – for the initial conditions of the ionization problem; $N_{e0} \approx 3.3 \cdot 10^{14}$ cm^{-3}, $T_{e0} \sim 1$ eV – for the initial conditions in the afterglow problem. The temperature and electron density at other times of the afterglow are given in Fig. 8.8.

The results of the calculations. The parameters required for the calculations are given in Table 8.2. Figures 8.9 and 8.10 show the results of two calculations with different initial electron densities. These densities correspond roughly to the times $\Delta t = 70$ ns and $\Delta t = 15$ ns in the afterglow of the previous pump pulse in the experiments [2,51,59,60].

As can be seen from Figs. 8.9 *b* and 8.10 *b*, the bulk of the power supplied to the medium by Joule heating is consumed under these conditions for the ionization of copper. Only a small amount of energy is used for gas heating due to elastic collisions and ionization of neon.

The electron temperature has a maximum near the point in time, for which the power introduced into the medium is compared with the power expended on ionization. This takes place in the region of growth of the beam current. The presence of a maximum in the electron temperature is crucial for achieving lasing conditions (see below).

When the neon density is high (our case) the elastic collisions with the neon atoms at all ionization stages (with the exception of the initial stage) dominate over Coulomb collisions with the electrons and ions, i.e. collisions with neon atoms make the main contribution to the resistivity of the plasma. Calculations show that at low neon densities the Coulomb collisions dominate during much greater period of time.

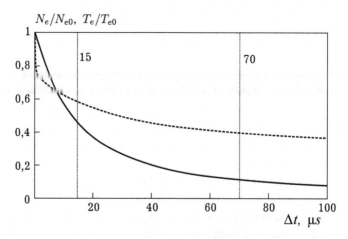

Figure 8.8. The time dependence of the density (solid curve) and electron temperature (dashed curve) in the afterglow: $N_{e0} = 3.3 \times 10^{14}$ cm^{-3}, $T_{e0} = 1$ eV.

Table 8.2. The parameters used in the calculations

The density of neon	$N_{Ne} = 1.5 \cdot 10^{18}$ cm^{-3}
The density of the copper vapour	$N_{Cu} = 2 \cdot 10^{15}$ cm^{-3}
The gas temperature	$T_g = 2800$ K
The length and diameter of the plasma column	$L = 40$ cm, $d = 2$ cm
Peak current and current density	$J_{max} = 300$ A; $j_{max} = 95.5$ A/cm^2
The initial density and electron temperature in the afterglow	$N_{e0} = 3.2 \cdot 10^{14}$ cm^{-3} $T_{e0} = 0.8$ eV
Prepulse electron density and temperature for the delays in the afterglow of 100 μs, 70 μs and 15 μs, respectively	$N_{e0} = 2.6 \cdot 10^{13}$ cm^{-3}; $N_{e0} = 3.7 \cdot 10^{13}$ cm^{-3}; $N_{e0} = 1.5 \cdot 10^{14}$ cm^{-3};

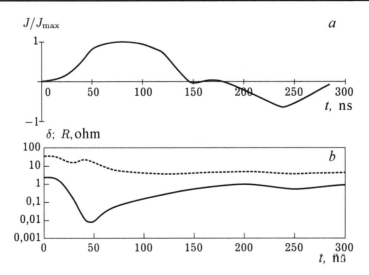

Fig. 8.9. Time dependence of the current and the characteristics of the plasma during the heating pulse: *a*: solid line – the current passing through the tube $J(t)$, dependence is taken from experiments [2,51,59,60], *b*: solid line – $N_e(t)$, dotted line – $T_e(t)$; initial electron density $N_{e0} = 3.7 \times 10^{14}$ cm^{-3}, roughly equivalent to $t = 70$ ns for the afterglow (see Fig. 8.8).

This determines the resistance of the plasma in the initial moments of time.

The above simple estimate (8.5) is confirmed by direct calculations. For example, at $T_{ecr} = 2$ eV, $j_{max} = 95$ A/cm^2, $N_{Cu} = 2 \cdot 10^{15}$ cm^{-3} for the critical value of the electron density we have $N_{ecr} = 1.6 \cdot 10^{14}$ cm^{-3}.

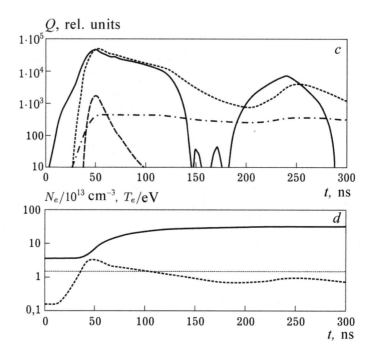

Figure 8.9. *Continued.* Time dependence of the current and the characteristics of the plasma during the heating pulse: *c*: solid line – power, introduced to the medium (Joule heating $j^2(t)/\sigma$); dotted line – power expended on copper ionization Q_{iCu}; dashes – power expended for ionization of neon Q_{iNe}; dot-dash – the power consumed for heating the gas in the elastic collisions $Q_{\Delta T}$; *d*: solid curve – the ratio of the contribution to the conductivity of plasma of Coulomb collisions to the contribution of elastic collisions with neon $\delta = (K_{Ne}(T_e) \cdot N_{Ne}/2k_{ei}(T_e)Ne)$; dotted line – the resistance of the plasma column *R*. The initial electron density $N_{e0} = 3.7 \times 10^{14}$ cm^{-3} corresponds to approximately $t = 70$ ns for the afterglow (see Fig. 8.8).

The calculation for the initial electron density $N_{e0} = 1.5 \cdot 10^{14}$ cm^{-3} (see Fig. 8.10) shows that the electron temperature is really not sufficient for lasing.

According to the above calculations of the afterglow (see Fig. 8.8), focused on experiments [2,51,59,60], the electron density at 15 μs after the pulse is $N_{e0} = 1.5 \cdot 10^{14}$ cm^{-3}, which is close the critical value. Therefore, as shown by experiments with double pulses (see Fig. 6 in [51]), when applying the second pumping pulse 15 μs after the end of the first pulse the population of the resonant level is less than the population of the metastable during the entire pump pulse. If the second pulse is applied after 70 ns, the initial electron density $N_{e0} = 3.7 \cdot 10^{13}$ cm^{-3} is less than critical. In this case, according to the same measurements the inversion does take place.

This does not mean that the high initial electron density is in the experiments [2,51,59,60] the only reason for failure of lasing at a small delay in the double pulse. The high initial population of the metastable states also worsens the inversion conditions. However, it is clear that even af the population of metastable states equal to zero the inversion may stall due to the high residual electron density.

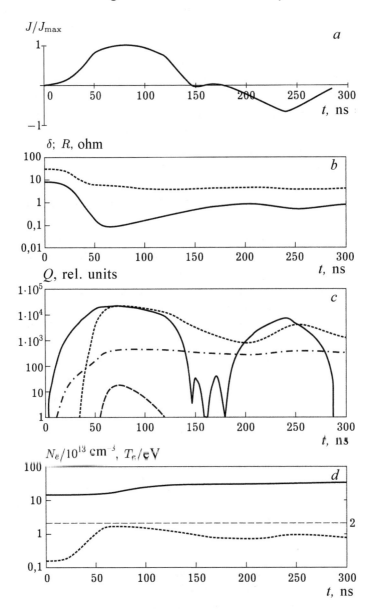

Figure 8.10. Same as in Fig. 8.9. The initial electron density $N_{e0} = 1.5 \times 10^{14}$ cm^{-3} corresponds approximately to $t = 15$ ns for the afterglow (see Fig. 8.8).

Detailed kinetic model [5–7]

Section 8.2 provides a detailed kinetic model of the active medium, which describes the time variation of the volume-averaged values of the level populations of the copper atom and neon, the density of the copper and neon ions, the electron temperature and laser radiation intensity on the green and yellow lines of copper.

Testing of the model. We experienced difficulties when testing the model using the experimental. The fact is that although the total number of experimental works on the CVLs is very large (see, e.g., in [1,2]), comprehensive measurements of some parameters of the active medium were carried out only in the works of Petrash [51,59–63], a group of Piper [16] and in Hogan's thesis [64]. However, the data in these studies are also not sufficient for complete analysis. For example, in [16] not measurements were taken of the initial prepulse concentrations of electrons and copper atoms in the ground state. In comparisons with the calculations and experimental data from this work, in our calculations these concentrations were chosen equal to the estimated values used in calculations [16] (see below). The time dependences of the lasing power were not measured.

In [60,63] no measurements were taken of the time dependences of the concentrations of the upper and lower states for the yellow lines, in addition measurements were made only in the small signal gain mode.

However, when using the experimental dependence of the current on time we obtained a good agreement of the populations of the levels of copper, the electron density and lasing power on the 510 and 578 nm lines with the available experimental data [16,62–64].

Comparison with the calculated data of the model by Karman and others [16]. The model was tested by comparing the different simulation characteristics, calculated on the basis of the model described above, with the theoretical time dependences of the voltage across the discharge gap, the reagent concentrations of the active medium, the lasing power on the 510 and 578 nm lines, and also the electron temperature from [16].

The calculations used the second of the specified versions of the model using the experimental time dependence of the current passing through the discharge gap. This dependence has the form shown in Fig. 8.11 a. The calculated time dependence of the voltage across the discharge gap (Fig. 8.11 a) has the same shape of the curve as in calculations [16], but exceeds it by about a factor of 2.

Temporal behavior of the reactant concentrations. The calculation in the present model slightly overestimates the population of resonance levels and lowers the population of metastable levels as compared with the data in [16] (Fig. 8.11 b, c).

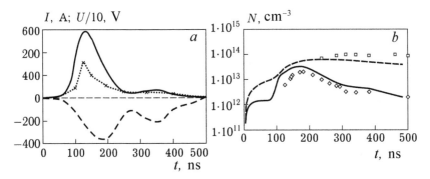

Figure 8.11. Comparison with the calculated data of the model by Karman and others [16] *a*: dependence of the current in the discharge gap (dashed line) on time in the experiments [16]. This relationship is used to set the pumping in the following calculations. Comparison of the calculated time dependence of the voltage in the discharge gap by the present model (solid line) with the calculated dependence (icons) in [16], *b*: comparison of the calculated dependences of the concentration of resonance ($P_{3/2}$) (solid line) and metastable ($D_{5/2}$) (dashed line) levels of the copper atom on time by the present model with experimental (points) dependences of [16]. The calculated and experimental curves [16] are in good agreement with each other so the former are not given.

In [16] experimental measurements were taken only of the time dependence of the concentrations of resonance and metastable states of copper, and the calculated curves of the same work are in good agreement with them. However, there are also calculated dependences of the concentrations of other reactants which are also interesting to compare, because both models are very close. This comparison for the time dependence of the concentration of excited states of copper $Cu(5^2S_{1/2})$, Cu**, as well as concentrations of the ground states of the copper atom and ion for the two models is shown in Fig. 8.11 d.

Lasing pulses. The form of the calculated time dependences of the radiation power at different wavelengths differs significantly (Fig. 8.11 e) in the calculations by the present model and the model in [16]. In the present model there is a delay in the lasing pulse at the 578 nm line compared with the line at 510 nm, in the model in [16] there is no such delay. The lasing pulse durations are about the same in both cases.

In the experiment, the emitted energy was 0.51 + 0.25 mJ, where the first number refers to the 510 nm line, and the second – to the 578 nm line. The model in [16], as the authors themselves point out, gives a value of 0.62 + 0.33 mJ, our model – the value of 1.28 + 0.94 mJ. Here we should note two things. First, if we use the data in [16] on the calculated output power, the value of the estimated radiated energy must be 0.37 + 0.19 mJ and not 0.62 + 0.33 mJ. Second, the initial concentration of copper atoms in [16] in

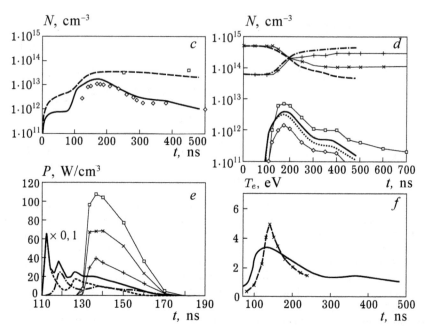

Figure 8.11. *continued.* Comparison with the calculated data of the model by Karman et al [16] *c*: Comparison of the calculated concentration dependence of the resonance $(P_{1/2})$ (solid line) and metastable $(D_{3/2})$ (dashed line) levels of copper atoms on time according to the experimental model (icons) and the dependences in [16]. Calculated and experimental dependences [16] are in good agreement with each other, whereby the former are not presented. *d*: Comparison of the calculated dependence of the concentration of copper ions Cu⁺ (dash-dotted line), as well as the ground Cu (dashed line) and excited levels of Cu (5 $S_{1/2}$) (dotted line) and Cu** (solid line) on time represented by the model with the calculated dependences of [16] (symbols, respectively direct crosses, oblique crosses, diamonds, squares), *e*: Comparison of the calculated time dependences of the specific total lasing power (solid line) and the lasing on the lines 510 (dashed line) and 578 nm (dash-dotted line), calculated from the present model (power divisible by 10 are given), with the calculated dependences of the specific powers (symbols: squares – total lasing, oblique crosses – lasing at the 510 nm line and direct crosses – lasing at 578 nm) of the work [16]. *f*: Comparison of the calculated time dependence of the electron temperature (solid line) by the presented model with the calculation (points) [16].

the calculations was chosen to be $5 \cdot 10^{14}$ cm⁻³ (Fig. 8.11 d). Since in this study we compare our estimates with the calculated data in [16], then we have chosen in the calculations the same value for the initial concentration of the copper atoms (as well as for the initial electron density (copper ions) used in the calculations [16]). However, the authors [16] pointed out that the wall temperature of the gas discharge chamber was 1790 K, the vapour pressure at a temperature of $1.85 \cdot 10^{15}$ cm⁻³, which means the average concentration of copper atoms in the cross section should be about $1 \cdot 10^{15}$ cm⁻³. Note also that since neither the initial concentration of copper atoms, the initial concentration of copper ions nor temporary forms of lasing power pulses were measured in the experiments, the comparison of the energy radiated in each line can be provide only a small amount of information.

The electron temperature. Qualitatively the behaviour of the time dependence of the electron temperature in our model and the electron temperature corresponding to the low-energy part of the distribution function [16] coincide (Fig. 8.11 *f*). The width of the dependence in our case is slightly greater and the maximum value is smaller compared with the dependence in [16].

Comparison with the results of G.G. Petrash et al. [60,63]. *Description of the laser.* The measurements [60,63] of the time dependence of the reagent concentrations and amplification of the probe signal in copper vapour lasers were performed for a standard industrial-type gas discharge tube of the UL-101 type. The active zone of the tube was 40 cm long and 2 cm in diameter. Neon pressure was about 300 Torr, the temperature of the tube walls T_w varied from 1400°C to 1650°C. The maximum power achieved at T_w = 1570–1590°C. The discharge tube was excited with a constant power of 2 W received from a rectifier, at a pulse repetition frequency of 10 kHz.

Time dependences. Comparison was made for the wall temperature T_w = 1590°C [63]. As in the previous paragraph, the pumping of the medium was calculated using the experimentally measured time dependence of current passing through the active medium (Fig. 8.12 *a*).

The concentration of electrons. The time dependence of the electron concentration was not measured. In the calculations, the chosen concentration was such that the voltage in the tube, determined by the resistance of the discharge plasma did not exceed the experimental by more than two-fold (Fig. 8.12 a). The same ratio was also obtained in the comparison between the calculated voltage in our model and the model in [16] (see above). For this value of the initial electron density ($2 \cdot 10^{14}$ cm^{-3}) we obtained the correct ration (of 4) of the concentration of the resonance levels $P_{3/2}$ in the first maximum to the concentrations of these same levels in the second maximum in time (Fig. 8.12 b). According to estimates in [60,63], the electron concentration at the end of the discharge pulse is approximately $(6-7) \cdot 10^{14}$ cm^{-3}. The calculated time dependence of the electron density is shown in Fig. 8.12 c. It is seen that the calculated value at the end of the pulse is in good agreement with these estimates.

Density of atomic states. In contrast to [16], in [60,63] measurements were taken on the concentration of resonance and metastable levels only for the 510 nm transition (Fig. 8.12 b). But in [63,60] there are data on the measurement of the density of copper in the ground state (Fig. 8.12 c), missing in [16]. This information is valuable because the prepulse concentration of copper atoms on the axis of the tube can, depending on the temperature of the walls and other parameters for pipes with a diameter of 2 cm, differ from the concentration at the wall of the laser tube by up to fivefold [62], and its theoretical definition may be an additional source of error.

Weak signal gain. In [60, 63] the time dependence of the gain of the probe signal was also measured. This dependence was compared with the time dependence of the weak signal gain, which was calculated from the model (Fig. 8.12 *d*).

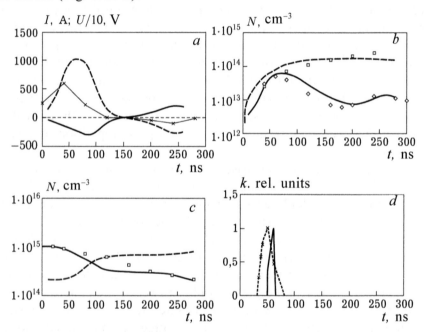

Figure 8.12. Comparison with the results of G.G. Petrash et al. [60,63]: *a*: experimental time dependence of the value of the current flowing through the medium (solid line) [63]. This relationship is used to define pumping in the following calculations. Comparison of the calculated time dependences of the voltage produced by the active resistance of the plasma discharge by the present model (dashed line) with the experimental dependence (icons) in [63]. *b*: comparison of the calculated time dependences of the concentrations of resonance ($P_{3/2}$) (solid line) and metastable ($D_{5/2}$) (dashed line) levels by the present model with the experimental (points) dependences of [63]. *c*: Calculated time dependence of electron concentration (dashed line). Comparison of the calculated time dependences of the concentration of copper atoms in the ground state by the present model (solid line) and experimental (symbols) dependence of [63]. *d*: Comparison of the calculated time dependence of the coefficient of the weak signal gain (solid line) with the experimental time dependence (dashed line, circles) of the amplification of the probe signal [63].

Table 8.3. Parameters of the laser under test [64]

Plasma tube diameter	42 mm
Plasma tube length	1500 mm
The buffer gas	neon
Buffer gas pressure	25 mbar
The pulse repetition frequency	6500 Hz
The voltage in the storage capacitor	16 kV
Power pumping (based on stored energy)	58 W
The average laser output power	40 W
Plasma tube wall temperature	1700 K
Storage capavity	7 nF
Peaking capacity	2 nF

The experimental dependences in Figure 8.12 correspond to the tube axis. For the optimum temperature of the walls of the discharge tube T_w = 1590 C measurements on the tube radius showed that the vast majority of the concentrations of various states of copper is almost uniform along the radius of the tube (the difference does not exceed a factor of 2), which agrees with the conclusions of [16] where it is reported that a substantial inhomogeneity begins to manifest itself at a pulse repetition frequency of v > 20 kHz. The experimental measurements show that the large difference in the inhomogeneity (2 to 5 times) can occur only for the population of the ground state at the time of maximum depletion of this level and for the population of the metastable $D_{5/2}$ in the prepulse period. During the excitation and lasing period the dependences of the population of the upper $P_{3/2}$ and the lower $D_{5/2}$ levels of the laser transition at the 510 nm line are weakly inhomogeneous.

Comparison with Hogan's results [64]. The study [64] is probably the only paper in which measurements were the temporal and spatial dependences of the most complete set of concentrations of various reagents and simultaneously the lasing power at the 510 and 578 nm lines. The measurements were made during lasing. A model of a copper vapour laser Oxford Laser CU-25, modified specifically for this study, was used. The known parameters of this laser are shown in Table 8.3. They do not include the parameters of the resonator. In comparing the calculations with the experimental data the inverse lifetime of a photon in the resonator (see 8.2) $\gamma = c\ln (1/(r_1 r_2))/2l = 6.44 \cdot 10^8$ s^{-1} was adjusted so that the calculation reproduced a pronounced vibrational structure of the radiation power at 510 nm, which took place in the experiment (Fig. 8.13 b). In addition, the equation of radiative transfer in zero-dimensional approximation took into account the radiation losses in the resonator – wI, where $w = 8.5 \cdot 10^8$ s^{-1} was chosen so that the calculated output power corresponded to the experimental power (40 W). We note immediately that these parameters – γ and w do not affect the dependences of the concentrations of various reagents on time discussed below.

Time dependences. As in comparison with the experiments [16,60,63] (see above), pumping in the model was calculated using the dependence of current on time presented in the experiment (Fig. 8.13 a). In [64] it is

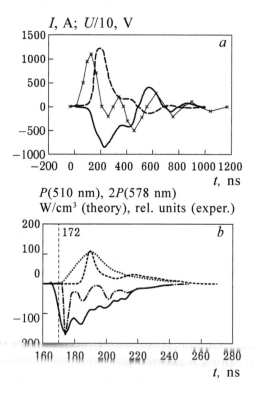

Figure 8.13. Comparison with the results of Hogan [64]: *a*: experimental time dependence of the current flowing through the active medium (solid line) [64]. This relationship is used to set the pumping the following calculations. Comparison of the calculated time dependence of the voltage produced by the resistance of the discharge plasma using the present model (dashed line) with the experimental dependence of the tube voltage (circles) from [64]. *b*: Comparison of the calculated time dependence of lasing power on the 510 nm line (dash-dotted line) in the present model (W/cm^3) with the experimental dependence (solid curve, rel. units) in [64] and the calculated dependence of the double lasing power at 578 nm (dashed curve) for the present model (W/cm^3) with the experimental curve (dotted line, rel. units) in [64].

noted that the lasing starts after about 200 ns after the beginning of the current pulse, and the current value in the first 100 ns is very small and the criterion by which to gauge whether the current pulse started or not is not clarified. When using the experimental dependence of the current on time the lasing pulse in the calculation starts at 172 ns (Fig. 8.13 b). When combining the first maxima of the time, we see that in this case the beginning of the lasing in the experiment must comply with the time of 164 ns. In [64] the zero time value was represented by the start of lasing. Therefore, when comparing the results further, we assume that at time $t = 164$ ns in the calculation corresponds to the time $t = 0$ in the experiments, i.e. in comparison the experimental curves are shifted by 164 ns. The time lag in the lasing pulse on the 578 nm line relative to the 510 nm line, obtained in the model, coincides with the experimental values (Fig. 8.13 b).

Voltage. The calculated time dependence of voltage in the active resistance of the discharge plasma on the time and the experimental time dependence of voltage in the discharge tube are shown in Fig. 8.13 a. Generally speaking, these voltages may not coincide exactly, since the discharge tube also has some inductance and capacitance. This comparison is purely illustrative. The figure shows that the tube should in fact have a capacitance or in the experimental curves there is some shift in time, because if we accept that in addition to the active resistance the tube has only inductance, it is unclear why the experimental voltage after 164 ns begins to fall, because the derivative of current with respect to time has not yet changed its sign.

The population of working levels. There is also a noticeable shift in time between the calculated and experimental time dependences of the concentration of the metastable levels (Fig. 8.13 c). The maximum experimental values of the concentration of metastable levels of $D_{3/2}$ were 50% lower than calculated. The temporary forms of the calculated and experimental dependences of the concentrations resonance levels coincide (Fig. 8.13 d), but the maximum experimental value is 2.6 times lower in comparison with the calculated level for the $P_{3/2}$ and 2.7 times lower than the calculated level $P_{1/2}$. In [63] it is noted that the author was not quite sure regarding the time dependence of the concentrations of the $D_{3/2}$ levels, since the measurement of populations of the $D_{3/2}$ levels by the method of hooks used in the experiments is much more complex than measuring the populations of other levels. With respect to time shifts of the dependences in Fig. 8.13 c, they seem to be associated with inaccuracies in the calibration of the time scale of the experiment. Indeed, Figs. 8.13 e and 8.13 f show the experimental and calculated time dependence of the populations of the resonant and

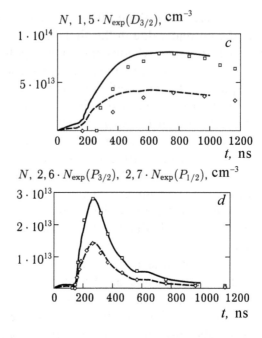

Figure 8.13. Comparison with Hogan's results [64] *c*: Comparison of the calculated time dependences of the concentration of metastable $(D_{5/2})$ (solid line) and $(D_{3/2})$ (dashed line) levels of the presented model with the experimental dependences (squares and diamonds respectively) from [64]. For the level $(D_{3/2})$ the experimental values are approximately one and a half times less than the calculated values. *d*: Comparison of the calculated time dependences of the concentrations of resonance $(P_{3/2})$ (solid line) and $(P_{1/2})$ (dashed line) levels by the present model with experimental dependences (respectively squares and diamonds) from [64]. The experimental values are about 2.6 and 2.7 times, respectively, lower than the calculated values for the levels $P_{3/2}$ and $P_{1/2}$

metastable levels. The concentrations of the metastable levels in Figure 8.13 e are multiplied by the ratio of the statistical weights (2/3) of the resonance and metastable levels, i.e. the intersection of the curves in the figure corresponds to the zero gain coefficient in the laser medium. For example, Fig. 8.13 f shows that if the calibration of the time line in the experiment is chosen correctly, the lasing at the 578 nm line should not begin after 170 ns (Fig. 8.13 b) but earlier than 164 ns (Fig. 8.13 f), and should not end at 250 ns (Fig. 8.13 b) but later than t = 290 ns (Fig. 8.13 f). The concentrations of the metastable levels in Figure 8.13 f are multiplied by the ratio of the statistical weights (1/2) of the resonance and metastable levels, i.e. the intersection of the curves in the figure corresponds to the zero coefficient of the gain in the laser medium.

Concentration of the ground state of copper and electrons. Burnout of copper in the calculation is 1.5 times faster than in the experiment, and the electron concentration at the end of the pulse

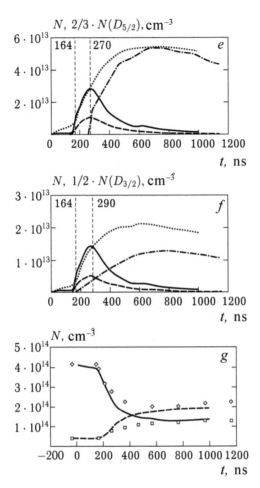

Figure 8.13. *continued.* Comparison with Hogan's results [64]: *e*: experimental [64] (dashed line, dash-dotted line) and calculated (solid line, dotted line) time dependence of the concentration of resonance ($P_{3/2}$) and metastable ($D_{5/2}$) levels. *f*: experimental [64] (dashed line, dot-dash) and calculated (solid line, dotted line) time dependences of the concentrations of the resonance ($P_{1/2}$) and metastable ($D_{3/2}$) levels. *g*: Comparison of the calculated time dependences of the concentration of the copper atoms in the ground state by the present model (solid line) with the experimental curve (diamonds), and the time dependence of electron concentration by the present model (dashed line) with the experimental dependence (squares) of [64].

reaches the value 1.5 times greater than in the experiment (Fig. 8.13 g). It should be noted that the electron density was quite accurately measured at the front of the excitation pulse and during the collapse. During the time when the electron density is close to saturation, the author [64] found it difficult to give an exact value of the electron concentration.

When comparing the calculated data with experimental data presented above, the experimental measurements correspond to the measurements at the discharge axis. A tube with a diameter of 4.2 cm was used in the experiments. The distribution of populations of different states of copper in the tube was close to uniform, the ratio of the maximum and minimum values for most of the diameter of the tube (except for the near-wall region of ~1–2 mm) was approximately 1.5. The prepulse electron concentration distribution in the tube cross section was almost uniform, after the excitation pulse the concentration on the tube axis exhibited a pronounced maximum, and $n_e(r = 0 \text{ mm})/n_e(r = 17.5 \text{ mm}) \approx 2$, $n_e(r = 0 \text{ mm})/n_e(r = 20 \text{ mm}) \approx 4$. In [64] it is stated that the maximum on the axis of the tube can be caused by the contribution of regions near the electrodes with pure neon where the discharge is shunted and the electron concentration is different from zero only in a small region near the axis of the discharge. These areas could not be excluded from the measurements, as n_e was determined using the radiation propagating in the longitudinal direction. For a tube with a diameter of 4.2 cm the delay in the beginning of the lasing pulse on the axis as compared with the beginning of the lasing pulse at the wall is maximum for the 510 nm line and is less than 7 ns. The pulse intensity varies along the radius of the tube by no more than 1.5 times.

Effect of initial density of electrons and metastable copper atoms on the failure of lasing
Electrical circuit. The question of the existence of the critical initial electron density was investigated using one of the following three-contour electrical circuits (Fig. 8.14, see also [53]). This circuit corresponds to the following equations for the currents and voltages:

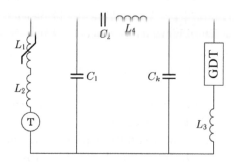

Figure 8.14. Diagram of the electrical circuit that was used in the calculations. GDT – gas discharge tube, T – thyratron, inductance $L_{10} = 27.2$ µH, $L_2 = 1.5$ µH, $L_3 = 3.2$ µH, $L_4 = 20$ nH, capacitance $C_1 = 1.5$ nF, $C_2 = 1$ nF, $C_k = 0.235$ nF.

$$-(L_1+L_2)\frac{dI_1}{dt}=I_1R_{th}-U_1, \qquad \frac{dU_1}{dt}=-\frac{I_1+I_2}{C_1},$$

$$-L_4\frac{dI_2}{dt}=-U_1+U_3-U_2, \qquad \frac{dU_2}{dt}=-\frac{I_2}{C_2},$$

$$-L_3\frac{dI_3}{dt}=I_3R_d-U_3, \qquad \frac{dU_3}{dt}=\frac{I_2-I_3}{C_k}.$$

Here I_1, I_2, I_3 are the currents in the first, second and third circuit, respectively (the current I_3 flows through a discharge tube with a resistance $R_d(N_e, T_e)$); U_1, U_2, U_3 is the voltage across the respective capacitors.

The equations for voltage and current in the electrical circuit were solved together with the kinetics equations which were used to determine the resistance of the tube from the temperature and electron density: $R_d = \rho l/S$, where $l = 40$ cm is the length of the tube, $S = \pi R^2$ is its cross section (the tube radius $R = 1$ cm). The voltage in the tube was calculated from Ohm's law: $U_d = R_d I_3$.

The initial conditions for the circuit were taken by the capacitor voltage $U_1 = U_2 = 14$ kV; after firing the thyratron and polarity change of the voltage in the first capacitor, the total voltage across the two capacitors was about 28 kV. It was assumed that at the initial time there are no currents:

$$I_1(0) = I_2(0) = I_3(0) = 0.$$

The time dependence of the resistance of the thyratron was expressed through the time dependence of the current $I_1(t)$, flowing through the thyratron,

$$R_t(t) = R_0 + R_1 \cdot \exp(-4 \cdot I_1(t)/I_0),$$

using the following parameters: $R_0 - 3$ Ohm, $R_1 - 2$ MOhm, $I_0 - 1$ A. The time dependence of the variable inductance is also expressed through the time dependence of the current $I_1(t)$,

$$L_1(t) = L_{10} \cdot \exp(-0.1(I_1(t)/I_0)^4).$$

The following values were used for constant inductances and capacitances: $L_{10} = 27.2$ μH, $L_2 = 1.5$ μH, $L_3 = 3.2$ μH, $L_4 = 20$ nH, $C_1 = 1.5$ nF, $C_2 = 1.$ nF, $C_k = 0.235$ nF. The following values were assumed in the calculations: $N_{Cu} = 0.6 \cdot 10^{15}$ cm^{-3}, $N_{Ne} = 1.62 \cdot 10^{18}$ cm^{-3}.

Disruption of lasing with increasing initial electron density. In [52] the author evaluated the critical density (4.3) through the experimental values of peak current j_{max}. It was assumed that the lasing is disrupted

when the electron temperature during the pulse does not exceed the
critical value of $T_{ecr} \approx 2$ eV.

The fact that the higher initial electron density reduces the maximum
electron temperature during the pump pulse, and this leads to the
disruption of the inversion, is illustrated by calculations based on our
detailed model (Figs. 8.15, 8.16). Indeed, at a relatively low initial
electron density $N_e(0) = 4 \cdot 10^{13}$ cm^{-3}, the electron temperature at the
peak is significantly higher than T_{ecr} and the population inversion takes
place (Fig. 8.15). At a higher initial density $N_e(0) = 4 \cdot 10^{14}$ cm^{-3}, he

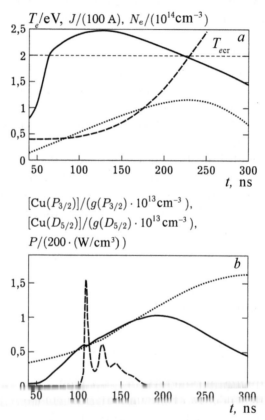

Figure 8.15. The dependences on the plasma parameters and the populations levels
at a low initial density of electrons (ns): *a*: electron temperature T_e (eV, solid line),
the current through the discharge tube, J (in units of 100 A, dotted line); the electron
density N_e (in units of 10^{14} cm^{-3}, the dashed curve), *b*: population of the resonant level
$[Cu(P_{3/2})]/g(P_{3/2})$ (units of 10^{13} cm^{-3}, solid line); the population of the metastable
level $[Cu(D_{5/2})]/g(D_{5/2})$ (units of 10^{13} cm^{-3}, dotted line); specific total power of the
laser radiation in two lines P (units of 200 W/cm^3, dashed curve). Here $g(P_{3/2}) = 4$,
and $g(D_{5/2}) = 6$ is the statistical weight of respective states, initial electron density
$4 \cdot 10^{13}$ cm^{-3}, the initial density of metastable levels $1 \cdot 10^{13}$ cm^{-3}, the initial density of of
Cu atoms in the ground state $0.6 \cdot 10^{15}$ cm^{-3}, the concentration of neon $1.62 \cdot 10^{18}$ cm^{-3}.

electron temperature, even at the maximum, is much lower than the T_{ecr} and the population inversion does not occur (Fig. 8.16).

In order to find the critical electron density, a series of calculations with a change in values of the initial electron density was carried out. The initial density of the metastable atoms was assumed to be zero. This was done in order to demonstrate that the high initial density of electrons can disrupt the generation, even in the absence of the metastable atoms. Different initial electron densities correspond to different frequencies of the pump pulse repetition frequency (see [52]). As might be expected, according to [52] the lasing fails when the initial electron density exceeds a critical value $N_{ecr} \approx 2 \cdot 10^{14}$ cm^{-3} (see Fig. 8.17).

Compare this result with the estimate (8.5). To calculate the conditions under consideration, we have: $j_{max} = 67.5$ A/cm^2, $N_{Cu} = 0.6 \cdot 10^{15}$ cm^{-3}, $N_{Ne} = 1.62 \cdot 10^{18}$ cm^{-3}. Hence, at the critical electron density from (8.5),

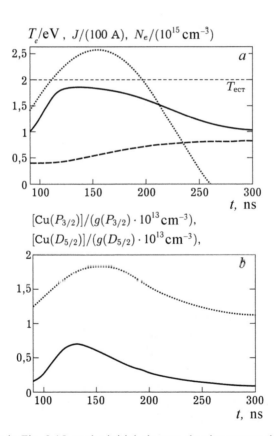

Fig. 8.16. Same as in Fig. 8.15, at the initial electron density greater than the critical value of $4 \cdot 10^{14}$ cm^{-3}. The initial density of metastable atoms were set equal to zero.

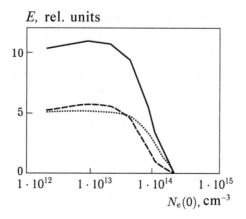

Fig. 8.17. Dependence full laser pulse energy E (rel. units). On 510 lines (dashed line) and 578 (dotted line) nm, and the total energy generation (solid line) of the initial electron density (cm^{-3}). The initial population of the metastable atoms were set equal to zero

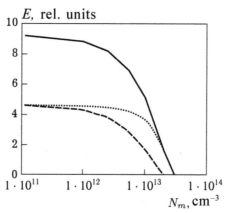

Figure 8.18. The dependence of the output energy E (rel. units) on lines 510 nm (dashed line) and 578 nm (dotted line), and the total energy generation (solid line) of the initial population of the metastable N_m (cm^{-3}). The initial electron density was assumed to be 4×10^{13} cm^{-3},

we have a value $N_{ecr} = 2.27 \cdot 10^{14}$ cm^{-3}. It agrees well with calculations based on the detailed model ($2 \cdot 10^{14}$ cm^{-3}).

Thus, these results confirm the conclusion made in [52] on the existence of the critical prepulse electron density above which the lasing fails. Moreover, the estimated values of the critical electron density are consistent with calculations based on the detailed model.

Disruption of lasing with increasing initial density of metastable atoms. Another series of calculations demonstrated the possibility of failure of lasing due to the high density of metastable atoms. In this series the initial values of the population of metastable atoms $N_{D_{5/2}}(t =$

0), $N_{D_{3/2}}$ $(t = 0)$ in the states $D_{5/2}$ and $D_{3/2}$ was varied. For simplicity, the initial densities of metastable atoms in the $D_{5/2}$ and $D_{3/2}$ states were set equal to: $N_{D_{5/2}}$ $(t = 0) = N_{D_{3/2}}$ $(t = 0) = N_m$. The initial electron density was $N_e(0) = 4 \cdot 10^{13}$ cm^{-3}.

Calculations show that the lasing stops at a sufficiently high initial density of metastable atoms $N_m \approx (2 \div 3) \cdot 10^{13}$ cm^{-3} (see Fig. 8.18). This value is close to the initial electron density. The calculations also showed that with a decrease in the initial electron density by 10 times (to $N_e(0) = 4 \cdot 10^{12}$ cm^{-3}), the critical density of the metastable atoms remained the same and therefore far exceeded $N_e(0)$. However, under normal circumstances, the afterglow population levels (including the metastables) is much smaller than the electron density.

It follows that the restriction on the initial density of metastable atoms, at least in the conditions considered in this case, are less important than the restrictions on the initial electron density. Indeed, if the initial electron density is low, the number of the metastable states will be small. If the initial electron density is high, even if there are no the metastable states, the lasing fails.

The results of experiments [59] were used in [51, 55] as the main evidence that the most important limitation on the pulse repetition rate is the initial density of the metastable atoms. Indeed, according to these studies at high initial densities of metastable atoms the inversion disappears. However, this does not imply that the failure of the inversion occurred only because of the high initial density of the metastable atoms. To directly demonstrate this, we carried out calculations in which the initial density of metastable atoms disrupted the inversion, and then reduced the initial density of electrons and again there was an inversion. For example, when the initial electron density was $8 \cdot 10^{13}$ cm^{-3} the lasing at the 510 nm line fails when the initial density of the metastable atoms reached the value $1 \cdot 10^{13}$ cm^{-3}; with a decrease in the electron density to $4 \cdot 10^{13}$ cm^{-3} and the same initial density of the metastable atoms the lasing took place.

A series of calculations was also carried out in which the initial electron density was varied, and the initial density of the metastable atoms was taken in accordance with the Boltzmann distribution for the initial electron temperature. The initial electron temperature $T_e(0)$ was taken equal to the temperature $T_e(t)$ for the moment t of the afterglow at which the given electron density $N_e(t)$ corresponds to the initial electron density for the pump pulse $N_e(0)$. The dependence $N_e(t)$, $T_e(t)$ in the afterglow is given in [52]. The calculations of the afterglow, carried out within the framework of the model presented here, are consistent with the results in [52]. The effect of ambipolar diffusion and thermal

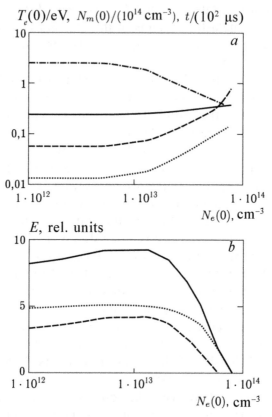

Figure 8.19. The dependence of the initial parameters of the plasma (*a*) and lasing energy (*b*) of the initial electron density (cm^{-3}): *a*: solid line – initial electron temperature (eV), dashed line – the initial population of the metastable state $D_{5/2}$ for the 510 nm line (10^{14} cm^{-3}); dotted line – the initial population of the metastable state $D_{3/2}$ for the 578 line (10^{14} cm^{-3}); dot-dash – the time points (100 µs) at which the electron concentration on the abscissa is realized as the solution of the equations for the electron temperature and concentration in the afterglow; *b*: lasing energy *E* (ref. units): dashed line – the 510 line; dotted line – the 578 nm line; solid line – the total lasing energy.

conductivity of electrons on the dependences of $N_e(t)$, $T_e(t)$ for tubes with a radius about 1 cm is less than 5%.

The results of calculations with the non-zero initial density of metastable atoms are shown in Fig. 8.19. Lasing breaks down at lower (2.5–3 times) initial electron densities than in the case of the zero initial density of the metastable atoms. For these pumping conditions and the density of the copper atoms the restrictions associated with the initial density of metastable atoms are more significant.

Conclusion

The results can be summarized as follows.

1. For a given pumping system there is a critical initial electron density above which the lasing of a CVL stops [52].
2. The calculations also show that there are critical initial densities of the metastable atoms. However, at a sufficiently low initial electron density the critical initial density of metastable atoms at which there is a breakdown of lasing is higher than the initial electron density. It follows that to increase the frequency of the laser pulse repetition we must first reduce prepulse electron density.

The question of the existence of the critical electron density is important because easing of the restrictions related to the prepulse densities of electrons and metastable atoms should be realized in different ways. Let us discuss this.

Due to the restrictions posed by the prepulse density of metastable atoms, we note the following. As discussed here, the relaxation time of metastable atoms under the influence of electron collisions is much shorter than the time between the pump pulses. The high prepulse density of metastable atoms is not connected with the fact that they do not have time to decay before the next pulse, and is associated with a relatively high prepulse electron temperature, due to which the metastable atoms are populated by electron excitation from the ground state. Therefore, when considering the restrictions associated with prepulse density of metastable atoms, we should not talk about the 'residual' density of metastable atoms but we should consider that this restriction is associated both with the relaxation time of electron temperature and the fact that the gas temperature is high. The gap between the gas and the electron temperature can be reduced in many ways, for example, increasing the density of the buffer gas and replacing neon with helium, as well as using small molecular additives. Perhaps it makes sense to try to provide a high density of copper at low gas temperatures, using pulsed evaporation.

The restrictions on the prepulse electron density should be softened by another procedure. The rate of decrease of the electron density between pulses is much lower than the decrease of temperature. Even at an electron temperature equal to the gas temperature, the recombination time can be a significant limitation on the pulse repetition rate. To increase the pulse repetition frequency we should increase the peak current density and decrease the energy input to the plasma whilst increasing the buffer gas density.

8.5. Kinetically enhanced lasers)

Introduction of hydrogen in the active medium of a copper vapour laser under certain experimental conditions leads to a significant increase in lasing efficiency [2,65]. The increase of the lasing characteristics of metal vapour lasers with small amounts of hydrogen was recently observed in a large number of studies (see, for example [66–75]). At this point, the lasers containing a small (a few percent) admixture of hydrogen to the buffer gas are the most effective (reaching 1–3%) at the level of the average power of 100–200 W [2]. Improved laser characteristics for a copper vapour laser were noted in [66,67] and in the study of CuBr vapour lasers [68–70] and CuCl lasers [71].

Comparison of the characteristics of copper vapour lasers of identical size with the addition to the active medium of hydrogen molecules (H_2), bromine (Br_2), pure hydrogen bromide (HBr) and its mixture with hydrogen (H_2) showed that the introduction of impurities of molecular bromine leads to a reduction of the lasing parameters, while the introduction of molecules of hydrogen bromide can more than double the lasing power [76,77]. This work stimulated the creation of metal vapour lasers (see, e.g., [78]), where the active medium of standard copper vapour lasers contained, together with hydrogen, hydrogen halides, e.g. hydrogen chloride (HCl). Lasers of this new species are called the kinetically enhanced lasers. In [78] in investigating the frequency–energy characteristics of a given type of copper vapour laser the authors already an almost threefold increase in the lasing power with the addition of HCl. In [79,80] the hydrogen chloride addition allowed to reach the 100 kHz excitation pulse repetition frequency, with an average output power of 3 W. Increasing the pulse repetition frequency was another positive aspect of adding HCl. High pulse repetition frequencies is required in principle in the problems of precision micromachining of materials, navigation tasks, sensing of the atmosphere, etc. [74]. Thus, a significant increase in the frequency of pulse repetition is possible either with the use of special design features of the active element of the laser, or the introduction to the active medium of additives modifying the kinetics. The use of HCl additives allowed the authors of [81] to obtain from one meter of the active medium (the diameter of the aperture of the laser was 32 mm) more than 100 W average output power or 130 W/liter.

The proposal to use additives of cesium atoms to improve the energy characteristics of copper vapour lasers was first suggested in [82]. It is not clear whether is it really cesium additives that lead to an increase in the lasing characteristics of the laser. For example, in

[83] the introduction of additives of about $10^{15}-10^{16}$ cm^{-3} reduced the lasing energy, and in [84] at much lower concentrations (about $10^{13}-10^{14}$ cm^{-3}) the lasing energy increased 1.3–1.4 times.

Later, the influence of additives of cesium was studied, but not to the active medium of the pure copper vapour laser, and to lasers on the chloride vapours [80,85], bromide vapours [86], and copper vapours and a HyBrID laser [87]. But in this case it is also not yet possible to clearly answer the question of the effect of additives of cesium. In the experiments in [85–87] the authors observed an increase in laser energy with the addition of cesium but in the experiments [80] no such increase was detected.

A detailed analysis of the mechanisms of change in the kinetics after adding hydrogen chloride, cesium, bromine and hydrogen bromide to the active medium of the CVL can be found in [28–31,88].

8.5.1 Additions of hydrogen

The development of kinetic models of copper vapour lasers with added hydrogen has been the subject of several studies [28,29,36,72,73]. Analysis of the performance improvement of the lasers with hydrogen additions was made on the basis of [28,29].

Description of kinetic model

To determine the mechanisms of the effect of hydrogen addition on lasing characteristics a kinetic model was developed using the previously constructed model for pure copper vapour lasers [5–7] (see section 8.2). We discuss briefly additions to the model made to account for hydrogen and hydrogen-containing reagents. We also take into account the ground H, excited H* atomic states, the ions H$^+$, H$^-$, in the case of molecular hydrogen the ground state H$_2$ was taken into account, the state of the vibrationally excited levels $v = 1-5$, the molecular ions H$_2^+$ and H$_3^+$. Also considered was the ground state of the CuH molecules. The model took into account the elastic cooling of electrons, the inelastic processes (direct and stepwise electron-impact ionization of atomic and molecular hydrogen, impact and dissociative recombination of atomic and molecular hydrogen ions, the excitation/de-excitation by electron impact of atomic hydrogen, excitation/de-excitation of vibrational levels by the electrons of the hydrogen molecule (up to $v = 5$ inclusive), VV-relaxation in molecular hydrogen, the quenching of metastable and resonance levels of the copper atom by vibrationally excited hydrogen molecules, the quenching of metastable levels of the copper atom by the negative ion of hydrogen molecules with the formation of a CuH molecule); dissociative attachment of electrons to atomic and molecular hydrogen, the triple ion–ion recombination involving the negative hydrogen ion, association and dissociation of hydrogen molecules.

The heat balance equation takes into account the power density spent on ionization and excitation of hydrogen-containing components (these values are expressed in terms of the population of excited atomic and molecular states, they take into account the processes of excitation and de-excitation of the levels considered in the model), the terms Q_{wall} and $Q_{\Delta T}$ now have the form

$$Q_{wall} = \frac{5.41 \cdot 10^4 T_e^{1.5} N_e}{R^2 \left(\sigma_{eNe} N_{Ne} + \sigma_{eH_2} N_{H_2} \right)}, W/cm^3$$

is the power of the heat sink on the wall; T_e is in eV, $\sigma_{eNe}(T_e)$ is the transport cross section for elastic electron collisions with neon atoms (in units of 10^{-16} cm^2), it is weakly dependent on temperature in the range 2 eV and equals approximately $1.5 \cdot 10^{-16}$ cm^2, $\sigma_{eH_2}(T_e)$ is the cross section of electron collisions with hydrogen (in units of 10^{-16} cm^2) and is about $1.5 \cdot 10^{-15}$ cm^2, and the tube radius R is in cm;

$$Q_{\Delta T} = 2 \cdot [(m_e/m_{Ne}) \cdot k_{Ne} \cdot N_{Ne} + (m_e/m_{Cu}) \cdot k_{ei} N_e + $$
$$ + (m_e/m_{H_2}) \cdot k_{H_2} N_{H_2}] \cdot N_e \cdot (T_e - T_g)$$

is the power density spent on cooling of electrons due to elastic collisions with neon atoms, copper ions and hydrogen molecules; k_{Ne}, k_{H_2}, k_{ei} is the rate of elastic collisions of electrons with neon atoms, hydrogen molecules and ions; N_{Ne} is the density of neon atoms, N_{H_2} is the concentration of molecular hydrogen; m_e is the electron mass, m_{Ne}, m_{Cu}, m_{H_2} is the atomic mass of neon and copper and the hydrogen molecule, respectively.

The non-stationary equations for the concentrations of various reagents, the balance equation for the electron temperature, radiation intensity, and (if necessary) the Kirchhoff equations for electrical circuits (a total of 39 equations) were solved self-consistently using the PLAZER software package (for details see section 8.2, [28,29]). In total, the model took into account ~200 kinetic reactions.

Mechanisms of the effect of hydrogen impurity
We list the assumptions put forward in various papers about the mechanisms of the influence of hydrogen impurities on the work of metal vapour lasers:

M1. In [65] there was no significant increase in the lasing characteristics of copper vapour lasers with hydrogen additives. However, this work seems to be the first work in which it was concluded that the admixture of hydrogen leads to an increase in the optimum pulse repetition frequency. It was noted that the main mechanism of recombination of the plasma is ambipolar diffusion, and the optimal pulse repetition frequency is increased by increasing the three-body recombination with the introduction of hydrogen.

M2. According to [5–7,52,54,89,90] the decrease of the prepulse concentrations of electrons and copper atoms in the metastable state increases the energy and frequency characteristics of metal vapour lasers. Additions of hydrogen can provide these changes: a large cross section of elastic [66,67] collisions with hydrogen enables the effective cooling of electrons can effectively cool in the interpulse period, leading to a decrease in the concentration of copper atoms in the metastable state; the electron cooling, in turn, leads to an increase in the rate of recombination of positive copper ions, which leads to a decrease in the prepulse electron density.

M3. The prepulse electron density in the active medium can also be reduced by other reasons [68–70]:
  1) because of the electron attachment to atomic hydrogen;
  2) due to the dissociative attachment of the electrons to the hydrogen molecules (in particular, to the vibrationally excited molecules);

M4. Improved matching of the gas discharge tube (GDT) with the excitation circuit will allow for more efficient excitation of the active medium and, therefore, provide a more efficient device. This improvement, according to [67–70, 72, 73], is due to the lower initial conductivity of GDT.

M5. In [70,91] it is stated that the introduction of hydrogen in the active medium lowers the extent of depletion of the ground state of copper during the excitation pulse, combined with more rapid recovery of copper in the interpulse interval, especially at the initial stage (first few microseconds). The higher recovery rate of the concentration of copper atoms in the ground state will increase the energy output in the optimum mode. Note that in [68–70] the mechanisms described in M.3–M.5 were discussed with respect to CuBr lasers.

M6. Improvement of the lasing characteristics of a CVL with the introduction of hydrogen is associated with an increase in the concentration of copper atoms in the ground state in the active medium due to the higher wall temperature of the GDT [73]. Apparently, the temperature increase of the GDT is associated with an increase of the energy supplied into the discharge. Note that this mechanism may be the cause of the negative and positive feedback: the increase of the power input leads to an increase in wall temperature, which in turn increases the concentration of copper, because of which, by reason of increasing the frequency of collisions, increases the resistance of plasma, etc., which ultimately may result in a break in lasing.

M7. According to [73] the increase of the thermal conductivity of the active medium with the addition of hydrogen decreases the gas temperature at the axis of GDT, which leads to a decrease in the concentration of copper atoms in the metastable states (the lower laser level) and, accordingly, to increase of the laser radiation power. At present views differ with respect to changes in the thermal conductivity of the plasma after addition of hydrogen. Some studies [73,92] show the variation of the radial profile of gas temperature, other studies indicate that these changes are minimal [69]. It should be noted that the improvement in the uniformity of the radial distribution of the gas temperature is beneficial to the operation of the laser, since it is easier to 'tackle' overheating of the active medium.

M8. In [67] the decrease of the total radiated power with increasing hydrogen concentration above the optimal value is attributed to increasing energy loss in inelastic processes of excitation and ionization of molecular hydrogen. In [73] it is stated that when adding an excessive amount of hydrogen reduces the concentration and temperature of the electrons, which slows down the improvement in the laser characteristics.

M9. According to [93,94], the improvement of the laser parameters may be influenced by an increase in the quenching rate of metastable copper atoms by the vibrationally excited hydrogen molecules with the formation of a CuH molecule.

Table 8.4. The parameter values of the GDT used in the calculations

Parameter	1 [73]	2 [72]
Length of active medium (cm)	90	150
GDT diameter (cm)	4	3.8
Repetition frequency of excitation pulses (Hz)	5000	12000
Neon pressure (Torr)	30	36

A detailed comparison of the theoretical calculations with the experimental and calculated data from [72,73], the most detailed studies known to the authors of this book, is presented below. The simulation results are in satisfactory agreement with these works, so the identification of mechanisms of the influence of hydrogen impurity on the kinetics of the active medium was carried out using the experimental conditions of operation of the lasers in one of these papers, namely [72] (see Table 8.4).

Analysis of the assumptions
Evaluation of the significance of different processes in the calculations is greatly simplified by using flows of these reactions (see for example [95]). The reaction flux in the time interval between the times t_1 and t_2 is defined as

$$F(t_1,t_2) = \int_{t_1}^{t_2} k[N_1]...[N_n] dt,$$

where k is the reaction rate, N_i are the reactants taking part in the reaction. The relationship of these functions directly tells us about the role of reactions in a given interval of time in the appearance or disappearance of any reactant. If we are interested in the contribution of any reaction in the energy balance of the electrons, this contribution is determined by the product of the flux of the given reaction by the amount of energy transferred to electrons by the elementary act defined in this reaction.

A1. In [65] it is argued that the main contribution to the decrease in electron density is provided by ambipolar diffusion. According to the calculations, for example, in the experimental conditions in [72] the ambipolar diffusion is greatly inferior to the bulk recombination in the process of reducing the concentration of electrons. The relaxation of the plasma in pulsed metal vapour lasers is determined by three body recombination. This fact was noted long ago in studies of metal vapour lasers, for example, in [2, 51]. On the other hand, in [65] it is noted that the introduction of atomic hydrogen is preferred to molecular. According to the results of our simulations, the main contribution to the cooling of the electrons comes from the dissociation and vibrational excitation of molecular hydrogen during the excitation pulse (A.8). The specific contribution of the elastic cooling of electrons by molecular hydrogen and atomic hydrogen should be about the same. However, increasing the lasing pulse repetition frequency with the addition of hydrogen is effectively connected with a decrease in electron density due to more intensive cooling of the electrons. As can be seen from Fig. 8.20 a, at the initial stage of the interpulse period the presence of the hydrogen addition gives a much higher rate of decrease of the electron temperature, which in turn leads to more rapid decrease in electron density. This mechanism is most important at high excitation pulse repetition frequencies (see A.6).

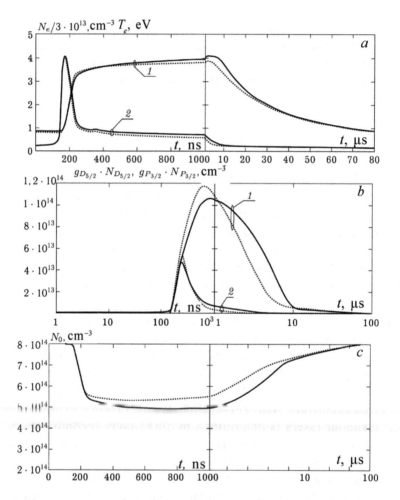

Figure 8.20. Time dependence of *a*: concentration (*1*) and temperature (*2*) of the electrons during the excitation pulse and the interpulse interval. *b*: the level populations $D_{5/2}$ (*1*) and $P_{3/2}$ (*2*) during the excitation pulse. *c*: concentration of copper atoms in the ground state for the excitation pulse and the interpulse interval. (Solid curve – $N_{Cu} = 8 \cdot 10^{14}$ cm^{-3}, $N_{H_2} = 0\%$; dotted line – $N_{Cu} = 1 \cdot 10^{15}$ cm^{-3}, $N_{H_2} = 2\%$). GDT parameters are given in Table 8.4, the values of the plasma reagents, used in the calculations are in Table 8.6 (for $N_{H_2} = 0\%$) and 8.7 (for $N_{H_2} = 2\%$).

Table 8.5. Cooling fluxes of electrons in the interpulse period: F (3 μs, 80 μs) to 2% of hydrogen impurity. In order to fully eliminate the influence of the excitation pulse, t_1 was set equal to 3 μs, the end of interpulse period corresponds to 80 μs (f = 12 kHz)

Reaction	Energy [eV]	Flux (cm^{-3})	
Ne$^+$ + 2·e → Ne** + e	3.1	1.41·10^{13}	1.41·10^{13}
Coulomb cooling	$-5.43·10^{-4} × (T_e - T_g)$	6.272·10^{19}	6.272·10^{19}
Cu($P_{3/2}$) + e → Cu + e	3.817	2.1·10^{13}	2.1·10^{13}
Cu($P_{1/2}$) + e → Cu + e	3.786	1.138·10^{13}	1.138·10^{13}
Cu + e → Cu($D_{5/2}$) + e	1.389	5.139·10^{14}	5.17·10^{13}
Cu($D_{5/2}$) + e → Cu + e		4.618·10^{14}	
Cu + e → Cu($D_{3/2}$) + e	1.642	2.06·10^{14}	1.32·10^{14}
Cu($D_{3/2}$) + e → Cu + e		3.38·10^{14}	
Cu($D_{5/2}$) + e → Cu($D_{3/2}$) + e	0.253	1.233·10^{15}	1.1·10^{14}
Cu($D_{3/2}$) + e → Cu($D_{5/2}$) + e		1.122·10^{15}	
Cu($P_{3/2}$) + e → Cu($P_{1/2}$)+ e	0.0313	4.395·10^{15}	8·10^{12}
Cu($P_{1/2}$) + e → Cu($P_{3/2}$)+ e		4.403·10^{15}	
Cu$^+$ + 2·e → Cu** + e	1.58	6.542·10^{13}	6.542·10^{13}
Electronic heat conductivity	1	1.601·10^{14}	1.601·10^{14}
H* + e = H + e	10.2	2.724·10^{12}	2.724·10^{12}
H$_2$ + e = H$_2$(v=1) + e	0.5	8.36·10^{14}	1.105·10^{14}
H$_2$(v = 1) + e = H$_2$ + e		9.465·10^{14}	
H$_2$(v = 1) + e = H$_2$(v=2) + e	0.5	1.203·10^{14}	2.76·10^{13}
H$_2$(v = 2) + e = H$_2$ (v=1) + e		1.479·10^{14}	
H$_2$ + e = H$_2$(v = 2) + e	1	1.394·10^{13}	1.394·10^{13}
H$_2$(v – 3) + e = H$_2$(v = 2) + e	0.5	2.070·10^{13}	2.878·10^{13}
H$_2$ + e = H$_2$ + e	$-5.43·10^{-4} × (T_e - T_g)$	1.07·10^{15}	1.07·10^{15}

A2. In [67] the authors put forward a mechanism of the effect of hydrogen on the lasing characteristics of copper vapour lasers, based on the fact that the hydrogen molecule has a large cross section of elastic collisions with electrons, and this circumstance, according to the authors of [67], should lead to improved output parameters due to cooling of the electrons.

Numerical experiments suggest that the effect of this mechanism on the kinetics in negligible (Table 8.5) (the GDT parameters are presented in Table 8.4, the values of the initial concentration of the reagents of the active medium are listed in Table 8.7).

When comparing the reaction fluxes it should be mentioned that due to the low energy transferred in elastic collisions, the contribution of the elastic reaction to the reduction of the cooling temperature of the electrons will be weaker than the contribution of the other reactions, in spite of its greater importance in comparison with other fluxes. The fact is that in the balance equation for the electron temperature (8.2), the contribution of an inelastic process is $E_{ik}k_{ik}N_eN_i$, where E_{ik} is the energy transfer in inelastic collisions of electrons (concentration N_e) with the atomic (molecular) component (concentration N_i), k_{ik} is the rate of this reaction. The contribution of elastic collisions is $3\delta k_{el}(T_e-T_g)N_eN$, where $\delta = m_e/m$ is the ratio of the electron mass to the mass of an atom or molecule (concentration N), k_{el} is the rate of the elastic process, T_e, T_g are respectively the electron temperature and gas temperature. It may be seen that the ratio of the elastic contribution to the contribution of inelastic reactions is equal to the ratio of reactions by a factor $\delta(T_e-T_g)/E_{ik}\ll1$.

We have already discussed the question of reducing the prepulse electron concentration due to cooling we will discuss this issue in greater detail. The beneficial influence of the hydrogen addition in M.3 is also associated with a decrease in the prepulse electron concentration. As already noted, earlier studies [5–7,52,54,89] have shown that the value of the prepulse electron concentration has a direct effect on the characteristics of the laser, in particular, the achievable frequency and power parameters. On this basis, we would expect the positive effect of reducing the prepulse electron concentration. However, as studies have shown (see section 'Discussion of prepulse values' and section 'Impact of the excitation pulse repetition frequency'), the positive effect of this mechanism becomes important only at high-frequency operation of the laser. At low excitation pulse repetition frequencies the main contribution to the increase in radiation energy is mediated through the increase in the concentration of copper in the active medium (see A.4). Moreover, if the introduction of hydrogen did not lead to the processes described in A.4, the resulting reduction in the prepulse electron concentration would be unable to compensate

Table 8.6. The initial values of the plasma reagents (concentrations – in cm³/s, T_e – in eV) at 12 kHz ($N_{Cu} = 8 \cdot 10^{14}$ cm⁻³) ($N_H = 1 \cdot 10^{15}$ cm⁻³ for all cases where N_{H_2} is not 0). The calculations used to model the kinetics the gas temperature $T_g = 0.146$ eV. The initial values of plasma reagents were obtained as a result of the self-consistent calculation. E_t – the total specific lasing energy, E_{510} – specific lasing energy at a wavelength of 510.6 nm, E_{578} – specific lasing energy at a wavelength of 578.2 nm (J/cm³). $N_{Cu\ min}$ – the minimum concentration of copper during the excitation pulse. Lasing energy for different gas temperatures ($T_g = 0.146$–0.215 eV) varies by about 1% if the calculations take the same density of the copper atoms in the ground state into account (see Table 8.6 for the gas temperature $T_g = 0.215$ eV)

$T_g = 0.146$ eV	$N_H=0$ $N_{H_2}=0$ (cm⁻³)	$N_{H_2}=1.65\cdot10^{15}$ (cm⁻³)	$N_{H_2}=3.3\cdot10^{15}$ (cm⁻³)	$N_{H_2}=1.65\cdot10^{16}$ (cm⁻³)	$N_{H_2}=3.3\cdot10^{16}$ (cm⁻³)	$N_{H_2}=4.95\cdot10^{16}$ (cm⁻³)
N_{Ne^+}	$6.25\cdot10^{12}$	$6.13\cdot10^{12}$	$5.96\cdot10^{12}$	$4.25\cdot10^{12}$	$2.8\cdot10^{12}$	$2.02\cdot10^{12}$
N_{Cu^+}	$1.96\cdot10^{13}$	$1.9\cdot10^{13}$	$1.79\cdot10^{13}$	$1.2\cdot10^{13}$	$8.35\cdot10^{12}$	$6.7\cdot10^{12}$
$N_{D_{5/2}}$	$7.54\cdot10^{11}$	$7.38\cdot10^{11}$	$6.31\cdot10^{11}$	$1.94\cdot10^{11}$	$9.65\cdot10^{10}$	$7.4\cdot10^{10}$
$N_{D_{3/2}}$	$1.06\cdot10^{11}$	$1.01\cdot10^{11}$	$8.66\cdot10^{10}$	$2.1\cdot10^{10}$	$8.23\cdot10^{9}$	$5.33\cdot10^{9}$
$N_{Cu\ min}$	$4.91\cdot10^{14}$	$5.07\cdot10^{14}$	$5.2\cdot10^{14}$	$5.73\cdot10^{14}$	$5.98\cdot10^{14}$	$6.16\cdot10^{14}$
T_e	0.171	0.17	0.165	0.1465	0.146	0.146
N_{H^-}	–	$1.63\cdot10^{9}$	$2.01\cdot10^{9}$	$2.36\cdot10^{9}$	$1.8\cdot10^{9}$	$1.54\cdot10^{9}$
N_{CuH}	–	$7.98\cdot10^{10}$	$1.19\cdot10^{11}$	$6.35\cdot10^{10}$	$1.83\cdot10^{10}$	$7.75\cdot10^{9}$
$N_{H_2}\ (v{=}1)$	–	$7.7\cdot10^{13}$	$1.39\cdot10^{14}$	$2.8\cdot10^{14}$	$1.85\cdot10^{14}$	$1.09\cdot10^{14}$
$N_{H_2}\ (v{=}2)$	–	$2.9\cdot10^{11}$	$4.77\cdot10^{11}$	$4.08\cdot10^{11}$	$8.01\cdot10^{10}$	$1.49\cdot10^{10}$
E_t (J/cm³)	$3.00\cdot10^{-6}$	$2.95\cdot10^{-6}$	$2.92\cdot10^{-6}$	$2.72\cdot10^{-6}$	$2.45\cdot10^{-6}$	$2.2\cdot10^{-6}$
E_{510} (J/cm³)	$2.14\cdot10^{-6}$	$2.1\cdot10^{-6}$	$2.09\cdot10^{-6}$	$1.95\cdot10^{-6}$	$1.77\cdot10^{-6}$	$1.61\cdot10^{-6}$
E_{578} (J/cm³)	$0.86\cdot10^{-6}$	$0.85\cdot10^{-6}$	$0.833\cdot10^{-6}$	$0.765\cdot10^{-6}$	$0.674\cdot10^{-6}$	$0.59\cdot10^{-6}$
$T_g = 0.215$ eV	$N_H=0$ $N_{H_2}=0$	$N_{H_2}=1.65\cdot10^{15}$	$N_{H_2}=3.3\cdot10^{15}$	$N_{H_2}=1.65\cdot10^{16}$	$N_{H_2}=3.3\cdot10^{16}$	$N_{H_2}=4.95\cdot10^{16}$
$N_{Cu}^+\,N_{Ne}^+$	$3.3\cdot10^{13}$	$3.2\cdot10^{13}$	$3.06\cdot10^{13}$	$2.08\cdot10^{13}$	$1.3\cdot10^{13}$	$9.2\cdot10^{12}$
$N_{D_{5/2}}$	$6.7\cdot10^{12}$	$6.6\cdot10^{12}$	$5.6\cdot10^{12}$	$1.7\cdot10^{12}$	$8.6\cdot10^{11}$	$6.6\cdot10^{11}$
$N_{D_{3/2}}$	$9.5\cdot10^{11}$	$9.1\cdot10^{11}$	$7.7\cdot10^{11}$	$1.9\cdot10^{11}$	$7.4\cdot10^{10}$	$4.8\cdot10^{10}$
$N_{Cu\ min}$	$4.9\cdot10^{14}$	$5.1\cdot10^{14}$	$5.2\cdot10^{14}$	$5.7\cdot10^{14}$	$6.\cdot10^{14}$	$6.2\cdot10^{14}$
T_e, eV	0.225	0.224	0.223	0.22	0.217	0.216
N_H^-	–	$2.5\cdot10^{9}$	$3.1\cdot10^{9}$	$3.6\cdot10^{9}$	$2.7\cdot10^{9}$	$2.3\cdot10^{9}$
$N_{Cu\,H}$	–	$2.2\cdot10^{11}$	$3.2\cdot10^{11}$	$1.7\cdot10^{11}$	$5.1\cdot10^{10}$	$2.1\cdot10^{10}$

N_{H_2} (v=1)	–	$8.4 \cdot 10^{13}$	$1.5 \cdot 10^{14}$	$3.1 \cdot 10^{14}$	$2.02 \cdot 10^{14}$	$1.2 \cdot 10^{14}$
N_{H_2} (v=2)	–	$5.6 \cdot 10^{11}$	$9.2 \cdot 10^{11}$	$7.9 \cdot 10^{11}$	$1.5 \cdot 10^{11}$	$2.87 \cdot 10^{10}$
E_l	$3 \cdot 10^{-6}$	$2.9 \cdot 10^{-6}$	$2.92 \cdot 10^{-6}$	$2.72 \cdot 10^{-6}$	$2.4 \cdot 10^{-6}$	$2.2 \cdot 10^{-6}$
E_{510}	$2.1 \cdot 10^{-6}$	$2.1 \cdot 10^{-6}$	$2.1 \cdot 10^{-6}$	$1.9 \cdot 10^{-6}$	$1.7 \cdot 10^{-6}$	$1.6 \cdot 10^{-6}$
E_{578}	$0.9 \cdot 10^{-6}$	$0.8 \cdot 10^{-6}$	$0.82 \cdot 10^{-6}$	$0.8 \cdot 10^{-6}$	$0.7 \cdot 10^{-6}$	$0.6 \cdot 10^{-6}$

for the negative effect of inelastic losses due to the addition of hydrogen (see M.8 and Fig. 8.20 *a*).

A3. Dissociative electron attachment to molecular hydrogen has practically no influence on the improvement of the lasing characteristics of the laser. This reaction occurs with low efficiency, both due to the small cross section ($10^{-21} \div 10^{-20}$ cm^{-2}), and quite a significant threshold (3.724 eV for the ground and 3.224 eV for the first vibrational level of the hydrogen molecule) for this reaction. The rate of the reaction with atomic hydrogen is of the order 10^{-28} cm^3/s.

A4. In experimental studies [67–70] it was noted that when hydrogen is injected into the laser active medium there are changes in the electrical processes in the discharge circuit, caused by changing of the conductivity of the plasma. In [67], this effect was reduced to the increase of the discharge power which, respectively, caused an increase in the wall temperature (gas) and the concentration of copper in the active medium of gas discharge tubes (GDT). In the simulation it was possible to take into account the heating of the active medium with the introduction of hydrogen. For this purpose, in the heat balance equation for gas temperature it is simply necessary to consider the energy input into the gas from each of the reactions included in the model. As mentioned in section 8.1, the balance equation for the gas temperature was not solved. But it is clear that the introduction of hydrogen into the mixture, the gas temperature is greater than the gas temperature without the addition of hydrogen due to, for example, reactions of the elastic cooling of electrons by hydrogen molecules and the dissociation of hydrogen by an electron impact

$$H_2 + e \rightarrow 2 \cdot H + e.$$

Thus, the introduction of hydrogen alters the initial concentration of copper atoms in the ground state. To test this hypothesis, we increased the concentration of the copper atoms in the active medium without addition of hydrogen (but GDT

Table 8.7. The initial values of the reactants of plasma at a frequency of 12 kHz ($N_{Cu} = 1\cdot10^{15}$ cm^{-3}) ($N_H = 1\cdot10^{15}$ cm^{-3}) (see caption to Table 8.6)

	$N_H = 0$, $\rho_{H_2} = 0$ (cm^{-3})	$N_{H_2} = 1.65\cdot10^{15}$ (cm^{-3})	$N_{H_2} = 3.3\cdot10^{15}$ (cm^{-3})	$N_{H_2} = 1.65\cdot10^{16}$ (cm^{-3})
N_{Ne^+}	$4.86\cdot10^{12}$	$4.75\cdot10^{12}$	$4.66\cdot10^{12}$	$3.41\cdot10^{12}$
N_{Cu^+}	$2.12\cdot10^{13}$	$2.05\cdot10^{13}$	$1.96\cdot10^{13}$	$1.32\cdot10^{13}$
$N_{D_{5/2}}$	$9.33\cdot10^{11}$	$8.8\cdot10^{11}$	$7.18\cdot10^{11}$	$2.23\cdot10^{11}$
$N_{D_{3/2}}$	$1.31\cdot10^{11}$	$1.23\cdot10^{11}$	$9.7\cdot10^{10}$	$2.39\cdot10^{10}$
$N_{Cu\ min}$	$5.91\cdot10^{14}$	$6.49\cdot10^{14}$	$6.63\cdot10^{14}$	$7.25\cdot10^{14}$
T_e	0.171	0.17	0.167	0.1465
N_{H^-}	–	$1.65\cdot10^9$	$2.06\cdot10^9$	$2.46\cdot10^9$
N_{CuH}	–	$9.92\cdot10^{10}$	$1.53\cdot10^{11}$	$8.17\cdot10^{10}$
N_{H_2} ($v=1$)	–	$7.8\cdot10^{13}$	$1.42\cdot10^{14}$	$2.96\cdot10^{14}$
N_{H_2} ($v=2$)	–	$3.06\cdot10^{11}$	$5.17\cdot10^{11}$	$4.72\cdot10^{11}$
E_t (J/cm^3)	$3.63\cdot10^{-6}$	$3.56\cdot10^{-6}$	$3.54\cdot10^{-6}$	$3.31\cdot10^{-6}$
E_{510} (J/cm^3)	$2.58\cdot10^{-6}$	$2.53\cdot10^{-6}$	$2.52\cdot10^{-6}$	$2.37\cdot10^{-6}$
E_{578} (J/cm^3)	$1.05\cdot10^{-6}$	$1.03\cdot10^{-6}$	$1.02\cdot10^{-6}$	$0.942\cdot10^{-6}$

wall temperature (and gas temperature) were not changed) and conducted a series of calculations. The result was an even greater increase in radiation energy than when hydrogen was added (Tables 8.6 and 8.7). Thus, the increase of the power output was probably due to the increase of the GDT temperature (gas temperature) and the concentrations of copper in the active medium in the experiments.

In the same paper [67], an experiment was conducted with a pure copper vapour laser in which the GDT temperature was forced to increase to that obtained with the admixture of hydrogen. However, the increase of the power output was only a few percent of the improvement obtained by the introduction of molecular hydrogen additives. Analysis of this result shows that by 'forced' heating of the GDT we not only change the concentration of copper (this what the experimenters in the work [67] tried to achieve) but at the same time we raise the level to which the electron temperature relaxes in the interpulse period, thus defining the equilibrium concentration of the copper atoms in the metastable state. Consequently, the forced increase in the gas temperature results in the simultaneous occurrence of two competing processes: an increase in the density of copper in the active medium and an increase in prepulse population of the metastable level of the copper atoms. From this we can conclude that increasing the concentration of copper atoms in the metastable state did not result in any significant increase in the power output when using a laser without hydrogen additions. Indeed, increasing the concentration of copper in the ground state due to the increased energy input (increasing the gas temperature), with the introduction of hydrogen in the active medium is accompanied, according to calculations, by a decrease in the prepulse values of the concentration of copper atoms in the metastable state due to quenching by the vibrationally excited hydrogen molecules (values of the prepulse concentrations of metastable copper atoms at the same time differ by 12–14% for 2% hydrogen addition), the total energy of radiation increases by 18% –from $3.00 \cdot 10^{-6}$ J/cm^3 per pulse up to $3.54 \cdot 10{-6}$ J/cm^3. The efficiency is increased by 20% – from 0.421% to 0.505%, see Tables 8.6 and 8.7 and Fig. 8.20 *b*.

Thus, according to the calculations, the power output with the addition of hydrogen was increased mainly due to the increase of gas temperature and the concentrations of copper in the active medium, while reducing the value of the prepulse population of the metastable levels by their quenching by hydrogen molecules in the experiments (see also section 'Testing the model').

A5. see A.4, A.6, A.8.

A6. In [69] the addition of hydrogen resulted in a decrease the extent of depletion and faster recovery of the prepulse concentrations of copper atoms in the ground state. In our numerical experiments we also observed this effect. Table 8.8 (see also Fig. 8.20 *a*) gives the main fluxes, changing the population of the ground level of the copper atom. It is evident that the introduction of hydrogen additives lowers the fluxes, leading to an overall increase in the minimum value to which the concentration of copper atoms in the ground state is reduced. The hydrogen addition has an indirect effect on the value of the population of the ground state of copper by changing the concentration and temperature of the electrons because Table 8.8 indicates that the absolute values show that the reaction involving hydrogen have no substantial effect on the concentration of the copper atoms in the ground state. The above table shows that molecular hydrogen 'captures' the energy

Table 8.8. Depletion flows of ground level of the copper atom within the excitation pulse F (0, 500 ns). Selection arguments t_1, t_2 of function F is due to the pulse duration of the pump current. (GDT parameters are given in Table 8.4, the initial values of the plasma concentrations of the reactants – in Tables 8.6–8.7).

Reaction	Flux (cm³)					
	0% H$_2$ N_{Cu}= 0.8·10^{15} cm^{-3}		2% H$_2$ N_{Cu}= 0.8·10^{15} cm^{-3}		2% H$_2$ N_{Cu} = 1·10^{15} cm^{-3}	
Cu + e → Cu ($P_{3/2}$) + e	1.515·10^{14}	8.399·10^{13}	1.402·10^{14}	7.959·10^{13}	1.671·10^{14}	9.314·10^{13}
Cu ($P_{3/2}$) + e → Cu + e	6.751·10^{13}		6.061·10^{13}		6.689·10^{13}	
Cu +e → Cu ($P_{1/2}$) + e	7.232·10^{13}	3.956·10^{13}	6.689·10^{13}	3.743·10^{13}	7.977·10^{13}	4.379·10^{13}
Cu ($P_{1/2}$) + e → Cu +e	3.276·10^{13}		2.946·10^{13}		3.598·10^{13}	
Cu + e → Cu ($D_{5/2}$) + e	1.293·10^{14}	9.436·10^{13}	1.229·10^{14}	8.844·10^{13}	1.568·10^{14}	1.115·10^{14}
Cu ($D_{5/2}$) + e → Cu +e	3.494·10^{13}		3.446·10^{13}		4.553·10^{13}	
Cu +e → Cu ($D_{3/2}$) + e	1.137·10^{14}	5.317·10^{13}	1.068·10^{14}	4.828·10^{13}	1.351·10^{14}	5.879·10^{13}
Cu ($D_{3/2}$) + e → Cu +e	6.053·10^{13}		5.852·10^{13}		7.631·10^{13}	
Cu +e → Cu* + e	1.191·10^{13}	1.182·10^{13}	1.107·10^{13}	1.1·10^{13}	1.317·10^{13}	1.308·10^{13}
Cu* + e → Cu +e	8.66·10^{10}		7.283·10^{10}		8.643·10^{10}	
Cu +e → Cu$^+$ + 2·e	1.916·10^{13}	1.916·10^{13}	1.842·10^{13}	1.842·10^{13}	2.135·10^{13}	2.135·10^{13}
Cu ($D_{5/2}$)+ Cu → Cu + Cu	1.63·10^{10}	1.63·10^{10}	1.656·10^{10}	1.656·10^{10}	2.606·10^{10}	2.606·10^{10}
Cu ($D_{3/2}$) + Cu → Cu + Cu	4.13·10^{9}	4.13·10^{9}	4.118·10^{9}	4.118·10^{9}	6.407·10^{9}	6.407·10^{9}
Ne* + Cu → Cu$^+$ + Ne +e	3.612·10^{12}	3.612·10^{12}	3.712·10^{12}	3.712·10^{12}	4.033·10^{12}	4.033·10^{12}
Cu ($P_{3/2}$) → Cu	2.687·10^{12}	2.687·10^{12}	2.492·10^{12}	2.492·10^{12}	2.919·10^{12}	2.919·10^{12}
Cu ($P_{1/2}$) → Cu	1.368·10^{12}	1.368·10^{12}	1.27·10^{12}	1.27·10^{12}	1.49·10^{12}	1.49·10^{12}
NeCu$^+$ + e → Cu + Ne	5.53·10^{10}	5.53·10^{10}	5.09·10^{10}	5.09·10^{10}	5.793·10^{10}	5.793·10^{10}
CuH + H$_2$ → H$_2$(v =3)+ Cu	–	–	7.934·10^{10}	7.934·10^{10}	9.898·10^{10}	9.898·10^{10}
Cu ($P_{1/2}$) + H$_2$ → Cu + H$_2$	–	–	1.493·10^{12}	1.493·10^{12}	1.752·10^{12}	1.752·10^{12}
Cu ($P_{3/2}$) + H$_2$ → Cu + H$_2$	–	–	2.892·10^{12}	2.892·10^{12}	3.391·10^{12}	3.391·10^{12}

of the excitation pulse for the dissociation and excitation of the first vibrational levels of the hydrogen molecule. Thus, the excitation of the copper atoms requires a smaller flux compared with the case of the absence of hydrogen, see also A.4.

A7. Changes in the electronic conductivity with the introduction of hydrogen addition were determined in a series of numerical calculations. Their results suggest that, despite the significant total cross section of electron scattering by molecular hydrogen, hydrogen has no significant contribution to the change of electronic thermal conductivity of hydrogen due to its low content, compared with the buffer gas (the flux of electronic thermal conductivity with the introduction of 2% hydrogen additive varies by less than 1%).

Table 8.9. Cooling fluxes of electrons F (0.500 ns) in the excitation pulse (see caption to Table 8.8)

Reaction	Flux (cm^{-3})			
	0% H$_2$ N_{Cu} = 0.8·10^{15} cm^{-3}		2% H$_2$ N_{Cu} = 0.8·10^{15} cm^{-3}	
Ne + e → Ne* + e	1.756·10^{14}	1.494·10^{14}	1.791·10^{14}	1.59·10^{14}
Ne* + e → Ne +e	2.614·10^{13}		2.007·10^{13}	
Ne + e → Ne$^+$ + 2·e	4.157·10^{12}	4.157·10^{12}	4.295·10^{12}	4.295·10^{12}
Ne* + e → Ne** + e	2.926·10^{14}	1.592·10^{14}	2.016·10^{14}	1.177·10^{14}
Ne** + e → Ne* + e	1.334·10^{14}		8.392·10^{13}	
Cu + e → Cu ($P_{3/2}$) + e	1.459·10^{14}	8.063·10^{13}	1.139·10^{14}	7.314·10^{13}
Cu ($P_{3/2}$) + e → Cu +e	6.527·10^{13}		4.076·10^{13}	
Cu +e → Cu($P_{1/2}$) + e	6.962·10^{13}	3.792·10^{13}	5.421·10^{13}	3.431·10^{13}
Cu ($P_{1/2}$) + e → Cu +e	3.17·10^{13}		1.99·10^{13}	
Cu +e → Cu ($D_{5/2}$) + e	1.301·10^{14}	1.301·10^{14}	9.005·10^{13}	9.005·10^{13}
Cu +e → Cu ($D_{3/2}$) + e	1.132·10^{14}	1.132·10^{14}	7.586·10^{13}	7.586·10^{13}
Cu ($D_{3/2}$) + e → Cu +e	6.199·10^{13}	6.199·10^{13}	4.444·10^{13}	4.444·10^{13}
Cu +e → Cu* + e	1.148·10^{13}	1.148·10^{13}	9.217·10^{12}	9.217·10^{12}
Cu +e → Cu$^+$ + 2·e	1.853·10^{13}	1.853·10^{13}	1.699·10^{13}	1.699·10^{13}
Cu ($D_{5/2}$) + e → Cu* + e	1.809·10^{13}	1.809·10^{13}	7.318·10^{12}	7.318·10^{12}
Cu ($P_{3/2}$) + e → Cu* + e	3.689·10^{13}	3.689·10^{13}	2.173·10^{13}	2.173·10^{13}
Cu ($P_{3/2}$) + e → Cu ($P_{1/2}$) + e	1.366·10^{16}	3.·10^{13}	8.531·10^{15}	1.6·10^{13}
Cu ($P_{1/2}$) + e → Cu ($P_{3/2}$) + e	1.369·10^{16}		8.547·10^{15}	
H$_2$ + e → H + H +e	–	–	6.931·10^{13}	6.931·10^{13}
H$_2$ + e → H$_2^+$ + 2·e	–	–	1.044·10^{13}	1.044·10^{13}
H$_2$ + e → H$_2$(v = 1) + e	1.01·10^{14}	1.01·10^{14}	6.407·10^{14}	6.407·10^{14}

A8. In [67], one of the possible negative factors of the change of the kinetics of copper vapour lasers with hydrogen additives were assumed to be energy losses due to the inelastic excitation and ionization of molecular hydrogen. A detailed analysis of this hypothesis was carried out. It turned out that during the excitation pulse the molecular hydrogen injected into the active medium has a significant effect on the kinetics of the processes. In particular, the reactions:

$$H_2 + e \rightarrow H + H + e$$

and

$$H_2 + e \rightarrow H_2(v = 1) + e,$$

exert, within a few hundred nanoseconds, a significant effect on the electron temperature (see Table 8.9 for reaction fluxes). In this case, there is also a decrease of electron concentration (see Fig. 8.20 *a*). With the increase in the percentage of the hydrogen addition the role of the dissociation processes and excitation of the first vibrational levels of molecular hydrogen becomes more significant, leading to a general change in the kinetic processes occurring in the active medium of the laser. By the end of the interpulse period at low excitation pulse repetition frequencies there is no significant decrease in the electron temperature and, accordingly, the copper concentration in the metastable state due to quenching of the vibrationally excited levels of the hydrogen molecules by the electrons. This slows down the decay rate of the electron density, which negatively affects the rate of recombination of the plasma. Thus, at low (about 10 kHz) excitation pulse repetition frequencies the prepulse concentrations and the electron temperature decrease slightly (by about 10–15% (see Fig. 8.20 *a*). By increasing the excitation pulse repetition rate the decrease of the prepulse values of electron concentration and temperature can, however, lead to positive effects (see the preceding paragraph and section 'The influence of the excitation pulse repetition frequency').

Interception of the energy used into the excitation of copper with the introduction of hydrogen and illustrated above (see also A.6) negatively affects the characteristics of the copper vapour laser. However, the optimum improvement of the lasing characteristics due to the effect described in A.4 is greater than the negative effects, discussed in this section.

A9. In [93] the authors proposed and investigated the mechanism of quenching of the copper atoms in the metastable state due to interaction with vibrationally excited hydrogen molecules. The effect of the process on the change in the population of metastable levels of the copper atom was analyzed by the calculation of reaction fluxes that change the concentration of copper atoms in the metastable state. As an example, Table 8.10 shows the reaction fluxes for the $D_{5/2}$ level. It may be noted that the quenching of the metastable level of the copper atom by the vibrationally excited hydrogen molecules is one of the most important processes affecting its population, especially in the beginning of the interpulse period, and has previously been noted in [94]. The action of this mechanism at low excitation pulse frequencies was described above, see A.4. Detailed

Table 8.10. Fluxes of reduction of the population level $D_{5/2}$ in the interpulse period, F (3 µs, 80 µs) (see caption to Table 8.5)

Reaction	Flux (cm⁻³)	
$Cu + e \rightarrow Cu\ (D_{5/2}) + e$	$5.139 \cdot 10^{14}$	$5.21 \cdot 10^{13}$
$Cu\ (D_{5/2}) + e \rightarrow Cu + e$	$4.618 \cdot 10^{13}$	
$Cu\ (D_{5/2}) + e \rightarrow Cu\ (P_{3/2}) + e$	$6.206 \cdot 10^{10}$	$5.58 \cdot 10^{12}$
$Cu\ (P_{3/2}) + e \rightarrow Cu\ (D_{5/2}) + e$	$5.643 \cdot 10^{12}$	
$Cu\ (D_{3/2}) + e \rightarrow Cu\ (D_{5/2}) + e$	$1.122 \cdot 10^{15}$	$1.11 \cdot 10^{14}$
$Cu\ (D_{5/2}) + e \rightarrow Cu\ (D_{3/2}) + e$	$1.233 \cdot 10^{15}$	
$Cu\ (D_{5/2}) + e \rightarrow Cu^* + e$	$7.525 \cdot 10^{10}$	$6.418 \cdot 10^{10}$
$Cu^* + e \rightarrow Cu\ (D_{5/2}) + e$	$1.107 \cdot 10^{10}$	
$Cu\ (D_{5/2}) + e \rightarrow Cu^{**} + e$	$5.758 \cdot 10^{6}$	$8.301 \cdot 10^{9}$
$Cu^{**} + e \rightarrow Cu\ (D_{5/2}) + e$	$8.307 \cdot 10^{9}$	
$Cu\ (D_{3/2}) + Cu \rightarrow Cu\ (D_{5/2}) + Cu$	$1.72 \cdot 10^{12}$	$1.72 \cdot 10^{12}$
$Cu\ (D_{5/2}) + Cu \rightarrow Cu + Cu$	$2.35 \cdot 10^{11}$	$2.35 \cdot 10^{11}$
$Cu\ (P_{3/2}) + Cu \rightarrow Cu\ (D_{5/2}) + Cu$	$6.752 \cdot 10^{12}$	$6.752 \cdot 10^{12}$
$Cu^{**} \rightarrow Cu\ (D_{5/2})$	$7.468 \cdot 10^{12}$	$7.468 \cdot 10^{12}$
$Cu\ (P_{3/2}) + hv \rightarrow Cu\ (D_{5/2}) + 2 \cdot hv$	$3.081 \cdot 10^{7}$	$3.081 \cdot 10^{7}$
$Cu\ (D_{5/2}) + hv \rightarrow Cu\ (P_{3/2})$	$1.046 \cdot 10^{9}$	$1.046 \cdot 10^{9}$
$Cu\ (P_{3/2}) \rightarrow Cu\ (D_{5/2}) + hv$	$1.385 \cdot 10^{9}$	$1.385 \cdot 10^{9}$
$Cu\ (D_{5/2}) + H_2(v = 1) \rightarrow CuH + H$	$2.533 \cdot 10^{13}$	$2.533 \cdot 10^{13}$
$Cu\ (D_{5/2}) + H_2(v = 2) \rightarrow CuH + H$	$1.693 \cdot 10^{12}$	$1.693 \cdot 10^{12}$
$Cu\ (D_{5/2}) + H_2(v = 3) \rightarrow CuH + H$	$2.407 \cdot 10^{11}$	$2.407 \cdot 10^{11}$

consideration at high frequencies will be given below in section 'Effect of the excitation pulse repetition frequencies'.

Testing of the model

The model was tested in experiments [72,73] These works contain the most detailed (known to us) results of theoretical and experimental studies of the effect of hydrogen additives on the characteristics of the copper vapour lasers. The model was tested by comparing various dependencies, calculated using the model described above, with the theoretical and experimental dependences of the electron and the copper atoms in the metastable state, the electron temperature, the energy and parameters of the laser pulse, as well as the electrical characteristics. The calculations were carried out using the experimentally measured time dependence of the current flowing through the gas discharge tube during the excitation pulse. According to [72,73] and our calculations (see for example the section 'Excitation pulse') the time dependence

Table 8.11. The initial plasma concentration of the reactants (for [73]). With the exception of the concentration of copper in the ground state and neon parameters were calculated self-consistently within the model

The concentration of neon	1×10^{17} cm^{-3}
The concentration of copper	1.8×10^{15} cm^{-3}
The initial concentration of copper ions	9.85×10^{12} cm^{-3}
The initial population of level $D_{3/2}$	2.34×10^{10} cm^{-3}
The initial population of level $D_{5/2}$	2.43×10^{11} cm^{-3}
The initial concentration of vibrationally excited hydrogen for $v = 1$	1.02×10^{13} cm^{-3}
The initial concentration of vibrationally excited hydrogen for $v = 2$	2.68×10^{9} cm^{-3}
The initial concentration of CuH molecules	3.07×10^{9} cm^{-3}
The initial concentration of negative hydrogen ions	2.81×10^{8} cm^{-3}

of the current varies slightly with the introduction of the hydrogen impurity (for concentrations close to optimal (few percent)).

The initial concentration of the copper atoms in the calculations corresponded to the initial concentration shown in [73]. The initial values of the other reagents were calculated in the model. These values were calculated based on iterations. The initial concentrations of the reactants were defined and this was followed by calculating their values for the beginning of the next pulse. The iterations were stopped when the difference between initial and final values was less than 1%. This procedure can be used because of the weak dependence of the form of the current on time when the hydrogen impurity is added.

Comparison with the results of [73]. The parameters of the gas discharge tube that was used in [73] and the corresponding values of reagent concentrations of the active medium are given in Tables 8.4 and 8.11 respectively.

Voltage. Voltage calculations were performed using the plasma resistance calculated in our work (see 'The effect of hydrogen additives on the electrical characteristics'). In the absence of hydrogen the time dependence of voltage is in good agreement with both the experimental and calculated dependences given in [73] (Fig. 8.21 *a*). When 1% of hydrogen is added the voltage in the GDT computed using our model, was about 1.5 times higher than calculated in [73], although very close in shape (Fig. 8.21 *a*). Despite the small amount in comparison with the concentrations of the main agents of the active medium, hydrogen significantly increases the maximum resistance (see Fig. 8.24 *a* and formula (8.6)). Therefore, the resulting discrepancy can be explained by the fact that perhaps the authors of [73] did not consider in their calculations the contribution to the conductivity due to the hydrogen addition.

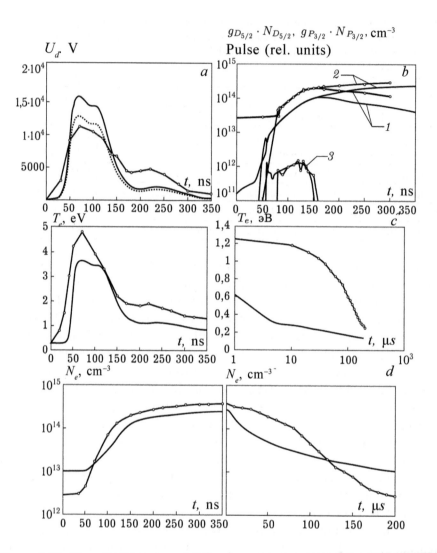

Figure 8.21. Time dependences a: GDT voltage (circles – experiment, dashed line – calculation without hydrogen in plasma conductivity, solid curve – taking into account the contribution of hydrogen) [73]. b: the populations of levels $P_{3/2}$ (1) and $D_{5/2}$ (2), lasing pulse (3) (circles – calculation in [73]), the solid curve – this work). c: electron temperature during the excitation pulse and in the interpulse period (circles – calculation [73], the solid line – results of this simulation). d: the electron density during the excitation pulse and in the interpulse period (circles calculations in [73], solid curve – the results of this work.) GDT parameters are given in Table 8.4, values of plasma reagents used in the calculations – in Table 8.11

Concentration of metastable levels. In comparison with the time dependence of the population of the metastable level $D_{5/2}$, presented in [73], the calculation using the present model for the initial value gives lower values (Fig. 8.21 *b*). This is due to a lower level of the electron temperature at the end of the interpulse period in our calculations. It should be noted that at the end of the interpulse period at a low pulse repetition frequency the population of the metastable level is close to equilibrium with the electron temperature and can be estimated (for example, for the $D_{5/2}$ level):

$$N_{D_{5/2}} = \frac{g_m}{g_0} N_0 \cdot \exp\left(-\frac{E_{D_{5/2}}}{T_{em}}\right).$$

In [73], the electron temperature at the end of the interpulse interval is ~0.25 eV. Thus, at a copper concentration in the ground state equal to $1.8 \cdot 10^{15}$ cm^{-3}, which was used in [73], the prepulse value of the population of the $D_{5/2}$ level must relax to a level of ~$2 \cdot 10^{13}$ cm^{-3}, which is two times lower than the value given in [73].

Temperature and electron density. The time dependences of the electron density and temperature were also compared (see Fig. 8.21 *c*, 8.21 *d*). As in the calculations in [73] the maximum electron concentration decreases with increase of the hydrogen addition. According to calculations, there is a delay in the start of growth of the electron concentration as a function of time (see Fig. 8.21 *d*). However, in contrast to the results in [73] where this delay increases with the percentage of hydrogen in the active medium, in the calculations using the model presented here this effect is not observed. When comparing the dependence of the electron temperature it can be noted that in our calculations the maximum electron temperature during the excitation pulse had a lower value (Fig. 8.21 *c*). In addition, in [73], the electron temperature with or without the hydrogen addition relaxed to about the same level. At the same time, the initial values of the populations of metastable levels were significantly different (for the case with the hydrogen addition the prepulse concentration of the copper atoms in the metastable state was two times higher than at 1% hydrogen addition). This is contrary to the above reasoning about the equilibrium population of levels at the end of the interpulse interval. The time dependence of the electron temperature in the interpulse period was also different. In our calculations, there is a clear separation of the relaxation period into two phases: a sharp decline in the first few microseconds and then a slow decline during the rest of the interpulse period. This is typical of the copper vapour lasers [2, 51]. At the same time in [73] the authors observed the opposite relationship: a slow decline at the beginning but at the end of the relaxation interval a marked decrease in the electron temperature (Fig. 8.21 *c*).

Indeed, the recombination rate depends on the concentration and, in particular, the electron temperature, so the intense recombination begins when the electron temperature reaches a certain value. However, in the process of recombination there also occurs recombination heating of electrons, which leads to a slower decay of the electron temperature. As a result, at the beginning of the interpulse period the electron temperature decreases rapidly, and then, with the beginning of recombination heating, the decline slows down sharply and the relaxation changes to the quasi-stationary phase which is determined by the balance of energy input to the electrons as a result heating and cooling them in elastic collisions. This fact has long been known in the physics of plasma lasers [8].

An increase in the prepulse electron concentration with the introduction of hydrogen was reported in [73]. Thus, according to [73], the addition of hydrogen increases both the concentration of metastable states and the electron concentration and this impairs the prepulse conditions. This behaviour is in contradiction with the results of our calculations.

Lasing pulse. The duration of the lasing pulse, calculated using the model described here, is almost equal to the duration compared to the pulse parameters given in [73], but has a smoother structure with a sharp leading edge, where the 'center of gravity of the pulse' is shifted to its end (Fig. 8.21 *b*).

Comparison with Piper' results [72]. Although the work [72] as a whole is devoted to the study of lasers with HCl additives, it compares the experimental and theoretical studies of a laser with a hydrogen addition: time dependences of the concentration and temperature of the electrons, the density of copper in the ground state, the plasma resistance of the GDT, the population of the metastable level.

In [72] calculations and the experiments were performed with and without the hydrogen addition. In our calculations we used a higher

Table 8.12. Comparison of the lasing power with [72]. GDT parameters are given in Table 8.4, the values of reactant concentrations of the active medium is Tables 8.6 and 8.7. The value of prepulse concentration of copper in the ground state for pure neon corresponds to $8 \cdot 10^{14}$ cm^{-3}, and for 2% hydrogen addition to $1 \cdot 10^{15}$ cm^{-3} (see text)

	Pure neon	2% H$_2$
Experiment (total power)	59.6 W	67.2 W
Model (total power)	61.04 W	70.83 W
510.6 nm	43.48 W	50.42 W
578.2 nm	17.56 W	20.41 W

value of the prepulse concentration of copper in the ground state (see Fig. 8.22 *a*) compared with the calculations in [72]. The fact is that according to the calculations of the radial distribution of gas temperature [96] $Tg_0/Tg_w \sim 1.3\div1.5$ (depending on the percentage of hydrogen in the active medium). For the wall density of copper presented in [72] we obtained the value used in our calculations. At this value of the initial density of copper there was also a quite good agreement with the experimentally measured value of lasing power (see Table 8.12). Note that [72] gives only the near-wall gas temperature. Considerations from which the authors chose the initial concentration of copper in the ground state in the centre of GDT in the calculations were not mentioned.

The calculated time dependence of the electron concentration differs from that given in [72] as follows: the value of the prepulse electron concentration obtained in our calculations is smaller (by about a half),

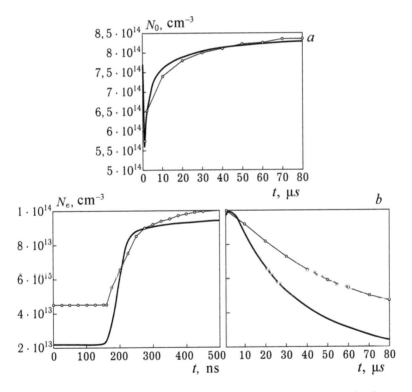

Figure 8.22. Time dependences of *a*: concentration of copper atoms in the ground state during the excitation pulse and the interpulse interval (circles – calculations in [72], solid curve – the results of this work, normalized to the values of [72]) (in the calculation in [72] the value $N_{Cu} = 0.83 \cdot 10^{14}$ cm^{-3} was used, in our calculation $N_{Cu} = 1 \cdot 10^{15}$ cm^{-3}, see text. *b*: the electron density during the excitation pulse and interpulse period (circles – calculations in [72], solid curve – the results of this work),

while the maximum value achieved during the lasing pulse is 5% lower. During the interpulse period in our calculations there was a higher rate of decrease of the electron concentration in comparison with the results of [72] (Fig. 8.22 *b*).

The temperature of the electrons during the excitation pulse has a very close resemblance to the data obtained in [72]. The dependence of the electron temperature during the interpulse period is similar to the curve of [72], but has a lower value (Fig. 8.22, *c*).

Also, there is a good agreement between the type of time dependence for the populations of the ground and metastable $D_{3/2}$ levels, but due to the higher concentration of copper in the active medium in our calculations compared with [72] the dependence of the population of the ground level is higher than that calculated in this work (Fig. 8.22 *a*, *d*).

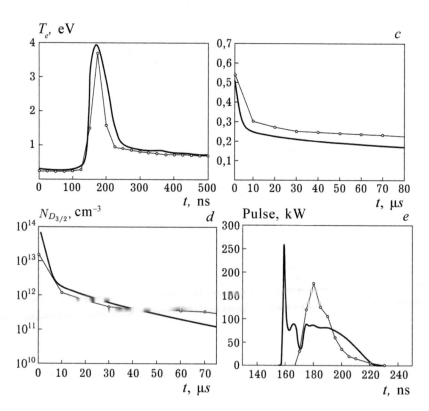

Figure 8.22. *continued.* Time dependences of: *c*: the electron temperature during the excitation pulse and the interpulse period (circles – calculation [72], solid curve – the results of this simulation). *d*: population level $D_{3/2}$ for interpulse interval (circles – calculation [72], solid curve – the results of this work. *e*: the laser pulse (circles –work [72], solid curve – the results of this work.) GDT parameters are given in Table 8.4, values of the plasma reactants used in the calculations are listed in Table 8.7.

Table 8.13 The initial values of the plasma reactants at a frequency of 50 kHz $(N_{Cu} = 8\cdot10^{14}$ cm$^{-3})$ $(N_H = 1\cdot10^{15}$ cm$^{-3})$. $(T_g = 0.215$ eV, see captions for Table 8.6)

	$N_H = 0,$ $N_{H_2} = 0$ (cm^{-3})	$N_{H_2} =$ $3.3\cdot10^{15}$ (cm^{-3})	$N_{H_2} =$ $1.65\cdot10^{16}$ (cm^{-3})	$N_{H_2} =$ $3.3\cdot10^{16}$ (cm^{-3})	$N_{H_2} =$ $4.95\cdot10^{16}$ (cm^{-3})
N_{Ne+}	$4\cdot10^{12}$	$4.02\cdot10^{12}$	$4\cdot10^{12}$	$3.7\cdot10^{12}$	$3.39\cdot10^{12}$
N_{Cu+}	$7.31\cdot10^{13}$	$6.95\cdot10^{13}$	$5.15\cdot10^{13}$	$3.8\cdot10^{13}$	$2.92\cdot10^{13}$
$N_{D_{5/2}}$	$3.62\cdot10^{12}$	$3.53\cdot10^{12}$	$1.75\cdot10^{12}$	$7.75\cdot10^{11}$	$3.96\cdot10^{11}$
$N_{D_{3/2}}$	$6.94\cdot10^{11}$	$6.92\cdot10^{11}$	$3.17\cdot10^{11}$	$1.15\cdot10^{11}$	$5.08\cdot10^{10}$
T_e	0.218	0.222	0.201	0.173	0.165
N_{H-}	–	$1.48\cdot10^{10}$	$2.8\cdot10^{10}$	$2.18\cdot10^{10}$	$1.55\cdot10^{10}$
N_{CuH}	–	$3.66\cdot10^{12}$	$4.2\cdot10^{12}$	$2.2\cdot10^{12}$	$1\cdot10^{12}$
$N_{H_2}(v=1)$	–	$4.4\cdot10^{14}$	$1.42\cdot10^{15}$	$1.65\cdot10^{15}$	$1.5\cdot10^{15}$
$N_{H_2}(v=2)$	–	$8.75\cdot10^{12}$	$1.87\cdot10^{13}$	$2.04\cdot10^{13}$	$8.47\cdot10^{12}$
E_t (J/cm^3)	$1.96\cdot10^{-6}$	$1.92\cdot10^{-6}$	$2.14\cdot10^{-6}$	$2.07\cdot10^{-6}$	$1.96\cdot10^{-6}$
E_{510} (J/cm^3)	$1.49\cdot10^{-6}$	$1.46\cdot10^{-6}$	$1.61\cdot10^{-6}$	$1.55\cdot10^{-6}$	$1.47\cdot10^{-6}$
E_{578} (J/cm^3)	$0.47\cdot10^{-6}$	$0.46\cdot10^{-6}$	$0.53\cdot10^{-6}$	$0.52\cdot10^{-6}$	$0.49\cdot10^{-6}$

A comparison of the lasing pulse shows that the pulse calculated in this work has a more pronounced vibrational structure at the start of lasing (Fig. 8.22 *e*).

Discussion of prepulse values and time dependences of plasma reagents

Depending on the hydrogen content in the active medium the prepulse values of the plasma reagents change (some of which are shown in Tables 8.13 and 8.14). The concentration of negative hydrogen ions, as well as the concentration of vibrationally excited hydrogen molecules, increases with the increase of the hydrogen addition. It is seen that with increasing hydrogen addition the prepulse concentrations of electrons and copper atoms in the metastable state decreases both for the case $N_{Cu} = 8\cdot10^{14}$ cm^{-3} and for the copper concentration corresponding to the increased gas temperature (see A.4) as a result of adding hydrogen $(N_{Cu} = 1\cdot10^{15}$ cm$^{-3})$. However, despite this, at low laser pulse repetition frequencies (the case of a repetition frequency of 12 kHz), while

Table 8.14. The initial values of the plasma reactants at a frequency of 100 kHz ($N_{Cu} = 8\cdot10^{14}$ cm^{-3}) ($N_H = 1\cdot10^{15}$ cm^{-3}). ($T_g = 0.215$ eV, see captions in Table 8.6)

	$N_H = 0$, $N_{H_2} = 0$ (cm^{-3})	$N_{H_2} = 1.65\cdot10^{16}$ (cm^{-3})	$N_{H_2} = 3.3\cdot10^{16}$ (cm^{-3})	$N_{H_2} = 4.95\cdot10^{16}$ (cm^{-3})
N_{Ne+}	$4.5\cdot10^{11}$	$1.19\cdot10^{12}$	$1.74\cdot10^{12}$	$1.92\cdot10^{12}$
N_{Cu+}	$1.25\cdot10^{14}$	$9.05\cdot10^{13}$	$6.95\cdot10^{13}$	$5.45\cdot10^{13}$
$N_{D_{5/2}}$	$7.14\cdot10^{12}$	$5.5\cdot10^{12}$	$2.22\cdot10^{12}$	$1.12\cdot10^{12}$
$N_{D_{3/2}}$	$1.54\cdot10^{12}$	$1.1\cdot10^{12}$	$4.12\cdot10^{11}$	$1.85\cdot10^{11}$
T_e	0.21	0.211	0.202	0.185
N_{H-}	–	$7.6\cdot10^{10}$	$7\cdot10^{10}$	$5.37\cdot10^{10}$
N_{CuH}	–	$2.82\cdot10^{13}$	$1.75\cdot10^{13}$	$1.01\cdot10^{13}$
N_{H_2} ($v = 1$)	–	$2.33\cdot10^{15}$	$3.2\cdot10^{15}$	$3.26\cdot10^{15}$
N_{H_2} ($v = 2$)	–	$7.05\cdot10^{13}$	$7.5\cdot10^{13}$	$5.06\cdot10^{13}$
E_l (J/cm³)	$5.83\cdot10^{-7}$	$1.17\cdot10^{-6}$	$1.4\cdot10^{-6}$	$1.43\cdot10^{-6}$
E_{510} (J/cm³)	$5.83\cdot10^{-7}$	$9.66\cdot10^{-7}$	$1.13\cdot10^{-6}$	$1.15\cdot10^{-6}$
E_{578} (J/cm³)	0	$2.04\cdot10^{-7}$	$2.66\cdot10^{-7}$	$2.8\cdot10^{-7}$

maintaining the concentration of copper atoms in the ground state constant, the total lasing energy decreases with an increase in the percentage of hydrogen in the active medium. This is due to the fact that the improvement of the prepulse conditions is accompanied by a decrease of the energy flux used for the excitation of copper atoms because of energy 'costs' of the dissociation and excitation of vibrational levels of molecular hydrogen (for details see A.4 and A.8).

The overall changes in the time behaviour of the various reagents of the active medium are shown in Figs. 8.20 *a–c* and 8.23 *a–b*.

Excitation pulse. As noted earlier (see A.8) the introduction of hydrogen addition to the active medium results in a reduction of the maximum values of temperature and electron concentration attainable within the excitation pulse. This is due to the fact that some of the energy supplied to the medium is spent on the dissociation and excitation of vibrational levels of the hydrogen molecule (see A.8). The lasing pulse is shifted slightly forward, staying in shape like the pulse generated in the absence of hydrogen (see Fig. 8.23 *a*). The addition of hydrogen decreases the resulting depletion of the populations of the ground level, while shifting forward the minimum point (Fig. 8.20 *c*).

Interpulse period. In discussing the effect of the hydrogen addition on the plasma parameters of the GDT the interpulse period can be divided into two parts. During the first phase, lasting a few microseconds, the addition of hydrogen leads is a significant reduction in the temperature and electron concentration (see Fig. 8.20 *a*), as well as the concentration of copper atoms in the metastable state (see Fig. 8.20 *b*), while increasing the rate of restoration of the concentration of copper in the ground state (see Fig. 8.20 *c*), which can lead to an improvement of the lasing characteristics at high excitation pulse repetition rates (see section 'Effect of excitation pulse repetition frequency'). Then comes the mode with the slow decay of the electron temperature and population of the metastable levels and the dependences for the cases with and without the addition of hydrogen become similar. A similar behaviour is also observed for the population of the ground level. The dip of the concentration of negative hydrogen ions at the beginning of the excitation pulse is associated with the reaction (see Table 8.15)

$$H^- + e \rightarrow H + 2 \cdot e.$$

The reaction

$$H^- + H \rightarrow H_2 + e$$

does not contribute to a sharp change in the ion concentration. The concentration of negative hydrogen ions at the maximum reaches a value of 10^{10} cm^{-3} and has no significant effect on the kinetics of the active processes occurring in the active medium (Fig. 8.23 *b*). This fact was also noted in [97] in connection with the supposed importance of this ion in the kinetics of the copper laser [69]. This figure also shows as an illustration the time dependences of the concentration of vibrationally excited molecules ($v = 1.2$) and CuH molecules.

Influence of the excitation pulse repetition frequency
Experiments with a high excitation pulse repetition frequency. For the above case with an excitation pulse repetition frequency of 12 kHz the positive effect of reducing the prepulse values of the concentration of the electrons and the copper atoms in the metastable state is nullified by the losses due mainly to the losses of the input energy in the dissociation and excitation of the first vibrational levels of the hydrogen molecule (A.8). However, in experimental studies [98–100] with a

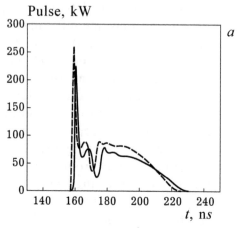

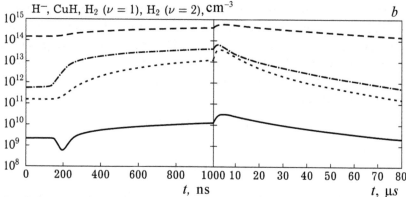

Figure 8.23. The time dependences of: *a*: the lasing pulse (solid curve – N_{Cu} = 8·10¹⁴ cm⁻³, N_{H_2} = 0%; dotted line – N_{Cu} = 1·10¹⁵ cm⁻³, N_{H_2} = 2%). *b*: the concentration of negative hydrogen ions H^- (solid curve), the CuH molecules CuH (dotted curve), vibrationally excited hydrogen molecules (v = 1 (dashed line), v = 2 (dot-dashed curve)) in the excitation pulses and the interpulse interval. The GDT parameters are in Table. 8.4, plasma reactants values used in the calculations – in Table 8.6 (for N_{H_2} = 0%) and 8.7 (for N_{H_2} = 2%).

copper vapour laser the use of hydrogen made it possible to achieve a high lasing pulse repetition frequency. Despite the fact that the studies [99, 100] refer to the lasers on copper bromide vapours, the mechanisms identified in the case of pure copper with the addition of hydrogen may also exert an effect in the active medium of the bromide vapour laser. The changes caused by the introduction of hydrogen additions in the case of high frequencies were analyzed. With this aim the operation of the laser was investigated at the excitation pulse repetition frequencies of 50 and 100 kHz at a constant pulse shape of the pump current. Tables 8.13 and 8.14 show that even at a constant concentration of

Table 8.15. Fluxes of reactions affecting the concentration of negative hydrogen ions, in the beginning of the excitation pulse F (0, 210 ns). The value of t_2 corresponds to the end of the concentration dip of H$^-$. GDT parameters are given in Table 8.4, plasma reactants values are in Table 8.6

Reaction	Flux (cm^{-3}) (2% H$_2$, N_{Cu} = 0.8·10^{15} cm^{-3})	
H$^-$ + e → H + 2·e	3.494·10^9	3.393·10^9
H + 2·e → H$^-$ + e	1.006·10^8	
H$_2$ + e → H$^-$ + H	1.442·10^9	9.643·10^8
H$^-$ + H → H$_2$ + e	4.777·10^8	
H$_2$ (v = 1) + e → H$^-$ + H	9.694·10^8	9.694·10^8
H$_2$ (v = 2) + e → H$^-$ + H	1.021·10^8	1.021·10^8
H$_2$ (v = 3) + e → H$^-$ + H	1.847·10^8	1.847·10^8

copper atoms in the ground state the lasing energy increases with the introduction of hydrogen, and for the case of a repetition frequency of 100 kHz the increase of radiation energy is greater. It should also be noted that when the frequency increases the maximum power is shifted towards a higher percentage of hydrogen (see Tables 8.13 and 8.14).

As noted above, the improvement of the laser characteristics with the addition of hydrogen occurs both due to the decrease of the prepulse concentration of electrons and the decrease of the prepulse concentrations of copper atoms in metastable states. Special calculations for a pulse repetition frequency of 100 kHz were carried out to determine the strength of the effect of these factors on this improvement. Self-consistent calculations were carried out for the parameters of the mixture shown in Table 8.7 and a hydrogen concentration of 16.5·10^{15} cm^{-3} to determine the initial concentrations of reagents, which included the concentration of electrons and metastable atoms. The specific lasing energy per pulse with the hydrogen addition was 1.17·10^{-6} J/cm^{-3}. Further, two calculations were carried out in the presence of hydrogen. In the first calculation of the prepulse electron concentration was taken equal to the value obtained in the absence of hydrogen. In the second – the prepulse concentration of metastable atoms were set equal to the values obtained in the absence of hydrogen. In the first case the radiation energy was about 1.88·10^{-7} J/cm^{-3}, and in the second 4.71·10^{-7} J/cm^{-3}. Thus, from these calculations we can conclude that the main contribution to the increase in lasing energy comes from the reduction of the prepulse electron concentration as a result of adding hydrogen to the active medium.

Of course, in a real experiment, the current amplitude and shape will vary somewhat with increasing excitation pulse repetition frequency, but such changes are quite difficult to take into account in the simulation, as the specific changes depend on the specific experimental conditions. In addition, the temperature of the gas medium, etc. can also vary. Incorporating all these factors would lead to a significant complication of the model and in the experiment there are rather scarce data on the magnitude of these changes while working with a laser with a hydrogen addition. Therefore, the current pulse in the calculations is considered constant, however, already from these results (Tables 8.13, 8.14) we can see the possibility of a significant increase of the lasing characteristics. Furthermore, even if all the above-mentioned factors were included in the model, their simultaneous consideration would lead to difficulties in understanding the situation because of the large number of these factors.

Effect of hydrogen additions on the electrical characteristics of the excitation circuit of the copper vapour laser

Excitation pulse. In the study of the impact of the hydrogen addition on the electrical characteristics of the excitation scheme, both during the excitation pulse and the interpulse period we used one of the following three-contour circuits (Fig. 8.7), used by us earlier [5–7]. Kirchhoff's equations are given in section 8.4.

The equations for voltage and current in the electrical circuit were solved together with kinetic equations which were then used to calculate the tube resistance from the determined temperature and electron density: $R_d = \rho l / S$, where $l = 150$ cm is the length of the tube, $S = \pi R^2$ is its cross-section (the tube radius $R = 1.9$ cm), ρ is the resistivity of the plasma. The voltage in the tube was calculated from Ohm's law: $U_d = R_d I_3$.

The initial conditions for the circuit was the capacitor voltage $U_1 = U_2 = 14$ kV. After firing the thyratron and changing the voltage polarity in the first capacitor, the total voltage across the two capacitors was about 28 kV. It was believed that at the initial time there are no currents: $I_1(0) = I_2(0) = I_3(0) = 0$.

The time dependences of the resistance of the thyratron and variable inductance as well as the values of constant inductances and capacitances in the calculations were identical with the dependences and values in section 8.4. The self-consistent calculations of the time dependences of various parameters of the active medium were carried out in conjunction with the Kirchhoff equations for the excitation circuit.

The resistivity of the plasma. Theoretically, the resistivity ρ is associated with the conductivity σ of the plasma by the ratio

$$\rho = 1/\sigma = mv/e^2 n_e,$$

where m – mass, e – electron charge, n_e – the concentration of electrons, v – the effective frequency of electron collisions (EFEC).

The frequency of elastic collisions. The effective frequency of electron–atom collisions in neon is

$$v(\text{Ne}) = [\text{Ne}]v\sigma_{tr}(\text{Ne}).$$

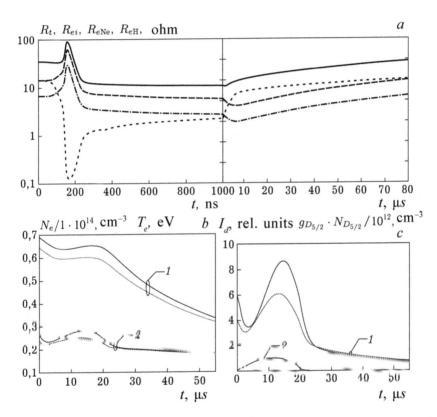

Figure 8.24. The time dependences of: *a*: plasma resistance (solid curve), the contribution of the Coulomb electron–ion collisions (dotted curve), electron–atom collisions (dashed line), the contribution of collisions with hydrogen (dot-dashed curve) during the excitation pulse and interpulse interval. *b*: concentration (*1*) and the temperature (*2*) of the electrons in the interpulse interval. *c*: population of the level $D_{5/2}$ (*1*) and current (*2*), flowing through the GDT, during interpulse interval. *b, c*: solid curve – $N_{Cu} = 8 \cdot 10^{14}$ cm⁻³, $N_{H_2} = 0\%$; dotted line – $N_{Cu} = 1 \cdot 10^{15}$ cm⁻³, $N_{H_2} = 2\%$, at $t = 0$ corresponds to 24.5 μs of the interpulse period. GDT parameters are given in Table 8.4, the values of the plasma reagents used in the calculations – in Table 8.6 (for $N_{H_2} = 0\%$) and 8.7 (for $N_{H_2} = 2\%$),

Taking into account the elastic collisions of electrons with copper atoms does not significantly alter the effective frequency of elastic electron–atom collisions.

The frequency of Coulomb collisions. The effective frequency of the Coulomb electron–ion collisions is

$$\nu(\text{coul}) = \frac{4\sqrt{3}\pi}{9} \frac{e^4 n_e \ln \Lambda}{m_e^{1/2} T_e^{3/2}},$$

where $\ln \Lambda$ is the Coulomb logarithm (≈ 10).

In the calculations ρ in a mixture of Ne–Cu was described by the expression

$$\rho \approx \frac{7.456 \cdot 10^{-2}}{\sqrt{T_e}} \left(\frac{1}{T_e} + 5.181 \cdot 10^{-3} \frac{N_{\text{Ne}}}{n_e} T_e \right) \text{ohm} \cdot \text{cm} \tag{8.6}$$

For mixtures containing hydrogen the electron collisions with hydrogen molecules were taken into account ρ in accordance with the full transport scattering cross section

$$\rho \approx \frac{7.456 \cdot 10^{-2}}{\sqrt{T_e}} \left(\frac{1}{T_e} + 5.181 \cdot 10^{-3} \frac{N_{\text{Ne}}}{n_e} T_e + 0.122 \frac{N_{\text{H}_2}}{n_e} T_e \right) \text{ohm} \cdot \text{cm}$$

From this expression it is clear that due to the large cross sections of interaction of the hydrogen molecule with the electron, even a small admixture of hydrogen can significantly alter the electrical characteristics of the active medium. To analyze this question, we conducted a series of calculations of the change in the conductivity of the plasma, current and voltage in the GDT. The GDT was a tube with the parameters corresponding to [72].

Figure 8.24 a shows that the hydrogen addition gives approximately the same relative contribution to the resistance of the plasma, both during the excitation pulse and in the interpulse period. In this case the current flowing during the excitation pulse through the gas-discharge tube is slightly reduced.

During the interpulse period the resistance of the plasma, defined by the contribution of hydrogen, increases monotonically, remaining less than the contributions of the Coulomb electron–ion and electron-atom collisions.

Effect of interpulse current. In the excitation circuits, discussed in this work, current passes through the GDT in the interpulse period which leads to an increase in the electron temperature and, consequently, slows down the relaxation of the plasma. We calculated the effect of the presence of current on relaxation of the basic reagents of the plasma medium and the radiation energy in the presence of hydrogen in the active medium and without hydrogen. Our calculations were based on the time dependence of the current flowing in the interpulse period through the discharge tube shown in [67]. In the simulation we used a sinusoidal current whose amplitude value is chosen based on the need to achieve the prepulse energy values of the storage capacitor $q^2/2C$, where C is the value of storage capacity, $q = \int_{t_1}^{t_2} i_{ch}(t)dt$, t_1 and t_2 – the beginning and end of the current flow through the discharge tube.

Accounting for the presence of interpulse current leads to an increase in prepulse values of the concentrations of the electron and the copper atoms in the metastable state and, consequently, to a decrease of the lasing power (see Fig. 8.24 b, c). Interpulse current reduces the rate of rise of the copper atoms in the ground state and the plasma resistance. It should be noted that when the interpulse current is taken into account the relaxation of vibrationally excited hydrogen molecules slows down. Increase of the percentage of hydrogen in the active medium leads to a decrease in the amplitude of the charging current (addition of hydrogen leads to a decrease in the electron concentration (see, e.g., Fig. 8.20 a, as well as A.2) and therefore to a decrease of the conductivity of the active medium), which was experimentally confirmed in [67], and has a positive effect on the prepulse values of the reagents of the active medium, increasing, in turn, the radiation energy.

Effect of prepulse conditions of the active medium on the output characteristics of the copper vapour laser

Changes of the prepulse concentration values of various reagents of the active medium, including the concentration of the electrons and the copper atoms in the metastable state can occur when the excitation pulse repetition frequency is changed and/or when different active substances are added to the active medium of the laser. In particular, increasing the excitation pulse repetition frequency may result in the accumulation of prepulse parameters of the active medium, which at a given frequency do not have time to relax completely. The magnitude of this effect is determined by the relaxation rate of the plasma parameters.

The existence of critical values of the prepulse concentration of the electrons [5–7,52,54,89] and copper atoms in the metastable state [5–7,52,54,101,102] was previously confirmed to for pure copper vapour lasers. It is of interest to clarify the influence of these factors on the lasing characteristics of the copper vapour laser with hydrogen additives, since, as discussed above, the hydrogen addition modifies the prepulse plasma conditions of the active medium. Moreover, this effect becomes more significant at higher excitation pulse repetition frequencies. This has been the subject of numerous experimental studies in recent years.

Let us separately consider the effect of prepulse concentrations of electrons and copper atoms in the metastable state (Fig. 8.25 a, b). In the above results the initial concentrations of the plasma reagents were obtained by self-consistent calculations by iteration. Here, prepulse concentrations of various reagents were self-consistently found the next conditions ($N_{Ne} = 1.25 \cdot 10^{17}$ cm^{-3}; $N_{H_2} = 0$, $3.3 \cdot 10^{15}$ cm^{-3}; $N_{Cu} = 8 \cdot 10^{14}$ cm^{-3}, $1 \cdot 10^{15}$ cm^{-3}; $f = 12$ kHz). After this the prepulse concentrations of the reactants were fixed, and changes were made without matching either in the prepulse concentration of electrons or the prepulse concentration of metastable atoms. Different initial values of the electrons and metastable atoms thus correspond to the various possible processes mentioned at the beginning of the paragraph.

Effect of prepulse electron concentration. The calculation was performed for three groups of parameters:

1) $p_{H_2} = 0\%$, $N_{Cu} = 8 \cdot 10^{14}$ cm^{-3};
2) $p_{H_2} = 2\%$, $N_{Cu} = 8 \cdot 10^{14}$ cm^{-3};
3) $p_{H_2} = 2\%$, $N_{Cu} = 1 \cdot 10^{15}$ cm^{-3}.

The latter calculation was carried out in order to stress the influence of increasing concentration of copper, which occurs with the introduction of hydrogen [67] (see A.4).

The presented dependences of laser energy on the initial concentration of electrons $E = f(N_e)$ show that the curve constructed for the case *2* is slightly shifted downward relative to the curve *1*, which leads to a decrease in the critical prepulse electron concentration. However, since, as previously noted (see A.2) the hydrogen addition reduces the value of prepulse electron concentration, especially at high frequencies, then we should expect that the failure of lasing will occur at a higher pulse repetition frequency (or more adverse conditions resulting in such values of the electron concentration).

If we now consider the case with the increased concentration of copper in the active medium (which corresponds to heating of the GDT with the introduction of hydrogen, see A.4), it may be noted that, despite the higher values of radiation energy in the optimal range of the prepulse electron concentration, the lasing energy decreased more rapidly and the critical value of the prepulse electron concentration becomes slightly smaller. It is also necessary to mention the fact that the failure of lasing the yellow line occurs at lower initial values of the electron concentration than for the green line.

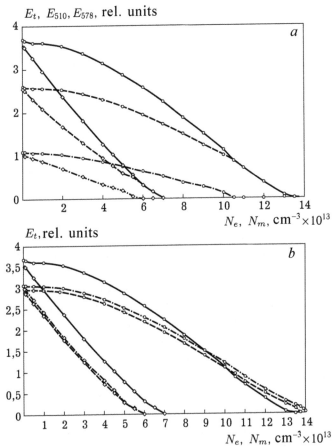

Figure 8.25. Dependences: *a*: total (solid line), at 510.6 nm (dashed curve) and at 578.2 nm (dot-dashed curve) lasing energy on the prepulse concentration of copper atoms in the metastable state and electrons (circles – the dependence $E = f(N_e)$, diamond – the dependence $E = f(N_m)$) for $N_{Cu} = 1 \cdot 10^{15}$ cm^{-3}, $N_{H_2} = 2\%$. *b*: the total lasing energy on the prepulse concentrations of copper atoms in the metastable state and electrons (circle – the dependence $E = f(N_e)$), diamond – the dependence $E = f(N_m)$) for $N_{Cu} = 1 \cdot 10^{15}$ cm^{-3}, $N_{H_2} = 2\%$ (solid curve) $N_{Cu} = 8 \cdot 10^{14}$ cm^{-3}, $N_{H_2} = 2\%$ (dashed curve), $N_{Cu} = 8 \cdot 10^{14}$ cm^{-3}, $N_{H_2} = 0\%$ (dot-dashed curve). GDT parameters are given in Table, 8.4, the values of plasma reagents used in the calculations – in Table 8.6 (for $N_{Cu} = 8 \cdot 10^{14}$ cm^{-3}) and 8.7 (for $N_{Cu} = 1 \cdot 10^{15}$ cm^{-3}).

The mechanism of constraints associated with the prepulse electron concentration has been studied in sufficient detail earlier for the pure copper vapour lasers ([5–7,52,89], section 8.4). Briefly, we note that it is due to the fact that for efficient excitation of the upper laser levels of the electron temperature must exceed a certain critical value, equal to ~1.7 eV. At lower temperatures more effective excitation is observed in the lower (metastable) laser levels and the population inversion is not achieved. Thus, for the available power supply with a specific power output we have a threshold value of the electron concentration above which the electron temperature can not reach the values required to operate the laser.

Effect of prepulse concentration of copper atoms in the metastable state. The effect of the prepulse concentration of copper atoms in the metastable state is due to the fact that at their high initial value, the population of the metastable levels can exceed the population of resonance levels during the excitation pulse, i.e. there is no population inversion. Earlier studies for pure copper [5–7,54,101,102] showed the presence of a limit above which the lasing is disrupted. Below we present a series of calculations with a hydrogen addition. Changes were made in the initial values of the populations of metastable levels $D_{5/2}$ and $D_{3/2}$, where they were taken to be $N_{D_{5/2}} = N_{D_{3/2}} = N_m$. Modelling was performed, as in the case of assessing the impact of the prepulse electron concentration, for the three groups of parameters (see the previous section):

Figure 8.25 a, b shows that the radiation energy becomes zero at a sufficiently high prepulse concentration of copper atoms in the metastable state is equal to $(6–7) \cdot 10^{13}$ cm^{-3}, which is only half the value for the critical concentration of electrons. The dependence $E = f(N_m)$ for the case of the hydrogen addition $N_{Cu} = 8 \cdot 10^{14}$ cm^{-3} is shifted downward relative to the curve plotted for the case without hydrogen. Moreover, for both cases the critical values of the initial population of the metastable levels are practically identical. When the calculations were carried out taking into account the change in the concentration of copper atoms in the ground state in the active medium the dependence $E = f(N_m)$ was situated above the curves plotted for $N_{Cu} = 8 \cdot 10^{14}$ cm^{-3}, and the critical initial concentration of metastable atoms increased.

In [103] it is concludes that the findings of [28,29] can not be justified, as the very model and simulation method are a clear step backward in comparison with previously published studies [16, 36, 50, 72, 73, 81]. In addition, it is argued that the conclusions about the mechanism of the influence of hydrogen additives made in our studies [28,29] differ from the findings of work "used a much more informed self-consistent model." These statements are erroneous and the general critical approval is not true [104].

Conclusion

To determine the effect of hydrogen additions to the active medium of copper vapour lasers we constructed a detailed kinetic model of the Ne–Cu–H$_2$ active medium.

A comparison with the available experimental data was made.

From this analysis it follows that the increase in the laser radiation power with the addition of molecular hydrogen has, apparently, a different nature at low and high pulse repetition rates.

At high frequencies ($f \gg 10$ kHz) the laser radiation power can be increased by reducing the prepulse concentrations of electrons and copper atoms in the metastable state, as well as by increasing the rate of reduction of the concentration of copper in the ground state of the copper atom.

When operating at low excitation pulsed repetition frequencies ($f \sim 10$ kHz) the mechanisms mentioned above are not sufficient. The laser radiation power increases due to the increase in the concentration of the copper atoms in the active medium by increasing the heating of the GDT with the introduction of hydrogen and the simultaneous decrease in the prepulse concentration of metastable atoms due to quenching of these levels by the hydrogen molecules in vibrationally excited states. In addition, the addition of hydrogen to the active medium decreases the value of current flowing in the interpulse interval through the discharge tube, which accelerates the flow of relaxation processes in the plasma.

Note also the negative effects of the introduction of hydrogen. It is the consumption of a significant amount of the energy, deposited into the plasma during the pump pulse, for the vibrational excitation and dissociation of molecular hydrogen, as well as a slowdown in the relaxation of the plasma in the interpulse period, due to the release of energy in the electronic component in the process of thermalization of vibrationally excited hydrogen molecules.

8.5.2. Hydrogen chloride additives

At the moment there are several studies [36,72,105], which simulated an active medium with the addition of $H_2 + HCl$, but only one of them, namely [72], attempted to analyze the possible changes that accompany the introduction of HCl. The improvement of the laser characteristics with hydrogen additives will be analysed on the basis of [30].

Description of kinetic model

To determine the mechanisms of the effect of the addition of hydrogen and HCl (a mixture of $Cu–Ne–H_2–HCl$) on the lasing characteristics a kinetical model was used for the previously constructed models for pure copper vapour lasers [5–7] (see section 8.2) and for copper vapour lasers with the addition of hydrogen [28,29] (see section 8.5.1). We mention briefly additions to the model made to account for hydrogen chloride and chlorine-containing reagents. In constructing the model with the addition of hydrogen chloride we took into account our experience in modelling the exciplex XeCl laser [10,13,14,17,106] and KrCl lamps [20–23]. We took into account the ground and the first two vibrational states of the HCl molecule, the ground states of the molecule and the molecular ion of chlorine and also the first excited electronic state of the chlorine molecule. For atomic chlorine the model takes into account the first lower excited state and ground states of the negative and positive ions. Also considered was the ground state of the CuCl molecule. For hydrogen the same states as in [28, 29] were considered. These papers also give plasma-chemical reaction rates with hydrogen-containing reactants (see also section 8.5.1). A total of 46 plasma reagents of the active medium were taken into account.

The kinetic model takes into account the following basic processes involving chlorine-containing reagents: vibrational excitation and quenching of the HCl molecule (up to $v = 2$ inclusive); electronic attachment/detachment to the chlorine atom; dissociative attachment to the molecules of HCl and CuCl; associative detachment from the negative chlorine ion; ion–ion recombination of a negative chlorine ion and the positive copper and neon ions, the association/dissociation of the molecules of HCl, CuCl and Cl_2; electronic excitation/quenching of the chlorine atoms, the direct and stepwise ionization of atoms and chlorine molecules and recombination of their ions, electronic excitation/quenching of the chlorine molecule; VV- and VT-relaxation of HCl molecules; quenching of metastable levels of the copper atom by the negative chloride ion to form a molecule of CuCl; recovery on the wall of the gas discharge tube of the molecules of H_2, Cl_2, HCl, etc. (for details see [30]).

In comparison with the model [28,29] (see also section 8.5.1) the heat balance equation takes into account the power density spent on the ionization and excitation of chlorine-containing components (these values are expressed in terms of the population of excited atomic and molecular states, they take into account the processes of excitation and de-excitation of the levels considered in the model), the terms and Q_{wall} and $Q_{\Delta T}$ now have the form

$$Q_{wall} = \frac{5.41\cdot10^4 T_e^{1.5} N_e}{R^2\left(\sigma_{eNe}N_{Ne}+\sigma_{eH_2}N_{H_2}+\sigma_{eHCl}N_{HCl}\right)}, \text{W/cm}^3$$

is the power of the heat sink on the walls (T_e – in eV, $\sigma_{eNe}(T_e)$ is the transport cross section for elastic collisions of the electron with the neon atom (in units of $10^{-16}\,\text{cm}^2$), it is weakly dependent on temperature in the range 2 eV and is about $1.5 \cdot 10^{-16}$ cm^2, $\sigma_{eH_2}(T_e)$ is the collision cross section of the electron with hydrogen (in units of 10^{-16} cm^2) approximately equal to $1.5 \cdot 10^{-15}$ cm^2, $\sigma_{eHCl}(T_e)$ is the transport cross section of electron collisions with HCl (in units of 10^{-16} cm^2), which is approximately $1 \cdot 10^{-15}$ cm^2, the tube radius R is in cm;

$$Q_{\Delta T} = 2 \cdot [(m_e / m_{Ne}) \cdot k_{Ne} \cdot N_{Ne} + (m_e / m_{Cu}) \cdot k_{ei} N_e +$$
$$+ (m_e / m_{H_2}) \cdot k_{H_2} N_{H_2} + (m_e / m_{HCl}) \cdot k_{HCl} N_{HCl}] \cdot Ne \cdot (T_e - T_g)$$

is the power density spent on cooling of electrons due to elastic collisions with the neon atoms, the copper ions and the hydrogen and HCl molecules; k_{Ne}, k_{H_2}, k_{HCl}, k_{ei} is the rate of elastic collisions of electrons with the neon atoms, the molecules of hydrogen, HCl and the ions; N_{Ne} is the density of the neon atoms, N_{H_2} is the concentration of molecular hydrogen; N_{HCl} is the concentration of the HCl molecules; m_e is the electron mass, m_{Ne}, m_{Cu}, m_{H_2}, m_{HCl} is the atomic mass of neon and copper and the hydrogen and HCl molecules.

The system of rigorous ordinary differential equations for describing the time dependences of the plasma reactant concentrations of the active medium, the heat balance equation for the electron temperature and the intensity of radiation was self-consistently solved using the PLAZER software package (for details see sections 8.2, 8.5.1, [30]). A total of about 250 kinetic reactions was taken into account in the model.

Mechanisms of the effect of HCl additives

Consider the previously discussed mechanisms of the effect of HCl additives on the operation of a copper vapour laser [36,72,77–81,90, 91,105,107,108]:

M1. In [72,80,81,90,91,105] it is concluded that the main change in the kinetics of the introduction of HCl additives is to increase the decay rate of the electron concentration in the interpulse period owing to the reactions of dissociative electron attachment to HCl molecules, which leads to a decrease of the prepulse concentration of electrons, creating more favourable operating conditions of the laser.

M2. A decrease of the prepulse electron concentration increases the resistance of the plasma during the excitation pulse [72,79,91, 108] leading to a decrease in the maximum value of the GDT current, while increasing the voltage across the GDT.

M3. It is noted [72,81,91] that the introduction of HCL is accompanied by an increase of the concentration of the copper atoms in the active medium with a simultaneous increase of the rate of recovery in the afterglow which, however, does not fully explain the observed increase in lasing energy.

M4. From the simulation results [72] it follows that the addition of HCl increases the maximum electron temperature, which should lead to an increase in lasing energy.

M5. According to [72,105] the rate of decrease of the population of metastable levels of the copper atom in the initial phase of the interpulse period decreases due to a drop in the electron density, which reduces the efficiency of electron de-excitation.

M6. In [72] the simulation results are used to conclude that there is a small energy loss of the electrons in collisions with chlorine-containing reagents, compared with the losses in collisions with the components of copper, hydrogen and neon.

M7. In [36,72] it is suggested that the influence of HCl additions on the kinetics of the active medium during the excitation pulse is small, due to the low concentration.

M8. According to [72], in the initial stage of the interpulse period [72] the rate of decline in the electron temperature is lower due to the decrease of the number of positive ions in the plasma and, consequently, decrease of the number of elastic Coulomb collisions, leading to a decrease in the electron temperature. Decrease of the concentration of the positive ions in the plasma is due to the intense ion–ion recombination.

M9. According to [72,107,108] the addition of HCl into the active decreases the intensity of the plasma skin effect, which determines the development of the radial electric field and changes the spatial evolution of the optical gain factor, which improves the quality of the output beam. The same effect causes a shift of the 'centre of gravity' of the lasing pulse at the end.

M10. Improvement of the characteristics of the laser is a result of increasing the degree of dissociation of molecular hydrogen [36]. In [36] the degree of dissociation of molecular hydrogen increased up to 50% because of the reactions

$$H_2(v=0.1)+Cl \rightarrow HCl(v=0-2)+H.$$

In [72] the degree of dissociation of molecular hydrogen was smaller, about 30%.

Below we compare in detail the numerical calculations based on the developed model with the experimental and calculated data from [72,80] which are the most comprehensive of the works known to the authors. The simulation results agree well with these works, so the analysis of the effect of HCl additives on the kinetics of the active

Table 8.16. The initial values of the plasma reactants at a frequency of 12 kHz (prepulse value of NH_{20}: $NH_{20} = 3 \cdot 10^{15}$ cm^{-3}). The initial values of plasma reagents were obtained as a result of the self-consistent calculation. Parameters of GDT are in Table 8.19. The calculations were performed for gas temperatures $T_g = 0.146$ and 0.215 eV. E_t – the total specific lasing energy, E_{510} – specific lasing energy at a wavelength of 510.6 nm, E_{578} – specific lasing energy at a wavelength of 578.2 nm (J/cm^3). $N_{Cu\,min}$ – the minimum concentration of copper during the excitation pulse. Lasing energy for different gas temperatures ($T_g = 0.146$–0.215 eV) varies by about 1% if the calculations take the same density of the copper atoms in the ground state into account.

Reagent	$N_{HCl} = 0$, (cm^{-3})	$N_{HCl} = 5 \cdot 10^{13}$, (cm^{-3})	$N_{HCl} = 1 \cdot 10^{14}$, (cm^{-3})	$N_{HCl} = 1.5 \cdot 10^{14}$, (cm^{-3})	$N_{HCl} = 2 \cdot 10^{14}$, (cm^{-3})	$N_{HCl} = 3 \cdot 10^{14}$, (cm^{-3})
\multicolumn{7}{c}{Gas temperature for all cases $T_g = 0.146$ eV}
Cu	$0.8 \cdot 10^{15}$	$0.9 \cdot 10^{15}$	$1 \cdot 10^{15}$	$1.1 \cdot 10^{15}$	$1.1 \cdot 10^{15}$	$1.4 \cdot 10^{15}$
I, A	840	760	680	610	520	360
Cu$^+$	$2.34 \cdot 10^{13}$	$1.33 \cdot 10^{13}$	$1.1 \cdot 10^{13}$	$8.41 \cdot 10^{12}$	$8.33 \cdot 10^{12}$	$5.08 \cdot 10^{12}$
Ne$^+$	$6.33 \cdot 10^{12}$	$6.22 \cdot 10^{12}$	$4.84 \cdot 10^{12}$	$3.88 \cdot 10^{12}$	$3.03 \cdot 10^{12}$	$1.8 \cdot 10^{12}$
$Cu_{D_{5/2}}$	$1.5 \cdot 10^{12}$	$6.81 \cdot 10^{11}$	$4.14 \cdot 10^{11}$	$2.34 \cdot 10^{11}$	$2.19 \cdot 10^{11}$	$1.07 \cdot 10^{11}$
$Cu_{D_{3/2}}$	$2.55 \cdot 10^{11}$	$9.59 \cdot 10^{10}$	$5.11 \cdot 10^{10}$	$2.46 \cdot 10^{10}$	$2.22 \cdot 10^{10}$	$7.36 \cdot 10^{9}$
H	$1.28 \cdot 10^{15}$	$2.7 \cdot 10^{15}$	$3.35 \cdot 10^{15}$	$3.08 \cdot 10^{15}$	$2.86 \cdot 10^{15}$	$2.39 \cdot 10^{15}$
H$_2$	$2.25 \cdot 10^{15}$	$1.55 \cdot 10^{15}$	$1.31 \cdot 10^{15}$	$1.46 \cdot 10^{15}$	$1.39 \cdot 10^{15}$	$1.78 \cdot 10^{15}$
H$^-$	$2.23 \cdot 10^{9}$	$7.72 \cdot 10^{8}$	$5.54 \cdot 10^{8}$	$3.88 \cdot 10^{8}$	$3.48 \cdot 10^{8}$	$1.66 \cdot 10^{8}$
Cl	–	$8.73 \cdot 10^{12}$	$2.5 \cdot 10^{13}$	$3.1 \cdot 10^{13}$	$3.14 \cdot 10^{13}$	$4.24 \cdot 10^{13}$
CuH	$9.72 \cdot 10^{10}$	$5.51 \cdot 10^{9}$	$1.24 \cdot 10^{9}$	$6.35 \cdot 10^{8}$	$5.7 \cdot 10^{8}$	$2.59 \cdot 10^{8}$
CuCl	–	$1.08 \cdot 10^{12}$	$9.42 \cdot 10^{11}$	$1.16 \cdot 10^{12}$	$1.1 \cdot 10^{12}$	$1.66 \cdot 10^{12}$
H$_2$ ($v = 1$)	$1.07 \cdot 10^{14}$	$2.4 \cdot 10^{13}$	$1.06 \cdot 10^{13}$	$9.23 \cdot 10^{12}$	$8.99 \cdot 10^{12}$	$8.19 \cdot 10^{12}$
H$_2$ ($v = 2$)	$7.79 \cdot 10^{11}$	$2.93 \cdot 10^{10}$	$3.78 \cdot 10^{9}$	$5.19 \cdot 10^{8}$	$4.35 \cdot 10^{8}$	$1.18 \cdot 10^{8}$
H$_2$ ($v = 3$)	$2.39 \cdot 10^{10}$	$1.24 \cdot 10^{9}$	$2.44 \cdot 10^{8}$	$1.06 \cdot 10^{8}$	$9.62 \cdot 10^{7}$	$5.24 \cdot 10^{7}$
HCl ($v = 0$)	–	$3.8 \cdot 10^{14}$	$7.45 \cdot 10^{13}$	$1.14 \cdot 10^{14}$	$1.21 \cdot 10^{14}$	$2.45 \cdot 10^{14}$
HCl ($v = 1$)	–	$2.88 \cdot 10^{12}$	$4.23 \cdot 10^{12}$	$4.91 \cdot 10^{11}$	$1.63 \cdot 10^{12}$	$4.82 \cdot 10^{17}$
HCl ($v = 2$)	–	$1.05 \cdot 10^{11}$	$1.3 \cdot 10^{11}$	$1.22 \cdot 10^{11}$	$1.09 \cdot 10^{11}$	$6.19 \cdot 10^{10}$
T_e	0.191	0.177	0.163	0.153	0.149	0.146
E_{510} (J/cm^3)	$2.4 \cdot 10^{-6}$	$2.61 \cdot 10^{-6}$	$2.74 \cdot 10^{-6}$	$2.75 \cdot 10^{-6}$	$2.66 \cdot 10^{-6}$	$2.35 \cdot 10^{-6}$
E_{578} (J/cm^3)	$1.06 \cdot 10^{-6}$	$1.2 \cdot 10^{-6}$	$1.24 \cdot 10^{-6}$	$1.17 \cdot 10^{-6}$	$1.19 \cdot 10^{-6}$	$0.99 \cdot 10^{-6}$
E_t (J/cm^3)	$3.465 \cdot 10^{-6}$	$3.81 \cdot 10^{-6}$	$3.98 \cdot 10^{-6}$	$4.02 \cdot 10^{-6}$	$3.85 \cdot 10^{-6}$	$3.33 \cdot 10^{-6}$
P, W	70.7	77.78	81.25	82.044	78.6	67.98
P_{exp}, W	67.2	–	–	86.4	–	–
\multicolumn{7}{c}{Gas temperature for all cases $T_g = 0.215$ eV}
Cu	$0.8 \cdot 10^{15}$	$0.9 \cdot 10^{15}$	$1 \cdot 10^{15}$	$1.1 \cdot 10^{15}$	$1.1 \cdot 10^{15}$	$1.4 \cdot 10^{15}$

I, A	840	760	680	610	520	360
Cu^+	$2.5 \cdot 10^{13}$	$1.4 \cdot 10^{13}$	$1.2 \cdot 10^{13}$	$9.1 \cdot 10^{12}$	$8.9 \cdot 10^{12}$	$5.5 \cdot 10^{12}$
Ne^+	$6.8 \cdot 10^{12}$	$6.6 \cdot 10^{12}$	$5.2 \cdot 10^{12}$	$4.1 \cdot 10^{12}$	$3.2 \cdot 10^{12}$	$1.9 \cdot 10^{12}$
$Cu_{D_{5/2}}$	$4.5 \cdot 10^{12}$	$2.7 \cdot 10^{11}$	$2 \cdot 10^{12}$	$1.1 \cdot 10^{12}$	$1.6 \cdot 10^{12}$	$7.9 \cdot 10^{11}$
$Cu_{D_{3/2}}$	$7.6 \cdot 10^{11}$	$3.8 \cdot 10^{11}$	$2.5 \cdot 10^{11}$	$1.2 \cdot 10^{11}$	$1.1 \cdot 10^{11}$	$5.1 \cdot 10^{10}$
H	$1.2 \cdot 10^{15}$	$2.6 \cdot 10^{15}$	$3.2 \cdot 10^{15}$	$2.9 \cdot 10^{15}$	$2.7 \cdot 10^{15}$	$2.3 \cdot 10^{15}$
H_2	$2.3 \cdot 10^{15}$	$1.6 \cdot 10^{15}$	$1.3 \cdot 10^{15}$	$1.5 \cdot 10^{15}$	$1.4 \cdot 10^{15}$	$1.8 \cdot 10^{15}$
H^-	$2.5 \cdot 10^{9}$	$8.5 \cdot 10^{8}$	$6.1 \cdot 10^{8}$	$4.3 \cdot 10^{8}$	$3.8 \cdot 10^{8}$	$1.9 \cdot 10^{8}$
Cl	–	$8.6 \cdot 10^{12}$	$2.4 \cdot 10^{13}$	$3 \cdot 10^{13}$	$3.1 \cdot 10^{13}$	$4.1 \cdot 10^{13}$
CuH	$2.6 \cdot 10^{11}$	$1.4 \cdot 10^{10}$	$3.3 \cdot 10^{9}$	$1.7 \cdot 10^{9}$	$1.5 \cdot 10^{9}$	$6.9 \cdot 10^{8}$
$CuCl$	–	$1.8 \cdot 10^{12}$	$1.5 \cdot 10^{12}$	$1.9 \cdot 10^{12}$	$1.9 \cdot 10^{12}$	$2.7 \cdot 10^{12}$
$H_2\ (v = 1)$	$1.1 \cdot 10^{14}$	$2.6 \cdot 10^{13}$	$1.1 \cdot 10^{13}$	$1 \cdot 10^{13}$	$9.8 \cdot 10^{12}$	$8.9 \cdot 10^{12}$
$H_2\ (v = 2)$	$1.5 \cdot 10^{12}$	$5.7 \cdot 10^{10}$	$7.3 \cdot 10^{9}$	$1.1 \cdot 10^{9}$	$8.4 \cdot 10^{8}$	$2.3 \cdot 10^{8}$
$H_2\ (v = 3)$	$3.6 \cdot 10^{10}$	$1.8 \cdot 10^{9}$	$3.7 \cdot 10^{8}$	$1.6 \cdot 10^{8}$	$1.4 \cdot 10^{8}$	$7.9 \cdot 10^{7}$
$HCl\ (v = 0)$	–	$4.1 \cdot 10^{13}$	$8.1 \cdot 10^{13}$	$1.2 \cdot 10^{14}$	$1.3 \cdot 10^{14}$	$2.6 \cdot 10^{14}$
$HCl\ (v = 1)$	–	$3.4 \cdot 10^{12}$	$4.9 \cdot 10^{12}$	$5.7 \cdot 10^{12}$	$5.4 \cdot 10^{12}$	$5.6 \cdot 10^{12}$
$HCl\ (v = 2)$	–	$1.3 \cdot 10^{11}$	$1.6 \cdot 10^{11}$	$1.5 \cdot 10^{11}$	$1.3 \cdot 10^{11}$	$7.9 \cdot 10^{10}$
T_e, eV	0.238	0.235	0.233	0.227	0.224	0.22
E_{510} (J/cm³)	$2.4 \cdot 10^{-6}$	$2.6 \cdot 10^{-6}$	$2.7 \cdot 10^{-6}$	$2.8 \cdot 10^{-6}$	$2.6 \cdot 10^{-6}$	$2.3 \cdot 10^{-6}$
E_{578} (J/cm³)	$1. \cdot 10^{-6}$	$1.2 \cdot 10^{-6}$	$1.2 \cdot 10^{-6}$	$1.2 \cdot 10^{-6}$	$1.2 \cdot 10^{-6}$	$1. \cdot 10^{-6}$
E_t (J/cm³)	$3.4 \cdot 10^{-6}$	$3.8 \cdot 10^{-6}$	$3.9 \cdot 10^{-6}$	$4. \cdot 10^{-6}$	$3.8 \cdot 10^{-6}$	$3.3 \cdot 10^{-6}$
P, W	70.6	77.8	81.2	82	78.6	68
P_{exp}, W	67.2	–	–	86.4	–	–

Table 8.17. Fluxes of variation of the concentration of copper ions changes during the excitation pulse, F (0, 500 ns) with the HCl addition at constant pumping conditions ($N_{HCl} = 1.5 \cdot 10^{14}$ cm⁻³, $N_{Cu} = 0.8 \cdot 10^{15}$ cm⁻³, $I_{max} = 810$ A). (GDI parameters are shown in Table 8.19, the values of initial concentrations of reagents of the active medium obtained in self-consistent simulations – Table 8.16)

Reaction	Flux (cm⁻³)
$Cu + e \rightarrow Cu^+ + 2 \cdot e$	$2.015 \cdot 10^{13}$
$Cu^{**} + e \rightarrow Cu^+ + 2 \cdot e$	$1.862 \cdot 10^{13}$
$Cu\ (P_{3/2}) + e \rightarrow Cu^+ + 2 \cdot e$	$7.105 \cdot 10^{12}$
$Cu^* + e \rightarrow Cu^+ + 2 \cdot e$	$6.978 \cdot 10^{12}$
$Ne^* + Cu \rightarrow Cu^+ + Ne + e$	$4.924 \cdot 10^{12}$
$Cu\ (P_{1/2}) + e \rightarrow Cu^+ + 2 \cdot e$	$4.38 \cdot 10^{12}$
$Cu\ (5\ S_{1/2}) + e \rightarrow Cu^+ + 2 \cdot e$	$3.756 \cdot 10^{12}$
$Cu\ (D_{5/2}) + e \rightarrow Cu^+ + 2 \cdot e$	$2.439 \cdot 10^{12}$
$Cu\ (D_{3/2}) + e \rightarrow Cu^+ + 2 \cdot e$	$1.271 \cdot 10^{12}$
$Cu^+ + Cl^- + Ne \rightarrow Cu\ (P_{1/2}) + Cl + Ne$	$9.319 \cdot 10^{11}$
$Cu^+ + Cl^- + Ne \rightarrow Cu\ (P_{3/2}) + Cl + Ne$	$9.319 \cdot 10^{11}$

Table 8.18. Part 1. Fluxes of electrons cooling in the pump pulse: F (0, 500 ns) without hydrogen chloride ($N_{Cu} = 0.8 \cdot 10^{15}$ cm^{-3}, $I_{max} = 810$ A). (GDT parameters are shown in Table 8.19, the values of initial concentrations of reagents of the active medium obtained by self-consistency calculations – Table 8.16)

Reaction	$N_{HCl} = 0$ cm^{-3}			
	Flux (cm^{-3})	Resultant flux	Energy	Energy flux
Pumping	$5.924 \cdot 10^{17}$	$5.924 \cdot 10^{17}$	–	–
Ne + e → Ne* + e	$1.344 \cdot 10^{14}$	$1.075 \cdot 10^{14}$	−16.6	$-2.234 \cdot 10^{15}$
Ne* + e → Ne + e	$2.69 \cdot 10^{13}$		16.6	
Ne* + e → Ne** + e	$2.974 \cdot 10^{14}$	$1.495 \cdot 10^{14}$	−1.9	$-2.84 \cdot 10^{14}$
Ne** + e → Ne* + e	$1.479 \cdot 10^{13}$		1.9	
Cu ($P_{1/2}$) + e → Cu ($P_{3/2}$) + e	$1.465 \cdot 10^{16}$	$4 \cdot 10^{13}$	−0.0313	$-1.252 \cdot 10^{12}$
Cu ($P_{3/2}$) + e → Cu ($P_{1/2}$) + e	$1.461 \cdot 10^{16}$		0.0313	
Cu + e → Cu ($P_{3/2}$) + e	$1.555 \cdot 10^{14}$	$8.571 \cdot 10^{13}$	−3.817	$-3.271 \cdot 10^{13}$
Cu ($P_{3/2}$) + e → Cu + e	$6.979 \cdot 10^{13}$		3.817	
Cu + e → Cu ($P_{1/2}$) + e	$7.422 \cdot 10^{13}$	$4.028 \cdot 10^{13}$	−3.786	$-1.525 \cdot 10^{14}$
Cu ($P_{1/2}$) + e → Cu + e	$3.394 \cdot 10^{13}$		3.786	
Cu ($D_{3/2}$) + e → Cu ($D_{5/2}$) + e	$1.194 \cdot 10^{14}$	$4 \cdot 10^{12}$	0.253	$-1.012 \cdot 10^{12}$
Cu ($D_{5/2}$) + e → Cu ($D_{3/2}$) + e	$1.19 \cdot 10^{14}$		−0.253	
Cu + e → Cu ($D_{3/2}$) + e	$1.228 \cdot 10^{14}$	$4.776 \cdot 10^{13}$	−1.642	$-7.842 \cdot 10^{13}$
Cu ($D_{3/2}$) + e → Cu + e	$7.504 \cdot 10^{13}$		1.642	
Cu + e → Cu ($D_{5/2}$) + e	$1.42 \cdot 10^{14}$	$9.613 \cdot 10^{13}$	−1.389	$-1.335 \cdot 10^{14}$
Cu ($D_{5/2}$) + e → Cu + e	$4.587 \cdot 10^{13}$		1.389	
Cu + e → Cu$^+$ + 2·e	$1.988 \cdot 10^{13}$	$1.988 \cdot 10^{13}$	−7.726	$-1.465 \cdot 10^{14}$

Table 8.18. Part 2. Cooling fluxes of the electrons for the pump pulse: F (0, 500 ns) with the HCl addition at constant pumping conditions ($N_{HCl} = 1.5 \cdot 10^{14}$ cm^{-3}, $N_{Cu} = 0.8 \cdot 10^{15}$ cm^{-3}, $I_{max} = 810$ A). (GDT parameters are shown in Table 8.19, the values of initial concentrations of the reagents of the active medium obtained by self-consistency calculations – Table 8.16)

Reaction	$N_{HCl} = 1.5 \cdot 10^{14}$ cm^{-3}			
	Flux (cm^{-3})	Resultant flux	Energy	Energy flux
Pumping	$7.427 \cdot 10^{17}$	$7.427 \cdot 10^{17}$	–	–
Ne + e → Ne* + e	$2.396 \cdot 10^{14}$	$2.018 \cdot 10^{14}$	−16.6	$-3.345 \cdot 10^{15}$
Ne* + e → Ne + e	$3.775 \cdot 10^{13}$		16.6	
Ne* + e → Ne** + e	$4.103 \cdot 10^{14}$	$2.149 \cdot 10^{14}$	−1.9	$-4.083 \cdot 10^{14}$
Ne** + e → Ne* + e	$1.954 \cdot 10^{14}$		1.9	
Cu ($P_{1/2}$) + e → Cu ($P_{3/2}$) + e	$1.387 \cdot 10^{16}$	$4 \cdot 10^{13}$	−0.0313	$-1.252 \cdot 10^{12}$
Cu ($P_{3/2}$) + e → Cu ($P_{1/2}$) + e	$1.383 \cdot 10^{16}$		0.0313	

Cu $(D_{3/2})$ + e → Cu $(D_{5/2})$ + e	$1.103\cdot10^{15}$	$4\cdot10^{12}$	0.253	$-1.012\cdot10^{12}$
Cu $(D_{5/2})$ + e → Cu $(D_{3/2})$ + e	$1.099\cdot10^{15}$		-0.253	
Cu + e → Cu $(P_{3/2})$ + e	$1.459\cdot10^{14}$	$7.983\cdot10^{13}$	-3.817	$-3.047\cdot10^{14}$
Cu $(P_{3/2})$ + e → Cu + e	$6.607\cdot10^{13}$		3.817	
Cu + e → Cu $(P_{1/2})$ + e	$6.964\cdot10^{13}$	$3.75\cdot10^{13}$	-3.786	$-1.42\cdot10^{14}$
Cu $(P_{1/2})$ + e → Cu + e	$3.214\cdot10^{13}$		3.786	
Cu + e → Cu $(D_{3/2})$ + e	$1.179\cdot10^{14}$	$4.863\cdot10^{13}$	-1.642	$-7.985\cdot10^{13}$
Cu $(D_{3/2})$ + e → Cu + e	$6.927\cdot10^{13}$		1.642	
Cu + e → Cu$^+$ + 2·e	$2.015\cdot10^{13}$	$2.015\cdot10^{13}$	-3.89	$-1.556\cdot10^{14}$
Cu + e → Cu $(D_{5/2})$ + e	$1.365\cdot10^{14}$	$9.434\cdot10^{13}$	-1.389	$-1.31\cdot10^{14}$
Cu $(D_{5/2})$ + e → Cu + e	$6.884\cdot10^{13}$		1.389	

medium was carried out by the example of modelling the experimental conditions [72,80].

Analysis of previously discussed mechanisms of the effect of HCl additions

A1. The positive effect of HCl additions in M1 associated with a decrease in the prepulse electron concentration.

Since the development of copper vapour lasers, researchers have been trying to determine the root cause of restrictions on their frequency and energy characteristics. The main points of view today are: a high value of the prepulse population of the metastable levels of the copper atom [51,55] and a high prepulse concentration of electrons [5–7,52,54,89,90]. In general, these two factors are interrelated, but nevertheless, earlier studies [5–7,52,54,89,90] showed that the prepulse electron concentration is a major restricting factor in determining the limiting characteristics of this class of lasers. From this we can expect a positive effect of reducing the prepulse electron

Table 8.19. The GDT parameter values used in the calculations

Parameter	1 [81]	2 [80]
The length of the active medium (cm)	150	58
GDT diameter (cm)	3.8	2.5
Repetition frequency of excitation pulses (kHz)	12	50
Neon pressure (Torr)	36	40
Copper concentration (cm^{-3})	$1.1\cdot10^{15}$	$1.72\cdot10^{15}$

Table 8.20. Fluxes of depletion of the ground level of the copper atom during the excitation pulse F (0, 500 ns). (GDT parameters listed in Table 8.19, the values of the initial plasma reactant concentrations obtained by self-consistency calculations – Table 8.16). Conditions of excitation of the active medium in both cases are the same ($N_{Cu} = 0.8 \cdot 10^{15}$ cm^{-3}, $I_{max} = 810$ A)

Reaction	Flux (cm^{-3})			
	$N_{HCl} = 0$ cm^{-3}		$N_{HCl} = 1.5 \cdot 10^{14}$ cm^{-3}	
$Cu + e \rightarrow Cu\,(P_{3/2}) + e$	$-1.555 \cdot 10^{14}$	$-8.571 \cdot 10^{13}$	$-1.459 \cdot 10^{14}$	$-7.983 \cdot 10^{13}$
$Cu\,(P_{3/2}) + e \rightarrow Cu + e$	$+6.979 \cdot 10^{13}$		$+6.607 \cdot 10^{13}$	
$Cu + e \rightarrow Cu\,(D_{5/2}) + e$	$-1.42 \cdot 10^{14}$	$-9.613 \cdot 10^{13}$	$-1.365 \cdot 10^{14}$	$-9.434 \cdot 10^{13}$
$Cu\,(D_{5/2}) + e \rightarrow Cu + e$	$+4.587 \cdot 10^{13}$		$+4.216 \cdot 10^{14}$	
$Cu + e \rightarrow Cu\,(D_{3/2}) + e$	$-1.228 \cdot 10^{14}$	$-4.776 \cdot 10^{13}$	$-1.179 \cdot 10^{14}$	$-4.863 \cdot 10^{13}$
$Cu\,(D_{3/2}) + e \rightarrow Cu + e$	$+7.504 \cdot 10^{13}$		$+6.927 \cdot 10^{13}$	
$Cu + e \rightarrow Cu\,(P_{1/2}) + e$	$-7.422 \cdot 10^{13}$	$-4.028 \cdot 10^{13}$	$-6.964 \cdot 10^{13}$	$-3.75 \cdot 10^{14}$
$Cu\,(P_{1/2}) + e \rightarrow Cu + e$	$+3.394 \cdot 10^{13}$		$+3.214 \cdot 10^{13}$	
$Cu + e \rightarrow Cu^+ + 2 \cdot e$	$-1.988 \cdot 10^{13}$	$-1.988 \cdot 10^{13}$	$-2.015 \cdot 10^{13}$	$-2.015 \cdot 10^{13}$
$Cu + e \rightarrow Cu^* + e$	$-1.227 \cdot 10^{13}$	$-1.227 \cdot 10^{13}$	$-1.137 \cdot 10^{13}$	$-1.137 \cdot 10^{13}$
$Ne^* + Cu \rightarrow Cu^+ + Ne + e$	$-3.163 \cdot 10^{12}$	$-3.163 \cdot 10^{12}$	$-4.924 \cdot 10^{12}$	$-4.924 \cdot 10^{12}$
$Cu\,(P_{3/2}) \rightarrow Cu$	$+2.351 \cdot 10^{12}$	$+2.351 \cdot 10^{12}$	$+2.384 \cdot 10^{12}$	$+2.384 \cdot 10^{12}$
$Cu\,(P_{1/2}) \rightarrow Cu$	$+1.199 \cdot 10^{12}$	$+1.199 \cdot 10^{12}$	$+1.217 \cdot 10^{12}$	$+1.217 \cdot 10^{12}$

Table 8.21. Fluxes of changes in the concentration of the $D_{5/2}$ levels of the Cu atom in the interpulse period, F (3, 80 μs). To eliminate fully the influence of the excitation pulse, t_1 was set equal to 3 μs, the end of the interpulse period corresponds to 80 μs ($f = 12$ kHz) (GDT parameters are given in Table 8.19, the values of the initial plasma reactant concentrations obtained by self-consistency calculations are in Table 8.16). Conditions of excitation of the active medium in both cases are the same ($N_{Cu} = 0.8 \cdot 10^{15}$ cm^{-3}, $I_{max} = 810$ A)

Reaction	Flux (cm^{-3})			
	$N_{HCl} = 0$ cm^{-3}		$N_{HCl} = 1 \cdot 10^{14}$ cm^{-3}	
$Cu\,(D_{3/2}) + e \rightarrow Cu\,(D_{5/2}) + e$	$+1.052 \cdot 10^{15}$	$+3.9 \cdot 10^{13}$	$+1.259 \cdot 10^{15}$	$+3.9 \cdot 10^{13}$
$Cu\,(D_{5/2}) + e \rightarrow Cu\,(D_{3/2}) + e$	$-1.017 \cdot 10^{15}$		$-1.22 \cdot 10^{15}$	
$Cu\,(D_{5/2}) + e \rightarrow Cu + e$	$-1.059 \cdot 10^{14}$	$+3.93 \cdot 10^{13}$	$-8.748 \cdot 10^{13}$	$+4.243 \cdot 10^{13}$
$Cu + e \rightarrow Cu\,(D_{5/2}) + e$	$+7.097 \cdot 10^{13}$		$+4.505 \cdot 10^{13}$	
$Cu\,(P_{3/2}) \rightarrow Cu\,(D_{5/2})$	$+7.024 \cdot 10^{12}$	$+7.024 \cdot 10^{12}$	$+7.832 \cdot 10^{12}$	$+7.832 \cdot 10^{12}$
$Cu^{**} \rightarrow Cu\,(D_{5/2})$	$+8.764 \cdot 10^{12}$	$+8.764 \cdot 10^{12}$	$+2.343 \cdot 10^{12}$	$+2.343 \cdot 10^{12}$
$Cu\,(P_{3/2}) + e \rightarrow Cu\,(D_{5/2}) + e$	$+6.503 \cdot 10^{12}$	$+6.332 \cdot 10^{12}$	$+4.857 \cdot 10^{12}$	$+4.231 \cdot 10^{12}$
$Cu\,(D_{5/2}) + e \rightarrow Cu\,(P_{3/2}) + e$	$-1.71 \cdot 10^{11}$		$-6.258 \cdot 10^{11}$	

$Cu^* + e \rightarrow Cu\,(D_{5/2}) + e$	$+6.479 \cdot 10^{11}$	$+6.01 \cdot 10^{11}$	$+3.462 \cdot 10^{12}$	$+3.233 \cdot 10^{12}$
$Cu\,(D_{5/2}) + e \rightarrow Cu^* + e$	$-4.697 \cdot 10^{10}$		$-2.286 \cdot 10^{11}$	
$Cu\,(D_{5/2}) + H_2\,(v{=}1) \rightarrow CuH + H$	$-4.105 \cdot 10^{12}$	$-4.105 \cdot 10^{12}$	$-3.375 \cdot 10^{12}$	$-3.375 \cdot 10^{12}$
$Cu\,(D_{5/2}) + H_2\,(v{=}2) \rightarrow CuH + H$	$-4.907 \cdot 10^{11}$	$-4.907 \cdot 10^{11}$	$-3.745 \cdot 10^{11}$	$-3.745 \cdot 10^{11}$
$Cu\,(D_{5/2}) + H_2\,(v{=}3) \rightarrow CuH + H$	$-5.182 \cdot 10^{10}$	$-5.182 \cdot 10^{10}$	$-3.914 \cdot 10^{10}$	$-3.914 \cdot 10^{10}$
$Cu\,(D_{5/2}) + Cl \rightarrow CuCl + e$	–	–	$-1.078 \cdot 10^{11}$	$-1.078 \cdot 10^{11}$

concentration. Indeed the addition of the optimum HCl concentration of 0.1% from the viewpoint of the lasing energy resulted, according to the model, in more than halving the prepulse electron concentration (see Table 8.16 and Fig. 8.26 a). We can also note a slight decrease in the electron concentration during the excitation pulse obtained in the simulations, both at

Table 8.22. Part 1. Fluxes of electron cooling in the interpulse period: F (3 µs, 80 µs) without a mixture of hydrogen chloride ($N_{Cu} = 0.8 \cdot 10^{15}$ cm^{-3}, $I_{max} = 810$ A). To eliminate the influence of the excitation pulse, t_1 was set equal to 3 ms, the end of interpulse period corresponds to 80 µs ($f = 12$ kHz) (GDT parameters are given in Table 8.19, the values of the initial reactant concentrations of the active medium obtained from the self-consistency calculations – Table 8.16)

Reaction	$N_{HCl} = 0$ cm^{-3}			
	Flux (cm^{-3})	Resultant flux	Energy	Energy flux
$H_2 + e = H_2\,(v = 1) + e$	$7.541 \cdot 10^{14}$	$5.4 \cdot 10^{13}$	-0.5	$-2.7 \cdot 10^{13}$
$H_2\,(v = 1) + e = H_2 + e$	$7.001 \cdot 10^{14}$		$+0.5$	
$Cu\,(D_{3/2}) + e \rightarrow Cu\,(D_{5/2}) + e$	$1.052 \cdot 10^{15}$	$3.5 \cdot 10^{13}$	$+0.253$	$+8.871 \cdot 10^{11}$
$Cu\,(D_{5/2}) + e \rightarrow Cu\,(D_{3/2}) + e$	$1.017 \cdot 10^{15}$		-0.253	
Elastic scattering	$7.006 \cdot 10^{19}$	$7.006 \cdot 10^{19}$	$-8 \cdot 10^{-5} \times (T_e{-}T_g)$	–
$Cu\,(P_{3/2}) + e \rightarrow Cu\,(P_{1/2}) + e$	$5.106 \cdot 10^{15}$	$1 \cdot 10^{13}$	$+0.0313$	$+3.13 \cdot 10^{11}$
$Cu\,(P_{1/2}) + e \rightarrow Cu\,(P_{3/2}) + e$	$5.096 \cdot 10^{15}$		-0.0313	
$Cu\,(D_{3/2}) + e \rightarrow Cu + e$	$9.406 \cdot 10^{13}$	$6.458 \cdot 10^{13}$	$+1.642$	$+1.06 \cdot 10^{14}$
$Cu + e \rightarrow Cu\,(D_{3/2}) + e$	$2.948 \cdot 10^{13}$		-1.642	
$Cu\,(D_{5/2}) + e \rightarrow Cu + e$	$1.059 \cdot 10^{14}$	$3.493 \cdot 10^{13}$	$+1.389$	$+4.851 \cdot 10^{13}$
$Cu + e \rightarrow Cu\,(D_{5/2}) + e$	$7.097 \cdot 10^{13}$		-1.389	
$Cu^+ + 2 \cdot e \rightarrow Cu^{**} + e$	$7.735 \cdot 10^{13}$	$7.735 \cdot 10^{13}$	$+1.58$	$+1.222 \cdot 10^{14}$
$Cu\,(P_{3/2}) + e \rightarrow Cu + e$	$2.435 \cdot 10^{13}$	$2.435 \cdot 10^{13}$	$+3.786$	$+9.219 \cdot 10^{13}$
$Ne^+ + 2 \cdot e \rightarrow Ne^{**} + e$	$1.931 \cdot 10^{13}$	$1.931 \cdot 10^{13}$	$+3.1$	$+5.986 \cdot 10^{13}$

Table 8.22. Part 2. Fluxes of electron cooling in the interpulse period: F (3 μs, 80 μs) with a mixture of hydrogen chloride at constant pumping conditions ($N_{HCl} = 1.5 \cdot 10^{14}$ cm^{-3}, $N_{Cu} = 0.8 \cdot 10^{15}$ cm^{-3}, $I_{max} = 810$ A) . To eliminate the influence of the excitation pulse, t_1 was set equal to 3 μs, the end of interpulse period corresponds to 80 μs ($f = 12$ kHz) (GDT parameters are given in Table. 8.19, the values of the initial reactant concentrations of the active medium obtained from the self-consistency calculations – Table 8.16)

Reaction	$N_{HCl} = 1.5 \cdot 10^{14}$ cm^{-3}			
	Flux (cm^{-3})	Resultant flux	Energy	Energy flux
HCl + e = HCl ($v = 1$) + e	$1.135 \cdot 10^{15}$	$2.805 \cdot 10^{14}$	-0.37	$-1.038 \cdot 10^{14}$
HCl ($v = 1$) + e = HCl + e	$8.545 \cdot 10^{14}$		$+0.37$	
Cu ($D_{3/2}$) + e → Cu ($D_{5/2}$) + e	$1.259 \cdot 10^{15}$	$3.9 \cdot 10^{13}$	0.253	$+9.867 \cdot 10^{12}$
Cu ($D_{5/2}$) + e → Cu ($D_{3/2}$) + e	$1.22 \cdot 10^{15}$		-0.253	
Cu ($D_{3/2}$) + e → Cu + e	$9.807 \cdot 10^{13}$	$7.185 \cdot 10^{13}$	1.642	$+1.18 \cdot 10^{14}$
Cu + e → Cu ($D_{3/2}$) + e	$2.622 \cdot 10^{13}$		-1.642	
Cu ($P_{3/2}$) + e → Cu ($P_{1/2}$) + e	$4.067 \cdot 10^{15}$	$1.3 \cdot 10^{13}$	0.0313	$+4.069 \cdot 10^{11}$
Cu ($P_{1/2}$) + e → Cu ($P_{3/2}$) + e	$4.054 \cdot 10^{15}$		-0.0313	
Cu ($D_{5/2}$) + e → Cu + e	$8.748 \cdot 10^{13}$	$4.243 \cdot 10^{13}$	1.389	$+5.894 \cdot 10^{13}$
Cu + e → Cu ($D_{5/2}$) + e	$4.505 \cdot 10^{13}$		-1.389	
H$_2$ + e = H$_2$ ($v = 1$) + e	$2.087 \cdot 10^{14}$	$3.57 \cdot 10^{13}$	-0.5	$-1.785 \cdot 10^{13}$
H$_2$($v = 1$) + e = H$_2$ + e	$1.73 \cdot 10^{14}$		$+0.5$	
Cu ($P_{3/2}$) + e → Cu + e	$1.937 \cdot 10^{13}$	$1.937 \cdot 10^{13}$	1.642	$+3.18 \cdot 10^{13}$
HCl ($v = 1$) + e = HCl ($v = 2$) + e	$1.43 \cdot 10^{14}$	$9.689 \cdot 10^{13}$	-0.37	$-3.585 \cdot 10^{13}$
HCl ($v = 2$) + e = HCl ($v = 1$) + e	$4.611 \cdot 10^{13}$		$+0.37$	
Ne$^+$ + 2·e → Ne** + e	$1.271 \cdot 10^{13}$	$1.271 \cdot 10^{13}$	3.1	$+3.94 \cdot 10^{13}$

a constant pump current pulse and the concentration of copper atoms in the active medium without the HCl and at the current and the concentration corresponding to operation with the HCl addition. As shown in Table 8.17 for the reaction fluxes (see section 8.5.1), the ion–ion recombination as well as the reaction of dissociative attachment to the molecules of HCl and CuCl during the excitation pulse have no significant effect on the behaviour of the electron concentration. Reducing the maximum electron concentration is associated with a lower prepulse value for the given HCl addition.

A2. In [72,79,91,108] it is reported that the introduction of HCl results in a change in the electrical processes in the discharge circuit, caused by a decrease in the plasma conductivity. While maintaining a constant discharge current, this should increase the energy supplied the discharge (see Table 8.18). However, it is clear that even with a constant energy supplied

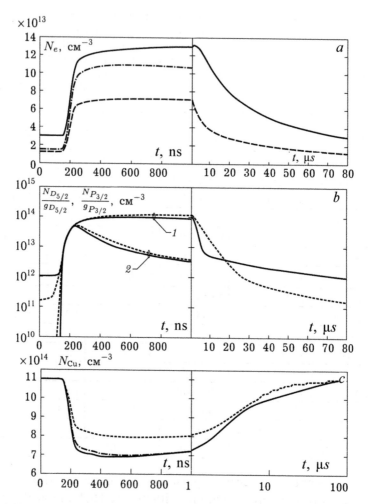

Figure 8.26. Time dependences of: *a*: the electron concentration for excitation pulses and interpulse interval (solid curve corresponds to initial concentrations $N_{HCl} = 0\%$, $N_{Cu} = 0.8 \cdot 10^{15}$ cm^{-3} and a maximum current of GDT $I_{max} = 810$ A, dotted $- N_{HCl} = 1.5 \cdot 10^{14}$ cm^{-3}, $N_{Cu} = 1.1 \cdot 10^{15}$ cm^{-3}, $I_{max} = 640$ A, dot-dash curve $- N_{HCl} = 1.5 \cdot 10^{14}$ cm^{-3}, $N_{Cu} = 0.8 \cdot 10^{15}$ cm^{-3}, $I_{max} = 810$ A). *b*: population levels $P_{3/2}$ (*1*) and $D_{5/2}$ (*2*) (solid line $- N_{HCl} = 0\%$; dotted $- N_{HCl} = 1.5 \cdot 10^{14}$ cm^{-3}). *c*: concentration of copper atoms in the ground state during the excitation pulse and interpulse interval (solid curve $- N_{HCl} = 0\%$, $N_{Cu} = 0.8 \cdot 10^{15}$ cm^{-3}, $I_{max} = 810$ A, dotted line $- N_{HCl} = 1.5 \cdot 10^{14}$ cm^{-3}, $N_{Cu} = 1.1 \cdot 10^{15}$ cm^{-3}, $I_{max} = 640$ A, dash-dot curve $N_{HCl} = 1.5 \cdot 10^{14}$ cm^{-3}, $N_{Cu} = 0.8 \cdot 10^{15}$ cm^{-3}, $I_{max} = 810$ A),

into the discharge the addition of HCl to the active medium, gas temperature will be higher than the gas temperature without HCl additions due to, for example, the following reactions

$$H + Cl_2 \rightarrow HCl(v = 0-2) + Cl,$$

which, accordingly, will increase the concentration of copper in the ground state. However, simply increasing the GDT wall temperature (and hence the concentration of the copper atoms in the active medium) leads to an undesirable increase in the prepulse electron temperature, which relaxes to the gas temperature, and, accordingly, the prepulse population of metastable levels of the copper atom. While increasing the concentration of copper in the ground state with the introduction of HCl in the active medium is accompanied, according to calculations, by a decrease of the prepulse values of the concentrations of copper atoms in the metastable states (the values of the prepulse concentrations of copper atoms in the metastable states decrease quite considerably (several times, see Table 8.16).

For the mixtures containing hydrogen and HCl we have no experimental data on the magnitude of the effective frequency of electron collisions, thereby electron collisions with hydrogen and HCl molecules are taken into account out in accordance with the full transport scattering cross section [28,29] (see also § 8.5.1) :

$$\rho = \frac{1}{\sigma} \approx \frac{7.5 \cdot 10^{-2}}{\sqrt{T_e}} \left(\frac{1}{T_e} + 5.2 \cdot 10^{-3} \frac{N_{Ne}}{N_e} T_e + 0.122 \frac{N_{H_2}}{N_e} T_e + \right.$$
$$\left. + 1.3 \cdot 10^{-2} \frac{N_{HCl}}{N_e} T_e \right) \text{Ohm} \cdot \text{cm}.$$

Table 8.25. Fluxes of changes in the concentration of copper ions during the afterglow F (3, 80 μs) in the case of the HCl addition under constant pumping conditions ($N_{HCl} = 1.5 \cdot 10^{14}$ cm^{-3}, $N_{Cu} = 0.8 \cdot 10^{15}$ cm^{-3}, $I_{max} = 810$ A) (GDT parameters are listed in Table 8.19, the values of the initial plasma reactant concentrations obtained by self-consistent calculations are in Table 8.16)

Reaction	$N_{HCl} = 1.5 \cdot 10^{14}$ cm^{-3}
	Flux (cm^{-3})
$Cu^+ + 2 \cdot e \rightarrow Cu^{**} + e$	$1.869 \cdot 10^{13}$
$Cu^+ + Cl^- + Ne \rightarrow Cu\ (P_{1/2}) + Cl + Ne$	$1.563 \cdot 10^{13}$
$Cu^+ + Cl^- + Ne \rightarrow Cu\ (P_{3/2}) + Cl + Ne$	$1.563 \cdot 10^{13}$
$Cu^+ + 2 \cdot Ne \rightarrow NeCu^+ + Ne$	$3.022 \cdot 10^{12}$
Ambipolar diffusion	$7.256 \cdot 10^{11}$
$Ne^* + Cu \rightarrow Cu^+ + Ne + e$	$2.719 \cdot 10^{11}$

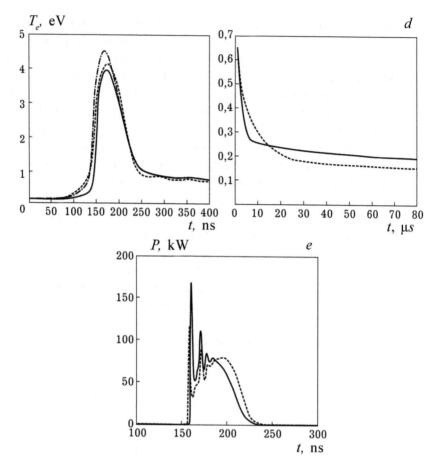

Figure 8.26. *continued.* Time dependences of: *d*: the electron temperature during the excitation pulse and the interpulse interval (solid curve – N_{HCl} = 0%, N_{Cu} = 0.8·10^{15} cm^{-3}, I_{max} = 810 A; dotted curve – N_{HCl} = 1.5·10^{14} cm^{-3}, N_{Cu} = 1.1·10^{15} cm^{-3}, I_{max} = 640 A, dash-dot curve N_{HCl} = 1.5·10^{14} cm^{-3}, N_{Cu} = 0.8·10^{15} cm^{-3}, I_{max} = 810 A). *e*: the lasing pulse (solid curve – N_{HCl} = 0%; dotted curve – N_{HCl} = 1.5·10^{14} cm^{-3}). GDT parameters are given in Table. 8.19, the values of the plasma reagents used in the calculations – in Table. 8.16,

As can be seen from the above expression for the resistivity of the plasma the HCl addition has an indirect effect on the change in the conductivity of the active medium due to its low (in comparison with hydrogen and neon) concentration, changing the time dependence of temperature and, above all, the electron density.

A3. Along with a decrease in the prepulse electron concentration there is also a decrease of the prepulse values of the populations of the copper atoms in the metastable states while increasing the concentration of copper in the active medium; this may also favorably affect the laser characteristics (see Fig. 8.26 b).

Additional calculations were carried out to assess the significance of these processes. Self-consistent calculations were carried out for the parameters of a $Cu-Ne-H_2$ mixture of the laser, shown in Table 8.19, to determine the initial concentrations of reagents and lasing energy, which was (at both wavelengths) $3.83 \cdot 10^{-6}$ J/cm^3. This was followed by a number of additional calculations in which the concentration of copper atoms in the ground state and the electron concentration and the concentration of copper atoms in the metastable states were taken equal to those obtained by self-consistent calculations for the $Cu-Ne-H_2-HCl$ active medium. For the first calculation the lasing energy was $5 \cdot 10^{-6}$ J/cm^3, for the second $- 3.95 \cdot 10^{-6}$ J/cm^3, for the third $- 3.9 \cdot 10^{-6}$ J/cm^3. It is seen that the main contribution to the increase in lasing energy when adding HCl at low excitation pulse repetition frequencies (12 kHz) is due to the increase in the working concentration of copper in the active medium.

In [70,81,91] the introduction of HCl resulted in a decrease in depletion and faster recovery of the prepulse concentrations of copper atoms in the ground state. This effect was also observed when calculating using the developed model. Table 8.20 (see also Fig. 8.26 c) gives the main fluxes, changing population of the ground level of the copper atom during the excitation pulse. It may be seen that the addition of HCl is accompanied by a decrease of the values of the main fluxes that affect the population of the ground state of the copper atom, which determines the reduction in the degree of depletion during the excitation pulse. However, as shown in Table 8.20, the reactions involving chlorine-containing reagents during the excitation pulse have no significant effect on the population of the ground state of copper, i.e., the HCl addition has an indirect effect by changing the time dependence of temperature and, above all, the electron concentration (see Fig. 8.26 and Table 8.20).

A4. In [72] there is an increase of the maximum electron temperature when adding HCl in the active medium.

In our calculations there was no substantial increase in the electron temperature (see Fig. 8.26 d). The influence of the HCl addition on the time dependence of the electron temperature during the excitation pulse was clarified by the analysis of the energy fluxes under constant pump conditions and the concentration of copper atoms in the active medium (see Table 8.18). It is seen that due to the increased resistance of plasma the total energy flux deposited in the plasma during the pump pulse increases. This leads to an overall increase in other fluxes of the plasma-chemical reactions. However, it should be noted that the reactions involving chlorine-containing components in general do not have a significant effect on the electron temperature change during the excitation pulse (see Table 8.18), although, of course, the dissociation and excitation of the first vibrational levels of the HCl molecule use some part of the input energy into the plasma.

A5. In the simulation in [72,105] the authors noted slowing of the decline of the concentration of copper atoms in the metastable states during the early afterglow due to the decrease of the electron concentration in the active medium with the introduction of HCl.

This effect was also observed in our work. However, due to the higher concentration of non-dissociated (see subsection 'Testing the model') HCl, leading to a smaller concentration of electrons, the results of our calculations predict a stronger slowdown in decline of the population of metastable levels of copper, thereby reducing their quenching by electrons, which is also illustrated by the decrease in the fluxes of electron de-excitation of the $D_{5/2}$ level (see Table 8.21 for a decrease in $D_{5/2}$ in the interpulse period.) The contribution of reactions

$$Cu + Cl^- \rightarrow CuCl + e,$$
$$Cu(D_{3/2}) + Cl^- \rightarrow CuCl + e,$$
$$Cu(D_{5/2}) + Cl^- \rightarrow CuCl + e,$$
$$Cu + HCl(v = 2) \rightarrow CuCl + H$$

in quenching of the metastable levels is small. In addition, it should be noted that some slowdown is provided by a decrease in the intensity of the quenching of the copper atoms in the metastable states by vibrationally excited hydrogen molecules whose concentration in the active medium decreases markedly when HCl is added (see Table 8.16). Slowing of the decline of the concentration of copper atoms in the metastable states during the early afterglow can be critical in the transition to high frequencies of the excitation pulses.

A6. The simulation results [72] indicate a small value of the electron energy losses in collisions with chlorine-containing reagents, compared with the losses in collisions with components of copper, hydrogen and neon.

Indeed, as seen from the table of the energy fluxes during the excitation pulse, in pumping under the same conditions (same dependence of the current on time and the initial concentration of copper atoms in the ground state) for the active media of Cu–Ne–H$_2$ and Cu–Ne–H$_2$–HCl in during the excitation pulse the reactions involving chlorine-containing reagents have no significant effect on the time dependence of the electron temperature (see Table 8.18).

A7. The authors of [36,72] suggests that during the excitation pulse, the influence of the HCL addition on the kinetics of the active medium is small due to the low concentration.

In general, the influence of HCl additives through the occurrence of plasma chemical reactions in the active medium of the laser during the excitation pulse is really insignificant. This is confirmed by our calculations (see Tables 8.17, 8.18, see also A.3, A.4, A.6), but, as already noted, the introduction of HCl significantly altered the excitation conditions,which leads to a change in the pumping process.

A8. According to the results in [72], in the initial stage of the interpulse period there is a slower decline in the electron temperature due to the reduction of positive ions in the plasma and, consequently, reduction of the number of elastic Coulomb collisions, leading to a decrease in the electron temperature. Reducing the concentration of positive ions in the plasma is due to the intense ion–ion recombination.

We have also observed this effect (see Fig. 8.26 *b* and Table 8.22 for cooling in the interpulse period). It is seen that in the kinetics of the laser with HCl additives the contribution of elastic Coulomb collisions in the afterglow to a decrease in the electron temperature decreases significantly compared to the contribution in Cu–Ne–H$_2$ and Cu–Ne lasers, where it is one of the main processes. It is also necessary to note that the process of vibrational excitation of hydrogen molecules for electron cooling is less important(see Table 8.22). This process is noticeably inferior to the process of vibrational excitation of HCl molecules that have a slightly lower value of the vibrational quantum. The process of vibrational excitation of HCl molecules is one of the most significant processes in cooling of the electrons in the afterglow of a laser with the HCl addition (see Table 8.22).

A9. When adding HCl the experiments revealed [72,107,108] a shift of the 'centre of gravity' of the lasing pulse to its end.

Figure 8.26 e shows the lasing pulses without additives and with the HCl addition. It is seen that the simulation results predict the effect that is due to the change of the time dependence of population inversion with the introduction of HCl.

A10. Study [36] indicates an increase in the proportion of dissociated hydrogen when HCl is added to the laser.

In this work, the degree of dissociation of molecular hydrogen with the introduction of HCl also increases from 25% for copper vapour lasers without HCl up to 50% with the addition of HCl (see Table. 8.16). This occurs through the reactions:

$$H_2(v=0) + Cl \rightarrow HCl(v=0) + H, \tag{8.7}$$

$$H_2(v=1) + Cl \rightarrow HCl(v=0) + H, \tag{8.8}$$

$$H_2(v=1) + Cl \rightarrow HCl(v=1) + H, \tag{8.9}$$

in which molecular hydrogen is 'transferred' into atomic. The much lower degree of dissociation of molecular hydrogen, derived in [72], is connected, apparently, with these rates corresponding to temperatures around 300 K, while the rates at the operating temperature (~1700 K) are much larger (see below). Note that due to the greater degree of dissociation of the vibrationally excited hydrogen molecules in the active medium the addition of HCl greatly reduces the amount of molecular hydrogen, which is reflected both as a decrease in the electron temperature in the afterglow and in the quenching of the copper atoms in the metastable states (see also 'Testing of the model').

The model was tested by comparing the calculated data with experimental results [72, 80]. These works contain the most detailed (known to us) results of theoretical and experimental studies of the effect of HCl additives on the lasing characteristics of the copper vapour lasers. Testing was conducted by comparing various dependences, calculated using the model described above, with the theoretical and experimental dependences of the densities of the electrons and the copper atoms in the ground and metastable states, the time dependence of electron temperature, the energy and lasing pulse parameters, as well as the electrical characteristics.

The initial concentrations of copper atoms in the calculations corresponded to the initial concentrations given in [72, 80]. The initial values of the other reagents were calculated in the model. These values were calculated based on iterations. The initial concentrations of the reactants were defined and their values were then calculated at the beginning of the next pulse. The iterations were stopped when the difference between the initial and final values was less than 0.5%. The amplitude of the discharge current in the GDT for the optimal HCl addition was reduced by approximately 25%. No simulation of the pumping circuit was carried out due to the lack of data on its structure and component values and the experimentally measured dependences were used. The effect of intermediate values of HCl for which no experimental dependences were available was calculated using the linear approximation for the amplitude of the discharge current, while the form of the time dependence of the discharge current remains the same.

On the recovery of hydrogen and HCl concentrations. Description of the restoration of the molecules of H_2, HCl is one of the major issues considered in the simulation of the laser. During the excitation pulse, a significant part of the molecular reagents dissociate and should recover in the interpulse period. Let us therefore study these issues in detail.

Recovery on the walls. The cross section for the capture processes for the potentials of the type $U = C_n/r^n$ is equal to [109]:

$$\sigma = \pi \cdot n \cdot (n-2)^{(2-n)/n} \times \left(\frac{C_n}{3 \cdot T_g} \right)^{\frac{2}{n}}.$$

Then the cross section for the reactions

$$H + H \rightarrow H_2, \tag{8.10}$$

$$Cl + Cl \rightarrow Cl_2, \tag{8.11}$$

$$H + Cl \rightarrow HCl. \tag{8.12}$$

can be estimated on the basis of van der Waals interaction potential $U = C_6/r^6$ for atoms. If the coefficient C_6 is used for evaluating hydrogen (6.5 atomic units [241]), then at a gas temperature of the order of 0.15 eV the cross section has a value approximately equal to $1.55 \cdot 10^{-15}$ cm^2.

The upper limit values of the reaction rates for processes (8.13–8.15) in the recovery of the molecules on the walls

$$H + H + Wall \rightarrow H_2 + Wall, \tag{8.13}$$

$$Cl + Cl + Wall \rightarrow Cl_2 + Wall, \tag{8.14}$$

$$H + Cl + Wall \rightarrow HCl + Wall \tag{8.15}$$

can be estimated as product k'_n, where n is the number of the process,

$$k' = \sigma \cdot v = 8.15 \cdot 10^{-16} \left(\frac{1}{T_g(\text{eV})} \right)^{1/3} \cdot \sqrt{\frac{3 \cdot T_g}{m}} =$$

$$= 1.41 \cdot 10^{-9} \cdot \frac{T^{1/6}(\text{eV})}{\sqrt{\mu}} (\text{cm}^3 / s) \qquad (8.16)$$

by the ratio $\Delta l/l$, which characterizes the probability of finding a particle at a distance Δl from the wall of the gas discharge tube of width l, μ is the reduced mass of the molecule under consideration.

The value of $\Delta l/l$ in the order of magnitude is 10^{-8}, which gives the values of process rate coefficients (8.13–8.15):

$$\text{H} + \text{H} + \text{Wall} \rightarrow \text{H}_2 + \text{Wall } k_1^w \approx 2 \cdot 10^{-17} \cdot T_g^{1/6}(\text{cm}^3 / s),$$

$$\text{Cl} + \text{Cl} + \text{Wall} \rightarrow \text{Cl}_2 + \text{Wall } k_2^w \approx 1.4 \cdot 10^{-17} \cdot T_g^{1/6}(\text{cm}^3 / s),$$

$$\text{H} + \text{Cl} + \text{Wall} \rightarrow \text{HCl} + \text{Wall } k_3^w \approx 0.97 \cdot 10^{-18} \cdot T_g^{1/6}(\text{cm}^3 / s).$$

Based on the values of the determined coefficients of the reactions it can be concluded that the processes (8.13–8.15) in the active element of a copper vapour laser with HCL additives have no significant effect on the kinetics of the processes.

Three-dimensional reconstruction. The occurrence of two-body reactions (8.10–8.12) is made difficult due to the fact that the energies of the initial and final states are different. The necessary energy can be carried away by a third particle. We estimate the three-particle velocity according to Thomson's theory. Then the rate will be equal to the product of the value (8.16) by the probability of the presence of the atom of the third particle in the volume determined by the interaction radius at which this interaction is of the order of the temperature of the gas medium

$$k_{\text{II}}^V = \mu \cdot \lambda^{1/6} \cdot P$$

where

$$P = \Delta V \cdot N_B,$$

and at the same time $\Delta V = \frac{4}{3} \cdot \pi \cdot R_{\text{int}}^3$ and $\frac{\alpha}{R_{\text{int}}^6} \sim T_g.$

At temperatures of about 0.15 eV the interaction radius varies in the range (3÷6) angstroms. The range of probabilities is from $1 \cdot 10^{-5}$ to $1 \cdot 10^{-4}$, which gives the magnitude of the velocity at a gas temperature of the order of 0.15 eV from $3 \cdot 10^{-14} \cdot T_g^{1/6}$ to $3 \cdot 10^{-13} \cdot T_g^{1/6}$ for the recovery of the hydrogen molecule.

In this paper, for the reactions

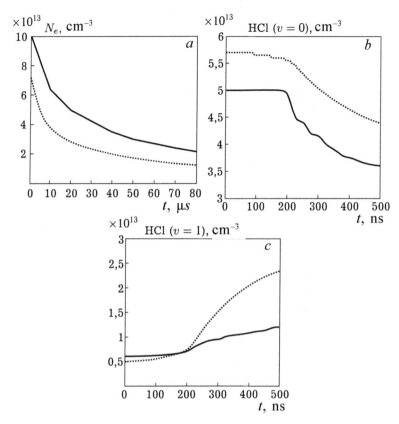

Figure 8.27. Time dependences of: *a*: the electron concentration in the interpulse period (solid line – calculation [72], dashed line – the results of this work). *b*: concentration of HCl ($v = 0$) during the excitation pulse and interpulse period (solid line – calculation [72], dotted line – the results of this simulation). *c*: concentration of HCl ($v = 1$) for excitation pulse and the interpulse period (solid line – calculated in [72], dotted line – the results of this simulation),

$$H + H + Ne \rightarrow H_2 + Ne$$
$$Cl + Cl + Ne \rightarrow Cl_2 + Ne$$
$$H + Cl + Ne \rightarrow HCl + Ne$$

the coefficients of the reaction rate equal to:

$$H + H + Ne \rightarrow H_2 + Ne \; k_4^V \approx 2.73 \cdot 10^{-13} \cdot T_g^{1/6} \, (cm^3 / s);$$
$$Cl + Cl + Ne \rightarrow Cl_2 + Ne \; k_5^V \approx 1.91 \cdot 10^{-13} \cdot T_g^{1/6} \, (cm^3 / s);$$
$$H + Cl + Ne \rightarrow HCl + Ne \; k_6^V \approx 1.32 \cdot 10^{-14} \cdot T_g^{1/6} \, (cm^3 / s);$$

are used. Note that, however, the main contribution to the recovery of HCl molecules made by the reactions (8.7)–(8.9).

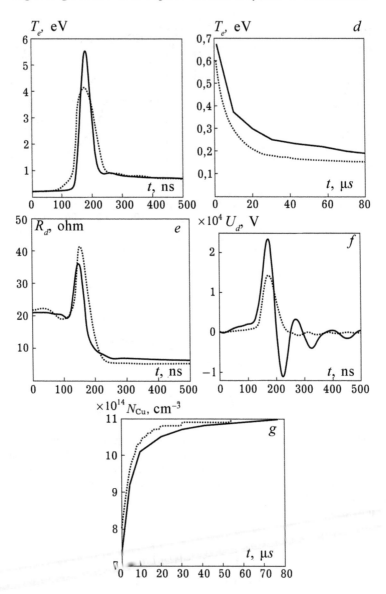

Figure 8.27. *continued.* Time dependences of: *d*: the electron temperature during the excitation pulse and the interpulse period. *e*: resistance of plasma. *f*: the voltage in GDT. *g*: the concentration of copper atoms in the ground state during the interpulse interval. The GDT parameters are in Table 8.19, the values of plasma reagents used in the calculations are in Tab. 8.16. *d*, *e*, *f*, *g* – solid line – calculation [72], dashed line – the results of this simulation.

Thus, the recovery of the HCl molecule in the interpulse period is mainly due to the reactions

$$H + H + Ne \rightarrow H_2 + Ne,$$
$$H_2(v = 0.1) + Cl \rightarrow HCl(v = 0.1) + H.$$

Comparison with the results of [72]. The parameters of the gas discharge tube that was used in [72] and the corresponding values of the reagent concentrations of the active medium are shown in Tables 8.19 and 8.16. The calculated time dependence of the electron concentration is distinguished from that given in [72] by the following: the value of prepulse electron concentration (see Fig. 8.27 a) is smaller (about 1.5 times), while the maximum value attained during the excitation pulse is 25% less. It should be noted that the time dependences of the electron concentration in the interpulse period are similar and differ only by a numerical factor. The observed differences can be explained by different rates of reactions (8.7)–(8.9) used in this paper and in [72]. The authors of [72] used a value of $2 \cdot 10^{-11} \cdot \exp(-0.1865/T_g)$ (T_g in eV) obtained in [110] for the reaction (8.7) at 300 K, while in this work this value is $3.6 \cdot 10^{-10} \cdot \exp(-0.06195/T_g) \cdot T_g^{-0.294}$, recommended in [111]. The latter value at the temperature of the active medium of the copper vapour laser is significantly different from the first. In a theoretical paper [111], the rates of these reactions are calculated in a wide temperature range and the same at 300 K with the data of [110]. For this reason, there are also large differences in the concentrations of HCL molecules in the ground and first vibrational state, as well as the concentration of the negative Cl ions, but in general their behaviour is similar (see Fig. 8.27 b, 8.27 d). Note that in the experiment, the excitation conditions of the mixture with and without the HCl additive were somewhat different. When HCl was used the voltage in the storage capacitor was lower. In this case the current through the GDT turned out to be also smaller. In the calculations for the mixture with and without the HCl addition (Fig. 8.27), we used the experimental dependence of current on time corresponding to these mixtures.

The behaviour of the values of electron temperature during the excitation pulse was similar to that in [72]. However, they differed at the maximum in time and had a larger time interval exceeding the value of 1.7 eV, when conditions were more favourable for the excitation of the resonance levels of the copper atom (Fig. 8.27 e). In the afterglow the electron temperature and the time dependences of the electron temperature were similar. When comparing the concentrations of copper atoms in the metastable states one can note a significant difference in the magnitude of prepulse values. The point is that at the excitation pulse repetition frequency of 12 kHz at the end of the afterglow the value of the population of metastable levels can be very accurately estimated using the Boltzmann distribution with an electron temperature corresponding to the end of the relaxation period. The simulation results presented in this paper (see Fig. 8.27 e) and described in [72] predict a decrease in the prepulse electron temperature when adding HCl. However, the population of metastable levels in [72] with the HCl addition does not decrease but increases.

The lasing pulse, calculated in this study, compared to the results in [72], is wider and its amplitude is smaller.

The time dependence of the plasma resistance in the GDT is similar to that presented in [72], although it has a slightly larger amplitude value (see Fig. 8.27 f). The voltage in the GDT calculated on the basis of the developed model is approximately twice smaller than the experimental value (see Fig. 8.27 g), although, as already noted, there is good agreement in the time dependences of the resistivity of the plasma (see Fig. 8.27 f). The observed difference in the amplitude values of the voltage in the GDT in comparison with those of [72] may be due to the neglect in our model of the voltage drop across the electrodes (according to [112] this may lead to a more than 30% voltage drop) and the near electrode gaps. The calculated time dependence of the population of the ground state of the copper atom coincides within the experimental error with that measured in [72] (see Fig. 8.27 h). The increase of the average output power in comparison with the experiment [72] obtained as a result of modelling for the Cu–Ne–H$_2$ laser is somewhat larger, whereas for the simulation results for the Cu–Ne–H$_2$–HCl laser predict a somewhat smaller increase in the average radiation power (Table 8.16). It should be noted that these simulations indicate the optimal value of the HCl additive concentration in the range $1 \div 2 \cdot 10^{14}$ cm^{-3}, which is generally consistent with [72]. Note also that the optimum concentration of the HCl injected into the active medium may change in modification of the excitation conditions.

Comparison with the results of [80]. The work [80] shows the experimental results of measurements of the electron concentration of the copper atoms in the ground and metastable states and electrons. Currently this work is the only one which shows this set of parameters for a copper vapour laser with the HCl addition. The main drawback is its lack of measurements of the concentrations the HCl and hydrogen additions in the active medium. This leads to some uncertainty in the excitation conditions. In the simulation we used the estimated values of these reagents which are given by the authors of [79,80], namely: the hydrogen concentration was assumed to be 2% of the concentration

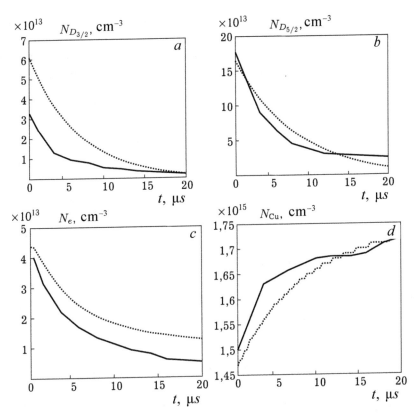

Figure 8.28. Time dependences of: *a*: population level $D_{3/2}$ for interpulse interval. *b*: population level $D_{5/2}$ for interpulse interval. *c*: the electron concentration for interpulse interval. *d*: concentration of copper atoms in the ground state during the interpulse interval. The GDT parameters are shown in Table 8.19, values of the plasma reactants are given in the text. *a, b, c, d* – solid line – experiment [80], dashed curve – the results of this work.

of neon, and the HCl concentration –0.15%. Parameters of the gas discharge tube that was used in [79] are shown in Table 8.19.

The calculated time dependence of the concentration of copper atoms in the $D_{3/2}$ state (see Fig. 8.28 *a*) compared with the experimentally measured dependence has, within the experimental error, the same prepulse value but its maximum value is twice as high. On the other hand, for the concentration of copper atoms in the $D_{5/2}$ state (see Fig. 8.28 *b*) the calculations based on the proposed model predicts a slightly smaller maximum value along with a smaller prepulse concentration, although in general it has a similar behavior (see Fig. 8.28 *c*). The calculated time dependence of the electron concentration for the experimental conditions [80] is higher than the measured dependence throughout the afterglow. The value of the prepulse electron concentration differs by more than half from the experimentally measured value. The differences in the time dependences of the electron concentration for that matter, and the differences in the behaviour of the populations of metastable levels are most likely due to some differences in the HCl and hydrogen concentrations estimated in [80] and actually present in the GDT plasma. Comparison of the time dependence of the concentration of the copper atoms in the ground state shows that the population calculated on the basis of the model is in agreement, within experimental error, with the experimentally measured dependence(see Fig. 8.28 *d*).

It should be noted that, as in the experiment, if no HCl was added there was no lasing at a frequency of 50 kHz. The main reason for the lack of inversion in the medium without HCl additives was the high prepulse concentration of the electrons and the available energy was not sufficient for heating them. When HCl was added inversion appeared as a result of a sharp decrease in the prepulse electron concentration. Thus, the HCl additive can experimentally realize higher lasing pulse repetition frequencies in a copper vapour laser, which is important for practice.

Discussion of prepulse values and time dependences of plasma reactants of the laser active medium

The prepulse values, as also the time dependences, of the plasma reagents vary depending on the percentage of HCl in the active medium. Table 8.16 presents the prepulse values of some of the most important reagents of the active medium of copper vapour lasers with the HCl addition. The analysis of Table 8.16 shows that the increase in the percentage of HCl in the active medium leads to a monotonic decrease in the prepulse concentrations of the electrons and copper atoms in metastable states. Also increases monotonically the prepulse concentration of almost all chlorine-containing reagents, except for the vibrationally excited HCl molecules whose concentration decreases because of the decrease of the prepulse electron temperature to the equilibrium value to which these values tend. It should also be noted that the introduction of HCl is accompanied by a decrease in the number of the vibrationally excited hydrogen molecules, which reduces their impact on the population of metastable levels in the reactions

$$Cu(D_{3,5/2}) + H_2(v = 1,2,3) \rightarrow CuH + H$$

and the behaviour of the electron temperature by the reactions

$$H_2(v = n) + e \rightarrow H_2(v = n+1) + e.$$

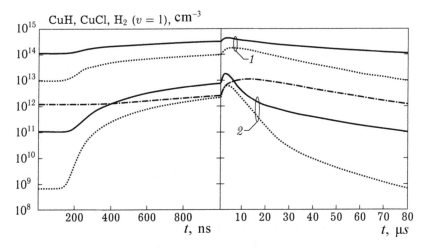

Figure 8.29. Time dependence of the concentration of molecules of H_2 ($v = 1$) (*2*) molecules CuH (*1*) and CuCl (dot–dash) for the excitation pulse and interpulse interval for $N_{HCl} = 1.5 \times 10^{14}$ cm^{-3}. The solid curve – $N_{HCl} = 0\%$; dotted line – $N_{HCl} = 1.5 \times 10^{14}$ cm^{-3}. Options are GDT Table. 8.19, the values of plasma reagents used in the calculations – in tab. 8.16

As already noted (see A.10) the addition of HCl is accompanied by an increase in the degree of dissociation of molecular hydrogen. The degree of dissociation increases up to a value determined by the reduction of the energy input to the discharge when adding high HCl concentrations. The overall changes in the behaviour of various reagents are shown in Figs. 8.26 a–d, and 8.29.

Excitation pulse. As noted earlier (see M.1) the addition of HCl is accompanied by some reduction in the maximum electron concentration during the excitation pulse with a simultaneous decrease of the prepulse value. Adding HCl leads to a slight increase of both the maximum value of the electron temperature and the duration of the period during which the electron temperature exceeds the value of 1,7 eV at which the rates of excitation of the resonance and metastable levels of the copper atom become equal. The degree of depletion of the ground level of the copper atom decreases with a slight shift of the minimum point towards lower values on the time axis (see Fig. 8.26 c). The intensity of the CuH molecule formation in the active medium increases as a result of the enhanced vibrational excitation of hydrogen molecules (see Fig. 8.26 d). At the end of the excitation pulse there is a somewhat greater concentration of copper atoms in the metastable states due to an increase in the electron temperature (see Fig. 8.26 b). The lasing pulse starts a little earlier for the case of the HCl addition and has a weaker vibrational structure, and the 'centre of gravity' of the pulse is shifted to the end (see Fig. 8.26 *e*), which, as noted in [107], improves its quality.

Interpulse period. As with the analysis of the influence of hydrogen additions [28,29] we divide the interpulse period into two phases. During the first phase, lasting about 20 μs, the most significant difference is observed in the electron concentration in the Cu–Ne–H$_2$ and Cu–Ne–H$_2$–HCl lasers. In this case, the electron temperature, as the concentration of copper atoms in metastable states, for the active medium with the HCl addition is higher (see A.3 and A.4). The difference in the concentrations of copper atoms in the ground state in the active media of Cu–Ne–H$_2$ and Cu–Ne–H$_2$–HCl in the afterglow varies non-monotonically (Fig. 8.26 *c*). It should be noted that in [79,80] namely the use of HCl greatly increased the excitation pulse repetition frequency due to the fact that it is possible to considerably reduce the prepulse electron concentration (see A.1 and Fig. 8.26 a). Thus, we conclude that it was a significant decrease in the electron concentration by adding HCl that allowed the authors of [79,80] to achieve laser operation at high excitation frequencies (around 50 kHz). Despite the higher values during the first phase of the interpulse period the prepulse electron temperature and, accordingly, prepulse population of the metastable levels of the copper atom in the case of the HCl addition is smaller as a result of cooling, including the energy loss in the vibrational excitation of HCl molecules. Note the essential role of the processes of ion–ion recombination to decrease the concentration of copper ions in the afterglow (see Table 8.23), in particular, in comparison with the role of three-body recombination.

Effect of prepulse conditions of the active medium on the output characteristics of the copper vapour laser

The influence of prepulse values of the concentration of the electrons and the copper atoms in metastable states was also analyzed. The prepulse plasma parameters may vary due to many causes, in particular when the excitation pulse repetition frequency changes or the active substances that affect the excitation or effective relaxation of the plasma during the afterglow are added. The change of the prepulse parameters will be determined by the specific experimental conditions and plasma parameters.

The characteristics of lasers based on vapours of pure copper and copper with the addition of hydrogen depend on the critical values of the concentrations of the electrons and the copper atoms in the metastable states [5–7,28,29,51,52,54,55,89,90,102]. Consider the influence of these factors on the lasing characteristics of the copper

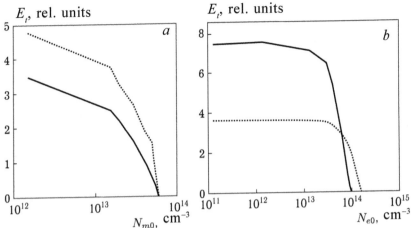

Figure 8.30. Dependences *a*: the total lasing energy generation on prepulse concentrations of copper atoms in the metastable states for $N_{HCl} = 0$ cm^{-3} (solid curve), $N_{HCl} = 1.5 \cdot 10^{14}$ cm^{-3} (dashed curve) for the same values of efficiency (see text). *b*: the total lasing energy in the prepulse electron concentration for $N_{HCl} = 0$ cm^{-3} (solid curve) $N_{HCl} = 1.5 \cdot 10^{14}$ cm^{-3} (dashed curve) for the same values of efficiency (see text). Prepulse concentration of copper atoms for the levels of $D_{3/2}$ and $D_{5/2}$ are taken to be those obtained on the basis of the Boltzmann distribution for the electron temperature corresponding to the prepulse concentrations of electrons. GDT parameters are given in Table 8.19, values of the plasma reagents used in the calculations –are in Table 8.16

vapour laser with HCl additives, because changes of the prepulse conditions caused by the introduction of HCl can alter the critical values at which the lasing fails.

We consider separately the effect of the prepulse concentrations of electrons and copper atoms in metastable states carrying out a series of calculations. After the self consistent determination of the initial plasma parameters these values were fixed and then we changed either the prepulse electron concentration or the prepulse concentration of the copper atoms in metastable states when another prepulse concentration was unchanged. To obtain more information, the energy and lasing efficiency at different concentrations of HCl were multiplied by the same factors. The values of the factors were chosen so that the product of these factors by the appropriate efficiency had the same value. We call the product of these factors by the lasing energy as the reduced energies.

Effect of prepulse electron concentration. In this case, the calculation was performed for the following plasma parameters:

1) $N_{HCl} = 0$;
2) $N_{HCl} = 1 \cdot 10^{14}$ cm^{-3}.

The reduced energy of laser radiation as a function of the initial electron concentration (Fig. 8.30 *b*) for the active medium with the HCl addition is significantly shifted downward relative to the energy calculated for pure copper, but the critical value at which the failure of generation increases slightly. Note, as mentioned in M.1, that adding HCl is accompanied by a sharp decrease in the prepulse electron concentration. Thus, in the case of a copper vapour laser with HCl the critical electron concentration will be achieved at a higher excitation pulse repetition frequency (or more adverse external conditions ensuring increased concentration of electrons). It should be noted that the failure of lasing for the yellow emission line occurs at lower values of the prepulse electron concentration compared to the failure on the green line.

Effect of prepulse concentration of copper atoms in metastable states. The change of the prepulse concentration of the copper atoms in the metastable states leads to a change in their time dependence both during the excitation pulse and the afterglow. With their high initial value the population of the metastable levels may exceed the population of resonance levels during the pump pulse, i.e. the population inversion does not occur [5–7,28,29,51,52,54,55,102]. Here, the dependence of the reduced lasing energy on the prepulse concentration of metastable atoms is obtained taking into account the presence of HCl in the active medium (Fig. 8.30 b). The dependence of the reduced energy generation on the prepulse concentration of metastable atoms in the presence of HCl is higher than the corresponding dependence for the case of pure copper (Fig. 8.30 a). Due to the fact that the critical values at which the breakdown of inversion takes place for the cases with and without HCl are close, and given that the introduction of HCl is accompanied by a significant decrease in the initial population of metastable levels of the copper atom, one can expect that the failure of lasing in the presence of HCl additives in the active medium will occur at high excitation pulse repetition frequencies as compared with the case of the absence of additives.

Conclusion

The effect of HCl additives on the operation of the copper vapour laser was clarified by constructing a detailed kinetic model of the active Cu–Ne–H$_2$–HCl medium.

The detailed analysis of the points of view available at the time of writing the book was carried out.

A good agreement of simulation results with available experiments was observed.

The analysis of the kinetics of recovery of the HCl molecule indicates basically that its recovery in the interpulse period is due to the reactions:

$$H + H + Ne \rightarrow H_2 + Ne,$$
$$H_2(v = 0.1) + Cl \rightarrow HCl(v = 0.1) + H.$$

When introducing HCl the energy losses due to the vibrational excitation of molecular hydrogen are reduced because of its greater degree of dissociation compared to the case of no HCl addition. Reducing the amount of molecular hydrogen decreases somewhat its positive impact, however, it is more than offset by the positive changes (decrease in the prepulse concentrations of electrons and copper atoms in the metastable states and increase of the density of copper atoms in the active medium of the laser) in the kinetics due to the presence of the HCl molecules in the active medium.

It is shown that the effect of HCl additives has a different nature at high (50 kHz) and low excitation pulse frequencies. When working with a low pulse repetition frequency of pumping the main contribution to the increase in lasing energy is an increase in the concentration of copper atoms in the ground state present in the active medium. This increase in the concentration of the copper atoms in the ground state is mainly a result of increased levels of energy input (due to the increase of the resistivity of the active medium) and of the decrease of the prepulse electron concentration which also contributes to the increase of lasing energy. At high excitation pulse repetition frequencies the most significant influence on the lasing characteristics is exerted by the reduced prepulse electron concentration caused by electron attachment to the HCl molecules. Note that a similar effect was detected previously for the addition of molecular hydrogen [28,29] (see also section 8.5.1).

The calculated optimum concentration of HCl additives agrees with the experimental optimum concentration and ranges from 0.1 to 0.2% of the concentration of the buffer gas for standard operating conditions of the majority of copper vapour lasers (medium pressure of 30 Torr).

Note that in addition to the positive factors, the introduction of HCl is accompanied by negative factors. These factors include the loss of energy due to vibrational excitation of HCl molecules and the excitation chlorine-containing component with increasing concentration in the active volume. In addition, it is the slowing down of the relaxation of metastable levels of the copper atom during the initial phase of the afterglow, which can adversely affect the laser output parameters when operating at very high excitation pulse repetition frequencies (more than 50 kHz). However, at optimal HCl concentrations the positive influence is much higher than the negative one.

8.6. Formation of high-quality radiation of the copper vapour laser in the master oscillator–amplifier system

The possible area of application of copper vapour lasers is constantly expending. For some applications, such as laser isotope separation [113–116], it is necessary to have high quality radiation with small divergence. The possibilities of using radiation for high-quality high-precision drilling and distribution in a waveguide (fibre coupling) are shown in [117], and its use for pumping solid-state lasers is also mentioned.

One way to obtain high-quality radiation consists in shaping it in an oscillator using unstable resonators. However, the possibilities of such a form of radiation are limited. As a rule, high-quality radiation is generated only by the end of the pulse, even when using unstable resonators with a large gain [118–123]. Another method is to form the desired beam divergence in the oscillator and then increased it in an amplifier or an amplifier cascade [117,124–128]. At the same time the energy of the oscillator does not play a special role. The main function of the oscillator (master oscillator) is basically only in the formation of the desired beam divergence. The possibilities of this approach are much broader than the first approach, however, when it is used there are additional complexities and questions. In amplifiers, along with increased quality of radiation the spontaneous emission is also amplified. Therefore, one of the questions is the question of timing of

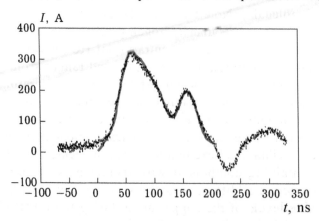

Fig. 8.31. The time dependence of the experimental current passing through the gas discharge tube (points) and the current used in the simulation (solid line). In the calculations, $t = 0$ was taken as the beginning of the excitation pulse current.

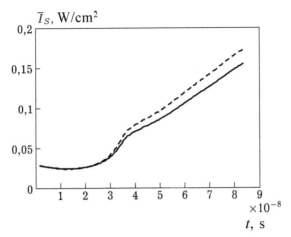

Figure 8.32. Calculated dependence of the time-averaged intensity of saturation. The solid curve corresponds to the 510 nm laser line, dashed – 578 nm. In the calculations the initial values for concentrations of the electrons and copper $[N_e] = 1.5 \cdot 10^{14}$ cm^{-3} and $[N_{Cu}] = 1 \cdot 10^{15}$ cm^{-3}, the concentration of neon in a mixture of $[N_e] = 1.62 \cdot 10^{18}$ cm^{-3}.

delivery of the high-quality radiation quality of the master oscillator to the amplifier. The second important issue is the question of the level of power supplied to the signal amplifier input sufficient for complete removal of the inversion amplifier to high-quality radiation.

The purpose of this section is to obtain data for the characteristics of LT-40Cu lasers (GNPP Istok) [75] using the resuls in [116].

The kinetic model
The model is described in [5–7,116] (see section 8.2).

The excitation of the medium. The calculations conducted in the description of excitation used the typical experimental time dependence of pump current of the CVL (Fig. 8.31) for the type LT-40Cu lasers.

Resonator. Based on the fact that the total power of the amplifier when it is not fed the amplified radiation (i.e. in the generator mode) is about 40 W we selected the values γ (6.16·10^8 s^{-1}) and w (10^8 s^{-1}).

Form of the amplified pulse. In modelling the time dependence $f(t)$ of the radiation applied to the input of the amplifier had the form of a trapeze with the leading edge duration of 2 ns, the trailing edge duration of 10 ns, and the total duratiuoin of 22 ns. According to experimental data this form close to the real form. For definiteness, we assumed that the oscillator creates a pulse in which the powers of the green and yellow lines correspond to each other as 55:45, the time dependences of the radiation of the green and yellow lines are the same and the yellow line emission is delayed by 8 ns with respect

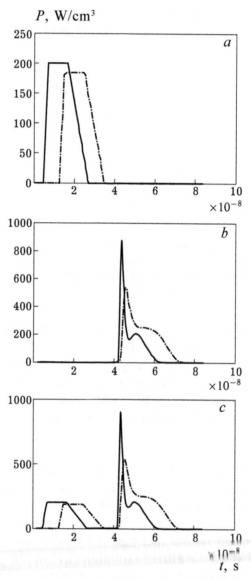

Figure 8.33. Dependence of the calculated power of high-quality (*a*), amplified spontaneous (*b*) and total (*v*) radiation in the pulse on the time of the start of sending the current pulse on the amplifier. The delay of the supply of radiation at the wavelength 510 nm to the amplifier from the master oscillator 4.75 ns. The solid curve corresponds to the 510 nm emission line, dot-dash – the 578 nm line.

to the green, which also roughly corresponds to the experimental data. To investigate the effect of the delay of the emission of amplified radiation with respect to the excitation pulse current of the amplifier, the calculations considered a shift of the beginning of the pulse applied to the amplifier (Figs. 8.33–8.36, respectively, delay 4.75, 15, 30 and 42 ns).

Normalization. The time dependence of the radiation power supplied to the amplifier can be expressed, therefore, by the functions

$$W_1 = W(\lambda = 510 \text{ nm}) = A_1 E_1 c Sf(t),$$
$$W_2 = W(\lambda = 578 \text{ nm}) = A_2 E_2 c Sf(t),$$

where

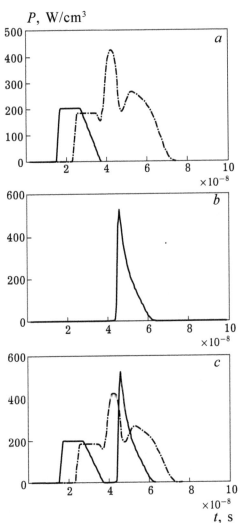

Figure 8.34. Dependences of the calculated power of high-quality (*a*), amplified spontaneous (*b*) and total (*c*) radiation in the pulse on the time of the start of sending the current pulse on the amplifier. The delay of the supply of input radiation at the wavelength of 510 nm to the amplifier from the master oscillator 15 ns. The solid curve corresponds to the 510 nm emission line, dot-dash – the 578 nm line.

$$\frac{A_1}{A_2} = \frac{55}{45}\frac{E_2}{E_1},$$

and E_1 and E_2 are the photon energies for the green and yellow lines of lasing (respectively, 2.428 and 2.143 eV). If the total radiated power supplied to the amplifier is W then the normalization factors A_i are

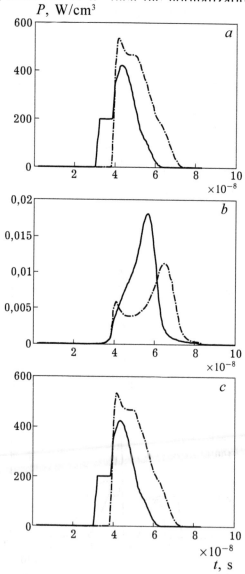

Figure 8.35. Dependences of the calculated power of high-quality (*a*), amplified spontaneous (*b*) and total (*c*) radiation in the pulse on the time of the start of sending the pulse on the amplifier. The delay of the supply of input radiation at a wavelength of 510 mm in the amplifier from the master oscillator 30 ns. The solid curve corresponds to the 510 nm emission line, dot-dash – the 578 nm line.

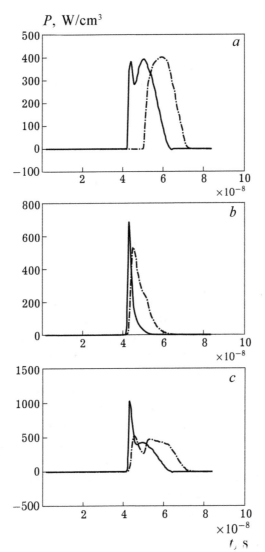

Figure 8.36. Dependences of the calculated power of high-quality (*a*), amplified spontaneous (*b*) and total (*c*) radiation in the pulse on the time of the start of sending the current pulse on the amplifier. The delay of the supply of input radiation at a wavelength of 510 mm in the amplifier from the master oscillator 42 ns. The solid curve corresponds to the 510 nm emission line, dot-dash – the 578 nm line.

the photon densities, determined from the relation

$$A_1 E_1 (1 + A_2 / A_1) Sc \int f(t) dt = W\delta, \qquad (8.17)$$

here S is the cross-sectional area of the laser beam (diameter of the active medium is equal to 2 cm), c is the speed of light, δ is the duty cycle (equal to 10^{-4} for the excitation pulse repetition frequency of 10 kHz).

Calculations were carried out in the approximation of the effective photon lifetime in the resonator (zero-dimensional approximation of the radiation in the resonator). In this approximation it is assumed that the radiation intensity is uniform over the volume of the resonator, i.e. homogeneous concentration of photons. The radiation power emitted from the cavity is given by

$$\hbar \omega n(t) V \cdot (\gamma - w),$$

where V is the volume of the active medium, $n(t)$ is the concentration of photons. By analogy with the presented expression the normalization factors will be determined by the equation

$$A_1' E_1 (1 + A_2 / A_1) V (\gamma - w) \int f(t) dt = W \delta,$$
$$A_1' / A_2' = A_1 / A_2,$$

(8.18)

Here A' has the same meaning as A – the photon density of the radiation, but are determined from the different ratios. Since the calculations were performed in the zero-dimensional approximation, for the normalization of the photon density we used the relation (8.18) and not (8.17). The fact is that the expression (8.17) gives the density of photons of freely propagating radiation. This photon density should be compared with the density of photons present in the cavity and corresponding to the zero-dimensional approximation after freely propagating amplified radiation enters the cavity. So, if the freely propagating radiation with power W enters the empty cavity, the power at its output should be the same W. Thus, the concentration of photons of external radiation with power W, applied to the amplifier after its entrance into the cavity will be determined according to expression (8.18). In this case, an empty (without the active medium) cavity with radiation will produce at the output the same intensity as the freely propagating radiation. Note that at the active medium length $l = 123$ cm and a diameter $d = 2$ cm, corresponding to the described lasers, the values A' are half the values of A.

Separation in laser radiation of photons of amplified radiation and amplified photons of spontaneous radiation. To describe the amplification of high-quality and spontaneous radiation, instead of equations (8.1) solution was found for four equations for intensities $I(\lambda)$ of the amplified spontaneous and $I_q(\lambda)$ of the amplified high-quality radiation with a small divergence. To the two equations (8.1) for $I(\lambda)$ we added two more equations for $I_q(\lambda)$ in which the contribution of the last term of (8.1) decreased by 10^5 times – the value corresponding to approximately the ratio of the square of the angle, corresponding to the geometrical divergence, to the square of the angle corresponding to the diffraction divergence. When calculating the time dependence of the concentrations of resonance and metastable states of copper, the reactions of the interaction of radiation with these states contained the total intensity (the sum of the intensities of high-quality and amplified spontaneous radiation).

Saturation intensity

Saturation intensity is defined as the intensity at which the contributions during the lifetime of the upper laser level from the kinetic processes taking place in the active medium, and from the presence of radiation in this radiation medium become equal. Accordingly, the expression for it has the form

$$I_s = \frac{\hbar\omega}{\sigma\tau},$$

where ω is the radiation frequency, σ is the stimulated emission cross section and τ is the lifetime of the upper state.

Here is a rough estimate of the saturation intensity. Radiative lifetimes of the laser levels are approximately 1 µs. The initial electron density is about 10^{14} cm^{-3}, the rate of de-excitation of resonant states by electrons to metastable states is approximately 10^{-8} cm^3/s, so the contribution of de-excitation by the electrons to τ also corresponds to the inverse lifetime equal to 1 µs. Over time, the concentration of

Table 8.24. Experimental data on the amplification of the weak signal and saturation energy for some of the active media of copper vapour lasers [129]

Tube diameter (cm)	Tube length (cm)	Mean power (W)	Frequency (kHz)	Amplification of the weak signal (cm^{-1})	Saturation energy (µJ/cm^2)	Reference
1.8	40	9.8	6	0.23	64	122
2.1	45	-	7	0.22	-	130
3	130	20	5.8	0.11	36.1	129
4	90	25	4.3	0.12	42	131
4.2	-	-	-	0.053	-	132
6	-	-	-	0.05–0.065	-	133
6	250	-	-	0.04	170	134
8	-	-	-	0.03–0.05	-	135
2	123	40	10	0.16-0.21	≈ 10	116, calculations

electrons increases. Given the contribution to τ of all other kinetic processes, we estimate τ as 10^{-8} s. Then, at the pulse repetition frequency of 10 kHz ($\delta = 10^{-4}$) we obtain that the time-averaged saturation intensity

$$\bar{I}_s = \frac{\hbar\omega}{\sigma\tau}\delta$$

is approximately 0.12 W/cm².

In the model, calculation of the dependence of τ on the real time t is not difficult, since in the solutions we know the time dependences of the concentration and temperature of the electrons for all the relevant reagents (Fig. 8.32).

We are interested in the values of the time-averaged saturation intensity at times when there is lasing (amplification) of the active medium. These times vary from about 40 to 70 ns. It is seen that the saturation intensity values are 0.07–0.12 W/cm².

Table 8.24, taken from [129], shows the saturation energies for various systems of copper vapour lasers, along with a brief description of these systems. The last line gives the information for the discussed lasers, evaluated according to the calculated data. Simulation of the considered lasers based on the model described above leads to the highest value of the gain $\alpha = 0.21$ cm⁻¹. During the laser pulse (~15 ns), the mean value of this coefficient is $\alpha = 0.16$ cm⁻¹. Taking the average frequency of radiation data from Table 8.24 equal to 5 kHz, we find that the time-averaged saturation intensities are in the range 0.18–0.8 W/cm², i.e., do not exceed 0.8 W/cm². For the data in line 7 the excitation pulse repetition frequency is not given. If we take

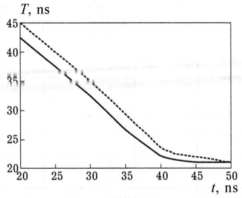

Figure 8.37. The dependence of the calculated duration of the high-quality radiation emitted from the amplifier on the delay of radiation input at a wavelength of 510 nm to the amplifier from the master oscillator with respect to the time of the beginning of the current pulse in the amplifier. The solid line corresponds to the line emission 510 nm, the dotted line – 578 nm. Full radiation power supplied to the amplifier 20 W.

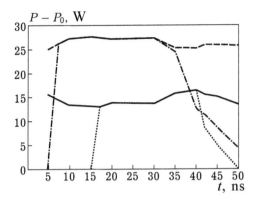

Figure 8.38. The dependence of the calculated excess radiation power leaving the amplifier on the delay in input radiation at a wavelength of 510 nm to the amplifier from the master oscillator in relation the start of the current pulse in the amplifier. The solid curve corresponds to the total power at the 510 nm emission line, dotted line – the high-quality radiation power for the same line, the dashed curve corresponds to the total power at the 578 nm emission line, dot-dashed – high-quality radiation power for the same line. Full power of radiation supplied to the amplifier 20 W.

that its value is not 5 and is 10 kHz, and then the estimated time-averaged saturation intensity would be about 1.7 W/cm^2. It is seen that the calculated saturation intensities of the lasers under consideration correspond to approximately the lower limit of the data in Table 8.24.

Effect of the delay of radiation supplied to the input of the amplifier on the characteristics of high-quality and amplified spontaneous radiation coming from the amplifier
Time dependences. For short delay times of the radiation supplied to the amplifier input with respect to the start of activation of the excitation current in the amplifier (Figs. 8.33–8.36), the applied radiation propagates in an unexcited medium and the amplifier generates the exclusively amplified spontaneous radiation [116]. The same result is obtained at longer delay times. In this case, the amplifier forms exclusively the amplified spontaneous emission, and only then the radiation supplied to the amplifier propagates through the medium.

Duration of high-quality radiation on the base. Starting at about 20 ns delay (and up to 40 ns), according to the calculations all of the radiation output of the amplifier at a wavelength of 510 nm becomes high-quality radiation (see, for example, Fig. 8.35). The duration of radiation on the base at both wavelengths with such a delay is equal to approximately twice the duration of the high-quality applied to the amplifier. By increasing the delay we obtain an approximately linear decay, which ends at 40 ns. The duration of the radiation on the base at both wavelengths is approximately equal to the duration of the high-quality radiation applied to the amplifier (Fig. 8.37).

Excess output power. The dependence of the excess output power on the delay was calculated for the intensities of radiation applied to the input of the amplifier in the range $W = 5$–40 W. These dependences are very weak, almost indistinguishable with the naked eye, so only one such relationship is presented – for $W = 20$ W (Fig. 8.38).

8.6.1. The dependence of the additional power, taken from the amplifier, on the pump power

We return to the main question that is of interest when considering the formation of high-quality radiation in the 'master oscillator–amplifier' system. What is the sufficient intensity I_{min} supplied to the amplifier to ensure that the output radiation would be entirely or almost entirely of high-quality? This value should not significantly exceed the saturation intensity. But it is not clear whether it can be lower than the saturation intensity. For this purpose, calculations were carried out of the dependences of the additional power taken from the amplifier, on the pump power at different delays of delivery of high-quality radiation, generated in the oscillator, to the amplifier [116]. These calculated

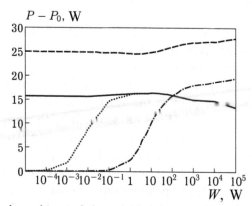

Figure 8.39. The dependence of the calculated excess radiation power emerging from the amplifier on the total power (on two lines) of the radiation supplied to the amplifier. The solid curve corresponds to the total power at the 510 nm emission line, the dotted line – the power of high-quality radiation for the same line, the dashed curve corresponds to the total power at the 578 nm line, dot-dashed curve – power of high-quality radiation for the same line. The delay in supplying input radiation at a wavelength of 510 nm to the amplifier from the master oscillator in relation to the start time of the current pulse in the amplifier is 40 ns.

curves (Fig. 8.39, 8.40) are very similar to the dependences obtained in other studies for two-pass amplifiers [117,124–126].

In the calculations in the zero-dimensional approximation we can not talk about any particular number of passes of radiation in the amplifier. Nevertheless, the calculated behaviour of the amplifier output power coincides with the behavior of output power for the case of two-pass amplifiers. If the saturation of the two-pass amplifier occurs gradually (curves of the dependence of the excess power of the amplifier on the power of supplied high-quality radiation are directed upwards, Fig. 3, 5, in [117]), then in the case of the single-pass amplifier the saturation curve resembles a linear relationship, or is even somewhat concave downward (Fig. 4 in [125], Fig. 3 in [117]). But this is not always the case. In some cases (Fig. 4 in [125], Fig. 1 in [124], Fig. 3 in [126]) the dependence for the single-pass amplifier is the same as for the two-pass amplifier in [117]. Note that in [126] this dependence for the two-pass amplifier increases at first, and then there is its weak decay. Despite the different nature of the approach to saturation, the very maximum output value is nearly independent of the number of passes of radiation in the amplifier (Fig. 3 of [117], Fig. 3 of [126]). Direct comparison with the results of experimental work is very difficult, because, as a rule, these works do not present all the date required for modelling. In the experiment [117] the diameter of the amplifier was 25 mm, and the saturation was already reached at $W = 0.1$ W (Fig. 3, 5, from [117]). In the experiment [124], the diameter of the amplifier was about 20 mm, and the saturation occurred at $W = 0.02$ W (Fig. 1 in [124]). With increasing W to 0.6 W the output power does not significantly increase, the increase of W by up to 30 times only leads to a 25% increase in output power. In the experiment [125] the diameter of the amplifier was 36 mm, and the saturation occurred for a kinetically enhanced amplifier at $W = 0.1–1$ W (Fig. 4 in [125]). For the single-pass amplifier in the same figure the maximum output power is not clear, however, the growth of output power with increasing input intensity from 1 to 10 W increased from 35 to 37 W, i.e. we can assume that in this case the saturation of the output power occurs at $W = 0.1–1$ W. In the experiment [126] the diameter of the amplifier was 20 mm, and the yield to saturation occurred at around $W = 0.06$ W (Fig. 3 of [126]). By increasing the supplied power by up to 30 times the output power increased by no more than 30%. For the single-pass amplifier the value of W was somewhat greater than for the two-pass amplifier, noting that low-quality radiation disappeared for a single-pass amplifier starting with $W' = 5 \cdot 10^{-4}$ W, and for the two-pass amplifier – from $W' = 10^{-3}$ W. Taking into account the values of W and the diameters

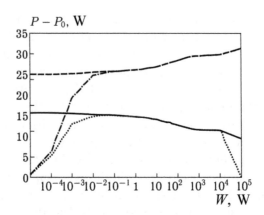

Figure 8.40. See the caption for Fig. 8.39. Delay in input of radiation at a wavelength of 510 nm to the amplifier from the master oscillator in relation to the start time of the current pulse in the amplifier is 30 ns.

Table 8.25. I_{min} values according to different studies and the calculations with respect to the considered lasers LT-40Cu. In the last two rows the word 'delay' means a delay in supplying the input radiation at a wavelength of 510 nm to the amplifier from the master oscillator relative to the start of the current pulse in the amplifier (for more details see text)

Tube diameter (mm)	Tube length (cm)	Frequency (kHz)	I_{min} (W/cm²)	Reference
25	100	12	0.022	117
20	80	10	0.007	124
36	150	12	0.0098–0.098	125
20	80	8	0.2	126
20	123	10	$3.2 \cdot 10^{-4}$ – $3.2 \cdot 10^{-3}$	Delay 30 ns, calculations, 116
20	123	10	0.032–0.32	Delay 40 ns, calculations 116

of the amplifiers, it can be seen that the experimental values of the intensities applied to the amplifier at which almost all of the output radiation was of high quality, are in agreement with the values of 0.022 [117], 0.007 [124], 0.0098–0.098 W/cm² [125], 0.02 [126], i.e. are in the range 0.007–0.022 W/cm². With a further large increase in W (tens of times or more) the output power from the amplifier increases by no more than 25–30%.

However, the delays used in these studies were not reported and no data on the dependence of the investigated dependences on delays were published. In laser isotope separation the most interesting line is usually 510 nm. For this line in the calculations the reduction of the delay from 40 ns to 30 ns results in about a 10–100 fold reduction in I_{min}, which amounts to from $3.2 \cdot 10^{-4}$ W/cm^2 ($0.32 \cdot 10^{-2}$ from I_s, $W = 0.001$ W) for 87% high-quality of the output radiation to $3.2 \cdot 10^{-3}$ W/cm^2 ($0.32 \cdot 10^{-1}$ from I_s, $W = 0.01$ W) for 98% high-quality light output (Fig. 8.40). With a further decrease of the delay to 20 ns I_{min} increases again, but does not exceed the values for the 40 ns delay. Thus, if the dependences of I_{min} on the delay are not optimized, then I_{min} can be estimated by the values less than I_{min} for the 40 ns delay, varying from $3.2 \cdot 10^{-2}$ W/cm^2 ($0.32 I_s$, $W = 0.1$ W) for 93% high-quality radiation at the output to 0.32 W/cm^2 (3.2 from I_s, $W = 1$ W) to 99.4% high-quality of the radiation output. The results of the above analysis are shown in Table 8.25.

Conclusion

One of the main issues arising from the implementation of the master oscillator–amplifier system is a question about the level of the intensity of the high-quality radiation I_{min} supplied to the amplifier and sufficient for obtaining also high-quality radiation at the amplifier output.

The value of I_{min} should not significantly exceed the saturation intensity I_s. According to the available experimental data, the saturation intensities are mostly in the range 0.18–0.8 W/cm^2 and, apparently, do not exceed a value of 1.7 W/cm^2. The theoretical value obtained in this study for the LT-40Cu laser, is 0.1 W/cm^2.

However, according to the calculations it appears that the values of I_{min} can be significantly reduced compared to I_s when choosing the optimum delay of the radiation applied to the amplifier input in relation to the beginning of the excitation current of the amplifier. According to experimental data I_{min} is in the range 0.007–0.022 W/cm^2, i.e., at least ten times less less than the saturation intensity I_s. According to the calculations the decrease of I_{min} compared with I_s can be up to $3 \cdot 10^{-3}$ and is $3.2 \cdot 10^{-4}$–$3.2 \cdot 10^{-3}$ W/cm^2 for producing 87% and 98% high-quality radiation at the output of the amplifier.

References

1. Soldatov A.N., Solomonov V.I., Gas-discharge lasers on self-terminating tran-sitiions in metal vapors. Novosibirsk: Nauka. 1985. P.149.
2. Batenin V.M, Buchanov V.V., Kazaryan M.A., Klimovskii I.I., Molodych E.I.,

Lasers self-terminating metal atoms. Moscow: Nauchnaya kniga. 1998. P.544.

3.   Little C.E. Metal Vapor Lasers Physics, Engineering & Applications. – John Willey & Sons Ltd. Chichester, UK. 1998. P.620.

4.   Encyclopedia of low-temperature plasma, Series B: Reference application, database, and database / Ch. Ed. Fortov VE Volume XI-4: Gas and plasma lasers. Ans. Ed. Yakovlenko SI - Moscow: Fizmatlit. 2005.P.822.

5.   Boychenko A.M., Yakovlenko S.I., Kvantovaya elektronika, 2002, 32, N2, P.172-8.

6.   Boichenko A.M., Yakovlenko S.I. . Laser Physics, 2002, 12, N7, P.1007-21.

7.   Boichenko A.M., Yakovlenko S.I., Critical prepulse electron density and metastable states in copper vapor lasers, 406-422. In the book: Encyclopedia of low-temperature plasma, Series B: Reference applications, databases, and database. Ed. Fortov V.E. Vol. XI-4: Gas and plasma lasers. Ed. Yakovlenko SI – Moscow: Fizmatlit. 2005, P.822 .

8.   Gudzenko L.I., Yakovlenko S.I., Plasma lasers. Moscow: Atomizdat. 1978.

9.   Derzhie V.I., Zhidkov A.G., SI Yakovlenko S.I.,The emission of ions in non-equilibrium dense plasma. Moscow: Energoatomizdat. 1986. P.160.

10. Yakovlenko S.I., (eds.) Plasma lasers in visible and near UV ranges (Tr. At GPI, T. 21). Moscow: Nauka, 1989. 142 P.

11. Yakovlenko S.I., Plasma Lasers. Laser Physics, Vol. 1, No. 6, P. 565-589, 1991.

12. Yakovlenko S.I., Gas and plasma lasers. Encyclopedia of Low-Temperature Plasma, Ed. V.E. Fortov. Introductory volume. IV. Moscow: MAIK Nauka Interperiodica, 2000, P. 262.

13. Boichenko A.M., Tarasenko V.F., Yakovlenko S., Laser Physics, 10, 1159 (2000).

14. Boichenko A.M., VF Tarasenko V.F., SI Yakovlenko Exciplex lasers and excimer molecules, 471-503. In the book: Encyclopedia of the low temperature plasma, Series B: Reference application, database and database, Ed. Fortov V.E., Volume XI-4: Gas and plasma lasers. Ed. Yakovlenko S.I. Moscow: Fizmatlit. 2005.

15. Yakovlenko S.I. Excimer and exciplex lasers. In: Gas Lasers, Ed. M. Endo, R.F. Walter. CRC Press, Taylor & Francis Group. 2007. P. 369-411.

16. Carman R.J., Brown D.J.W., Piper J.A., IEEE J. Quantum Electron., 30, 1876 (1994).

17. Boychenko A.M., et al., Kvantovaya elektronika 16, 278 (1989).

18. Boichenko A.M., Derzhiev V,I,, Yakovlenko S.I., Laser Phys., 2, 210 (1992).

19. Boichenko A.M., Karelin A.V., Yakovlenko S.I., Laser Phys., 5, 80 (1995).

20. Boychenko A.M., Racehorse V.S., Sosnin E.A., Tarasenko V.F., Yakovlenko S.I., Kvantovaya elektronika, 23, 344 (1996).

21. Boichenko A.M., Skakun V.S., Sosnin E.A., Tarasenko V.F., Yakovlenko S.I., Laser Phys., 10, 540 (2000).

22. Boichenko A.M., Yakovlenko S.I., Laser Phys., 2004, 14, N1, 1-14.

23. Boichenko A.M., Yakovlenko S.I., Modelling lamp sources radiation, 569-608. In the book: Encyclopedia of low-temperature plasma, Serie B: Reference application, database and database. Ed. Fortov V.E., Volume XI-4: Gas and plasma lasers. Ed. Yakovlenko S.I., Moscow: Fizmatlit. 2005, 822 P.

24. Radtsig A.A., Smirnov B.M., The parameters of atoms and atomic ions. Directory. Moscow: Energoatomizdat. 1986.

25. Massey B.E. Electron and ion collisions. Moscow: Publishing House of Foreign Literature. 1958.

26. Raiser Yu., The physics of the gas discharge. Moscow: Nauka. 1987. P.592 .

27. See chapter 6 of this book (The numerical study of pulsed metal vapour lasers).

28. Boichenko A.M., Evtushenko G.S., Zhdaneev G.S.,Yakovlenko S.I., Kvantovaya

elektronika. 2003. 33, N12, 1047-58.

29. Boichenko A.M., Evtushenko G.S., Yakovlenko S.I., Zhdaneev O.V., Laser Phys., 2003. 13, N10, 1231-1255.
30. Boichenko A.M., Evtushenko G.S., Yakovlenko S.I., Zhdaneev O.V., Laser Phys., 2004. 14, N7, 930-952.
31. Boichenko A.M., Evtushenko G.S., Yakovlenko S.I., Zhdaneev O.V., Laser Phys., 2004. 14, N6, 818-834.
32. Boichenko A.M., Evtushenko G.S., Yakovlenko S.I., Zhdaneev O.V., Laser Phys., 2004. 14, N7, 922-929.
33. Mnatsakanyan A.H., Naydis G.V., Stern N., Kvantovaya elektronika, 1978, 5, 597.
34. Carman R.J., J. Phys. D, 1991, 24, 1803.
35. Zaitsev V.V., Mashkov A.V., Goryansky E.S., Petrash G.G., Proc. IVth Conference: Molecular physics of nonequilibrium systems. 2000. 19.
36. Ivanov V.V., Klopovskii K.S., Mankelevich Yu. A., Motovilov S.A., et al., Proc. SPIE, 4747, 198.
37. Vriens L., J. Appl. Phys., 1974, 45, 1191.
38. Morgan W.L., Vriens L., J. Appl. Phys., 1980, 51, 5300.
39. Ligthart F.A.S., Keijser R.A., J. Appl. Phys., 1980, 51, 5295.
40. Cherninyani K. Theory and applications of the Boltzmann equation. Moscow: Mir, 1978, P.496 .
41. Smith K., Thompson R., Numerical modeling of gas lasers. Moscow: Mir, 1981, P.516.
42. Shkarovsky I., et al., Kinetics of plasma particles. Moscow: Atomizdat. 1969.
43. Boichenko A.M., Derzhiev V.I., Yakovlenko S.I., Laser Phys., 1992, 2, 210 (GPI Preprint No. 48, 1991).
44. Sereda O.V., The thesis for the degree of Ph.D., 1990, Moscow, P.152 .
45. Scheibner K.F., Hazi A.U., Henry R.J.W. . Phys. Rev. A., 1987, 35, 4869.
46. Msezane A.Z., Henry R.J.W. . Phys. Rev. A., 1986, 33, 1631.
47. Weinstein L.A.,, Sobel'man I.I., Yukov E.A., The cross sections for the excitation of atoms by electrons and ions. Moscow: Nauka, Home Edition physical and mathematical literature, 1973.
48. Fisher V., Bernshtam V., Golten H., Maron Y. . Phys. Rev. A., 1996. 53. 2425.
49. Loeb L., Main processes of electric discharges in gases. Moscow, Leningrad: Gos. Ed. Technical and Theoretical Literature, 1950.
50. Carman R.J., in: Pulsed Metal Vapour Lasers, Kluver Academic Publishers, Dordrecht, 203-214, (1996)
51. Petrash G.G., Laser Physics, 10, 994 (2000).
52. Yakovlenko S.I., Kvantovaya elektronika, 30, 501 (2000).
53. Yakovlenko S.I. . Laser Physics, 10, 1009 (2000).
54. Boichenko A.M., Evtushenko G.S., Yakovlenko S.I., Zhdaniev O.V., Laser Phys., 11, 580 (2001).
55. Petrash G.G., Kvantovaya elektronika, 30, 407 (2001)
56. Withford M.J., Brown D.J.W., Mildren R.P., Carman R.J., Marshall G.D., Piper J.A., Progress in Quantum Electronics, 28, 165-196 (2004).
57. Gurevich A.V., Pitaevskii L.P. . JETP 46, 1281 (1964).
58. Tkachev S.I., Yakovlenko C.R., Kr. Soobshch. Fiz. FIAN, No.7, 10 (1990).
59. Kazakov V., et al., Kvantovaya elektronika, 15, 2510 (1988).
60. Isaev A.A., Petrash G.G., Proc. SPIE (Metal Vapor Lasers and Applications: CIS Selected Papers) v. 2110, P.2, (1993).
61. Isaev A.A., Kazakov V.V., Forest M.M., Markova S.V., Petrash G.G., Kvanto-

vaya elektronika, 13, 2302 (1986).

62.   Isaev A.A., Mihkel'soo V.T., Petrash G.G., Peet V.E., Ponomarev I.V., Tresh-chalov A.B., Yurchenko N.I., Kvantovaya elektronika, 16, 1173(1989).

63.   Isaev A.A., Petrash G.G., Proceedings of the Lebedev Physics Institute, 212, 93 (1991).

64.   Hogan G., A Study of the Kinetics of Copper Vapour Lasers, Wolfson College, Oxford, A thesis submitted for the degree of Doctor of Philosophy, Trinity Term 1993 at the University of Oxford.

65.   Bohan P.A., Silant'ev V.I., Solomonov V.I., Kvantovaya elektronika, 1980.7. 1264.

66.   Huang Z.G, Namba J., Shimizu F., Japan J. Apllied Physics, 1986. 25.1677.

67.   Withford M.J., Brown D.J.W., Piper J.A., Optics Communications. 1994. 110. 699.

68.   Astadjov D.N., Sabotinov N.V., Vuchkov N.K. Optics Communications.1985. 56. 279.

69.   Astadjov D.N., Sabotinov N.V., Vuchkov N.K. IEEE Quantum Electronics. 1988. 24, 1927.

70.   Astadjov D.N., Isaev A.A., Petrash G.G., Ponomarev I.V., Sabotinov N.V., Vuchkov N.K., IEEE Quantum Electronics. 1992. 28. 1966.

71.   Sabotinov N.V., Vuchkov N.K., Astadjov D.N., Optics Communications. 1993. 95. 55.

72.   Carman R.J., Mildren R.P., Withford M.J., Brown D.J.W., Piper J.A., IEEE Quantum Electronics, 2000. 36. 438.

73.   Cheng C., Sun W., Optics Communications, 1997. 144. 109.

74.   Lyabin N.A., 2002 PhD Thesis, Moscow.

75.   Grigoryants A.G., Kazaryan M.A., Lyabin N.A., Copper vapor laser. Moscow: Fizmatlit. 2005, P.312 .

76.   Carman R.J., Withford M.J., Brown D.J.W., Piper J.A.. Proc. SPIE. 1996. 3092. 68.

77.   Withford M.J., Brown D.J.W., Carman R.J., Piper J.A.. Optics Commun., 1997. 135.164.

78.   Withford M.J., Brown D.J.W., Carman R.J., Piper J.A.. Optics Letters. 1998. 23. 706.

79.   Marshall G.D., Coutts D.W. . IEEE J. of Selected Topics in Quantum Electron. 2000. 6. 623.

80.   Marshall, G.D. 2003, PhD Thesis, Oxford.

81.   Mildren R.P., Withford M.J., Piper J.A. 2002, Conference Digest of the XIV International Symposium On Gas Flow & Chemical Lasers and High Power Laser Conference-GCL, HPL 2002. Wroclaw. Poland.

82.   Marius T. 1974, USA patent # 3.831.107.

83.   Isakov V.K., Kalugin M., Pavlov S.V., 1981 Abstracts, 10 Siberian Conference on Spectroscopy. Tomsk. 155.

84.   Belokrinitsky N.S.,et al., Institute of Physics. Preprint Ukrainian SSR. Kiev. 1988. No.6. 33.

85.   Masumura Yu., Ishikawa T., Saitoh H. Appl. Phys. Lett., 1994. 64. 3380.

86.   Sakata S., Oohori K., Higuchi M., IEEE J. Quantum Electron., 1994. 30. 2166.

87.   Ohzu A., Kato M., Maruyama Y., Appl. Phys. Lett., 2000. 76. 2979.

88.   Boichenko A.M., Evtushenko G.S., Yakovlenko S.I., Zhdaneev O.V., Laser Phys., 2004. 14. N6, 835-846.

89.   Bohan P.A., Metal vapor lasers with collisional de-excitation of the lower working conditions. Doctor Thesis. Sci. Sciences. IOA, Novosibirsk. 1988. 418 P.

90. Carman R.J., Mildren R.P. Withford M.J., Brown D.J.W., Piper J.A., Optics Communications. 1998. 157. 99.
91. Mildren R.P., Withford M.J., W.Brown DJ, Carman R.J., Piper J.A., IEEE Kvantovaya elektronika, 1998. 34. 2275.
92. Hayashi K., Iseki Y., Suziki S., Watanabe I., Noda E., Morimiya O., Jap. J. Applied Physics. 1992. 31. L1689.
93. Garcia-Prieto J., Ruiz M.E., Poulain E., Ozin G.A., Novaro O., J. Chem. Phys., 1984. 81. 5920.
94. Carman R.J., Mildren R.P., Piper J.A., Marshal G.D., Coutts D.W., Proc. SPIE, 4184, 2001.
95. Boychenko A.M., Tarasenko V.F., Fedeneev A.V., Yakovlenko S.I ., Kvantovaya elektronika. 1997. 24. 697.
96. Kushner M.J., Warner B.E., J. Appl. Phys. 1983. 54.2970.
97. Isaev A.A., Jones D.R., Little C.E., Petrash G.G., White CG., Zemskov K.I., IEEE J. Kvantovaya elektronika. 1997. 33.919.
98. Soldatov A.N., Fedorov V.F., Izv. VUZ. Fizika. 1983. 26. No.9, P.80-84.
99. Yevtushenko G.S., Petrash G.G., Sukhanov V.B., Fedorov V.F., Kvantovaya elektronika. 1999 28.220.
100. Evtushenko G.S., Shiyanov D.V., Fedorov V.F., Optika atmosfery i okeana. 2000. 13. 254.
101. Maltsev A.N, Preprint No.1 IRA Tomsk branch of the Academy of Sciences of the USSR.1982.
102. Petrash G.G., UFN. 1971. 105. 645.
103. Petrash G.G., Kvantovaya elektronika. 2005. 35. No.6. 576-7.
104. Boychenko A.M., Evtushenko G.S., Zhdaneev O.V., Yakovlenko S.I., Kvantovaya elektronika.2005.35.No.6. 578-80.
105. Carman R.J., Mildren R.P., Piper J.A., Marshall G..D, Coutts D.W., Proc. SPIE. 2001. 4184. 215.
106. Boychenko A.M. Yakovlenko S.I., Kvantovaya elektronika. 1992. 19.1172.
107. Brown D.J.W., Withford M.J., Piper J.A. . IEEE J. Quantum Electron., 2001. 37.518.
108. Withford M.J., Brown D.J.W., Carman R.J., Piper J.A.. Optics communications. 1998. 154.160.
109. Landau L.D., Lifshitz E.M., Mechanics. Moscow: Nauka. 1988.
110. Westenberg A.A., DeHaas N., J. Chem. Phys., 1968. 48. 4405.
111. Wilkins R.L., J. Chem. Phys., 1975. 63. 2963.
112. Astadjov D.N., Sabotinov N.V., J. Phys. D., 1997. 30. 1507.
113. Yakovlenko S.I., Kvantovaya elektronika, 25, 971 (1998), (Quant. Electr., 28 (11). 945-961 (1998).)
114. Tkachev A.N., Yakovlenko S.I., . Kvantovaya elektronika, 33 (7). 581-592(2003). (Quant. Electr., 33 (7), 581-592 (2003).)
115. Bohan P.A., Buchanov V.V., Zakrevskii D.E., Kazaryan M.A., Kalugin M.M., Prokhorov A.M., Fateev N.V., Laser isotope separation in atomic vapors. Moscow: Fizmatlit. 2004.
116. Boichenko A.M., Yakovlenko S.I., Laser Phys., 2005. 15. N11.1528-1535.
117. Coutts D.W., IEEE J. Quant. Electron. 2002. 38. 1217.
118. Zemskov A.I., Isaev A.A., Kazaryan M.A., Petrash G.G., Rautian S.G., Kvantovaya elektronika. 1974. 1. N4. 863-869.
119. Isaev A.A., Petrash G.G., Kazaryan M.A., Rautian S.G., Kvantovaya Elektronika. 1974. 1. N 6. 1379.
120. Isaev A.A., Petrash G.G., Kazaryan M.A., Rautian S.G., Shalagin A.M., Kvan-

tovaya elektronika. 1975. 2. N 6. 1125.

121. Isaev A.A., Petrash G.G., Kazaryan M.A., Rautian S.G., Shalagin A.M., Kvantovaya elektronika. 1975. 4. N 5. 1325-1335.

122. Hargrove R..S, Grove R., Kan T., IEEE J. Quant. Electron. 1979. 15.1228.

123. Coutts D.W. . IEEE J. Quant. Electron.1995. 31.330.

124 Kalugin M.M., Kuzminova E.N., Potapov S.E., Kvantovaya elektronika. 1981. 8. 1085.

125. Brown D.J.W., Withford M.J., Piper J.A. . IEEE J. Quant. Electron, 200.137. 518.

126. Karpukhin V.T., Konev J.B., Malikov M.M., Izv. AN Ser. Fiz., 2002. 66. 924.

127. Karpukhin V.T., Malikov M.M., Kvantovaya elektronika. 2003. 33. No.5.411-415.

128. Karpukhin V.T., Malikov M.M., ZhTF. 2005. 75. No.10. 69-72.

129. Behrouzinia S., Sadighi R., Parvin P., Appl. Opt., 2003. 42. 1013.

130. Kazaryan M.A., Matveev V.M., Petrash G.G.. Sov. J. Quant. Electron., 1984. 14. 631.

131. Warner B.E., in: New Developments and Applications in Gas Lasers, Carlson L.R. ed., Proc. SPIE 737.2 (1987).

132. Naylor G.A., Lewis R.R., Kearsley A.J., in: Gas Laser Technology, Sauerbrey R.A., et al. eds. Proc. SPIE. 894. 1101(988).

133. Lewis R.R., Maldonada G., Webb C.E., in Metal Vapor, Deep and Blue and Ultraviolet Lasers, Kim J.J., Kimball R., Wisoff P.J. eds., Proc. SPIE. 1041. 54 (1989).

134. Kimura H., Aoki N., Konagai C., Shirayama S., Miyazawa T., J. Nucl. Sci. Technol., 31. 34 (1994).

135. Coutts D.W., Piper J.A., in: Conference on Lasers and Electro-Optics, Vol. 10 of OSA Technical Digest Series (Optical Society of America. Washington, D.C., 1991), paper CFH5.

Index